Organ and Tissue Donation for Transplantation

Organ and Tissue Donation for Transplantation

Edited by

JEREMY R. CHAPMAN
MD (Camb) FRACP FRCP

Director of Renal Medicine, Westmead Hospital
Medical Director, Tissue Typing Laboratory,
NSW Red Cross Blood Bank, Sydney,
Australia

MARK DEIERHOI
MD

Division of Transplantation, University of Alabama,
Birmingham, Alabama,
United States of America

CELIA WIGHT
SRN HVCert

EDHEP Co-ordinator, Eurotransplant Foundation
Donor Action Secretariat, Harston, Cambridge,
United Kingdom

A member of the Hodder Headline Group
LONDON   SYDNEY   AUCKLAND
Co-published in the USA by Oxford University Press, Inc., New York

First published in Great Britain 1997 by
Arnold, a member of the Hodder Headline Group,
338 Euston Road, London NW1 3BH

Co-published in the United States of America by
Oxford University Press, Inc.,
198 Madison Avenue, New York, NY10016
Oxford is a registered trademark of Oxford University Press

© 1997 Arnold

All rights reserved. No part of this publication may be reproduced or transmitted in any form or by any means, electronically or mechanically, including photocopying, recording or any information storage or retrieval system, without either prior permission in writing from the publisher or a licence permitting restricted copying. In the United Kingdom such licences are issued by the Copyright Licensing Agency: 90 Tottenham Court Road, London W1P 9HE.

Whilst the advice and information in this book is believed to be true and accurate at the date of going to press, neither the authors nor the publisher can accept any legal responsibility or liability for any errors or omissions that may be made. In particular (but without limiting the generality of the preceding disclaimer) every effort has been made to check drug dosages, donation protocols and tests for transmission of disease; however it is still possible that errors have been missed. Furthermore, dosage schedules are constantly being revised and new side-effects recognised, new diseases are discovered and tests for transmissible disease are constantly being improved. The variations in disease frequency in different populations mean that local consideration must always be made with respect to the particular tests required to screen donors to prevent transmission of disease. For these reasons the reader is strongly urged to consult the drug companies' printed instructions before administering any of the drugs recommended in this book.

British Library Cataloguing in Publication Data
A catalogue record for this book is available from the British Library

Library of Congress Cataloging-in-Publication Data
A catalog record for this book is available from the Library of Congress

ISBN 0 340 61394 7

Typeset in 10/11 pt Times by
Mathematical Composition Setters Ltd, Salisbury, Wiltshire
Printed and bound in Great Britain by
St Edmundsbury Press, Bury St Edmunds, Suffolk

Contents

Foreword *Sir Peter Morris* vii
Preface ix
Contributors xi

1 **Transplantation** *Jeremy R. Chapman, Bill New* 1

PART 1 Organ and tissue donation in society

2 **Religious views on organ and tissue donation** *The Right Reverend Lord John Habgood, Antonio G. Spagnolo, Elio Sgreccia, Abdallah S. Daar* 23

3 **Public attitudes toward organ and tissue donation** *Karen L. Smith, Judith B. Braslow* 34

4 **Paid organ donation: towards an understanding of the issues** *Abdallah S. Daar* 46

5 **Ethics in organ donation and transplantation: the position of the Transplantation Society (1996)** *Sir Roy Calne* 62

PART 2 The process of organ and tissue donation

6 **Brain death** *Ian Y. Pearson* 69

7 **Legislation on organ and tissue donation** *Bernard M. Dickens, Sev S. Fluss, Ariel R. King* 95

8 **Transmission of disease by organ transplantation** *Svetlozar N. Natov, Brian J.G. Pereira* 120

9 **Organ recovery from cadaveric donors** *Mark Deierhoi* 152

10	**The living organ donor** Richard D.M. Allen, Stephen V. Lynch, Russell W. Strong	162
11	**Organ and tissue preservation** Vernon C. Marshall	200
12	**Organization of donation and organ allocation** Johan De Meester, Bernadette J.J.M. Haase-Kromwijk, Guido G. Persijn, Bernard Cohen	226
13	**Bone marrow donation** Kerry Atkinson	239
14	**Tissue banking** Stephen Cordner, Lynette Ireland	268
15	**The donor family experience: sudden loss, brain death, organ donation, grief and recovery** Sue C. Holtkamp	304

PART 3   Methods of increasing organ and tissue donation

16	**Transplant coordinators** Barbara A. Elick	325
17	**Informed or presumed consent legislative models** Paul Michielsen	344
18	**The Spanish experience in organ donation** Rafael Matesanz, Blanca Miranda	361
19	**The European Donor Hospital Education Programme (EDHEP)** Celia Wight, Bernard Cohen	373
20	**'Making the Critical Difference'™: Education, motivation, donation** Gigi Politoski, Jan Boller, Kathleen Casey	382
21	**The Partnership for Organ Donation: a strategic approach to solving the organ donor shortage** Carol L. Beasley, Jessica D. Blaustein	389
22	**Education in schools** Napier M. Thomson, Roy Knudson, Geoff Scully	400
23	**Publicity and marketing strategies** Miriam Ryan	412
24	**Unrelated bone marrow donor registries** Patricia A. Coppo	430
25	**Xenotransplantation – a solution to the donor organ shortage** David J.G. White	446
	Index	459

Foreword

There is a critical shortage of organs for transplantation throughout the world and in this book the authors have attempted to draw together all aspects of organ donation. The contributions are divided into three major sections, covering the attitude to organ donation in society in general, the process of organ and tissue donation and finally methods of increasing organ and tissue donation. The contributors to this volume are drawn from a wide spectrum of the transplantation community throughout the world, and express some interesting, and often challenging, views of thorny problems that we are faced with today.

No doubt readers will concentrate on one or other major section depending on their background. The first section on society's attitude to organ and tissue donation will be of interest not only to the transplant community but also to the public, whereas the second section on the actual methods of organ and tissue donation will primarily be of interest to the transplant team. The final section on methods of increasing organ and tissue donation will be of interest to all, not only to those within the transplant community but also to the lay public, including members of government who have to provide the money to enable various approaches to improve organ donation to be addressed.

Without question this is an important contribution to the problems of organ donation which should be not only a useful source of reference for the transplant community, but also a source of information about organ donation for the public and governments in particular.

Peter J. Morris
PhD FRCS FRS

Preface

Transplantation of organs and tissues from one individual to another, has become a technically successful procedure throughout the world. Survival of the graft at one year can be expected in more than 80–90 percent of transplants, ranging from the kidney to the heart, liver, pancreas and heart–lung. The numbers of patients benefiting from tissue donation have risen as increasing ingenuity of medical practice has extended to the use of heart valves, cornea and bone, while scientific advances in histocompatibility matching have broadened the options for patients with lethal diseases treatable by bone marrow transplantation. The numbers of patients with end stages of organ or tissue failure who are suitable for transplantation have thus risen progressively. By contrast the number of donors has remained relatively static worldwide.

The increasing range and number of organs and tissues that can be transplanted and number of patients on the waiting list has meant that availability of organ donors has surpassed the immunological problems as the major challenge for transplantation. The therapy is now limited more by the rate at which donors consent than the rate at which transplants fail. As a result, there have been many different approaches, both at the legislative and the individual level, to increase the donor rates.

This book collects detailed analyses of the problems surrounding organ and tissue donation. The solutions are at least as complex as the problems, with the diversity of approach standing testament to the difficulty of the task and the need for multi-disciplinary approaches. Communication of ideas may help and this book is proffered in the hope that the experiences of others may be useful to those who strive to meet the needs of patients who can be helped by a transplant.

Jeremy Chapman
Mark Deierhoi
Celia Wight

Contributors

Richard D.M. Allen MBBS FRACS   Clinical Senior Lecturer, University of Sydney; Director, National Pancreas Transplant Unit; Head, Department of Transplant Surgery, Westmead Hospital; Consultant Transplant Surgeon, New Childrens Hospital, Sydney Australia.

Kerry Atkinson MD FRCP FRACP   Associate Professor of Medicine and Senior Staff Specialist, St Vincent's Hospital, Sydney, Australia; Director of Clinical Transplantation, Systemix, Palo Alto, California, United States of America.

Carol L. Beasley MPPM   Managing Director, The Partnership for Organ Donation, Boston, Massachusetts, United States of America.

Jessica D. Blaustein   The Partnership for Organ Donation, Boston, Massachusetts, United States of America.

Jan Boller RN MSN   National Kidney Foundation, New York, United States of America.

Judith B. Braslow   Director, Division of Organ and Tissue Transplantation, US Department of Health and Human Services, Rockville, Maryland, United States of America.

Sir Roy Calne FRS   Professor of Surgery, Addenbrooke's NHS Trust, Cambridge, United Kingdom.

Kathleen Casey BA   Program Director, National Kidney Foundation, New York, United States of America.

Jeremy R. Chapman MD(Camb) FRACP FRCP   Director of Renal Medicine, Westmead Hospital; Medical Director, Tissue Typing Laboratory, NSW Red Cross Blood Bank, Sydney, Australia.

Bernard Cohen MEC   Director, Eurotransplant Foundation, Leiden, The Netherlands.

Patricia A. Coppo MS   Chief Operating Officer, National Marrow Donor Program, Minneapolis, Minnesota, United States of America.

Stephen Cordner MA(Lond) MB BS BMedSc FRCPA FRCPath   Professor of Forensic Medicine, Monash University; Director, Victorian Institute of Forensic Medicine, Southbank, Victoria, Australia.

Abdallah S. Daar D Phil(Oxon) FRCP(Lond) FRCS FRCSEd   Professor and Chairman, College of Medicine, Sultan Qaboos University, Muscat, Sultanate of Oman.

Mark Deierhoi MD   Associate Professor of Surgery, Division of Transplantation, University of Alabama, Birmingham, Alabama, United States of America.

Johan De Meester MD   Head Medical Affairs, Eurotransplant International Foundation, Leiden, The Netherlands.

Bernard M. Dickens PhD LLD   Professor of Medical Law, Faculty of Law and Faculty of Medicine, University of Toronto, Ontario, Canada.

Barbara A. Elick BSN RN CCTC/CPTC   Executive Director, The Transplant Center, University of Minnesota, Minnesota, United States of America.

Sev S. Fluss BSc(Edin) MS(Wisconsin) Dip Ag(Cantab)   Programme Manager for Human Rights, Office of the Executive Administrator for Health Policy in Development, World Health Organization, Geneva, Switzerland.

Bernadette J.J.M. Haase-Kromwijk   Head, General Affairs, Eurotransplant Foundation, Leiden, The Netherlands.

The Right Reverend Lord John Habgood PC MA PhD DD   Former Archbishop of York, Malton, North Yorkshire, United Kingdom.

Sue C. Holtkamp PhD   Licensed Professional Counselor and Certified Grief Therapist, Founder and Director of Something More Center for Loss and Transition, Chattanooga, Tennessee, United States of America.

Lynettee Ireland   Head, Donor Tissue Bank of Victoria, Victorian Institute of Forensic Medicine, Southbank, Victoria, Australia.

Ariel R. King MPH MIHM   Thunderbird: American Graduate School of International Management, Glendale, Arizona, United States of America.

Roy Knudson BSc   Education Officer, Australian Kidney Foundation, Melbourne, Victoria, Australia.

Stephen V. Lynch FRACS   Associate Professor of Surgery, University of Queensland, Deputy Director, Queensland Liver Transplant Service, Princess Alexandra and Royal Children's Hospitals, Brisbane, Australia.

Vernon C. Marshall FRACS FACS   Professor and Chairman, Monash University Department of Surgery; Chairman, Division of Surgery, Monash Medical Centre, Clayton, Melbourne, Australia.

Rafael Matesanz MD   Organización Nacional de Trasplantes, Madrid, Spain.

Paul Michielsen MD   Emeritus Professor of Medicine, Katholieke Universiteit, Leuven, Belgium.

Blanca Miranda MD PhD   National Transplant Co-ordinator, Organización Nacional de Trasplantes, Madrid, Spain.

Svetlozar N. Natov MD   Instructor in Medicine, Tufts University School of Medicine, New England Medical Center, Boston, United States of America.

Bill New PhD Senior Research Officer, King's Fund Policy Institute, London, United Kingdom.

Ian Y. Pearson FRCA FFI CANZCA Previously Director of Instensive Care and Consultant Anaesthetist, Westmead Hospital, Westmead, NSW, Australia.

Brian J.G. Pereira MD DM Associate Professor of Medicine, Tufts University School of Medicine; Staff Physician, New England Medical Center, Boston, United States of America.

Guido G. Persijn MD Medical Director, Eurotransplant International Foundation, Leiden, The Netherlands.

Gigi Politoski BA Director, Program Division, National Kidney Foundation, New York, United States of America.

Miriam Ryan BA MIPR Healthcare Communications Consultant, Ryan de Long, London, United Kingdom.

Geoff Scully Transplant Co-ordinator, Monash Medical Centre, Clayton, Melbourne, Victoria, Australia.

Elio Sgreccia Professor of Bioethics, Istituto di Bioetica, Università Cattolica del Sacro Cuore, Facoltà di Medicina e Chirurgia, 'Agostino Gemelli', Roma, Italy.

Karen L. Smith MS Public Health Analyst, Bureau of Health Resources Development, US Department of Health and Human Services, Rockville, Maryland, United States of America.

Antonio G. Spagnolo MD Senior Researcher, Istituto di Bioetica, Università Cattolica del Sacro Cuore, Facoltà di Medicina e Chirurgia, 'Agostino Gemelli', Roma, Italy.

Russell W. Strong FRACS Professor of Surgery, University of Queensland; Director, Queensland Liver Transplant Service, Princess Alexandra and Royal Children's Hospitals, Brisbane, Australia.

Napier M. Thomson MB BS MD FRACP Professor of Medicine and Head of Department of Medicine, Monash University; Director of Renal Medicine, Alfred Hospital, Prahran, Victoria, Australia.

David J.G. White FRCPath Lecturer in Surgery, Douglas House, Cambridge, United Kingdom.

Celia Wight SRN HVCert EDHEP Co-ordinator Eurotransplant Foundation; Donor Action Secretariat, Harston, Cambridge, United Kingdom.

CHAPTER 1

Transplantation

JEREMY R. CHAPMAN, BILL NEW

Introduction 1
Factors that influence transplantation rates 2
Transplantation of the kidney 12
Transplantation of other solid organs 15
Costs and effects 16
References 18

Introduction

It has been said that transplantation was the first major area in medicine where 'technology' changed the outcome of otherwise inevitable disease processes. It is not of course true, but transplantation has allowed medicine to challenge irrevocable failure of a vital organ. There is no doubt that between 1965 and 1997 there has been a transformation in the prospects of a person with end-stage kidney, liver, heart or lung failure. Bone marrow transplants have achieved the same dramatic reversal for many with haematological malignancies, while the improvement in morbidity from corneal, heart valve and bone transplantation has delivered a similar quantum for those patients.

It is important to examine the realities of that progress on a global front, as opposed to the public perceptions. The media have actively colluded with both doctors and patients to publicize the success of transplantation, driven by the need to increase organ donation and to demonstrate to the general community the benefits that can flow to the recipient from an organ donation. In the late 1990s the time has arrived to accept that organ donation is an ordinary medical therapy and not an extraordinary one. With that acceptance must also come an understanding of the normal outcomes of transplantation which, like every human endeavour, has positive and negative aspects.

In this chapter we review the factors that affect the rate of transplantation around the world and the outcomes of transplantation of organs and tissues.

Factors that influence transplantation rates

A simplistic view of transplantation activity around the world would suggest the number of donors available is the only factor that really affects the number of transplants. While this might be true of kidney transplants in some countries, it is not true in other countries or with most other organs and tissues. It is almost impossible to measure the true impacts of the many different factors which influence the rate of transplantation of each organ in each country, partly because they interact with each other. While we can analyse each factor independently it would be wise to remember that the truth is more complex (Figure 1.1). There is no point seeking donors if there is no one to perform the transplants, insufficient patients waiting, or inadequate facilities. There is clearly little point developing programmes to transplant an organ where the technical and scientific understanding is lacking, yet it is only through a development programme that the understanding and technical solutions can be achieved.

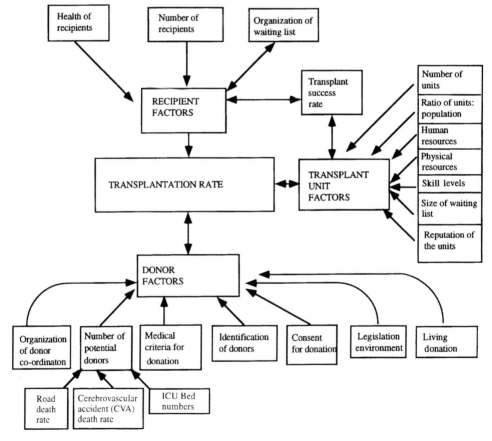

Figure 1.1 The factors that affect the rate of transplantation. These include those that relate to the recipients, the donors and the transplant units. The interaction between these factors determines how many transplants of a particular organ are performed.

Transplantation, perhaps more than other fields of medicine, has created well-known public individuals from the surgeon/scientist who have developed the programmes. Christian Barnard, Roy Calne, Tom Starzl and Victor Chang are among the names that have reached the public domain in this way.[1,2] An unanswered question is whether it is these individuals that create the transplant rates or are merely identifiable components of a complex process.

DONOR NUMBERS

The very early experimental transplantation programmes worked within a pioneering environment in which the availability of intensive care, capacity for long-term ventilation and ability to maintain the brain-dead patient, were all rudimentary, ensuring that suitable donors were few and far between. Brain death was not well understood as a concept, let alone established in law, such that cardiac standstill was the only defining moment from which to commence organ retrieval.

The development of renal transplantation from an experimental therapy to routine practice occurred progressively through the late 1960s and early 1970s. It was thus during this period that tension between the 'demand and supply' of donors developed. With relatively reliable and steadily improving results from renal transplantation, recipients became increasingly prepared to accept the option, while surgeons became increasingly prepared and able to offer the service. The limiting factor thus became the rate that suitable patients were identified as potential donors and consent gained from their bewildered relatives.

An early option for the pioneers of renal transplantation and then of bone marrow transplantation was to turn to the living relatives of patients in need and suggest that they donate. The first successful renal transplant was from an identical twin who gave a kidney to his brother dying of renal failure.[3] During the 1960s both living related and living unrelated donors gave their kidneys for transplantation. This alternative to cadaveric donation represented the most effective method of both providing well functioning kidney grafts and of releasing transplant programmes from the legal complexities of brain versus cardiac death and the perceived shortages of cadaveric donors.

In the 1990s the survival of both patients and grafts after all forms of solid organ transplant has risen to the point where, for the majority of straightforward first graft recipients, loss of the graft is an unusual event. Donor numbers thus provide the most important limitation on the ability to deliver a healthy outcome to patients with organ failure, yet there has actually been a decline in donor numbers in some countries.

In 1990, the United Kingdom Transplant Service Support Authority reported 16.6 cadaveric kidney donors per million of population (pmp) but fewer ever since (15.3, 15.8, 15.0, 15.2, 15.8 and 1991 to 1995). Similar changes have been seen in a number of other European countries (Figures 1.2 and 1.3). This has occurred despite implementation of a number of initiatives designed to improve donation rates during this period of time. In the United Kingdom, analysis of Department of Transport's data on road traffic accidents has showed a significant decline in deaths during 1991 coinciding with the introduction of new road safety laws. Similar road safety initiatives taken elsewhere have also had the effect of reducing the size of the pool of potential donors in those countries. The reduction in death from cerebrovascular accident (CVA) has also been dramatic. In Australia, for example, between 1967 and 1993, death rate from CVA fell by 73% in men and 77% in women between the ages of 20 and 69. In absolute terms

4 Transplantation

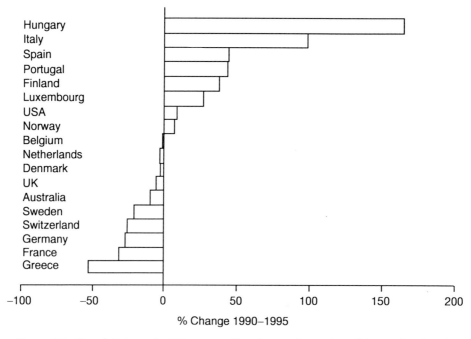

Figure 1.2 Trends in transplantation rates. The changes in number of donated cadaveric kidneys are shown for countries reporting data routinely.

this means that there were 5,000 fewer deaths from CVA in Australia during 1993 compared to 1967 (or 270 pmp). There is also considerable variation in death rate per 100,000 population from CVA between countries, Australia and the United States having rates in 1992 that were only 62% of the rate in Austria, while Portugal, for example, had twice that rate.

The response to low donation rates has been to try to understand the process of donation better so that rate limiting steps can be addressed; to accept as donors those who have pre-existing medical conditions such as diabetes and high blood pressure; to accept older donors; and to seek an alternative in non-human organs.

How many donors are being missed? This question is central to any strategy designed to increase cadaveric organ donation and despite considerable research there remains a great deal of uncertainty over the answer. If one asks this question of those who work in the emergency departments, intensive care units and neurosurgical units, they will almost without exception believe that organ donors are seldom, if ever, missed. If one asks this question of those who work in transplant and dialysis units, they will almost without exception believe that at least half of all potential donors are missed. How can one get at the truth? In order to understand that there is and can be no certain truth to this question one needs to understand the processes of brain injury and interventions that are possible for each individual patient. In Figure 1.4 a flow diagram attempts to show the complexity of possible outcomes for a patient who dies after a brain injury irrespective of the type of injury. Those patients who die from cardiac standstill on the left-hand side of the diagram are not potential brain-dead organ donors since, despite all possible medical intervention, their heart has ceased to beat. This group of patients

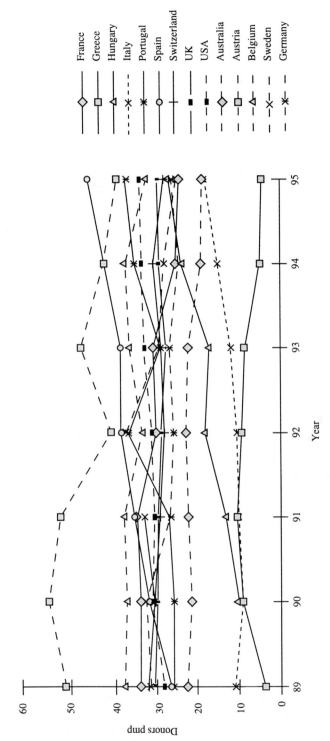

Figure 1.3 Percentage change in cadaveric donation between 1990 and 1995. The change in the rate of donation over the five year period 1990 to 1995 highlights the large number of countries with highly developed medical systems in which there has been either no change in donor rates or a decline.

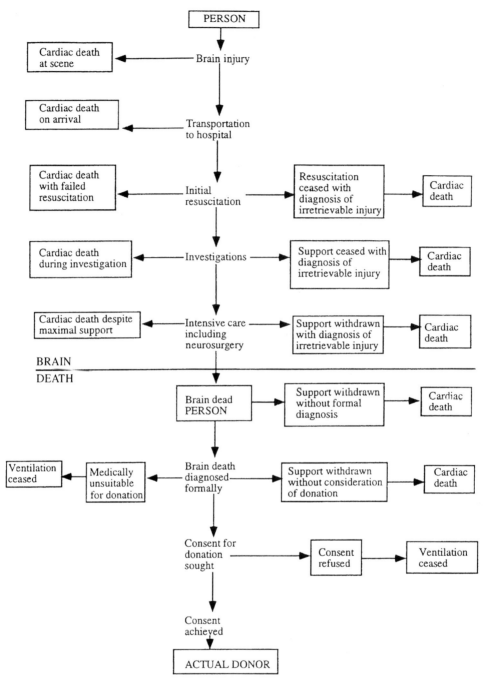

Figure 1.4 Outcome after serious brain injury. The possible events that follow initial brain injury in a patient who eventually dies as a result of the injury. The patient whose heart stops irretrievably at any point is depicted to the left of the flow diagram, while the patient whose active treatment ceases and who subsequently suffers cardiorespiratory arrest moves to the right of the diagram. Below the line the patient is brain dead, though the diagnosis may or may not be made formally.

might, however, be considered as non-heart beating donors as discussed below. Those patients who exit via the right-hand side of the diagram are potential brain-dead organ donors, with increasing certainty that they could truly have been both brain dead and organ donors the further down the flow diagram that they exit. Once the patient has actually become brain dead, that diagnosis may or may not be made formally, consent may or may not be sought and they may or may not be medically suitable as donors.

In any attempt to measure the number of patients that *could have* become actual donors most studies have started with the people who have actually died following brain injury from any cause. When one does that, either retrospectively or prospectively, there are two fundamental problems. The first is that unless the patient has been formally declared brain dead, it is impossible to be sure if they would have been brain dead if events had followed a different course. The second problem is that it is not possible to know what might have happened if resuscitation or cardiorespiratory support had been continued at maximal levels. In a proportion of patients, maximal support would simply have resulted in the patient dying from cardiac standstill anyway. In all of the studies of potential donor numbers it is thus important to remember that 'truth' does not exist, there are only best guesses.

Studies of the potential numbers of organ donors have, by their nature, to be retrospective since all start by reviewing events leading up to death of a patient. The methodology of most of the studies has been to review the case records of groups of patients who have died in defined settings such as in intensive care units or in particular hospitals during fixed periods of time. Some of the studies have endeavoured to review the events very soon after death and have thus established the population prospectively. An alternative strategy has been to examine the numbers of deaths due to causes which lead to organ donation, such as road trauma with head injury, and then to compare trends in those factors with trends in actual donations.

In the United Kingdom a national audit of potential donor numbers was based upon death in intensive care units.[4] There were 37 patients pmp who both died and were considered possibly to have satisfied the criteria for brainstem death. Only 22 pmp actually had brain death confirmed and donation occurred in 13.6 pmp. The single most important cause of failing to donate in those confirmed to be brain dead was refusal of consent. One of the limitations of this audit was that it did not include patients who died in the emergency departments and on general wards, implying that the figures should be seen as conservative estimates. Other studies have taken the wider focus of all deaths in defined hospitals (Table 1.1) and have arrived at higher figures for the potential numbers of organ donors.[5-12] In the Australian studies,[8,13] the disparity between estimates of up to 50 donors pmp and the reality of 10–13 actual donors pmp, was a source for conflict between intensive care specialists and those involved in transplant units. In analyses of these studies it is possible to confirm everyone's prejudices about the lack of donation. Transplant units 'can point the finger of blame' at intensive care specialists who fail to identify donors. On the other hand intensive care specialists can demonstrate that almost everyone who actually reaches an intensive care unit (ICU), becomes brain dead and does not succumb to cardiac arrest, is considered for organ donation. Those who believe non-beating heart donation will solve their problems can gain support from the large numbers that appear in the gap between possible and actual donation.

It was estimated that, in the United Kingdom, there would be nearly an additional 40 patients pmp each year in whom interventional ventilation, rather than death from respiratory depression could be considered. The proportion of those patients who could eventually donate might be as low as one quarter[15] and the risk that many might

Table 1.1 Estimates of the potential number of cadaveric organ donors

Location of study (Ref.)	Year of data collection	Potential donors pmp per year
USA[5]	1987	38–55
Spain[6]	1987	40
USA[7]	1988	51
Australia[8]	1989–92	48
England[4]	1989	22–37
USA[9]	1991	65
USA[10]	1992	44
USA[11]	1990–93	55
Spain[12]	1993	53

eventually be left in a persistent vegetative state has influenced attitudes. The barriers of relatives' consent and the relevance of using intensive care facilities for this purpose have not been widely tested because the legality of this practice has not been clarified in the United Kingdom and the initial study has been stopped.

The inescapable conclusion from these studies of potential donors is that there are unlikely to be more than 50 brain-dead donors per million population under any circumstances and that this is likely to be, from the recipients' perspective, an optimistic assessment. A different way of looking at that figure is that the maximum potential donor rate is in the region of 0.75% of all deaths, 2–4% of hospital deaths and up to 15% of deaths in the setting of intensive care. These figures thus set an estimate of the maximum rates that might be achievable in any given hospital, region or country, irrespective of the legal framework for consent (see Chapter 7).

The barriers to actual donation, defined by these studies, fall into two main categories. The first is lack of identification of a patient as a potential donor. In some instances this lack of identification is simply a lack of thought or lack of knowledge of criteria for acceptance of donation. In many instances, however, the normal process of care of the patient leads to a decision to withdraw active management at a time when the prognosis is obvious and inescapable, such as after massive head injury. If there is no purpose, other than organ donation, it is appropriate to place a patient in an intensive care unit? The second barrier is gaining consent to proceed to donation, with as many as 60% of families refusing in some studies. The true rate of family refusal has to be judged only from those studies where information was sought from intensive care staff in person soon after the death. In that circumstance the knowledge that one or more family member, or the patient themselves during life, had indicated refusal, may be apparent even though not recorded in the notes.

Analysis of the risk factors for dead patients not being considered for organ donation despite appearing to be suitable, has been undertaken in Australia.[16] The study was based on a prospective analysis of case notes and discussion within one week of the death with the senior clinician involved with care. Multifactorial analysis demonstrated that the patients who tended not to be asked were older, with the chance of being asked about organ donation halving between age 0–49, and 50–59 and halving again for 60–69 years. Those whose families were not asked about donation tended to die outside of an ICU, from causes other than trauma or intracranial haemorrhage, and finally they tended to die more quickly (within 12 hours of admission to hospital) than those whose families were approached about donation. Put the other way round, this study showed that the 'conventional donor' who is young, has died from trauma or intracranial haemorrhage, is in an ICU and is sufficiently stable to maintain cardiac output for more than 12 hours, almost always did get considered.

DEVELOPMENT OF THE CLINICAL SCIENCE OF TRANSPLANTATION

The evolution of the science of organ and tissue transplantation has also been an important determinant of both the rate of transplantation and its success. All forms of transplant have been caught in an impasse early in their development where only patients about to die are prepared to accept the uncertainty of an untried treatment, but those are the patients in whom success is most difficult to achieve. Perhaps the only organ where this impasse was circumvented quickly was the kidney, because of the ability to support the patient by dialysis, both before and after transplantation. This has meant that the patient need not be in extreme ill-health at the time of the operation and the organ need not function immediately. Acute renal failure due to reversible damage of the tubules can be managed for days or weeks by dialysis, while even total loss of the kidney does not lead to death of the patient. By contrast a patient with a year or two of life expected from their own heart, was in the early days, hardly likely to embark upon an operation in which their heart was removed and they became dependent upon both the immediate and long-term function of a transplanted heart. In that situation only those patients so ill that they were confined to a hospital intensive care unit with only a few weeks of life, were prepared to embark upon the then hazardous journey of cardiac transplantation. For this reason the cardiac transplantation programmes, started with such energy and popular enthusiasm in the mid-1960s, almost all fell into disrepute, bringing a backlash of approbation on those who undertook the surgery. Cardiac transplantation was only rescued by the long-term application of scientific method by the Stanford team led by Norman Shumway.

Bone marrow transplants provide a different example of the difficulty of applying an unknown therapy to the treatment of serious disease. Bone marrow like the heart, liver and lungs must start to function quite quickly if the patient is to survive. It is possible to support the function of the marrow artificially by blood and platelet transfusions. However, unless the graft commences production of red cells, white cells and platelets within two or three weeks, most patients die of overwhelming infection. The patients prepared to undergo this hazardous course should be fit and well and their haematological malignancy should be at its ebb. These are just the patients who cannot reasonably accept a high chance of immediate death in return for a small chance of long-term survival. There has been no substitute for the long and arduous process of scientific study and development, to change the outlook of bone marrow transplantation from predictably poor to predictably good and with that has come the increase in rate.

Some forms of transplantation remain at an impasse for one reason or another. For example, one might controversially place pancreas transplantation at this point. There is little doubt that pancreas transplants, especially when performed at the same time as a renal transplant, are a technically successful operation with graft survival rates similar to those seen in other forms of transplantation. The question is not whether it can be done, but whether it is actually worth doing with respect to the outcome of diabetes and its secondary complications. Those patients that are most suitable for the combined kidney/pancreas procedure with 20 or 25 years of diabetes and diabetic nephropathy also have the most established secondary complications, with only very limited room for improvement as a result of normal glycaemic control. The patients without secondary complications, in whom prevention would be a realistic aim, are seldom suitable for the unknown risk–benefit ratios of a transplant.

NUMBER OF TRANSPLANT CENTRES AND THEIR ORGANIZATION

There are a few gigantic transplant centres, mostly in the United States of America, where large numbers of surgeons, physicians and support staff transplant patients from large populations or draw patients from great distances, such as the Pittsburgh Center. Success with basic science has driven the size of some centres, but this has only really been accomplished where there is independence from cadaveric donation. The best example is probably the Fred Hutchinson Cancer Center in Seattle which continues to dominate the world of bone marrow transplantation. With these exceptions the transplantation rate in any country with a developed medical system is basically proportional to the number of transplant centres and to their geographical distribution (Table 1.2). Some cities have been characterized more by the number of transplant centres than by the number of transplants, demonstrating that geography and population

Table 1.2 The relationship between the number of transplant centres and the number of each type of transplant

Organ/tissue		Number of centres	Number of transplants per annum 1992	1993	1994	Mean number of transplants per centre per annum
Kidney	USA	220	9 976	10 625	10 548	47
	non-USA	349	14 040	14 034	12 305	39
Bone marrow	USA	96	2 790	2 934	2 958	30
	non-USA	225	3 762	3 829	3 759	17
Heart	USA	144	2 153	2 247	2 241	15
	non-USA	106	1 946	1 991	1 587	17
Heart/lung	USA	43	47	53	78	1.4
	non-USA	43	171	118	130	3.2
Lung	USA	59	550	646	550	30
	non-USA	44	328	363	388	8
Liver	USA	92	3 110	3 483	3 574	37
	non-USA	104	2 547	2 569	2 484	24
Pancreas	USA	39	94	137	117	3
	non-USA	41	38	28	19	0.7
Kidney/pancreas	USA	70	530	707	740	9
	non-USA	61	171	172	140	2.6

density act to impose limits on transplant rates, irrespective of how many units there are. There are nine renal transplant units in the New South Wales, Australia, eight of them in Sydney, and it is hard to believe that the increasing number delivers an increase in the transplant rate. Some will argue that proximity of a renal transplant unit directly affects the donor rate in a hospital, others will suggest that this represents a failure in the organization of organ donation and is an argument for properly resourcing organ donation processes, rather than resourcing a proliferation of transplant units.

The majority of countries utilize more than 95% of all kidneys offered for donation. There is, however, greater variability and a higher level of non-procurement of other organs. In some instances the reason for not retrieving an organ despite medical suitability and consent is the lack of a sufficiently close transplant unit, explaining the high 'wastage' rate of lungs. The increase in number of liver and heart transplant units and the sizes of their respective waiting lists has, in most countries, ensured that offers of those organs are almost always accepted, though regional variation is still seen.

The size of a transplant unit has a complex effect on the rate of transplantation. A very small unit, measured in terms of the number of transplants performed each year, will usually only have a small list of patients waiting. If one examines liver transplantation for example, a small unit may often have no suitable recipients for a particular donor offer because of the need to match for size and blood group. Unless there is an efficient organization for sharing between centres to create one large waiting list, there could be many liver donor offers refused and thus a reduction in the overall transplant rate. The same phenomenon can be seen with other organs (such as the lung) that share the characteristics of short storage times and stringent size, blood group or other compatibility criteria. At the opposite extreme, the global medical collaboration in unrelated bone marrow transplantation has ensured that the potential donations from more than three million volunteers are available to almost every patient irrespective of their location and the particular transplant unit.

Size of the unit, in terms of the number of surgeons, physicians, nurses and beds, has an impact on the transplant rate. It may become, for small units, a physical impossibility to undertake more than one donation and transplantation simultaneously or even consecutively. With the advent of multiorgan donation and multiple organ transplant services, a large hospital facility is required to undertake organ donation with up to eight hours of operating-suite time, followed simultaneously by a liver transplant in one theatre, a heart in another, two single lungs or one double lung in one or two more, and then perhaps a kidney–pancreas, kidney and two corneal transplants. A total of 40 hours of operating time spread across a minimum of three complete theatre suites and usually in the middle of the night, presents considerable problems. The logistics of size and the development of multiple transplant teams in single hospitals thus have the potential to reduce dramatically the actual transplant rate of solid organs. The individuals involved in the programmes are not indestructible, though the work rate in some centres implies immortality for surgeons. Small programmes, based upon a single experienced surgeon, are unable to iron out the peaks and troughs of activity that are the consequence of the random timing of death. It is for all these reasons that most countries have been driven towards the development of co-operation.

ORGANIZATION OF DONOR CO-ORDINATION

UNOS (United Network for Organ Sharing), Eurotransplant, Scandiatransplant, ONT (Organización Nacional de Trasplantes) and UKTSSA (United Kingdom Transplant

Support Services Authority) are all examples of the ways in which different countries or combinations of countries have sought to organize the process of sharing the solid organs donated. The drive for formation of these organizations came with a desire to maximize renal transplant outcome by tissue matching as well as to minimize wastage of kidneys of rare blood groups. A further invigorating force during the mid-1980s was the development of multiorgan donation and the need to meet the complex demands of the early liver, heart and lung programmes. In parallel with the development of such co-operative programmes, the co-ordination of donation at the hospital intensive care and operating suite needed 'professionalization'. Different countries have adopted different models for achieving this. For example, a series of organ donor procurement agencies was established in the USA with geographic responsibilities for co-ordination. In some countries donor co-ordination has remained within the transplanting units and hybrid arrangements are seen in others. There is even greater variety in the way in which corneal and other tissue banks have been organized (see Chapter 14). There can be little doubt that the transplantation rate in each country has been greatly influenced by the successes and failures of the chosen schemes. It would, however, be difficult to judge which model of organization leads to high transplant rates (Figures 1.2 and 1.3). There is no universal formula for success, since it is possible to find examples where one scheme leads to high transplant rates and short waiting lists in one place, but low transplant rates and long waiting lists in another.

Transplantation of the kidney

Renal transplantation first started as an experimental procedure in the 1950s,[17] though the first actual attempt was much earlier[18] and the decade of the 1950s saw only limited success, with the exception of identical twin transplants.[3] It was thus in the early 1960s with the advent of Azathioprine, in combination with corticosteroids, that transplantation became a viable option for a limited number of recipients.[19] There are still patients alive with grafts performed in the first half of the 1960s and though they are few in number, they do demonstrate that prolonged normal function can result from renal transplantation. The 1960s provided the basis of the explosion in clinical renal transplantation during the 1970s by determining the relationships between failure and blood group incompatibility[20] and the positive lymphocytotoxic crossmatch test.[21] The early groundwork for tissue type matching was also predominantly done in the 1960s, driving the science of histocompatibility and boosting development of the discipline of immunology.

Transplants in the 1990s came from both living and cadaveric donors with a surprising variation in the incidence of each between countries. There remain quite significant international differences in the prevalence of transplanted patients compared to those who remain on dialysis. In some countries, such as in the USA and Japan, dialysis is highly organized and therapy offered as an option almost irrespective of age, while in others either the incidence of renal disease is lower or the financial support is not available. In India, for example, only a small proportion of patients can be treated by dialysis and a renal transplant thus offers the most accessible form of long-term therapy.

The actual rates of transplantation expressed as a number of renal transplants per million population (pmp) are shown in Figure 1.5, updated since originally presented by the King's Fund Institute,[22] and show that the highest combined cadaveric and living related donor programmes achieve in the order of 40–45 renal transplants pmp. Norway reaches this figure by having a relatively large living related donor programme while Spain and Austria have almost no living donation but a high ratio of cadaveric donation. At the other end of the scale there are countries such as Australia and Denmark with fully developed transplant systems providing universal coverage to their populations, but with low rates of transplantation (25 pmp). There are also countries where for financial, religious, cultural or simply organizational reasons, cadaveric organ donation has been infrequent e.g. Saudia Arabia, Greece and Japan.

It is interesting to see the possible effect of the donor rate on the size of the waiting list of people in need of a renal transplant and the percentage of dialysing patients waiting. In Australia and New Zealand 37% and 34% of the dialysis population respectively, are on the active transplant list. Data from ANZDATA[23] are shown in Figure 1.6 demonstrating the percentages at each age group deemed to be appropriate for transplantation. The ratio of patients on the waiting list to patients transplanted each year provides a measure of the mean waiting time for a kidney and varies considerably between countries. In Norway the ratio is a low 0.7 so that the entire waiting list turns over in approximately $7\frac{1}{2}$ months, while in Australia there is a ratio of 3.0, implying a three-year waiting time. Both the USA and UK had a ratio of about 2 at the end of 1992, while Spain has been able to reduce its ratio progressively to less than 1 over the past few years. The approach both to dialysis therapy and to transplantation are dependent upon the average waiting time. Clearly in Australia with its very long waiting times, pre-dialysis transplantation cannot be achieved and all patients must expect to spend some years on dialysis unless they have a living related donor. In Norway, on the other hand it is frequently possible to transplant patients before dialysis is needed.

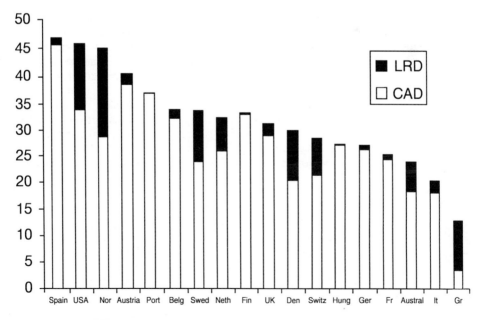

Figure 1.5 Renal transplantation rates. The rates of cadaver (CAD) and living related donor (LED) renal transplants are shown for several countries, demonstrating that the use of living related donation varies considerably.

Success rates of renal transplants have risen progressively since the early 1970s, with most programmes showing a step up of between 10% and 20% at the time of introduction of cyclosporine which occurred between 1981 and 1984 (Figure 1.7). There has been a major reduction in first cadaver graft loss from acute rejection such that it is now of equal importance to graft thrombosis and death with a functioning graft.[24] In the long term the two major issues affecting outcome are death, usually from coronary artery disease, and 'chronic' rejection (Figure 1.8). The focus of attention of many clinical nephrologists has thus turned to cardiac risk factors in transplanted patients and the definition and management of chronic graft dysfunction.

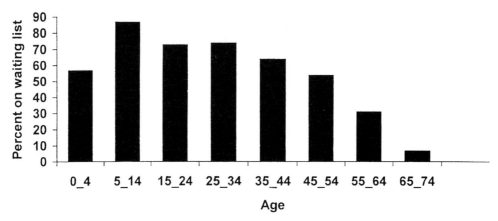

Figure 1.6 Age-specific proportions of dialysis patients awaiting transplantation. The percentage of patients being dialysed in Australia who are actively awaiting renal transplantation varies depending upon the patients' age. Approximately three-quarters of patients under 35 are thus deemed suitable for, or wish to have, a renal transplant.

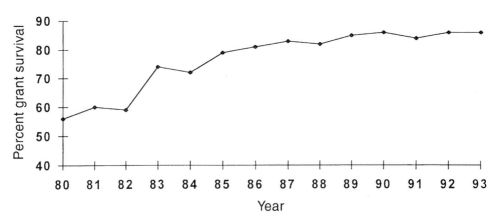

Figure 1.7 Renal transplant survival rates in Australia. The one-year actuarial graft survival for first cadaveric renal transplants is shown, based upon the year in which patients were transplanted in Australia.

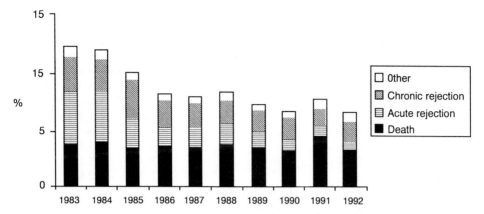

Figure 1.8 Causes of renal graft failure. The causes of graft loss in recipients of renal transplants in Australia are expressed as a percentage of the grafts at risk during that year. This shows that acute rejection has become a much less significant problem than chronic rejection and patient death.

Transplantation of other solid organs

The 1960s provided an era of exploration and discovery in transplantation during which many of the journeys ended on rocky shores. The uncharted territory included patient suitability, surgical technique, tissue matching, post-transplantation monitoring and immunosuppression. It is thus perhaps not surprising that, like the early geographers of our world, the early explorers in transplantation had systematically to set about charting the major features of the terrain.

The rate of progress during the 1960s and 1970s seen in renal transplantation was not echoed by other solid organ transplants. The number of active heart and liver transplant units was small until the 1980s and then grew explosively as use of Cyclosporine provided the turning point for achieving acceptable outcomes. The current status of solid organ transplants is shown in Table 1.3, which describes the number of each form of transplant in the main transplanting countries expressed per million population. In Table 1.4 the estimated total global numbers of transplants are shown by continent for 1995. The sum total of 51,964 solid organ transplants in 1995 represents a 4.5% increase over 1994, kidneys increasing by 4.4%; livers by 6.3%, lungs by 7.6% and pancreas by 9%, while heart increased by only 0.8% and heart lung decreased by 0.4%. The number of transplants actually decreased in Asia by 9.7% over this period while in North America and Europe there was an increase of 8.2 and 2% respectively, though much of this change occurred through variation in the rate of utilization of donor offers rather than changes in donor numbers.

The main indications for transplantation are shown in Table 1.5, though there is much variation from country to country and the particular mixture of diseases is also heavily influenced by the age range treated. Congenital biliary atresia is, for example, the diagnosis for 60% of paediatric and 0.3% of adult patients receiving a liver transplant.

Table 1.3 Rates of cadaveric organ transplantation in 1995 expressed as the number of transplants per million population (pmp). (*Source*: Council of Europe)

1995	Kidney	Liver	Heart	Heart/lung	Lung	Pancreas
Australia	19.4	6.6	5.3	0.7	3.7	0.7
Austria	39.1	14.7	14.4	–	3.8	1.0
Belgium	32.1	12.7	9.8	1.8	1.1	1.8
Canada	23.0	11.1	6.8	–	3.0	0.8
Denmark	21.9	8.9	5.7	0.5	3.5	–
Eurotransplant	27.0	8.2	6.4	0.3	1.1	1.0
Finland	32.8	6.2	5.2	1.2	0.8	–
France	24.0	11.1	7.4	–	1.4	0.8
Germany	26.0	7.3	6.1	0.3	1.0	0.7
Greece	4.2	0.7	1.0	–	0.1	–
Hungary	27.0	1.0	0.3	0	0	0
Italy	18.3	7.1	6.8	–	0.7	0.3
Netherlands	26.3	6.3	3.2	–	1.3	0.7
Norway	29.3	4.5	5.7	0.2	3.3	1.2
Portugal	36.8	7.0	0.5	–	–	–
Spain	46.0	18.1	7.2	–	1.1	0.6
Sweden	24.2	10.4	3.1	–	2.4	1.8
Switzerland	22.6	6.7	8.1	–	2.6	1.1
United Kingdom	28.9	11.3	5.6	0.9	1.8	0.5
USA	33.7	15.3	9.2	0.3	3.4	3.6

Table 1.4 Number of transplants performed worldwide in 1995. (*Source*: Sandoz Pharmaceuticals)

1995	Kidney	Liver	Heart	Heart/lung	Lung	Pancreas	Total
Europe	11 227	3 601	2 257	148	465	231	17 929
North America	12 611	4 240	2 541	75	944	934	21 345
South America	3 682	135	195	9	32	9	4 062
Asia	5 680	253	206	23	77	20	6 259
Middle East/Africa	2 165	69	116	1	14	4	2 369
Worldwide total	35 365	8 298	5 315	256	1 532	1 198	51 964

Costs and effects

One of the charges laid at the door of transplantation is that it is an expensive waste of money. Several different concepts are confused by those who make such charges and it is important to dissect those concepts to understand the issues. Transplantation, as a modern therapy, has needed to claim resources and in doing so those resources have had

Table 1.5 Indications for transplantation of solid organs. The percentages in brackets represent an approximate percent in adults during 1995; there are differences between countries and age groups

Kidney		
Glomerulonephritis	(50%)	
Polycystic kidney disease	(14%)	
Reflux nephropathy	(13%)	
Diabetic nephropathy	(10%)	
Renal vascular disease	(4%)	
Analgesic nephropathy	(2–10%)	
Liver		
Hepatitis/cirrhosis	(30%)	
Alcoholic cirrhosis	(20%)	
Primary biliary cirrhosis	(11%)	
Primary sclerosing cholangitis	(9%)	
Fulminant hepatic failure	(6%)	
Neoplasia	(5%)	
Metabolic diseases	(4%)	
Biliary atresia	(paediatric 60%)	
Heart		
Coronary artery disease	(45%)	
Cardiomyopathy	(45%)	
Valvular	(4%)	
Congenital	(1%)	
Heart/lung		
Primary pulmonary hypertension	(30%)	
Congenital	(28%)	
Cystic fibrosis	(15%)	
Emphysema	(5%)	
Lung	*Single lung*	*Double lung*
Emphysema	(42%)	(17%)
Interstitial pulmonary fibrosis	(16%)	(4%)
Alpha-1-antitrypsin	(16%)	(13%)
Primary pulmonary hypertension	(9%)	(10%)
Cystic fibrosis	(0.5%)	(36%)

to be measured, unlike those committed to 'traditional therapies' such as thoracic surgery for carcinoma of the lung or gastrointestinal surgery for bowel surgery. For those who seek to make comparisons it is important to understand the relative benefits of different types of treatment.

The quality adjusted life year (QALY) has been designed as a measure of the benefit that can be achieved by different treatments. Table 1.6 gives one set of such comparative benefits and makes the point that broad public health measures can, but do not always, yield very cost-effective results. On the other hand some therapies accepted as a norm, e.g. neurosurgery for cancer, are not such good value for money. In the realm of transplantation the cost of a life year is comparatively easy to quantify and so it is

Table 1.6 Cost per quality adjusted life year of a number of alternative health interventions. 1990/91 price[22]

Interventions	Cost per QALY (sterling)
Cholesterol testing and diet (40–69 g)	220
Neurosurgery for head injury	240
Pacemaker implant	1 100
Hip replacement	1 180
Coronary artery bypass, left main disease	2 090
Kidney transplant	4 710
Breast cancer screening	5 780
Heart transplant	7 840
Cholesterol testing and treatment (25–39 y)	14 150
Home haemodialysis	17 260
Coronary artery bypass, single vessel disease	18 830
CAPD	19 870
Hospital haemodialysis	21 970
Neurosurgery for malignant tumour	107 780

possible to produce quite accurate estimates, for example, of the relative costs of dialysis versus renal transplant. Transplantation is thus frequently now the treatment of choice by patients, their doctors and those who fund and administer health systems. The goal of this book is to understand what the real barriers are and how they may be overcome.

References

1. Starzl TE. *The Puzzle People*. University of Pittsburgh Press, Pittsburgh, 1992.
2. Barnard C, Pepper CB. *One Life*. George G. Harrap, London, 1969.
3. Merrill JP, Murray JE, Harrison JH, Guild WR. Successful homotransplantation of the human kidney between identical twins. *J Amer Medical Assoc* 1956, **160**: 277–282.
4. Gore SM, Taylor RM, Wallwork J. Availability of transplantable organs from brain stem dead donors in intensive care units. *Br Med J* 1991, **302**: 149–153.
5. Nathan HM, Jarreli BE, Broznik B et al. Estimation and characterization of the potential renal organ donor pool in Pennsylvania. *Transplantation* 1991, **51**: 142–149.
6. Espinel E, Deulofeu R, Sabater R, Manalich M, Domingo P, Rue M. The capacity for organ generation of hospitals in Cataluna, Spain. *Transplant Proc* 1989, **21**: 1419–1421.
7. Garrison RN, Bentley FR, Raque GH et al. There is an answer to the shortage of organ donors. *Surg Gyn Obstet* 1991, **173**: 391–396.
8. Hibberd AD, Pearson IY, McCosker CJ et al. Potential for cadaveric organ retrieval in New South Wales. *Br Med J* 1992, **304**: 1339–1343.
9. Siminoff LA, Arnold RM, Caplan AL, Virgin BA, Saltzer BA. Public policy governing organ and tissue procurement in the United States. *Am Intern Med* 1995, **123**: 10–17.
10. Evans RW, Orians CE, Ascher NL. The potential supply of organ donor. *J Amer Med Assoc* 1992, **267**: 239–246.

11. Gortmaker SL, Beasley C, Brigham LE et al. Organ donation potential and performance: size and nature of organ donor short fall. *Crit Care Med* 1996, **24**: 432–439.
12. Aranzabal J, Texeira JB, Darpon I et al. Capacidad generadora de organos en la CA del Pais Vasco. *Rev Esp Trasplantes* 1995, **4**: 14–18.
13. Thompson JF, McCosker CJ, Hibberd AD et al. The identification of potential cadaveric organ donors. *Anaesth Intens Care* 1995, **23**: 75–80.
14. Gore SM, Cable DJ, Holland AJ. Organ donation from intensive care units in England and Wales: Two year confidential audit of deaths in intensive care. *Med J* 1992, **304**: 349–355.
15. Feest TG, Riad HN, Collins CH, Golby MG, Nicholls AM, Hamad SN. Protocol for increasing organ donation after cerebrovascular deaths in a district general hospital. *Lancet* 1990, **335**: 1133–1135.
16. Chapman JR, Hibberd AD, McCosker CJ et al. Obtaining consent for organ donation in nine NSW metropolitan hospitals. *Anaesth Intens Care* 1995, **23**: 81–87.
17. Hume DM, Merrill JP, Miller BF, Thorn GW. Experiences with renal homotransplantation in the human: report of nine cases. *J Clin Invest* 1955, **34**: 327–382.
18. Voronoy YY. *Siglo Med* 1936, **97**: 296.
19. Calne RY. Organ transplantation from laboratory to clinic. *Br Med J* 1985, **291**: 1751–1754.
20. Gleason RE, Murray JE. Report from the transplant registry. *Transplantation* 1967, **5**: 343–359.
21. Patel R, Terasaki PI. Significance of the positive crossmatch test in kidney transplantation. *N Engl J Med* 1969, **280**: 735–739.
22. New B, Solomon M, Dingwall R, McHale J. A question of give and take. *Kings Fund Research Report 18*. Kings Fund Institute. London, 1994.
23. Australian and New Zealand Dialysis and Transplant Registry, 1983 to 1995. APS Disney (Ed.). Adelaide, South Australia.
24. Penny MJ, Nankivell BJ, Disney APS, Byth K, Chapman JR. Renal graft thrombosis. *Transplantation* 1994, **58**: 1–7.

PART 1

Organ and tissue donation in society

CHAPTER 2

Religious views on organ and tissue donation

**THE RIGHT REVEREND LORD JOHN HABGOOD,
ANTONIO G. SPAGNOLO, ELIO SGRECCIA, ABDALLAH S. DAAR**

Editors' introduction   23
    References   25
The Church of England   25
    Reference   27
The Roman Catholic Church   27
    References   29
Islam   29
    References   33

EDITORS' INTRODUCTION

Whenever individual attitudes to organ donation are studied, and especially when those attitudes are negative, there is a tendency to explain those attitudes in terms of the individuals' religious beliefs. Whenever the donation rates of a particular country are particularly low there is a similar tendency to turn to the religious persuasions of that country to find the answers. While it is usually possible to find a degree of divergence of view within a particular religion, there are very few religions for which organ or tissue donation conflicts with fundamental tenets. It is far more common for cultural, as opposed to religious, beliefs to contradict donation. In countries where the cultural beliefs dictate burial of the whole individual, for example, donation of organs such as the heart and kidneys may cause conflict. It is important to distinguish between the two issues primarily because religious views are defined by one process and cultural views by another. It is fallacious to believe that simply by the process of public understanding of a specific religious view, individual and cultural attitudes will follow. Equally, it is difficult for the individual to interpret the view of their religion without a lead. Failure to acknowledge the dilemmas and difficulties that modern medical therapies confront the individual with, would represent an abrogation of religious responsibility and most major religions have thus addressed the issue.

The definition of religious attitudes to organ and tissue donation has been tied to the progression of understanding of the potentials of transplantation. A therapy that is experimental, uncommon and of unproven benefit such as cardiac transplantation in the 1960s or baboon liver transplantation in the 1990s, attracts one level of analysis. A therapy that is standard, common and of proven benefit, joins the health measures that can together be seen as a society's fulfilment of providing a right to health for its members. The rate at which the different religions have thus needed to address the theological issues has also been partly dependent upon the stage of development of medical services. The converse of this is that interaction of religious stances, cultural views and medical technology has determined the delivery of service to each community. Where the religious and cultural views are not permissive, such as in Japan, lack of cadaveric organ donation has prevented delivery of transplantation therapy to large proportions of the community despite technical ability to do so.

During the late 1980s and early 1990s, there was a series of conferences which examined the ethical, religious and cultural aspects of transplantation. The first international congress on 'Ethics, Justice and Commerce in Transplantation: A Global View' was held in Ottawa, Canada in August 1989.[1] In June 1991 the Society for Organ Sharing met in Rome, Italy and elicited a special message on Organ Donation by His Holiness Pope John Paul II.[2] The third meeting was convened in Singapore in April 1992[3] and addressed amongst other topics the roles of religion in transplantation in developing countries. There have been other meetings in which religion and ethics have been placed side by side for discussion, but there have been no momentous shifts in the expounded views.

Despite slowly evolving analysis of religious view points on transplantation, it is the approach to the dead body that provides most religions with the underlying principles. In Japanese society the Shinto view of the dead body as an impure and dangerous object has been a basic concept for at least a thousand years. Injuring the dead body, for example by organ donation, has implications deeply embedded in belief and ritual,[4] that is not enhanced by a general public misunderstanding of brain death.[5] The Buddhist perceptions of impermanence of the relationships between body and mind, could lead to a positive view of organ donation both before and after death as an act of generosity which serves both as a compassionate act fulfilling one of the four guidelines for social living and as a condition for the realization of nirvana.[6] The Hindu religion has also a permissive approach to organ donation. The soul which lives irrespective of the death of the body and its cremation is supported by mythological use of human body parts to the benefit of humans and society. In India, it is thus the health priorities of sanitation and immunization that continue to impede transplantation in the public sector, though the adoption of cadaveric transplantation of organs other than the cornea has been slow also in the private sector.

While some religions draw on their understanding of human death to interpret the demands of organ donation and thus find broadly held views, other religions find diversity. The Jewish perspectives on organ donation cannot so comfortably be defined because of conflict between such views as the saving of life and prohibition of benefiting from a corpse. It is possible therefore to find a Jewish view that applauds both organ donation and transplantation,[7] but also to find community opposition.

The approaches of Muslim, Protestant and Roman Catholic religions are encompassed in the following sections. Each provide thoughtful essays outlining ways in which one of the problems of modern medicine can be seen.

References

1. Dosseter JB, Monaco AP, Stiller CR. Editors. Ethics, justice and commerce in transplantation: a global view. *Transplant Proc* 1990, **22**: 891–1054.
2. Pope John Paul II. Special message on organ donation. *Transplant Proc* 1991, **23**, XIII.
3. Daar AS, Lim S, Thompson JF. First International Congress on Transplantation in Developing Countries. *Transplant Proc* 1992, **24**: 1645–2133.
4. Namihira E. Shinto concept concerning the dead human body. *Transplant Proc* 1990, **22**: 940–941.
5. Aoki S, Akiba Y, Aranami Y, Ohkubo M et al. How to increase cadaver organ donation in Japan: a transplant recipient's approach. *Transplant Proc* 1992, **24**: 2064–2065.
6. Sugunasiri SHJ. The Buddhist view concerning the dead body. *Transplant Proc* 1990, **22**: 947–949.
7. Bulka RP. Jewish perspective on organ transplantation. *Transplant Proc* 1990, **22**: 945–946.

THE CHURCH OF ENGLAND

THE RIGHT REVEREND LORD JOHN HABGOOD

Anglicans share their basic beliefs and ethical principles with most other Christians. They often bring a particular style to ethical thinking, however, by placing strong emphasis on interdisciplinary study. Ethical judgements cannot generally be made on the basis of abstract principles alone, but need to be rooted in empirical realities and to draw on the insights from many disciplines including, of course, theology. The inclusion of chaplains in healthcare teams, when difficult ethical decisions have to be made, symbolizes this approach.

This willingness to stick close to the facts has encouraged a receptive attitude towards medical advances. Once the initial strangeness of transplant surgery was overcome, few Anglicans have had ethical difficulties with organ transplantation as such. Objections have mainly centred on the conditions under which organs are removed and transferred.

Basic ethical attitudes have been much influenced by the experience of blood transfusion. Titmuss's[1] famous study of blood transfusion as a 'gift relationship' has led to widespread recognition of the act of giving as important in itself, not only as a means of quality control, but for what it does morally to both donor and recipient. The same sense of moral value is also frequently expressed when tissues or organs are donated voluntarily, as in bone marrow or kidney transplants within families. Transplants from cadavers can also help relatives of the deceased to find a meaning in their death.

COMMERCIAL TRANSACTION

To turn such transfers into commercial transactions, however, radically alters their meaning, and is regarded by the Church of England as unacceptable, despite the obvious medical advantage in increasing the number of donors. To sell human organs would be to treat the human body as a commodity, thus dangerously reinforcing tendencies to interpret human life in more and more mechanistic terms. In Judaeo-Christian tradition respect for the body is an important part of respect for the person, and the ceremonies surrounding death bear witness to this understanding of the body as the bearer of personality. The buying and selling of body parts, like the buying and selling of persons, violates human dignity, and this devaluation of the person is perceived even when the sale is voluntary. Thus in Britain the sale of kidneys from living donors is prohibited. Blood, being a renewable resource, might in some circumstances be sold without attracting the same moral condemnation, but the weight of moral and medical opinion is in favour of voluntary donation.

ETHICAL DILEMMAS

Given an acceptable definition of brain death, the removal of tissues and organs from cadavers need not create ethical problems unless it interferes with the customary procedures for taking leave of the dying and paying respect to them. The haste with which transplant surgery needs to be performed can cause distress to relatives unless they are carefully prepared for it. The removal of organs from those whose hearts are still beating, while strongly desirable for medical reasons, can be traumatic both for medical staff and for relatives only just beginning to come to terms with the fact of death. There is no simple way of resolving the conflict of interest between those whose lives may depend on a transplant and those who want time to be with the dead and to distance themselves gradually from the death of loved ones. But it is important for medical staff to be constantly aware of the sensitivities involved.

PUBLIC ATTITUDES

Public attitudes may change as transplants become increasingly familiar, and the natural revulsion at some procedures may be overcome through education about the nature of death, and as the possibilities of giving additional meaning to death through organ donation are more widely perceived. It is essential, though, that donation should remain unmistakably voluntary, and for this reason the Church would be likely to oppose any move towards consent by contracting out rather than by contracting in.

USE OF TRANSGENIC ANIMALS

The use of transplants from transgenic animals would solve the problem of supply and avoid the ethical issues just discussed, but is likely to encounter strong emotional opposition. In strict logic those who eat meat and who rely on vaccines derived from animals cannot object in principle to the use of specially bred animal organs. There are, however, ethical questions to be asked about limits to the exploitation of animal life, especially when this entails breeding animals with specifically human genes. Crossing the species barrier between humans and, say, pigs may seem to reduce human distinctiveness, and with it the respect due to human life. But it is hard at present to know how to weigh such elusive considerations against the obvious advantages of animal transplants.

USE OF FOETAL TISSUE

Foetal tissue transplants have in general been ethically approved by church members, provided that decisions about abortion are kept quite separate from decisions about any subsequent use of such tissues. The use of foetal germ cells for cultivation as ova, and eventual transplantation into infertile women, has been opposed on the grounds that this might pose serious problems of identity for those conceived by this method. Germ cells are not like other tissues, and their use in research and therapy should remain strictly controlled.

Reference

1. Titmuss RM. *The Gift Relationship: from human blood to social policy.* Allen & Unwin, London, 1971.

THE ROMAN CATHOLIC CHURCH
ANTONIO G. SPAGNOLO, ELIO SGRECCIA

There are few moral objections within the Roman Catholic Church to organ transplantation. Apart from the sporadic negative reactions and vacillations of some Catholic moralists in the early years of transplants – mainly due to the limitations and dangers inherent in an experimental intervention – there was an enthusiastic basic approval from the Roman Pontiff in the fifties. He, on the occasion of a famous address to eye specialists and delegates of some concerned Italian associations, summed up Catholic teaching on transplants involving an organ (explicitly with regard to cornea, implicitly in the case of other structures) from a dead person.

At that time, Pius XII just stated that 'a person may will to dispose of his body and to destine it to ends that are useful, morally irreproachable and even noble, among them the desire to aid the sick and suffering'. So that, 'one may make a decision of this nature with respect to his own body with full realization of the reverence which is due it' and this decision should 'not be condemned but positively justified'.[1]

Also in the later magisterial teaching, to date, donation and transplant of organs are considered a 'service to life',[2] an intervention that 'shows its moral value and legitimizes medical practice',[3] a 'particularly praiseworthy example' of gesture of human sharing 'which build up an authentic culture of life'.[4]

There are, however, some conditions which must be observed, particularly those regarding donors, the organs donated and implanted, and the human act of donation. The Church refers for her teaching to some moral principles rooted in Sacred Scripture, Sacred Tradition and the Natural Law. The law of fraternal charity establishes that one's neighbour is to be regarded as another self and is based on the natural and supernatural unity of mankind.

We will distinguish between two main situations that can occur: the donation after death and the donation in life.

As the subject of this contribution is the 'donation' we exclude transplants from animals, as well as from anencephalic newborns and human foetuses from our considerations; nevertheless we would shortly recall that the Catholic moral teaching has a well-defined position on the above subjects.

Regarding *donation after death*, the Catholic moral theology reveals there is a lack of consistency in the ideas and fears of those opposed to these transplantations. On the contrary, we have already dealt with their worth and ethical validity recognized from the papal magisterium, in that they are legitimized by the Principle of Solidarity, which joins human beings, and by Charity, which prompts one to give to suffering brothers and sisters.[1,2] Besides, even if a corpse must always be respected as a human corpse, it no longer has the dignity of a subject and the value of a living person.

The questions which have arisen concerning this type of transplant derive therefore from factors other than transplantation itself.

One of these questions derives from the fear that in some cases the organ donor has not actually expired. This has been also recently denounced by Pope John Paul II, referring to 'more furtive, but no less serious and real, forms of euthanasia' that could occur 'when, in order to increase the availability for organs for transplants, organs are removed without respecting objective and adequate criteria which verify the death of the donor'.[4] Then, the removal of organs from a corpse is legitimate when the certain death of the donor has been determined. Hence, 'the duty of taking steps to ensure that a corpse is not considered and treated as such before death has been duly verified'.[3]

Much importance is now attached to the 'brain' criteria of death because the suitability of the organs to be removed is realized in the conditions made possible by life-support techniques. In order that a person be considered a corpse, it is enough that cerebral death of the donor be ascertained, which consists of the 'irreversible cessation of all cerebral activity'.[5] When total cerebral death is verified, in accordance with responsible and commonly accepted scientific criteria, it is licit to remove organs and also to surrogate organic functions artificially in order to keep the organs viable with a view towards a transplantation.

Another question concerns the authority over the body of the newly dead who did not bequeath his organs to others during his lifetime. As the medical intervention in transplants 'is inseparable from a human act of donation',[2] it is necessary to respect the consent of the nearest of kin because they can best interpret the wishes and attitudes of the newly dead: 'Organ transplants are not morally acceptable if the donor or those who legitimately speak for him have not given their informed consent'.[6]

Finally, there is the problem that ethically not all organs may be donated. It is a matter of organs, like the brain or the gonads, which are structurally connected with the personal and procreative identity respectively. They are organs 'which embody the characteristic uniqueness of the person, which medicine is bound to protect'.[3]

Regarding *donation in life*, it is clear that every organ or human tissue explanted from a living donor in some way impairs the corporeal integrity of the donor. Thus in the early years of this kind of transplant some eminent Catholic theologians argued that the Principle of Totality and Integrity – first formulated by Pius XII to prohibit eugenic sterilization[7] – would not justify the sacrifice of the person's bodily integrity for another, because one person is not related to another person as a means to an end or as a part of a whole. That initial rejection was founded on a wrong interpretation of a papal statement,[1] in which Pius XII would only remark that the Principle of Totality could never be used to justify organic transplantation between living human beings. But, while that principle could never be used to justify it, neither does it clearly exclude it. In fact, the later magisterium, appealing to the Principle of Charity, has legitimated this type of transplant to offering a chance of health and even of life itself to the sick who sometimes have no other hope: 'Organ transplants conform with the moral law and can be meritorious if the physical and psychological dangers and risks incurred by the donor are proportionate to the good sought for the recipient'.[6] 'Furthermore, the freedom of the perspective donor must be respected, and economic advantages should not accrue to the donor'.[8]

In conclusion, it must be recognized that with the advent of organ transplants the Roman Catholic Church is challenged to love her neighbour in new ways: 'even unto the end' (John 13:1). Yet this love has certain limits which cannot be transgressed – limits placed by human nature itself.

The possibility of 'projecting beyond death their vocation to love' should persuade persons 'to offer during life a part of their body, an offer which will become effective only after death'. This is 'a great act of love, that love which gives life to others'.[2] In this sense, 'Catholic health care institutions should encourage and provide the means whereby those who wish to do so may arrange for the donation of their organs and bodily tissue, for ethically legitimate purposes'.[8]

References

1. Pius XII. To the delegates of the Italian Association of Cornea Donors and the Italian Union for the Blind (May 14, 1956). In: *Acta Apostolicae Sedis*. Vatican City 1956, **48**: 462–465.
2. John Paul II. To the participants at the First International Congress on the Transplant of Organs (June 20, 1991). In: *Teachings of Jean Paul II*. Vatican City 1991, **XIV/1**: 1710–1712.
3. Pontifical Council for Pastoral Assistance to Health Care Workers. *Charter for Health Care Workers*. Vatican City 1995, ns. 83–91.
4. John Paul II. *Enciclical Letter 'Evangelium Vitae'*. Vatican City 1995, ns. 15 and 86.
5. Pontifical Academy of Sciences. Concluding Document on the artificial prolongation of life and the determination of the exact moment of death (October 21, 1985). In: Chagas C (ed.) *Pontificiae Academiae Scientiarum* Scripta Varia 1986, **60**: 113–114.
6. The Catechism of the Catholic Church, n. 2296.
7. Pius XII. Enciclical Letter 'Casti Connubi'. In: *Acta Apostolicae Sedis*. Vatican City 1930, **22**: 565.
8. National Conference of Catholic Bishops. Ethical and religious Directives for Catholic Health Care Services. Washington DC: US Catholic Conference Publishing Services, 1994. Directives 29–30 and 62–66.

ISLAM

ABDALLAH S. DAAR

Over a billion people in the world are Muslims; most live in developing countries, where rates of renal failure are high and transplantation is increasing. To most Muslims, religion plays a major role in their lives. It is not surprising, therefore, if questions related to religion often arise when discussing transplantation with Muslims and these are listed in the next section. Some of the basics are, of course, well-known to Muslims themselves, but in this day of increasing international contact, it may not be a Muslim physician or co-ordinator who has to discuss these matters with Muslim donors or recipients.

QUESTIONS RELEVANT TO ORGAN AND TISSUE DONATION

- What is Islam's conception of disease?
- Does Islam accept modern therapies such as transplantation? Does it encourage transplantation?

- What are Islam's teachings that would encourage donation?
- Is a Muslim a fatalist who need not seek therapy for major illness?
- What is the conception of the relationship of man to man, and of man to Allah?
- What is the meaning of life, and what is death? Who diagnoses death?
- If there is a day of judgement, and of resurrection, what happens if a part of the body is missing?
- Is it permissible to cut up a dead body to remove an organ?
- Have there been any official rulings about donation and transplantation? If so, what is the legal process?
- Should the donor sell his/her organs for transplantation?
- What is practised in Muslim countries?
- Looking ahead to xenotransplantation: what if the donor is the pig?

THE ANSWERS TO ISSUES MOST OFTEN ARISING

Nature of disease; obligation to seek treatment

Islam does not subscribe to the 'punishment' theory of disease causation. Diseases have causes other than retributive divine intervention. Life is sacred, and must be preserved whenever possible. To seek self-preservation through modern scientific treatment is therefore an obligation. Organ failure, just like any other disease, must be treated if possible; if the best method of treatment is transplantation, then that is not only acceptable, but must be encouraged and the authorities must provide for the therapy, provided this is not harmful to others. Primarily, however, disease must be prevented if possible and health is to be promoted.

Qualities conducive to transplantation

Islam encourages the following qualities which are supportive and conducive to organ donation and discourages exploitation and compulsion.

1. Altruism: the faithful are described in the Quran as: *'They give priority over themselves even though they are needy.'*
2. Generosity: *'In the society of the faithful, donations should be in generous supply and should be the fruit of faith and love of God and his subjects.'* Islamic Code of Medical Ethics.
3. Duty: *'The donation of body fluids or organs, such as blood transfusion to the bleeding or a kidney transplant to the patient with bilateral irreparable renal damage, ... is "Fardh Kifaya", a duty that donors fulfill on behalf of society'.* Islamic Code of Medical Ethics.
4. Charity: The Quran says *'And whosoever saves a human life it is as though he has saved all mankind.'* *'If the living are able to donate, then the dead are even more so; and no harm will afflict the cadaver if the heart, kidneys, eyes or arteries are taken to be put to good use in a living person. This is indeed a charity.'* Islamic Code of Medical Ethics.
5. Responsibility: Umar ibnul Khattab, the second Islamic Khalifa, decreed *'If a man ... dies of hunger ... the community should pay his "fidiah" (money ransom) as if they had killed him'.* *'The individual patient is the collective responsibility of society, which has to ensure his health needs by any means (while) inflicting no harm to others.'* Islamic Code of Medical Ethics.

6. Co-operation: *'The faithful in their mutual love and compassion are like the body ... if one member complains of an ailment, all other members will rally in response.'* Tradition of the Prophet. *'The faithful, to one another, are like the blocks in a whole building ... they fortify one another.'* Tradition of the Prophet.
7. Public health education: doctors are expected to play a major role in educating the public.

Allah, man and the gift of life

Man is encouraged to live in peace with fellow man: Muslims in a community constitute an 'Umma' – which requires mutual co-operation and support to thrive. While altruism is encouraged and minimal risk is acceptable, significant harm to oneself while alive is not encouraged in order to help another. The person is made up of the physical body and the soul. Not much knowledge has been given to man regarding the soul, except that Allah breathes His spirit into the foetus to make it a complete human being and death occurs when the soul has departed from the body; the exact moment of this occurrence is beyond man's perception. Life is a gift from God and it must not be forgotten that everything in life, including the body, belongs to God. Abuse of the body and suicide are therefore sinful.

Islamic law

Islamic law is the Shariah, which is at once permanent and yet capable of responding to every challenge, including issues in transplantation. The basis of Shariah is the Quran and the Sunna, i.e. the sayings, actions and rulings of the Prophet Mohammed (peace be upon him). The fundamental principles of jurisprudence when an issue arises that requires a decision are:

- Need and necessity are equivalent.
- Necessity allows 'prohibited' matters.
- Injurious 'harm' should be removed.
- Prevention of evil has priority over obtaining benefit.
- The greater benefit prevails over the lesser benefit.[1]

Rulings on transplantation by religious authorities

Based upon these principles and the foregoing attributes expected of a Muslim, the majority of Islamic legal scholars have concluded that transplantation of organs as treatment for otherwise lethal end-stage organ failure is a good thing. Donation by living donors and by cadaveric donors is not only permitted but is encouraged.

The senior Ulamaa (Scholars) Commission of Saudi Arabia in Fatwa (legal ruling) no. 99 dated 9.11.1402 (25 August 1982) resolved by majority the following:

> The permissibility to remove an organ or part thereof from a dead person for the benefit of a Muslim, should the need arise, should the removal cause no dissatisfaction, and should transplantation seem likely to be successful.

In Egypt, the Grand Mufti Gad al Haq decreed in Fatwa no. 1323 of 5 December 1979 that it is permissible to donate organs from the living provided no harm was done, the donation was done freely, in good faith and for the love of God and the human fraternity. He also sanctioned cadaver donation provided there was a will of testament or the consent of relatives of the deceased. In cases of unidentified corpses, an order from the Magistrate should be obtained prior to removal of the organs.

The Third International Conference of Islamic Jurists, at their meeting in Amman, Jordan, on 16 October 1986, in resolution no. 5, by majority, declared in favour of using brain death criteria to establish death, as follows:

> A person (is) considered legally dead, and all the Sharia's principles can be applied, when one of the following signs is established:
> 1. Complete stoppage of the heart and breathing, and the doctors decide that it is irretrievable.
> 2. Complete stoppage of all the vital functions of the brain, and the doctors decide it is irreversible, and the brain has started to degenerate.

It must be remembered that Islam is not monolithic and allows for differences of opinion in details. At various stages there have been minority opinions, requiring cessation of the circulation and respiration, for death to be diagnosed. Nevertheless, the Fourth International Conference of Islamic Jurists, meeting in Jeddah in 1988, endorsed all previous Fatwas, including the landmark Fatwa no. 5 (above) of 1986 which allowed cadaver donation and the use of brain death criteria. In addition, the 1988 Conference clearly rejected any trafficking or trading in human organs, and re-stressed the principle of altruism.[2]

Draft law on human organ transplants

Muslim Arab countries have actually taken the lead in trying to formulate international legal norms: the 12th session of the Council of Arab Ministers of Health meeting in Khartoum, Sudan, 14–16 March 1987, responding to a WHO call, promulgated the Unified Arab Draft Law on Human Organ Transplants (A.40/INF.DOC/6–30 April 1987). Article 1 states:

> Specialist physicians may perform surgical operations to transplant organs from a living or dead person to another person for the purpose of maintaining life, according to the conditions and procedures laid down in this law.

Article 5 states:

> Organ transplants may be performed from a dead body, provided that consent is obtained from the next of kin, under the following conditions:
> 1. Death has been definitely established by a committee of three specialist physicians, including a neurologist. The physician who is to perform the operation must not be a member of the committee.
> 2. The deceased, while alive, did not object to the removal of any organ from his body.

SUMMARY

In summary, disease must be prevented if possible, but when it develops, every available scientific method must be sought for its healing. Transplantation is considered a valid method of treatment, particularly to save lives. Both cadaver and live donor transplants are permitted, provided that there is no compulsion or exploitation and there is adequate informed consent. Organs should not be the object of trading or trafficking. Brain-death criteria to establish death have been accepted by the majority of Muslim Jurists. Individual scholars may hold different opinions; what I have described above is the majority opinion, as reflected in actual practice in Muslim countries, particularly in the Middle East, where transplantation is now practised in almost all the countries. Saudi Arabia performed the first cadaveric kidney transplantation in 1984[3] and has a very active cadaver programme based upon brain-death criteria. Oman has a similar Ministerial Decree, which has allowed us to use cadaver organs based on brain death criteria; Turkey has similar laws as do other countries, e.g. Kuwait. Not only kidneys, but now liver and heart transplants are being performed in Muslim countries.

If ever the day came when xenotransplantation was possible, this would easily be justified in Islamic law, even if the donor was a pig. I have discussed this extensively elsewhere[4] but one fact is enough to put this to rest. Over 900 years ago tooth and bone transplantation was practised in Muslim countries.[2] Porcine bones at that time gave better results than bones from other species for grafting, and despite the injunction against the eating of pig flesh, Islamic Jurists advocated the use of these porcine bones.

The question regarding waking up on the day of judgement/resurrection can be dealt with by the following logical argument: surely if Allah is powerful enough to be able to create a whole person in the first place, then Allah is more than capable of reconstituting him again if the need arose, for Allah is the creator of all the Universe, and he cherishes and cares for his creatures, especially man, whom he has designated his vicegerent on Earth.

References

1. Al Qattan M. The Jurisitic Ijtihad Regarding Transplantation of Organs in Organ Transplantation: Proceedings of a Symposium held in Riyadh 1984. MS Abomelha (Ed.), Medical Education Services, Oxford, 1984.
2. Al Bar MA. Islamic view on organ transplantation. In Proceedings of the 2nd International Conference of Middle East Society of Organ Transplantation, Kuwait, 11–15 March 1990. GM Abouna, MSA Kumar and AG White (Eds), Kluwer Academic Publishers, Dordrecht, Netherlands, pages 563–578, 1990.
3. Al Otaibi K, Al Khader A, Abomelha MS. First Saudi cadaver donation. *Saudi Medical Journal* 1985, **6**: 217.
4. Daar AS. Xenotransplantation and religion; the major monotheistic religion. *Xeno* 1994, **2**(4): 61–64.

Further reading

Daar AS. Current Practice and the Legal, Ethical and Religious Status of Post Mortem Organ Donation in the Islamic World. In *Organ Replacement Therapy; Ethics, Justice and Commerce*. W Land and JB Dossetor (Eds), Springer-Verlag, Berlin, pages 291–299, 1991.
Daar AS. Transplantation in Developing Countries. In *Kidney Transplantation, Principles and Practice*. 4th edn, PJ Morris (Ed.), Saunders, Philadelphia, pages 478–503.
Hathout H. Islamic basis for Biomedical Ethics. In *Transcultural Dimensions in Medical Ethics*. E Pellegrino, P Mazzarella, P Corsi (Eds), University Publishing Company, Frederick, Maryland, pages 58–72, 1992.
Islamic Code of Medical Practice. The Kuwait Document (1981). International Organisation of Islamic Medicine, 1st Edition Kuwait IOMS. See also Abdul Rahman, C Snine and Ahmed El Kadi (1989): Islamic Code of Medical Professional Ethics. In *Cross Cultural Perspectives in Medical Ethics: Readings*. Robert M Veatch (Ed.). Jones and Bartlett Publishers, Boston, pages 120–126. (Copies in Arabic and English can be obtained from the author of this section.
Sahin AF. Islamic Transplantation Ethics. *Transplant Proc* 1990, **20**(1): 1084–1088.

CHAPTER 3

Public attitudes toward organ and tissue donation*

KAREN L. SMITH, JUDITH B. BRASLOW

Introduction 34
Reluctance of minorities to donate 36
Brain death and presumed consent 40
Family discussion 41
Motor vehicle administrations 42
Conclusion 43
References 44

Introduction

Attitudes toward donation are complicated by many factors, which are reflected in public opinion polls conducted during the past seventeen years. Comparisons are difficult because of the lack of standardization in sampling methods, question formats, and polling techniques. Nevertheless, these polls have consistently shown that awareness of organ transplantation is very high among the general public and that organ donation is perceived as a socially desirable action.

Today, organ transplantation is no longer considered experimental and has been proven to increase greatly the quality and longevity of thousands of transplant patients. Currently, nearly 38,000 patients are awaiting life-saving and life-enhancing organ transplants. Unfortunately, while medical and pharmacological advances make it possible for greater numbers of people to recover from formerly fatal diseases and to live longer lives once they have received organs, the supply of organs has remained relatively static, thereby resulting in a critical shortage.[1]

* This chapter was written by K.L. Smith and J.B. Braslow in their individual capacities. The views expressed in this chapter do not necessarily reflect the view of the Department of Health and Human Services or the Federal Government. The authors acknowledge the assistance of Dr M.L. Gauikos in writing this chapter.

Societal attitudes including cultural values, religious beliefs and death anxiety that result in a reluctance to donate organs are cited as one reason for the shortage.[1] A greater understanding of the concerns and beliefs of the general public is necessary to effectively target public awareness efforts and ultimately, to reverse current trends and increase donation.

The success of organ donation depends on the altruism of the public. Research continues to show that the public, especially white and more highly educated survey respondents, supports organ donation in broad terms. In addition, surveys of both the general public and actual donor families indicate that family members are more likely not to donate organs and tissues if they do not know their relatives' wishes regarding donation and are asked to decide in the absence of this information.[2]

While required request laws have resulted in an increase in the number of families being approached for donation, some organizations report that a greater percentage of these families are refusing to consent to organ donation.[3]

THEORY OF REASONED ACTION BEHAVIOUR

Programmes designed to promote behaviour change are based on a number of models, theories and disciplines. Behavioural change models and theories outline the complex process of individual change. Trial of a new behaviour, such as family discussion and acceptance of organ donation intentions follows a sequence of events: awareness; understanding; interest; acceptance; personalization; decision-making related to the issue.

Carter[4] describes the Fishbein and Ajzen's 'Theory of Reasoned Action'[5] as the 'state of the art' for predicting specific behaviour intention as well as personal decision making in performing or not performing a behaviour. He states 'this theory can be used to explain virtually any behavior over which an individual has volitional control'. This model assumes behaviour intention as the mediator for all other factors influencing behaviour.

The Theory of Reasoned Action offers guidance to tracking the public's knowledge, attitudes and behaviours to assess whether and what changes are occurring in organ donation. It permits modifications of motivational strategies as the public views change. These motivational strategies can lead to effective mass media campaigns to increase awareness about the paucity of organ donors and establish it as a priority of concern. For organ donation, as for many other behaviour change issues, interactions between several people/organizations are required for the goal to be met.

PUBLIC EDUCATION

It is clear that while public education messages have been consistent in focusing on the beneficial and altruistic nature of donation and have been successful in reinforcing the value of donation to the white majority, they have failed to address the needs and concerns of substantial segments of the population. Public education efforts can affect donation rates only if they are targeted at those who are not already committed, including minorities and those with little formal education.

Currently, public education falls short of this goal. Its messages revolve around slogans about the need for organs or the moral worth of donation – matters undisputed by, but uninteresting to, the key target populations. The challenge of public education is to convey the dual message that organ and tissue donation protects rather than endangers the interests of potential donors and that the process itself is simple and unintimidating. Changes must be made both in the content of the messages conveyed and in the mechanisms used to transmit those messages to the appropriate people.[6]

The process of developing effective public education campaigns begins with a careful examination of the research conducted on various target populations. According to ethicist Arthur Caplan,[7] there are a variety of ways to measure public opinion about organ and tissue donation. One strategy is to determine the degree to which members of the general public are informed about various aspects of donation and transplantation. Another indicator is determining how willing individuals say they are to donate their own organs. An assessment can also be made of the reported willingness of the public to donate organs of their family members. In addition to prospective attitudinal scales, actual donation rates can be determined as can the percentage of people carrying organ donor cards.[7]

Although Caplan reports that according to public opinion surveys and polls conducted over a span of more than ten years, public education campaigns undertaken by various organ and tissue procurement organizations, community groups, government agencies and private foundations have been quite effective in increasing public awareness of transplantation (American Council on Transplantation and Task Force on Organ Transplantation, 1986), much work remains to be done to educate those with unfavourable attitudes toward donation as well as to bridge the gap between willingness to donate and actual donation.

In recent years, experts have begun to conduct studies with actual donor families in order to obtain insight and information that could lead to effective public education campaigns. This chapter will explore and attempt to consolidate much of the research that has been conducted regarding public attitudes toward donation and transplantation in an effort to define more clearly those issues which must be addressed in order to increase organ and tissue donation in this country.

Reluctance of minorities to donate

As minority populations experience rapid growth in the United States, it becomes increasingly important to recognize and understand cultural differences that influence these segments of the population, particularly in relation to health matters. African Americans, Hispanics, Asian/Pacific Islanders, and American Indians remain fairly unaware about organ and tissue donation even though data show an unabated need for donors from these subgroups. A LifeGift Organ Donation Center survey conducted in ten Houston clinics examined knowledge, attitudes and beliefs of minorities about organ and tissue donation. The survey revealed specific differences among minority groups, indicating the need for cultural awareness and sensitivity. Collectively, survey results suggested that minority groups are receptive to educational programmes and that the lack of donors may, in part, be due to a lack of information.[8] Jack Lynch, Hospital/Community Development Specialist of the Regional Organ Bank of Illinois contended:

> ... this segment of the population is unresponsive because it is being approached by the wrong people asking the wrong questions ... communication and perception lie at the root of the problem ... professionals in the transplant community have counted them out before they have been effectively informed ... (Lynch, 1990).

In a review of common reasons for the reluctance of minorities to donate organs for transplantation, it was found that major barriers stood between many African Americans, Hispanics, Asians/Pacific Islanders and Native Americans and their willingness to donate. These barriers include:

1. Distrust of the white majority group, of which most physicians and healthcare workers are a part.
2. Cultural differences in communications between ethnic minorities and healthcare personnel.
3. Belief that physicians would prematurely declare a minority patient dead in order to surgically remove an organ for transplantation to a patient of the majority group. This belief may be held more strongly by minority group members of low socioeconomic status, who also believe that the organ will likely be donated to an affluent patient.
4. Religion and/or superstition, such as some Native Americans' belief that the body must remain intact after death in order to become part of the great ancestral society.
5. Lack of awareness of the need for donated organs, which may be a result of the lack of exposure to information about organ donation and transplantation due to isolation from mainstream society.
6. Lack of faith in the success of transplantation, probably more prevalent among members of groups with low socioeconomic status, regardless of race or ethnicity.
7. Intense emotional state of the family, which may include hostility toward physicians or others who approach the family about donation. The frequency of this occurrence may be reduced if willing donors make their families aware of their wishes in advance.[9]

MEASURING ATTITUDES

In a telephone survey conducted in 1990 by Lieberman Research for the Association of Organ Procurement Organizations,[10] African Americans and Hispanics were oversampled to permit separate analysis of these groups and the total sample was weighted to reflect census estimates of the incidence of these minority groups in the population. The survey, conducted with 503 respondents, age 18 or older, concludes that general public education about the benefits of a strong organ donor programme can utilize appeals based on self-interest, altruism and transplant efficacy.

While the majority of the public believes that being an organ donor is 'the right thing to do' and would consider having a transplant themselves if needed (of the three appeals, altruism was the only one mentioned spontaneously as a reason for favourable attitudes toward being an organ donor), participation in an organ donor programme is inhibited by lack of general knowledge about donor and transplant programmes and specific knowledge about how or where to sign up.

Participation in organ donor programmes is also inhibited by public attitudes related to family, fairness, and medical trust. The study also concluded that public education efforts should include information on the costs of donation, that is, the fact that donor family members would not be charged for the surgical procedure to remove organs.

With regard to family issues, while almost everyone said they would respect family members' wishes to donate organs, most people, including many donor card signatories have never discussed organ and tissue donation with family or friends. Organ donation is adversely affected by this absence of family discussion, especially in light of the fact that almost a quarter of the respondents reacted negatively to the information that families must agree to donation at the time of death *even when a donor card has been signed.*

Fairness issues centre on the importance of communicating that matching organs to recipients is based on tissue typing only, not race, ethnic background, or socioeconomic status. The study found that, for the most part, the attitudes of African Americans and Hispanics differ from the general population in degree rather than direction. However, fairness issues, along with the issue of medical distrust, were cited more frequently in the AOPO study with African Americans having the highest level of mistrust about potential donors being declared brain dead too quickly.

Other factors which inhibit African American participation in organ donor programmes include: less confidence (than the general public) that their families would be willing to donate, resistance to the emotional benefits of organ donation and the positive benefits of transplantation, failure to believe that their religion permits organ donation, although there was a positive response from the information that being an organ donor would not preclude an open casket funeral.

Other factors which inhibit Hispanic participation in donor programmes, according to results from the AOPO study, include: less certainty that the public would follow their wishes, little knowledge about donor or transplant programmes or where to find donor cards that are culturally relevant, and uncertainty about whether their religion permits organ donation.

DEFINING ATTITUDES TOWARD ORGAN AND TISSUE DONATION

The amount of information available on Hispanic attitudes toward organ donation was greatly increased by the completion of a survey of Hispanic households in Northern California conducted by the Gallup Organization for Oscar Salvatierra, Executive Director of Pacific Transplant Institute and his colleagues with the Organ Procurement Transplant Service at the University of California, San Francisco. The survey consisted of telephone interviews of 505 Hispanic heads of household and was conducted in 1987.

The survey questionnaire was designed to measure and evaluate Hispanics' knowledge of and attitudes toward organ donation and related issues. While 82% of the entire sample said that they felt they were treated fairly when they went to a hospital, clinic, or physician, only 68% of the foreign-born respondents, 70% of the Spanish-dominant respondents, 72% of the low-income respondents, and 74% of the respondents with less than a high school education said they felt they were treated fairly at the hospital. The response to two related questions also indicates a lack of trust in physicians and in the health care system.[11]

The results of this study have far-reaching implications for the future of organ donation among Hispanics. Grass roots efforts to ensure that Hispanics feel comfortable with their hospital treatment and that of their relatives will have to be implemented before organ donation can be increased.

Attitudes among African Americans toward donating kidneys for transplantation were explored in a pilot project conducted by Dr Clive Callender et al.[3] Although the project consisted of a group discussion and was not a sample survey, it did reveal themes that have proven useful and have been substantiated by more scientific data. One general theme that surfaced was the lack of knowledge about kidney transplantation and the need for improved communication and education.

Another theme involved the lack of referral and follow-through on the part of healthcare personnel, as several project participants said they felt that the ability of the physician to impart (or not impart) information and explain the significance of the transplant was directly connected to their desire to donate. Some participants expressed scepticism about their treatment in a hospital where a kidney was needed, while others cited a religious or superstitious belief that, for them, precluded donation.

There were also several themes that had positive implications for donation, the strongest being empathy, or the desire to help a suffering individual. This was strongest in the case of a living donor. Another interesting positive factor was the belief that donation might compensate for an impure past life. In a sudden-death situation, the wishes of the deceased was a factor in the decision-making process.[3]

The relationship between someone having signed a donor card and a family's decision to donate has been established by a number of studies. Experts postulate that although healthcare personnel often hesitate to approach families because requests are often met with hostility and anger, as more people sign and carry donor cards the likelihood of a positive encounter increases. Dr Callender states that the most important role for organ donor cards is in stimulating family discussion.[12]

Surgeon General Antonia C. Novello, MD, MPH, echoes Horton and Horton's belief in the vital role of public education, writing in 1992, that public awareness campaigns of recent years have attempted to alleviate common misconceptions about donation and have stressed the importance of discussing donation with family members. Recalling the research that has been conducted, she reiterates what she believes are the major barriers to donation, namely, the fear of premature termination of care if it were known that the patient was a potential donor, uncertainty about the overall quality of care, and, to a lesser degree, religious concerns.[13]

Writing four years before the Surgeon General, Evans and Manninen[14] conclude that while surveys generally show that the public is both aware of and supportive of transplantation, they remain uncertain as to the merits of organ donation, and are even less willing to donate their own organs than those of a relative who has been pronounced brain dead.

This view was supported by Walker et al.[15] in a telephone survey of parents in the National Capital Region conducted to assess parents' attitudes toward donating their children's organs and to provide physicians with information that could help alleviate their concerns about approaching parents for consent. Of the 339 parents who agreed to answer questions after being given details about their child's 'death', 288 (85%) said they would be willing to donate their child's organs.

The degree of willingness was associated with the certainty of death, altruism and empathy toward other children in need of organs, a previous discussion about organ donation with a family member, and knowledge of an adolescent or adult child's attitude toward donation. Factors that inhibited the intention to donate included uncertainty of death, insufficient information from medical professionals and fear of mutilation.

The child's age was not significantly associated with intention to donate. The age, sex, socioeconomic status and religion of the parents did not significantly relate to the parents' intention to donate. Concordance between the results and actual donation rates in Canada and the United States supports the general applicability of the survey's findings.

Although most parents reported little personal experience or knowledge of issues related to transplantation, most still indicated a favourable attitude toward organ and tissue donation. Walker et al.[15] suggest that the combination of a favourable attitude and the lack of awareness, in particular the parents' overestimation of the frequency with which physicians initiate the request, contributes to the shortage of paediatric organs.

The authors write that apparently parents assume that they will be asked in appropriate donor situations. This implies that when they are not asked, which is often the case, they will not offer to donate because they do not think it is feasible.

The authors conclude that the positive feelings toward donation expressed by parents interviewed for this study may help to alleviate some of the concerns physicians have about the appropriateness of requesting consent for organ donation and provide them with greater confidence.[15]

Brain death and presumed consent

Public confidence in the concept of brain death and in the ability of physicians to declare death without conflict of interest, as demonstrated in the survey of parental attitudes, is essential to the future of transplantation. Once death has been established, the Uniform Anatomical Gifts Act (UAGA), enacted in every state and in the District of Columbia, authorizes the gift of cadaver organs and tissues. It allows individuals to arrange for the donation of their organs and tissues by signing a donor card or by making provisions in a will. For those instances where there are no indications as to wishes of the deceased, the UAGA allows family members to make the decision regarding donation.

Because only 15% of the potential donors actually donate, a number of laws have been enacted to help increase donation, some of which presume consent in the absence of any known objections. In about a dozen states, corneas (and sometimes pituitary glands, bone or skin) can be removed under the jurisdiction of a medical examiner without actual consent from family members. Although this system has produced thousands of corneas for transplantation, with few exceptions, there is still little public support for routine recovery of solid organs, such as kidneys or hearts. Currently, presumed consent for organs is not allowed in any state, although a number of European countries require it.[16]

The lack of support for presumed consent in the United States therefore requires that health professionals and others involved in organ donation and transplantation focus on education and communication regarding the issues surrounding organ and tissue donation. In March 1993, experts from across the country gathered to explore the findings and their implications for public and professional education of the largest survey ever conducted in the United States on public attitudes toward organ donation and transplantation. The survey was conducted by the Gallup Organization for The Partnership for Organ Donation.[17]

Survey results indicate that nearly nine out of ten Americans support the general concept of organ donation and virtually all agreed that it allows something positive to come from a person's death. Support for those concepts is positively correlated with higher levels of education. Predictably, support for the concept of organ donation is lower among non-whites than among white respondents. More than one-third of respondents (37%) reported they are very likely to donate their organs after death and an additional one-third (32%) reported they are somewhat likely to donate. There was virtually no difference between men and women.

African Americans and Hispanic respondents were also more likely than whites to indicate a desire to be buried 'intact'. Moreover, most respondents (85%) believe organ transplants extend the recipient's life and that the additional years are healthy ones. African American and Hispanic respondents tended to disagree with this statement.

Though the majority of respondents did not view organ transplantation as an experimental procedure and would accept a needed organ, nearly two out of five thought it was experimental. Nearly half of those who oppose organ donation indicated that they believed it was experimental. More than two-thirds of respondents said they believed that most people who needed transplants did not receive them, but 20% appeared to believe that the supply of organs was adequate for the demand. This misconception was particularly prevalent among those who opposed the concept of organ donation.

Family discussion

While a high percentage of respondents in The Partnership for Organ Donation study[17] (85%) said that they approve of organ donation, and virtually all agreed that it is important for families to discuss organ donation, less than half (42%) have themselves made a personal decision about donation of their organs. Even fewer (25%) have made a decision about the donation of family members' organs. Respondents who have made personal decisions regarding their own family members' organs tend to be more highly educated, and African American respondents were less likely to have made a decision than either white or Hispanic respondents.

Although the majority said they believe that organ donation may assist families in coping with the loss of a loved one, an additional 17% are unsure of the impact that organ donation has on the grieving process. And, nearly two out of five African American respondents disagree that organ donation helps families cope with their grief. This is a significantly higher level of disagreement than that found among white or Hispanic respondents.[17]

In a study conducted with actual donor families,[2] it was found that families have positive feelings about organ donation and their decision to donate in spite of the tragic circumstances surrounding their loved one's death. A majority of families in this study (80) stated that they understood the concept of brain death, while 14 families said they did not, even though they still consented to donate their loved one's organs. The major reasons given for a lack of understanding included:

- the donor did not 'appear dead' while being supported mechanically;
- the concept of brain death was vague or not well explained.

Most families (41) were approached about donation less than six hours after being informed about the death. Others reported time frames of six to twelve hours (17), twelve to twenty-four hours (21), greater than twenty-four hours (12), and 8 did not recall the time when approached. Some 85% felt that the time was appropriate for them.

Forty-six families were approached about donation by a physician (attending or resident). Twenty-two of the donations were family-initiated and 20 families were first approached by a transplant co-ordinator.

Families were also asked if being approached about donation caused additional stress at the time. Sixty-nine respondents said discussing organ donation caused no additional stress, although 23 indicated that it added some stress by forcing acceptance of the loss. Some said stress was caused by their belief that the donor's body would be 'violated' during organ recovery procedures. Most families said the initial stress subsided soon after the donation.[2]

Ongoing communication with donor families is important not only to help ensure that those families' issues and concerns are addressed, but to measure feelings and attitudes that can have important implications for educators and others who are striving to increase organ and tissue donation. Currently, The Partnership for Organ Donation, a non-profit research and consulting organization in conjunction with the Harvard School of Public Health is focusing its efforts on the attitudes of donor families on the reasons behind the decisions made both by donor and non-donor families with the hope of gaining insight that will lead to more effective ways to assist hospitals and organ procurement organizations increase donation by the way they handle the donation process.

Clearly the research shows that while a belief in altruism remains the basis for donation and that a majority of Americans subscribe to that belief, many other factors and issues must be addressed, aimed at key populations, in order for the gift of life to become reality for the thousands of transplant patients who so desperately need it.

Motor vehicle administrations

In the United States, the recording of one's intent to donate is typically carried out in two ways, by signing a donor card and by indicating intent to donate on the driver's licence. Some people employ one of these methods, some both. While donor card usage should not be discouraged, the driver's licence, together with other channels available through motor vehicle agencies, holds particular promise for generating positive changes in donation rates. Identifying potential donors and encouraging donation can both be facilitated through existing or emerging components of the motor vehicle system as discussed below.

Because of daily reliance on driving, and the importance of a driver's licence in providing identification for various purposes, it is likely that most Americans of driving ages have, and routinely carry, a licence. It would be difficult to make a similar statement about organ and tissue donor cards, or perhaps any other chattel. It can therefore serve as a fairly standardized form of donation intent, and one that is likely to be carried frequently by most Americans. All fifty states and the District of Columbia now have some method for stating intent to donate on the licence. In 1993, approximately 128,878,000 people aged 16 and over had a driver's licence.[18] This is an enormous number of individuals who all have the opportunity to declare their wishes to donate through an already existing system.

Furthermore, motor vehicle agencies (MVA) are perhaps the single institution in this country that have routine and periodic contact with a high proportion of the American public over the age of 16. This contact, whether in person, through the mail, or through other written materials, provides many potential avenues for encouraging donation. Throughout the United States, for example, MVAs, either on their own or in concert with organ procurement organizations, have implemented a variety of projects to encourage donation. Many states make some mention of donation in the driver's manual. Some send donation literature in the mail with renewal notices, while others make donation literature available in branch offices. Some states require MVA licensing clerks to remind applicants of the opportunity to indicate intent to donate on the licence. All of these types of activities are useful in increasing awareness and encouraging donation.

Additional mechanisms are necessary, however, to assure that 'good intentions' convert into actual donations. Specifically, in the event of death, it is important for surviving family members to be aware of the deceased's wishes with respect to donation. In an ideal world, family discussions about donation would have occurred before the death of a family member. In reality, this is unlikely the case in the majority of deaths today. Although a nationwide effort to promote family discussions has recently been introduced in the USA by The Coalition on Donation and The Advertising Council Inc., it may be years before family discussions become the norm. Because the driver's licence is such an important document and carried fairly routinely by millions of Americans, it seems more likely to be accessible at the time of death than other sources of information (i.e., wills or donor cards).

Even licences, however, may not be accessible at time of death. Many potential organ donors are trauma victims, such as those who die in automobile crashes. Often their wallet and driver's licences are locked up in the hospital safe or were never found at the crash site. An increasingly popular way to capture donation intentions and make this information available at death is by maintaining a registry of licencees and their donation wishes.

In approximately twenty states, the MVA currently maintains a registry where they record a driver's choice regarding donation as indicated on the licence. These registries vary in sophistication, ranging from a file system to a 24-hour accessible computerized registry such as those maintained by Maryland, Illinois, and Ohio. The Division of Organ Transplantation is working closely with the American Association of Motor Vehicle Administrators in hopes of eventually effecting a nationwide registry of donation information for drivers in all fifty states. The goal is for this system to be accessible to key hospital and procurement personnel on a 24-hour-a-day basis thereby apprising surviving family about the deceased's wishes regarding donation.

In summary, the motor vehicle system enjoys a particularly favourable position for increasing the supply of donor organs and tissues. In view of the large number of Americans who come into contact with this system on a regular basis, it holds considerable promise for: disseminating donation information, encouraging positive choices about future donation, recording licencee's donation choices, and making their donation intentions readily available at the appropriate time.

Conclusion

Although scientific advances in medicine, and the acceptance of cadaver organ transplantation has moved from the experimental stage to an accepted treatment for end-stage organ diseases, changes in societal attitudes have not kept pace with changes in transplant medicine, resulting in a continuing shortage of donor organs. Although numerous surveys conducted to gauge public opinion and awareness concerning organ and tissue donation and transplantation have indicated the organ shortage, overwhelming support for organ donation remains.

Theory is an important foundation of health behaviour and attitude change. The application of theoretical models, however, is often overlooked in the field of organ donation and transplantation. It should be included as an essential building block of organ donor programmes as they attempt to identify factors and barriers that might inhibit the donor process. For organ donation, an example of a barrier is religious beliefs that may predispose an individual against donation. Multidisciplinary efforts can help assure physicians, organ procurement co-ordinators, and/or designated requestors are accountable and understand the many facets that may have a tremendous effect on the consent rate of organ donation.

Thus, the issue at hand, is the message that is targeted at the public. It is vital that we understand existing attitudes of the general public about organ and tissue donation. Public education and the encouragement of family discussion must be targeted not only to the general public but also toward those segments of the population not presently supportive of organ and tissue donation. The categories of attitudes divide into three groups within the general public, namely, those committed to donation, those adamantly opposed and those who might change their opinions if they had more specific information. Many families are never asked to consider organ and/or tissue donation, or they are not asked in a culturally sensitive way that respects the grieving process.

Suggestions to increase the supply of donors are numerous. They include:

1. Increased public education.
2. Professional education targeted to physicians and nurses in hospitals.
3. Enforcement of 'required request' legislation.
4. Increased use of 'routine referral' to refer potential donors directly from the hospital admitting office to organ procurement agencies.
5. Public education materials on donation through motor vehicle agencies.
6. National advertising campaigns.

Public attitudes tend to vary over time and from society to society, depending on many known and unknown factors. It is evident that a greater understanding of the concerns and beliefs of the general public is necessary. Prottas and Batten[6] summarize it best that 'if we are going to be successful in increasing organ and tissue donation, changes both in the content of the messages conveyed and in the mechanisms used to transmit those messages to the appropriate people are needed'. We can create a formula for successful interaction with the public, and apply this statement to include the general public that: 'Altruism does not have to be sold: it only has to be encouraged in them'.

References

1. DeChesser AD. Organ donation: The supply/demand discrepancy. *Heart and Lung* 1986, **15**(6): 547–551.
2. Savaria DT, Rovelli MA, Schweizer RT. Donor family surveys provide useful information for organ procurement. *Transplantation Proceedings* 1992, **22**(2): 316–317.
3. Callender CO et al. Attitudes among blacks toward donating kidneys for transplantation: A pilot project. *Journal of the National Medical Association* 1982, **74**(8): 807–809.
4. Carter WB. Health behavior as a rational process: Theory of reasoned action and multiattribute utility theory. In *Health Behavior and Health Education: Theory, research, and practice*. K Glanz, FM Lewis, BK Rimer (Eds), Jossey-Bass Publishers, San Francisco, 1990.
5. Ajzen I, Fishbein M. Attitude-behavior relations: A theoretical analysis and review of empirical research. *Psychological Bulletin* 1977, **84**: 888–918.
6. Prottas JM, Batten HL. The willingness to give: The public and the supply of transplantable organs. *Journal of Health Politics and Law* 1991, **16**(1): 121–134.
7. Caplan A, Siminorff L, Arnold R, Virnig B. *Increasing organ and tissue donation: What are the obstacles, what are our options?* White Paper: Surgeon General's Workshop on Organ Donation, Washington, DC, 1991.
8. Health Resources and Services Administration, Department of Health and Human Services. Hospital liaison program: Advancing organ donation responsiveness within minority populations through education, Summary report, Houston, Texas, 1989.

9. Health Resources and Services Administration, 240-BMCHRD-11(9) Department of Health and Human Services. Identification of effective methods for minority education on organ donation and transplantation, Reference Manual, CSR, Inc., Washington, DC, 1991.
10. Association of Organ Procurement Organizations, Lieberman Research, Inc. A study of attitudes toward organ donation. Association of Organ Procurement Organizations (AOPO), Falls Church, Virginia, 1990.
11. Chapa J. *Hispanics and organ donation: Prospects, obstacles, and recommendations.* LBJ School of Public Affairs, University of Texas, White Paper: Surgeon General's Workshop on Organ Donation, Washington, DC, 1991.
12. Horton RL, Horton PJ. Knowledge regarding organ donation: Identifying and overcoming barriers to organ donation. *Social Science Medicine* 1984, **31**(7): 791–800.
13. Novello AC. From the Surgeon General US Public Health Service. *JAMA* 1992, **267**(2): 213.
14. Evans RW, Manninen DL. US public opinion concerning the procurement and distribution of donor organs. *Transplantation Proceedings* 1988, **10**(5): 781–785.
15. Walker JA et al. Parental attitudes towards pediatric organ donation: A survey. *Canada Medical Association Journal* 1990, **142**(12): 1383–1387.
16. Stryker J. Organ procurement, brain death, and public understanding. *Health Span* 1989, **6**(2): 9–14.
17. The Gallup Organization for The Partnership for Organ Donation. *The American public's attitude toward organ donation and transplantation.* Boston, MA, 1993.
18. National Safety Council. *Accident Facts*, 1994 Edition. National Safety Council, Itasca, Illinois, 1994.

CHAPTER 4

Paid organ donation: towards an understanding of the issues

ABDALLAH S. DAAR

Introduction   46
A historical perspective of the unfolding debate   47
Discussion of the issues   52
Conclusion   58
References   59

Introduction

We acknowledge that organ transplantation is one of the wonders of biomedical science; like a teenager, it is young, dynamic, successful, at times a little brash and certainly a little behind in its ability to cope with the societal issues its success has raised. In the 1990s its defining characteristic is the shortage of organs. This is becoming a serious problem for many countries, as shown in Table 4.1, which is a list of measures being discussed or adopted around the world to cope with the shortage. All of these measures are controversial, yet they bear discussion because they have been put forward by honest individuals who are trying to help sick patients. 'Paid organ donation' is an extremely complex subject that covers the spectrum from what is totally acceptable to what, for the profession, is totally unacceptable. Honest individuals and teams are working hard to define the acceptable, at the same time that many unscrupulous individuals and teams are exploiting the shortage to enrich themselves, using patients as the tools for enrichment.[31]

This chapter attempts to portray the complexities involved in this issue. It should help to distinguish the scrupulous from the unscrupulous. It acknowledges that the debate continues and must continue, because emotional or uninformed responses can only provide temporary solutions.

Table 4.1 Some controversial methods of increasing donation

A. **Cadaveric**
 1. Marginal donors
    (a) Hypertensives
    (b) Diabetics
    (c) Old age
 2. Consent manipulations
    (a) Presumed consent and variations
    (b) Mandated choice
 3. Perimortem issues
    (a) Non heart beating donors
    (b) Elective ventilation
    (c) Pittsburgh protocol (patients with advance directives)
 4. Financial incentives
    Futures market

B. **Living donors**
 1. Living unrelated kidney donors
 2. Paid unrelated kidney donors
 3. Living segmental liver donors
 4. Living segmental pancreas donors

A historical perspective of the unfolding debate

Having started with identical living twin transplants in the mid-1950s, transplantation moved rapidly to allotransplantation with the introduction in the early 1960s of chemical immunosuppression. In those early days, a significant proportion of living donation was from non-related individuals – this subject has been reviewed by Levey,[1] Tilney and Hollenberg[2] and Daar and Sells.[3] Non-related donation declined in importance with the advent of cadaver donation.

In 1970 the Transplantation Society felt that the fear of the buying and selling of organs was real enough and it was proclaimed that 'the sale of organs by donors living or dead is indefensible under any circumstances' (Committee on Morals and Ethics 1970).[4] This is an attitude that has certainly withstood the test of time *within the profession*. In the late 1970s and early 1980s, with the diffusion of knowledge and skills and the introduction of cyclosporine, transplantation spread to many countries outside of the USA and Europe. In the USA and Western Europe most of the transplants were from cadavers, a process facilitated by the adoption of brain-death criteria; in much of the rest of the world, however, cadaver donation was not yet developed (in fact, it is fair to generalize: it still is *not* developed).[5] However, the rapid expansion in transplantation at this time simply increased the demand for organs. In Germany, Count Rainer Rene Adelmann zu Adelmannsfelden, a 'specialist' in legal loopholes, set up an Organ Bureau offering to buy a kidney from those in financial need for US$30,000–40,000. Despite the absence of a law on transplantation in Germany, the profession adopted a code to ban any sales. In the USA Dr Jacobs of Virginia planned to set up a similar organization. The reaction was negative, and the US law subsequently closed this option. However, the US government's decision to get involved in issues related to end-stage renal failure, resulted in the adoption by Congress of the End Stage Renal Program, which has grown to cost the US government about US$9 billion per annum today. Poorer countries, of course, cannot afford such huge sums; in fact in India, for example, management of renal failure does not figure at all in the health budget. If an Indian patient developed end-stage renal failure and had no living relative to donate, he/she was allowed to die, because there was no affordable dialysis and no cadaver donation to fall back on. So one option that developed was the buying and selling of kidneys between unrelated living individuals.[6]

The model of paid donation that developed in Bombay was what was subsequently called 'rampant commercialism'.[7,8] It was totally unregulated; much of the money ended up in the hands of brokers; doctors and hospitals were in the business for the money, not for their patients' health and welfare; medical criteria of screening and selection were thus sacrificed for profit; and donors were used without attention to their subsequent health. We documented that such transplants resulted in high morbidity and mortality[9] and argued, on utilitarian grounds, that such transplants were unacceptable to us, because our patients, who had gone to Bombay against our advice, did have a choice of dialysis back home in the United Arab Emirates and Oman. Many patients, from many other countries, however, continued to go to Bombay for kidney transplants. The rest of the world criticized and condemned, and the media made much of this issue, which was, in retrospect, discussed rather naively in black and white terms.

INTRODUCTION OF A GREY COMPONENT

The condemnation by the West forced respectable units in India to stop and others went underground, with deterioration of results. Thinking individuals felt a need to discuss the issues, albeit timidly at first. A significant move was the talk given by Dr C.T. Patel of India at a 1987 conference in Pittsburgh to honour Dr Tom Starzl. Dr Patel argued eloquently that, under the circumstances prevalent in India, the exchange of money in the process of organ donation and transplantation was defensible. He said 'Kidney donation is a good act. It is the gift of life. The financial incentive to promote such an act is moral and justified'.[10] At the same meeting, Dr Francis Moore, obviously worried about the inequity of distribution that might result, introduced another element of greyness when he said 'Selling of kidneys from living donors, evidently a common practice in India, finds a negative response in our society *unless the recipients are chosen without respect to ability to pay*, i.e. some form of government subsidy (my italics).[11] The complexity of this subject can be seen even from these two apparently simple statements: notice the nuances from the above statements by Patel and by Moore detailed in Table 4.2.

THE ERA OF DEBATES AND DEFINITIONS

The (International) Transplantation Society, until 1988, had not had a major public discussion of this issue (or in fact of many other ethical issues) at any of its international congresses. Having adopted in 1985 a series of generally accepted guidelines, including a categorical ban on commerce (Transplantation Society 1985),[13] it felt no obvious need to discuss the matter further. Of course, in the meantime others[10,11] were

Table 4.2 Grey issues in payment for living organ donation

1. Payment does not necessarily mean 'buying' in the way generally understood
2. The payment can be seen to be an incentive
3. An incentive can be acceptable to increase the occurrence of a good act
4. The payment of the incentive does not diminish the goodness of the act
5. The act thus remains morally acceptable, and thus the payment is justified
6. The selling of organs 'finds a negative response in our society' (presumably by 'society' is meant the public. Actually, there is little evidence, whenever scientific polls have been carried out, to support the view that the public are against this.[12] The issue then, is that it is *society* that ultimately determines what is acceptable.[11] This raises the question: What about the profession? (For example, supposing that society said 'yes' – what would then justify the ban by the profession?)
7. Without some form of regulation and modulation, it is likely that only rich recipients will get kidneys, and the donors are going to be the poor.
8. Despite our acceptance of market forces as being the most efficient method of conducting business, we cannot allow the direction of organ exchange to be solely from poor to rich. This would be unacceptable.
9. 'Unless recipients are chosen without respect to ability to pay.' In other words, if arrangements could be made so that even the poor were able to pay for organs from living donors, this may make society's response less negative (or even positive?).
10. The government may have a role in subsidising the payments for those who would otherwise not be able to pay for kidneys. (Considering that the US Government alone spends about $9 billion per annum on the End Stage Renal Failure Program, it could be argued that the renal waiting list would be drastically reduced if some of the money went to pay living donors willing to part with their kidneys if they could receive money).
11. But what would the money be for? Would it be the purchasing price for the actual organ? Or would it be for the act of donation? Or for a service rendered? Or for the pain and inconvenience of undergoing a nephrectomy? Or for the loss of income? Or for the cost of investigations, surgery and hospitalization?

discussing it; 'commerce' did not stop; people were not satisfied with the absence of rational reasoning and arguments in favour of the ban; 'professional ethics' was being replaced by 'bioethics'; and the issue of scarce resources and their allocation was being claimed by 'societal ethics',[14] and, of course, the shortage was increasing and waiting lists were getting longer throughout the world.

And so the Scientific Committee of the XII International Congress of the Transplantation Society held a well-attended, rousing session on this question in Sydney in 1988. I presented our preliminary data regarding the transplants performed on our (Omani and UAE) patients in Bombay; I noted in passing that 'we have our prejudices, and these are to some extent coloured and informed by the prevailing conditions in the countries where we practise. While I believe there is a distinct difference between "rewarded gifting" and "rampant commercialism" I am worried about opening floodgates, for the consequences could be bad. And I ask myself: even if societal acceptance is the final arbiter of ethics where do we, as a profession, draw the line?'[7] Dr John Dossetor also distinguished between potentially acceptable and unacceptable forms of payment[15] – work that he developed later into a well-argued scheme that I will mention below.

Other significant contributions at that session included Peter Little's argument that the Transplantation Society was perhaps uninformed of realities in the poorer developing world and was also unrepresentative;[16] Dr Raj Yadav, from India, bitterly noted the unhelpful tendency of the Transplantation Society to condemn without making efforts even to visit India to talk of the issues and to encourage cadaver donation. He noted that key members of the Society were always too busy to visit India whenever he invited them.

Robert Sells introduced the work of the Madras group,[17] which claimed to have many checks and balances while performing paid organ transplantation. These included no use of brokers; strict medical and psychological evaluation of potential donors; fixed charges; transparent and fixed payments for donors; and a contractual commitment to care for the donors for a minimum of three years after donation.

THE OTTAWA CONFERENCE: A CLASSIFICATION OF THE ISSUES

As will now be obvious, the whole question was re-opened and became very confused. The confusion is perhaps less prevalent today as the issues for discussion have become clearer to a certain extent and some of the questions that need to be asked have been identified. Much of the progress is due to the continued efforts of dedicated individuals. Anthony Monaco, Calvin Stiller and John Dossetor organized a far-reaching multidisciplinary conference entitled 'Ethics, Justice and Commerce in Transplantation: A Global View'. It obviously covered many societal and technical aspects of transplantation as the field stood in 1989. In terms of the subject of paid organ donation, there were three main outcomes.

A classification of the practical and ethical issues was introduced[8] as an attempt to identify individual issues and categories to be discussed in depth without being confused with other issues and categories, see Table 4.3. (The classification was subsequently accepted by the Ethics Committee of the Transplantation Society as a basis for discussion.)

The Madras group vigorously and logically defended their practice of 'unconventional' organ donation, claiming results that were acceptable by Indian standards at that time.[18,19]

The Indian participants invited the Transplantation Society to send a delegation to India to look at the realities. Professor Richard Batchelor, the then President of the Society, chose Sells and Daar to undertake the task. The visit took place a few months later. We visited several units, talked to a large number of people involved, attended an annual conference of the Indian Society of Urology during which a major discussion, under the chairmanship of Dr Eli Friedman of New York, took place. The visit convinced us that the Indian medical profession was divided about the issue of paid kidney donation and would perhaps accept 'rewarded gifting' as a temporary measure before cadaver donation was established. We presented our findings and recommendations to the Council of the Transplantation Society, and subsequently to a general assembly of the Society at the XIIIth International Congress in San Francisco in 1990.[6] We noted that during our visit to India the Indian Society of Transplantation adopted a resolution declaring cadaver donation as the strategic aim of the Society. Our main recommendation was that the Society, rather than condemning, should try to help Indians develop cadaver transplantation, amongst other measures, by training Indian transplant surgeons and several members, notably Robert Sells, have since done so.

Table 4.3 Classification of ethical issues of living donor renal transplantation

Category	Name	Issue(s)	Gifting	Acceptable	Comment
I	Living related donor	Risk to donor	Preserved	Yes	Has always been acceptable. Risk to donor minimal
II	Emotionally related donor	Risk to donor. Results of transplantation	Preserved	Yes	Results are excellent
III	Altruistic donation	As II, plus question of motivation	Preserved	Yes, if truly altruistic	Rare to find genuine reports but they exist. Results same as II
IV	'Grey basket'	Many	?	?	Critical to this category is the context in which transplantation occurs: the aim of the enterprise is to benefit patients, not to enrich doctors, brokers or institutions. Many ideas are included
V	Rampant commercialism	Commerce in bodily parts. Risk to donor and recipient. Results of such transplants may not be acceptable. Cross-border transplantation (rich clients, poor donors)	Not preserved	Not acceptable	A major problem in some parts of the world. Not confined to developing countries. Dangerous according to our own 1990 results. Risk of dangerous infections. Broker, doctors, hospital receive bulk of payment instead of donor
VI	Criminally coerced organ procurement	A matter for the police but transplanters cannot ignore causes and effects	Not preserved	Not acceptable	No real convincing evidence as yet. Numerous reports of murders, kidnappings, and surreptitious organ removal in newspapers and TV programmes

OTHER COUNTRIES

Familiarity makes India a focus for discussion, but in fact other countries have also been reported as practising various forms of paid organ donation. These include Egypt, Iran, Iraq, Philippines, South Korea, Thailand and several Latin American countries.[20] A well publicized case in London, England, was the immediate cause for the hasty introduction of the British law on transplantation.[21]

THE MUNICH CONFERENCE OF 1990

Walter Land and John Dossetor organized another very successful and influential conference entitled 'Organ replacement therapy: ethics, justice and commerce'. The resulting book[22] is a landmark publication that should be read by anyone seriously interested in this and related issues. Perhaps the most significant additional dimension was L.R. Cohen's strong advocacy of the ethical virtues of a futures market in cadaveric organs.[23] His presentation was so radically different from the 'norm' that it was considered too wild even to comment upon in 1990.

Discussion of the issues

A basic assumption in medical ethics is that human life has value. It is in some ways sacred, and needs to be respected and saved whenever possible. Some of the factors entering the debate on paid organ donation are:

- The nature of life.
- Ownership of the body.
- Relations between individuals and their governments.
- Freedom and self determination and when these can be curtailed legitimately.
- Who makes decisions that have a societal, rather than a purely medical, dimension.
- What is the true basis for the various forms of organ donation.
- The spectrum of payments: compensation, incentives, purchase price, coercion.
- If people respond to religious beliefs and religious law, what do the major religions have to say about paid organ donation.
- Should deontological or utilitarian arguments prevail?
- Can 'rules' be made? What would the exceptions be? Will these be universally applicable?

THE CONTEXT OF THE DEBATE

There are six main issues which provide the context for this debate at the present time:

1. The bioethics establishment has essentially taken over major decision-making from the professional ethics establishment of the past.
2. Many members of the profession are not yet versed in the techniques of ethical debate. They have well-established intuitive responses that 'have stood the test of time', but they have enormous difficulties when challenged to explain the basis of their views.

3. Traditional awe and respect for the medical profession is being eroded daily. In this context, the profession's insistence that donation is based purely on altruism is placed in stark contrast to the enormous benefits the profession itself derives from transplantation.[24]
4. In today's world, traditional foundations for ethical decision-making hold little sway – 'moral strangers' can discourse only on the basis of mutual respect and limited democracy.[25]
5. The escalating cost of delivering quality tertiary health care will force decision-makers to contemplate the previously uncontemplated.
6. Altruism has been claimed to be the central and only acceptable motive for organ donation.[26] Several important insights about altruism have emerged in the debate recently. Altruism has been described as exaggerated as a motivating factor in much of human behaviour.[27] In the field of cadaver donation, those who consciously sign and carry donor cards can be assumed to be motivated by altruism. However, only a small proportion of cadaver organs are obtained from such donors; thus much of current cadaver donation does not reflect the donors' altruism at all (indeed the 'donor' is dead when the question arises). It has been argued by thoughtful ethicists that shared interest and purpose, rather than altruism, form a more solid basis for cadaver donation systems.[28] It has also been argued that incentives can be used to increase altruism, which is not thereby diminished in value and also that markets can be made altruistic.

ARGUMENTS AGAINST PAID ORGAN DONATION FROM THE PROFESSION

- Intuitive, unexplained.
    (a) Repugnant.
    (b) Repulsive.
    (c) Unethical.
    (d) Immoral.
- Theoretical bad consequences.
    (a) Will exploit the poor (as a class) and divide society.
    (b) The results will be poor if based on profit for the transplant establishment.
    (c) Will inhibit cadaver donation.
    (d) Will inhibit living related donation.
    (e) Will undermine any form of 'altruistic' donation.
- Surgical arguments.
    (a) Operation is not therapeutic for the donor.
    (b) A poor person who becomes a paid donor cannot give voluntary consent.
    (c) A poor person may sell vital organs (sacrificing his/her life) to help his/her family.
- Slippery slope. Any relenting, even of a minor degree, will inevitably lead to the most dire consequences. Therefore the subject should not be opened at all.
- Regulation. Arguments can be made that paid organ donation would be a good thing and help increase transplantation if it could be regulated; however, it can never be regulated. The risk is too high.

ARGUMENTS IN FAVOUR OF SOME FORM OF PAYMENT FOR ORGANS

- Self determination – a paramount value in secular Western societies since the Enlightenment.

- Autonomy. The decision to assume a risk should be made by the individual person concerned, not by the profession. If this makes him a 'vendor' instead of a 'donor' this is his business, so long as the interests of others are not prejudiced.
- Paid organ donation is justified ethically on utilitarian grounds because it will increase transplants and therefore the total amount of good for society. The onus is on those who oppose this view to demonstrate that this is not so. Perhaps a scientifically valid clinical trial is warranted.
- Voluntarism is more likely in the case of non-relatives since the donor is a free agent, alive, and is not compelled by family pressure. Voluntarism is an important element in legal consent.[29]
- Money is not necessarily dirty. Individuals freely make important decisions daily which reflect the need for money. The best illustration of this principle is the recent decision of the Council of the American Medical Association that a futures market to increase cadaver donation is ethically acceptable.[30]
- There are no compelling arguments against the sale of organs per se.[21,25] It is the potential *abuse* which is worrying. Therefore one can argue that one 'shouldn't throw the baby out with the bath water' but regulate the practice instead, and make it fair.
- 'What is less than best is not necessarily wrong'.[21]
- From the paid living donor's perspective, the act of selling his or her kidneys may well be altruistic if the money is desperately needed, say, to save the life of a member of his or her own family. The act is only self-degrading (Kant has argued that selling a part of the body reduces a 'person' to a 'thing') if the person evaluates the action as self-degrading.
- Living related donation is similarly open to abuse (moral bullying and other pressures). However, those of us involved in transplantation believe we can distinguish between willing and coerced donors and by our intervention we minimize abuse. The fear of abuse here has been dealt with rationally and the same methods can be applied in the case of non-related donors. It is also possible that in many family donations the donor is materially rewarded in one way or another. Transplant professionals have not dealt with this issue – partly because it does not attract much revulsion.
- Slippery slope arguments are philosophically unsound as a basis for public policy. Whatever risk is involved (e.g. alcohol and driving; nuclear power) is minimized by legislation, guidelines or regulations.
- Unequal bargaining situations ('the poor will be exploited') are common in everyday life. Exploitation is minimized in market economies by such checks and balances as minimum wage laws, protection by labour laws and social security systems.
- Previous attempts to commercialize transplants in the West were made only by individuals (Adelmann in Germany, Jacobs in the USA, Crockett in Britain) outside of, and reviled by, formal academic or healthcare institutions. They were quickly stopped by the response of the profession or by the law. There has, until now, been no instance by the profession or the government to organize and regulate any incentive or commercial arrangements for living donor or cadaver donor transplants.

Respectable members from the profession and from academia, however, have written and commented in favour of various forms of commercial arrangements that the authors obviously consider to be ethically appropriate. (Reviewed in reference no. 31.)

LIVING NON-RELATED DONATION AS AN ISSUE PER SE

Daar and Sells[3] reviewed this subject extensively, pointing out the historical precedents. We noted the almost illogical stand against living donation from unrelated individuals adopted by some European countries and even by some of the Nobel prizewinners in our field (Medawar, Dausset, Snell). Remnants of the opposition to any kind of living kidney donation, ostensibly driven by fear of commercialization, are still to be found in a few countries, notably in France.[32]

However, as has been pointed out,[32] the transplant profession is very susceptible to utilitarian arguments: as soon as a procedure or method is shown to give good results and it increases the overall perceived good, it becomes 'ethically justified' and is adopted. Living non-related donation has been practised in small numbers in reputable centres for a long time, with excellent results.[32,33] Terasaki and his group have now published the results of spousal transplants in the USA and, surprisingly, these are as good at three years as those in one haplomatched living related transplants, and significantly better than cadaver transplants, even when the latter are well matched.[34] There is little doubt that the numbers of such transplants will now increase. It is also important to note that over the past few years in the USA, there has been a marked increase in the rates of living donor kidney transplants at the same time that cadaver donor rates have increased: thus the argument that encouraging one method of donation will discourage another is without much foundation.

It can be safely concluded today that it is ethically acceptable to transplant between spouses, and it is becoming increasingly common to accept donations from genetically mismatched good friends. A statement such as this would have been considered heretical by some members of the profession only a few years ago.

Instead of paying attention to fearful, slippery slope scenarios of potential abuses, invalid consent, vulnerability, exploitation, commercialization, effect on cadaver donation etc., the focus has shifted to consideration of the excellent results, of regulation, and of the rights of donors and recipients to determine what is best for them. This is also a good illustration of the shift from 'professional ethics' to 'bioethics'.

THE RESPONSE TO THE FEAR OF COMMERCE

Professional/ethical response

No single body of professionals has accepted uncontrolled commerce as the basis for donation and transplantation. The following examples document the overwhelming negative response:

- 1971 Committee on Morals and Ethics of The Transplantation Society: 'The sale of organs by donors living or dead is indefensible under any circumstances.'
- 1985 Transplantation Society Guidelines: 'It must be established by the patient and the transplant team alike that the motives of the donor are altruistic and in the best interest of the recipient and are not self-serving or for profit. Active solicitation of living unrelated donors for profit is unacceptable. It should be clearly understood that no payment to the donor by the recipient, the recipient's relatives or any other supporting organization, can be allowed. However, reimbursement for loss of work earnings and any other expenses related to donation is acceptable.' Special Resolution: 'No transplant surgeon/team shall be involved directly or indirectly in the buying or selling of organs/tissues or in any transplant activity aimed at commercial gain to himself/herself or an associated hospital or institute. Violation of these guidelines by any member of The Transplantation Society may be cause for expulsion from the Society.'[13]

- World Health Organization: 'Such trade is inconsistent with the most basic human values and contravenes the Universal Declaration of Human Rights and the spirit of the WHO Constitution.'[35]
- European Health Ministers 1987: 'A human organ must not be offered for profit by any organ exchange organization, organ banking centre or by any other organization or individual whatsoever.'[36]
- Conference on Ethics, Justice and Commerce; A Global View. Ottawa August 20–24, 1989: 'The buying and selling of human organs and tissues for transplantation is unacceptable.'[37]
- Conference of Organ Replacement Therapy: Ethics, Justice and Commerce, Munich, December 11–14, 1990: 'The Congress supports maintaining a prohibition of commerce to obtain organs or tissues for allotransplant.'[22]

Similar sentiments have also been expressed by regional transplantation societies around the world.

Legal

A large number of countries have now included on their statute books laws banning commerce in organs for transplantation (see Chapter 7). These include countries with very little or no transplant activity. The WHO has recently published a book[38] which includes the legislative response to various issues of transplantation, including the question of protection of minors and other disadvantaged and vulnerable individuals, and the different types of consent.

Although transplantation has spread rapidly, many countries do not, as yet, have specific laws on transplantation. However, transplantation rarely operates in a legal vacuum. There are often laws from the past that regulate, for example, the legal status of dead bodies; the criteria of death have been developed regarding religious and public health concerns, for instance, that affect the transfer of body parts into living recipients.[39] Some developing countries, e.g. Argentina and South Africa, have had laws enabling transplantation for a long time; and more recently several developing countries have enacted specific organ transplant legislation. A major survey of organ transplant legislation in sixteen Latin American countries was recently published by the Pan American Health Organisation.[40] This review looked at the universally important transplant issues such as those above, and reviewed the 'required request' system, recipient selection, determination of death, conflict of interest, funding, donor compensation and international sharing of organs. In March 1987 the Council of Arab Ministers of Health adopted a Unified Arab Draft Law which, it is hoped, will be a model for both living donor and cadaver donor transplantation.[41,42]

Regarding the question of commercialism the World Health Organization, Geneva, has recently published a report on developments in the field of transplantation under its auspices between 1987 and 1991.[42] A large number of developing countries in the Americas, Europe, Eastern Mediterranean, Africa (e.g. Malawi, Zimbabwe), Western Pacific (Singapore) and South East Asia (Indonesia, Sri Lanka) already have some legislation; others have non-statutory measures (Hong Kong, Israel, Malta and Saudi Arabia, Sultanate of Oman), while in some jurisdictions (e.g. Israel and the Philippines) new legislation is either pending or in preparation[43] (see Chapter 7). The Indian Federal parliament has now passed the Transplantation of Human Organs Act, which has been ratified by some, but not yet all, of the states.[44] Egypt has adopted a resolution by the General Assembly of the Egyptian Society of Nephrology (effective as of 30 June

1992) authorizing living donor transplantation only between consanguineous relatives. (This resolution may not cover non-Egyptians.) Hong Kong issued on 27 March 1992 a Human Organ Transplant Bill prohibiting commerce and establishing a Human Organ Transplant Board (perhaps modelled on the British) to approve genetically unrelated living transplants, which would otherwise be illegal. In 1991 Israel inserted new provisions in its Public Health Ordnance that would ban commerce. Finally, the World Health Organization itself adopted in May 1992 its Guiding Principles on Human Organ Transplants[42,43] with important definitions and clarifications and any country wishing to develop transplant legislation would be advised to look to this for guidance. Furthermore, the Office of Health Legislation, World Health Organization, 1211 Geneva 27, Switzerland, has details on computer of most of the laws already enacted and its Chief, S.S. Fluss, co-author of Chapter 7, has offered to assist any country or organization in this matter.

THE 'GREY BASKET' CONCEPT

There are forms of living donation which are generally ethically acceptable: (a) living related; (b) emotionally related; and (c) altruistic strangers – see Table 4.3; other forms are obviously unacceptable (a) rampant commerce, i.e. without any regulation and meant to enrich the doctors and their colleagues at the expense of the patient and (b) criminal coercion. There are, however, a number of alternative arrangements that can be discussed and analysed to see whether they are acceptable or not. Before these are mentioned below, it is important to understand why a 'grey basket' category is needed and that it is essential to understand that not all living donation fits into the simplistic categories of good and bad.

- The shortage of organs persists.
- There are serious scholars and practitioners who will come up with ideas at the borderline of acceptability by current (or past) standards. Dismissing any such idea offhand is anti-intellectual and serves no useful purpose.
- Much of what the profession considers unacceptable is based on intuition and sometimes on emotion. Unless these views and opinions are challenged and clarified, they cannot form the basis for successful and enduring policies. They will not withstand analysis and challenge by such individuals as ethicists, economists, lawyers, philosophers, and religious law experts who are now part of the policy-making establishment and who are very often part of the bioethics establishment.
- It is to misunderstand the nature of ethics to believe that guidelines apply for all time and for all places. Recent events, particularly the question of a futures market for cadaver donation in the USA, demonstrate the rapid change that can take place in individual countries, let alone across different countries with different socioeconomic and cultural realities.
- It is crucial to understand the difference between upholding fundamental values on the one hand, and applying those values to individual circumstances. The former may be immutable (in a homogeneous society); the latter may be situational.
- Members of the profession, as part of their jobs, may hold various positions such as lawmakers, advisors, policy makers, members of institutional review boards and as individual physicians for individual patients. The basis of performance in these various roles may be different. What may be correct at a societal level, for instance, may not be good in individual doctor–patient relationships (the Hippocratic Oath,

CURRENT CONTENTS OF THE 'GREY BASKET'

- Compensated donation vs donation with incentive – this distinction primarily calls for a detailed analysis of just what is acceptable as genuine payment for the act of donation (lost income, hospital expenses, etc.) on the one hand; and an examination of the role of incentives and their limits.[41]
- Exploring the burden to benefit ratio in various socioeconomic circumstances implies evaluating the difference between availability and absence of dialysis as an alternative to transplantation and taking into account modulating values such as 'indirect altruism' and 'mandated philanthropy'.[45,46]
- A model, such as the 'Donor Trust'[47,48] would (a) separate the payment procedures from the donation/transplant procedures, (b) ensure fair compensation of donors and doctors, (c) eliminate exploitation of the poor, (d) ensure additional funds are available to the poor recipient who would otherwise not be able to afford to pay for a kidney donation and transplant. This model demonstrates one way in which regulation could be achieved.
- Payments to remove disincentives for a potential donor. Part of this is making sure that the donor does not lose out financially in the process of donation. Part of it may be payment of 'schmerzensgeld' (a legal term in German usage – it implies payment for 'suffering'). A donor may be entitled to schmerzensgeld because he does indeed undergo suffering, and unless we want to justify suffering on spiritual or theological grounds, it may be difficult to argue against schmerzensgeld. Furthermore, it does not diminish the altruistic nature of the donation and it may even be argued that it is morally wrong *not* to pay schmerzensgeld. To call this 'buying and selling by another name' is, in my opinion, simplistic.
- There are a number of other models already being practised by people who are happy to defend these practices publicly, whether we consider them ethical or not. These include the model in Madras[18,19] already mentioned above and the apparently institutionalized system in Iran.[49]
- The futures market idea proposed for cadaveric donation in the USA[30] is in my opinion the most daring and startling development in this debate. As a re-evaluation of the application of fundamental principles in the field of organ donation, it has no intellectual precedence.

Conclusion

The shortage of organs for transplantation not only continues but may be getting worse. Waiting lists around the world are certainly getting longer. There is no evidence that cadaver donation as currently practised will ever be adequate. In many countries, particularly those in the developing world, cadaver donation contributes either very little or nothing at all to the overall rate of transplantation.

Part of the problem of organ shortage is the immense success of transplantation itself. All breakthroughs of this nature result in the creation of many kinds of opportunity:

health for the patient; betterment of society; glory to the practitioners; economic savings for governments; and enrichment of the greedy. Laws and guidelines, however, have tended to follow rather than either precede or move together with the breakthroughs.

Over the past two decades a major concern of most members of the profession has been the safeguarding and increase in donation rates. Honourable members of the profession have come up with suggestions and solutions that are in many cases controversial. At the same time, there have been unscrupulous practitioners who have used the shortage to enrich themselves. The distinction between whose interests are primarily being served in transplantation may well underlie ethical acceptability, and it seems that the distinction is likely to affect the outcome of transplantation. Bad publicity has almost always resulted in a decrease of cadaver donor rates. The profession has always considered 'buying and selling' of organs as unacceptable, partly because of personal intuitive reasons (e.g. 'revulsion') and partly because of the fear of reducing cadaver donor rates. This question of 'commerce' has become one of the major issues for discussion. While it was confined to discussion within the profession there was not much debate – it was simply rejected.

The inclusion of various other interested groups, such as economists, lawyers, professional ethicists, religious authorities, social workers, anthropologists, etc., has meant that the issues are no longer black and white. Whether we like it or not, the debate now admits of some greyness. The conclusion in the end, if ever there is an end, may or may not be the original intuitive one adopted by the profession, but I believe that it is certainly necessary for us to go through the process of rational enquiry first. Members of the profession, particularly surgeons, are protected from discomfort by their own professional codes of conduct, their own beliefs and their own practices and behaviour. No surgeon need operate on a patient if he or she believes it is wrong to do so. However, surgeons and physicians now realize that decisions involving scarce resources have been taken out of their exclusive hands. Their own beliefs, the interests of their individual patients and the interests of society should ideally be congruent but in this messy world, unfortunately, they may not always be so.

References

1. Levey AS, Hou S, Bush HL, Jr. Kidney transplantation from unrelated living donors. *N Eng J Med* 1986, **314**: 914–916.
2. Tilney NL, Hollenberg NK. Use of living donors in renal transplantation. *Transplant Rev* 1990, **1**: 225–238.
3. Daar AS, Sells RA. Living non related donor renal transplantation – a reappraisal. *Transplantation Reviews* 1990, **4**(2): 128–140.
4. Committee on Morals and Ethics of the Transplantation Society. *Ann Int Med* 1970, **75**: 631–633.
5. Daar AS. Transplantation in developing countries. In *Kidney Transplantation, Principles and Practice*. PJ Morris, (Ed.) 4th edn, Saunders, Philadelphia, 1994, 478–503.
6. Daar AS, Sells RA. The problems of paid organ donation in India. Report on behalf of the Ethics Committee of the Transplantation Society to the President and Council of the Transplantation Society. 13th International Congress Transpl Soc, San Francisco, August 1990.
7. Daar AS. Ethical issues – a Middle East perspective. *Transplant Proc* 1989, **21**: 1402–1404.
8. Daar AS, Salahudeen AK, Pingle A, Woods HF. Ethics and commerce in live donor renal transplantation: Classification of the issues. *Transplant Proc* 1990, **22**: 922–924.

9. Salahudcen AK, Woods HF. Pingle A, Nur-el-Huda Suleyman M, Shakuntala K, Nandakumar M, Yahya T, Daar AS. High mortality amongst recipients of bought living unrelated donor kidneys. *Lancet* 1990, **336**: 725–728.
10. Patel CT. Live renal donation. A viewpoint. *Transplant Proc* 1988, **20**: 1068–1070.
11. Moore FD. Three ethical revolutions: ancient assumptions remodelled under pressure of transplantation. *Transplant Proc* 1988, **20** (Suppl. 1): 1061–1067.
12. Guttman A, Guttman RD. Attitudes of healthcare professionals and the public towards the sale of kidneys for transplantation. *J Med Ethics* 1993: 148–153.
13. Council of the Transplantation Society. Commercialisation in transplantation: the problems and some guidelines for practice. *Lancet* 1985, **ii**: 715–716.
14. Veatch RM. Theories of medical ethics: the professional model compared with the societal model. In *Organ Replacement Therapy: Ethics, Justice, Commerce*. W Land, JB Dossetor (Eds), Springer-Verlag, Berlin, 1991, 3–9.
15. Dossetor JB. Rewarded gifting: Is it ever ethically acceptable? *Transplant Proc* 1989, **24**(5): 2092–2094.
16. Little PJ, McMullin JP, McDonald A. Live donor renal transplantation in Iraq. *Transplant Proc* 1989, **21**(1): 1400–1401.
17. Sells RA. Ethics and priorities of organ procurement and allocation. *Transplant Proc* 1989, **21**: 1391–1394.
18. Reddy KC, Thiagarajan CM, Shunmugasundaram D, Jayachandran R, Nayar P, Thomas S, Ramachandran V. Unconventional renal transplantation in India. *Transplant Proc* 1990, **22**(3): 910–911.
19. Thiagarajan CM, Reddy KC, Shunmugasundaram, Jayachandran R, Nayar P, Thomas S, Ramachandran V. The practice of unconventional renal transplantation (UCRT) as a single centre in India. *Transplant Proc* 1990, **22**(3): 912–914.
20. Abouna GM, Sabawi MM, Kumar MSA, Samhan M. The negative impact of paid organ donation. In *Organ Replacement Therapy: Ethics, Justice and Commerce*. W Land, J Dossetor (Eds), Springer-Verlag, Berlin, 1991, 164–172.
21. Radcliffe-Richards J. From him that hath not. In *Organ Replacement Therapy: Ethics, Justice and Commerce*, W Land, J Dossetor (Eds), Springer-Verlag, Berlin, 1991, 191–196.
22. Land W, Dossetor JB (Eds). *Organ Replacement Therapy: Ethics, Justice and Commerce*. Springer-Verlag, Berlin, 1991.
23. Cohen LR. The ethical virtues of a futures market in cadaveric organs. In *Organ Replacement Therapy: Ethics, Justice and Commerce*. W Land, J Dossetor (Eds), Springer-Verlag, Berlin, 1991, 302–310.
24. Dickens BM. Human rights and commerce in health care. *Transplant Proc* 1990, **22**(3): 904–905.
25. Englehardt HT Jr. Is there a universal system of ethics or are ethics culture-specific? In *Organ Replacement Therapy: Ethics, Justice and Commerce*, W Land, J Dossetor (Eds), Springer-Verlag, Berlin, 1991, 147–153.
26. Suthanthiran M, Strom TB. Renal transplantation. *N Eng J Med* 1994, **331**: 365–376.
27. Childress JF. The body as property: some philosophical reflections. *Transplant Proc* 1992, **24**: 2143–2148.
28. Kleinman I, Lowy FH. Cadaveric organ donation: considerations for a new approach. *CMAJ* 1989, **141**: 107–110.
29. Land W. The problem of living donation: facts, thoughts and reflections. *Transplant Int* 1989, **2**: 168–179.
30. American Medical Association Council on Ethical and Judicial Affairs. Council Report: Strategies for cadaveric organ procurement. Mandated choice and presumed consent. *JAMA* 1994, **272**: 809–812.
31. Daar AS, Gutmann Th, Land W. Reimbursement, 'rewarded gifting', financial incentives and commercialism in living organ donation. In *Procurement and Preservation of Vascularised Organs*. Kluwer Academic Publishers, 1997, in press.
32. Daar AS. Living organ donation: Time for a donor charter (editorial). *Clinical Transplants* 1994, P Terasaki, M Cecka (Eds), UCLA, 376–380.

33. Young CJ, D'Allesandro AM, Sollinger HW, Belzer FO. Living related and unrelated renal donation: The University of Wisconsin perspective. *Clinical Transplants 1994*, P Terasaki, M Cecka (Eds), UCLA, 1995, pages 362–363.
34. Terasaki PI, Cecka M, Gjertson DW, Takemoto S. High survival rates of kidney transplants from spousal and living unrelated donors. *N Eng J Med* 1995, **333**: 333–336.
35. WHO (1987). 40th World Health Assembly. Resolution WHA 40.13. Geneva, May 1987.
36. European Health Ministers Council of Europe. Third Conference of European Health Ministers. Paris, 1987.
37. Ethics, Juistice and Commerce in Transplantation: A Report by the Program Committee. *Transplant Proc* **22**(3): 1054.
38. WHO. Legislative responses to organ transplantation. Kluwer Academic Publishers. Dordecht, 1993, ISBN 0-7923-2147-2.
39. Dickens BM. Legal aspects of transplantation – judicial issues. *Transplant Proc* 1992, **24**(5): 2118–2119.
40. Fuenzalida-Puelma HL. Organ transplantation: the Latin American legislative response. *Transplant Proc* 1990, **22**(3): 961–962.
41. Daar AS. Rewarded gifting and rampant commercialism in perspective: Is there a difference? In *Organ Replacement Therapy: Ethics, Justice and Commerce*, W Land, J Dossetor (Eds), Springer-Verlag, Berlin, 1991, 181–189.
42. WHO. Guiding principles. In *Human Organ Transplantation: a report on developments under the auspices of WHO (1987–1992)*. WHO, Geneva, 1991.
43. Fluss SS. Legal aspects of transplantation: emerging trends in international action and national legislation. *Transplant Proc* 1992, **24**(5): 2121–2122.
44. Nandan G. Transplantation of Human Organs Bill (India). *Brit Med J* 1994, **308**: 1657.
45. Dossetor JB, Manickavel V. Ethics in organ donation: contrasts in two cultures. *Transplant Proc* 1991, **23**: 2508–2511.
46. Dossetor JB. Rewarded gifting: is it ever ethically acceptable? *Transplant Proc* 1992, **24**(5): 2092–2094.
47. Sells RA. Towards an affordable ethic. *Transplant Proc* 1992, **24**(5): 2095–2096.
48. Sells RA. Some ethical issues in organ retrieval 1982–1992. *Transplant Proc* 1992, **24**(6): 2401–2403.
49. Simforoosh N, Amir Ansari B, Bassiri A, Gol S. Social aspects of kidney donations in 300 living related and unrelated renal transplants. In *Organ Replacement Therapy: Ethics, Justice and Commerce*. W Land, J Dossetor (Eds), Springer-Verlag, Berlin, 1991, 77–82.

CHAPTER 5

Ethics in organ donation and transplantation: the position of the Transplantation Society (1996)

SIR ROY CALNE

Introduction 62
Conclusion 66
References 66

Introduction

As organ transplantation has developed, a number of important ethical issues have arisen, particularly relating to the source of organs used for transplantation. Organs removed from both live and dead donors raise many important topics including the quality of the organ, who the donor was, or is, in the case of a live donor. Whether money has changed hands and other favours? There are also important moral questions as to the recipient. Whether priority should be given to patients according to age, usefulness to the community, dependence of a family, and medical state? How does one determine whether a patient is too ill or too well to receive an organ graft and if an organ transplant has failed, should a second transplant be offered to that patient, or to another patient on the waiting list? The customs vary in different countries, as does the availability of organs. In many developing countries, relatives will not give permission for them to be removed. Is it because they wish the body to be left intact after death, or because they do not entirely trust the doctors looking after their loved one and fear that brain death may be declared prematurely? The importance of the ethical aspect of transplantation resulted in the International Transplantation Society setting up an Ethics Committee who founded the guidelines which were adopted virtually unanimously.

Before 1985 the rules governing donation were largely intuitive, deriving from general bioethical principles and formulated by clinicians dedicated to saving lives threatened by terminal organ failure and anxious to make transplants available to as many deserving recipients as possible. Following the agreement and acceptance, by the profession, of the criteria for brain death in 1976 transplantations increased. With improved immunosuppresive agents, producing better results, renal transplantation became possible in most countries of the world. In the West kidney grafting began to move outside university hospitals to private care, with enthusiastic support from private insurers, particularly in the United States. This resulted in increased pressure to find organ donors. Wealthy patients started to advertise for donors, hospitals solicited for patients from other countries and cases appeared in newspapers where surgeons had transplanted kidneys from 'relatives' who were impoverished individuals who sold their kidneys to a fee-paying recipient in a Western centre.[1]

In September 1985 the Ethics Committee of the Transplantation Society published its guidelines for practice of clinical transplantation with particular emphasis on commercialism in the removal and transplantation of human organs. This publication was a landmark in the evolution of transplant ethics. The Council's *Lancet* paper in 1985 concludes by condemning all direct or indirect involvement in the buying and selling of organs and tissue, and of transplant activity aimed at commercial gain.[2]

The Ethics Committee of the Transplantation Society, 1990–1992 was appointed by President Tom Starzl at the XIIth International Meeting of the Society in San Francisco. The formal brief was to advise the Council on the ethical matters related to all aspects of transplantation. Two matters were identified as requiring urgent attention, the transplantation of organs removed post mortem from executed criminals and payment for organ transplants from living donors.

THE USE OF ORGANS FROM EXECUTED CRIMINALS

Newspaper reports on this practice had evoked horror and disbelief between laymen and professionals. The Council was advised of the following:

1. Medical involvement in executions to allow organ donation will dignify the process of legalized killing, execution technology will be modified to benefit transplantation, and organ donation could then be used as an argument by a government to further justify the death penalty.
2. The negative imagery of execution and positive imagery of transplantation should not be co-symbolized.
3. Since capital punishment is a violation of autonomy, how can voluntary informed consent be ever obtained?

PAYMENT FOR ORGANS

The resolution by the Transplantation Society in 1990 that a ban on payment for organs helps to develop cadaver transplantation in all countries was strongly supported. The Committee added the suggestion that while doctors in their own countries should decide on the incentive issue, the Transplantation Society should offer guidelines for suggestions of controls should they be asked for them.

Payment for cadaveric organs

The Committee rejected the 'economic imperative' as a reason to explore a future market. It was agreed that the question of payment for cadaveric organs must be approached from a humanitarian standpoint and not be forced by insurance lobbyists, lawyers or economists.

Education

An 'educational strategy' for the Transplantation Society was conceived. Third world doctors could enroll in approved training courses. Thus, a new generation of young doctors could learn the techniques of cadaveric transplantation, transplant medicine and chronic dialysis, but could observe those legal, social and religious principles that underpin cadaveric donation and its acceptance in most developed nations. It was reasoned that the secretariat of the Transplantation Society could act as a 'clearing house' to promulgate these fellowships and act as a register. Lack of available funds rendered this programme unworkable but the idea was felt to have merit and should remain a future consideration.

Living related donation

It was recognized that the profession may be more suspicious now than before that surreptitious payments are made between recipients' families and donors. The issue of living unrelated renal donation is of concern to many doctors as was the development of segmental pancreatic, hepatic and pulmonary transplants between living related donors.

XENOTRANSPLANTATION

The Ethical Committee continued to consider increasingly difficult ethical matters at length. Human xenotransplantation and animal studies in general were reviewed by the

Table 5.1 Recommendations on human xenotransplantation from the Ethics Committee of the Transplantation Society (1993)

1. The feasibility of human xenotransplantation should have been demonstrated before clinical trials by demonstrable success in an appropriate non-rodent animal xenograft model.
2. The clinical trials should be carried out:
    (a) by groups with a fully co-ordinated preclinical and clinical programme;
    (b) with local institutional and/or State or National Ethics Review Board approval;
    (c) with informed recipient consent.
3. The care and humane treatment of animals must always be of the highest standards.
4. All animal studies in transplantation must be approved by an institutional and/or State or National Ethics Review Board.
5. Endangered species must not be used.
6. Animals bred for the purpose of transplantation are the preferred source.
7. Research designed to diminish the need for use of animals in experimentation is to be encouraged.

These recommendations, presented by the Ethics Committee Chairman Professor Ross Sheil, were accepted by the Council of the Transplantation Society on 28 May, 1993

Committee. The Committee recommended that for human transplantation the use of xenografts, including those from non-human primates be acceptable in controlled circumstances. Further recommendations were made regarding animal studies (see Table 5.1).

POLICY STATEMENT FROM THE ETHICS COMMITTEE OF THE TRANSPLANTATION SOCIETY (1994)

Continuing allegations of immoral and criminal practices in the media stimulated a meeting in December 1993 between President, Professor Sir Roy Calne and Drs Bernard Cohen, Administrative Director of the Eurotransplant Foundation. A Memorandum was drafted on matters of ethical concern recently highlighted in the

Table 5.2 The Policy Statement

The Transplantation Society is the leading international professional society of doctors and scientists involved in the transplantation of tissues and organs. It will support the monitoring of transplantation activity in all countries, with the adherence of personnel involved to proper principles of practice.

The Transplantation Society recommends that all countries should:
1. Define death due to irreversible brain damage that is incompatible with life (brain death) and the method of its diagnosis. The concept of brain death should be enshrined in law with the requirement that its certification be by at least two appropriately qualified medical doctors who are independent and free from any direct interest in the transplant procedure.
2. Ensure effective monitoring of tissue and organ transplantation through:
    (A) The approval and certification of transplant centres and teams by relevant government and/or medical professional authorities
    (B) The requirement for:
        (i) signed and dated certification by the surgeon removing the tissues and organs, stating (a) that he/she has inspected and is satisfied with the certification of brain death and (b) which tissues and organs have been removed. Copies of this documentation must be forwarded with all tissues and organs. (ii) Full certification by the implanting surgeon stating that (a) he/she has inspected the retrieval documentation and deemed it satisfactory and (b) he/she has recorded the implantation of a stated organ or of stated organs into a specified individual or specified individuals. (iii) The accounting of tissues and organs that are not transplanted to relevant authorities.
    (C) The introduction by local authorities of methods for registering and tracking all donors and recipients.
    (D) The requirement for adequate standards in mortuaries and tissue banks with the implementation of regulation of their activities.
3. Enact legislation forbidding all commercial trafficking in tissues and organs.

With respect to more general issues the position of the Transplantation Society remains unchanged:
1. Organs and tissues should be freely given without commercial consideration.
2. Consent must be obtained.
3. Medical doctors and technicians should not be involved with obtaining organs from executed prisoners or in transplanting them.

media relating to organ transplantation. The joint Memorandum recommended governmental legislation and enforcement of codes of practice for national and international transplant procurement and allocation organizations. The recommendations covered transplant legislation, declaration of brain death, monitoring to prevent trafficking of organs and tissues and qualification of transplant centres and transplant teams.[4]

The XVth World Congress of the Transplantation Society took place in Kyoto, September 1994. Professor Sir Roy Calne requested the Ethics Committee to review the position of the Transplantation Society, before the Congress, regarding commercialism in transplantation. The joint Memorandum (Calne and Cohen 1993) was considered by Professor Ross Sheil, Chairman of the Society's Ethics Committee and used as part of the agenda for this Special Ethics Committee meeting which took place on 27 August 1994. The Policy Statement (Table 5.2) was approved by the Council of the Society on 28 August 1994 and ratified by the Members of the Business Meeting of the Society in Kyoto on 1 September 1994.[5]

This Policy Statement has reached all Society members and has also been promulgated through the World Health Organization and the scientific literature.

Conclusion

As organ transplantation advances so ethical matters may need to be reviewed, particularly, in organs transplanted from animals. There will be important guidelines to be established about which animal species are suitable, how the animals are managed and how the organs are offered to patients and the cost involved. These will be faced honestly and resolved to the best of the abilities of the experienced committee members.

References

1. Sells RA. Proceedings of the Ethics Committee of the Transplantation Society 1990–1992. *The Transplantation Society Bulletin*, Issue No. 1, August 1993, 4–7.
2. The Council of the Transplantation Society. Commercialism in Transplantation: The problems and some guidelines for practice. *Lancet*, September 28, 1985, 715–716.
3. Sheil AGR. Human xenotransplantation. *The Transplantation Society Bulletin*, Issue No. 1, August 1993, 8.
4. Calne RY. The President's Editorial. *The Transplantation Society Bulletin*, Issue No. 2, June, 1994, 1.
5. Sheil AGR. Policy Statement from the Ethics Committee of the Transplantation Society. *The Transplantation Society Bulletin*, Issue No. 3, June, 1995, 3.

PART 2

The process of organ and tissue donation

CHAPTER 6

Brain death

IAN Y. PEARSON

Introduction 69
The clinical condition 69
The acute brain injury 71
Definitions of brain death 74
The definitions 78
Brain death – concerns and issues 80
Conclusion 89
References 90

Introduction

Brain death is the state of bodily dysfunction in which a severe brain injury has caused an irreversible loss of the capacity for consciousness and irretrievable loss of all brainstem function, including the centres for respiratory, vasomotor and temperature control. Brain death is thus a condition that can occur only when artificial ventilation of the lungs is provided during and at the time of extinction of the brain cells. Even with that support cardiac arrest will occur within days unless further extraordinary support of the cardiovascular system is provided.

The clinical condition

THE DIAGNOSIS OF BRAIN DEATH

The diagnosis is made when, following a demonstrated and documented severe brain injury, and in the absence of drug effect or metabolic disturbance, all clinical evidence of consciousness and brainstem function is absent. Brain death can occur only if artificial ventilation is being provided at the time of cessation of brainstem function.

The condition of brain death may be apparent at first presentation following a severe brain injury, such as after arrival in the emergency room. Artificial ventilation will have been instituted by ambulance staff or by emergency room personnel as soon as reflex responses and breathing were noted to be diminishing or absent. In such a situation the intracranial event has been overwhelming, such as following massive intracerebral bleeding. Occlusion of incoming blood supply to the brain will have been immediate and absolute and the complete brain ischaemia that precedes brain death will have already occurred.

It is, however, more usual that brain death occurs during intensive care of the brain injured person, some 24 to 72 hours following the injury. The signs of brain death are obvious during routine nursing care of the person. Heart rate will become stable and will no longer alter in response to stimulation, and blood pressure will begin to fall. Core body temperature will fall. Pupil responses to light will be absent. There will be no response to routine nursing care of the eyes and mouth. There will be no cough upon tracheal suctioning. Spontaneous ventilatory movements, if previously obvious, will no longer occur. Somatic responses such as tachycardia or bradycardia, changes in blood pressure, or muscle responses will no longer occur in response to any noxious stimulus in the area of the cranial nerve distribution.

Brain death can almost always be predicted from knowledge of the severity of the brain injury, the time elapsed since the injury or further compounding injury, the rate at which these changes are occurring, and other evidence such as intracranial pressure both rising and becoming progressively less responsive to attempts to reduce its level.

The final evolution of those changes is usually gradual, occurring over a period of four to six hours. At this point of care the outcome will become increasingly obvious, irrespective of the extent and extremity of efforts to reverse the change. Within the context of competent routine care of the brain injured person the *diagnosis* is not made by chance observation by a senior nurse or physician, or at the time of extensive and exhaustive examination (at bedside rounds), but by the routine observation described. It is only formal *confirmation* of that diagnosis that then becomes the subject of a standardized protocol, whether that be legally required or be hospital or unit based practice.

THE CONFIRMATION OF BRAIN DEATH

Few countries have legal definitions of brain death that dictate tests of confirmation. Laws regarding confirmation of the diagnosis usually relate to requirements to be met before organ or tissue removal from the body and relate to the competence and suitability of those certifying death (see Chapter 7).

Standards and protocols for the confirmation of brain death have been created from guidelines provided by professional bodies such as The Royal Colleges in the United Kingdom[1] and the President's Commission (United States).[2] The extent and scope of tests to be used will vary from country to country, depending upon the perceived intention of legal definitions of brain death, and upon the acceptance by the medical community of the guidelines noted above. An outline of the tests which may be used for confirmation can be found on pages 75 to 78. More precisely detailed descriptions can be found in a number of recent publications.[3-5]

The acute brain injury

CAUSES OF BRAIN INJURY LEADING TO BRAIN DEATH

Traumatic brain injury is still the commonest cause of brain death in young males. The outcome of severe head injury (Glasgow Coma Score on admission of 8 or less) remains high at 31%,[6] 34%,[7] and 36%.[8] For both sexes brain death leading to organ donation is now most common as a result of spontaneous intracranial haemorrhage. Head trauma accounts for 44% in both sexes, although still 52% for males. Hypoxic brain injury is an uncommon cause (usually in situations such as respiratory arrest following acute asthma, 3%), and brain tumour even less common at 2%.[9]

THE PATHOPHYSIOLOGICAL PROCESSES UNDERLYING BRAIN DEATH

The pathological basis underlying brain death is the final occlusion of intracranial blood supply as it enters the cranium from the internal carotid and vertebrobasilar arteries. The mechanism by which this occurs is two-fold. A severe and overwhelming brain injury will lead to an increase in intracranial volume and pressure. If unrelieved, intracranial pressure will rise to a level exceeding incoming arterial pressure and blood supply to the whole brain will cease. In addition, haemorrhage and swelling cause distortion of the brain and compression of vessels against the edges and margins of structures such as the tentorium cerebelli and the foramen magnum (Figure 6.1).

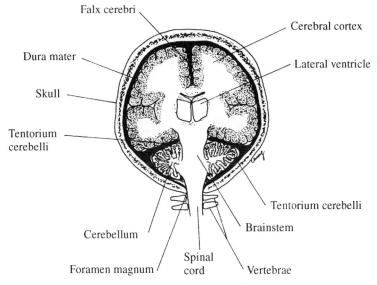

Figure 6.1 Cross section through the skull and upper part of the spinal cord seen from the back showing in black the fibrous falx cerebri and tentorium cerebelli which partition the intracranial compartment within which the brain is contained.

INTRACRANIAL VOLUME AND PRESSURE AND OCCLUSION OF BLOOD SUPPLY

The skull vault is often described as a closed box (Figure 6.2). After early childhood the bones of the skull are fused together in a rigid structure with apertures only for cranial nerves, blood vessels and for the spinal cord (foramen magnum). A substantial increase in intracranial volume caused by any space-occupying lesion within the closed cranium will result in an increase in intracranial pressure (ICP). Natural mechanisms of compensation do exist and provide initial relief. Cerebrospinal fluid (CSF) can be squeezed out of the cranium and into the subarachnoid space of the spinal cord. With increased pressure production of CSF ceases. A reduction in intracranial blood volume can also occur through cerebral vasoconstriction, partly through an increase in systemic blood pressure, and partly through spontaneous hyperventilation.

The intention of medical intervention is to accomplish the same reduction in ICP through the use of drugs such as intravenous mannitol, sedatives to reduce venous pressure and to reduce cerebral oxygen demand, direct CSF drainage, and the use of artificial hyperventilation. Surgical intervention is based upon removal of the offending volume. Removal of an extradural haematoma is the most effective intervention, the haematoma being most commonly caused by an extracerebral bleed, with little underlying brain injury. Removal of other haematomas is seldom as successful, because the haemorrhage is either caused by diffuse contusing forces, or spontaneous haemorrhage causes a diffuse brain injury itself. Surgical decompression of the closed skull vault for the relief of pressure is seldom successful as it does not alter the inexorable process of the underlying diffuse injury.

In any event, when natural and medical interventions fail to stop the rise in intracranial volume within the 'closed box' of the cranium, intracranial pressure will

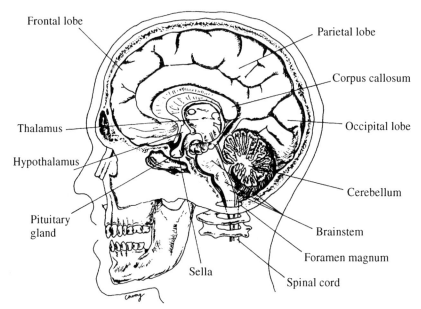

Figure 6.2 Cross-section through the skull and upper part of the spinal cord seen from the side.

increase ultimately to a level exceeding arterial pressure. Incoming blood supply is halted. Intracranial neuronal cells, deprived of nutrients, become structurally deformed, lose the ability to maintain internal processes, structure and cell wall integrity and the process of cellular dissolution, or necrosis, begins. Neuronal cells are uniquely sensitive to oxygen deprivation. It is accepted that an absence of blood supply for ten to fifteen minutes at normal brain temperature is sufficient to lead to diffuse and overwhelming brain tissue necrosis.

MECHANISMS OF VASCULAR OCCLUSION

Common descriptions of brain death, and most of the experimental work, are based upon the sudden and dramatic occlusion of intracranial blood flow by the mechanism referred to as 'coning'. In this situation a rapidly expanding supratentorial space-occupying lesion (or experimental balloon) forces the brain down through the tentorium cerebelli and against the edges of the foramen magnum. The distortion and compression occludes the blood vessels of the vertebrobasilar system and the carotid artery supply to the circle of Willis (Figure 6.3). Blood flow suddenly and immediately ceases. In this example the supratentorial injury may only be localized and circumscribed, with much of that tissue uninjured at the moment of occlusion. Ischaemia and subsequent necrosis of cerebral (supratentorial) tissue thus occurs at the same time as infratentorial tissue necrosis, that is, mostly *after* the loss of blood flow. In these circumstances massive and sustained increased pressure is sufficient to maintain that occlusion. Provided there

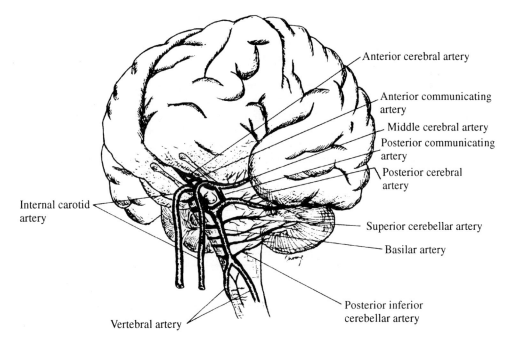

Figure 6.3 The underside of the brain showing the four main arteries (two internal carotid arteries and two vertebral arteries) which supply the brain.

is no release of pressure such as can occur with decompression from extensive skull fracturing, or by surgical exposure, necrosis of all intracranial tissue will be complete within minutes.

Usually, however, the process is quite different. The brain injury is usually more diffuse. The rate of rise of intracranial pressure is slower, partly through medical and surgical intervention. Although there are familiar patterns of injury caused by spontaneous bleeding, such as with aneurysmal rupture, in general the extent and situation of injury is random. Vasospasm may or may not occur, is extremely variable and the contribution to injury is uncertain.[10,11] Ischaemia and necrosis of brain tissue occurs more slowly and more diffusely. Because the primary injury is almost always supratentorial (that is, within the cerebral hemispheres) loss first occurs there, usually gradually and patchily at first. Steadily rising intracranial pressure slowly forces the brain downwards through the hard edges of the tentorium cerebelli, and then forces the brainstem downwards and against the edges of the foramen magnum, thus finally occluding incoming blood supply.

THE ACUTE BRAIN INJURY AND CELL NECROSIS

The forces that cause the initial injury cause bleeding from large vessels, such as dural and subdural arteries and veins. Shearing forces stretch and tear neuronal tissue. The same forces cause diffuse small vessel rupture and resulting contusions. Tissues supplied by those vessels are then deprived of blood flow. Oedema occurs through loss of integrity of vessel walls in the inflammatory process, and through the action of osmotically active particles as cellular breakdown ensues. Haemorrhage and oedema cause an increase in volume and pressure locally, thus further embarrassing blood supply. Vasospasm of larger vessels causes a further reduction in blood flow. Reductions in supply pressure and flow may be caused by other injuries to the body. Therapeutic manoeuvres such as hyperventilation to produce vasoconstriction, and intravascular volume restriction may all further embarrass blood flow and supply. Hypoxia, hypotension and raised venous outflow pressures are further aggravating factors. The cycle of cell loss and further oedema combines with the original and extending injury to further increase intracranial volume and pressure.

It is essential to emphasize that intracranial occlusion of blood supply occurs locally first at the level of the injury itself, and then globally as the increase in intracranial volume and pressure steadily exceeds the incoming blood supply pressure. **It is this knowledge of the process of an extending injury that is the basis for the expectation that following such an injury loss of brainstem function can confidently predict loss of the remaining brain tissue.**

Definitions of brain death

A full understanding of the clinical condition requires recognition of the variations in recognized definitions, the pathophysiological differences implied, and the wording of those definitions.

All existing definitions of brain death involve substantive differences and difficulties ranging from the use of the word 'death', to whether the condition involves all of the brain including the cerebral hemispheres or limits itself to the brainstem alone, and of the use of the words 'all' and 'function'. Legitimate concerns may be purely clinical, for example the anatomic difference implied by the term brainstem death and how that might differ from ischaemia and necrosis of all of the brain. Application of the word 'death' to the clinical condition has unnecessarily clouded an understanding of a very simple and very obvious sequence of pathological events. Public perception and understanding in an area of such extraordinary sociomedical importance is influenced by the definition chosen. Variations in definitions and the confusion attendant upon those differences inevitably leads to doubt, disbelief and suspicion. Teaching and explanation of the fundamental pathophysiological process is further hindered by a continuing lack of clarification.

FORMAL DEFINITIONS AND GUIDELINES

There have been a number of landmark publications of conclusions and guidelines of august medical bodies from which present practice is derived. All are based upon a knowledge of the processes causing intracranial occlusion of brain blood supply following a severe brain injury and the clinical manifestations of that process. All are founded upon clinical confirmation of the diagnosis of that state with varying recommendations for objective confirmatory tests.

> The Declaration of Sydney at the 22nd World Medical Assembly in Australia was adopted on 9 August 1968.[12] The statement remains just as current today: '...clinical interest lies not in the state of preservation of isolated cells but in the fate of a person. Here the point of death *of the different cells and organs* is not as important as the certainty that the process has become irreversible by whatever techniques of resuscitation that may be employed.' The declaration emphasized that it is the moment of acceptance of irreversibility of the processes leading to death which must be determined rather than the moment of death.

> The Statement issued by the Honorary Secretary of the Conference of Medical Royal Colleges and their Faculties in the United Kingdom on 11 October 1976[1] (the UK Colleges Statement).
>
> The basis of the statement are the words: 'It is agreed that permanent functional death of the brainstem constitutes brain death...'.
>
> The diagnosis should be considered when: 'There should be no doubt that the patient's condition is due to irremediable structural brain damage. The diagnosis of a disorder which can lead to brain death should have been fully established.' 'The patient is deeply comatose' – 'no suspicion that this state is due to depressant drugs' – 'primary hypothermia as a cause of coma should have been excluded' – 'metabolic and endocrine disturbances that can cause or contribute to coma should have been excluded.' 'The patient is being maintained on a ventilator because spontaneous respiration had previously become inadequate or had ceased altogether.'

The diagnosis is then confirmed by clinical assessment of brainstem responses as follows:

'All brainstem reflexes should be absent.

(a) The pupils are fixed in diameter and do not respond to sharp changes in the intensity of incident light.
(b) There is no corneal reflex.
(c) The vestibulo-ocular reflexes are absent.... .
(d) No motor responses within the cranial nerve distribution can be elicited by adequate stimulation of any somatic area.
(e) There is no gag reflex or reflex response to bronchial stimulation by a suction catheter passed down the trachea.
(f) No respiratory movements occur when the patient is disconnected from the mechanical ventilator for long enough to ensure that the arterial carbon dioxide tension rises above the threshold for stimulating respiration – that is the $PaCO_2$ must normally reach 6.7 kPa (50 mmHg).'

It also stated:

'It is customary to repeat the tests to ensure that there has been no observer error.' 'It is well established that spinal cord function can persist.... .' 'It is now widely accepted that electroencephalography is not necessary for diagnosing brain death.' 'It is recommended that it (body temperature) should be not less than 35 °C before the diagnostic tests are carried out.'

In October 1979 the Conference of Royal Colleges and their Faculties published the *Memorandum on the Diagnosis of Death*[13] which concluded that 'It is the conclusion of the Conference that the identification of brain death means that the patient is dead, whether or not the function of some organs, such as heart beat, is still maintained by artificial means.' The memorandum reinforced the fact that death is not an event but a process. It also emphasized that, **despite all previous history of cessation of heart beat as the traditional time of acceptance of death, brain death represents, in fact, 'the stage at which a patient becomes truly dead....'**

Cessation of heart beat, in this age of modern medicine, is no longer considered automatically irreversible, as the heart can often be restarted. In the reality of modern practice attempts to restart the heart and cardiopulmonary resuscitation are usually only abandoned when the period during which a circulation cannot be provided is accepted as well past the time of irrecoverable brain injury, usually at least thirty minutes. That same principle is applied by ambulance officers. All attempts at resuscitation will be applied to a warm body with a history of cardiac arrest of minutes, or to a cold body following immersion, if there exists any chance of resuscitation. It is only the certainty of an irrecoverable brain injury that precludes or stops attempts to restart the circulation.

Guidelines for the Determination of Death: Report of the Medical Consultants on the Diagnosis of Death to the President's Commission for the Study of Ethical Problems in Medicine and Biomedical and Behavioural Research[12] (The USA President's Commission – 1981).

'An individual who has sustained either (1) irreversible cessation of circulatory and respiratory functions, or (2) irreversible cessation of all functions of the entire brain, including the brainstem, is dead.'

For the latter to be diagnosed:

1. CESSATION is recognized when – (a) Cerebral functions are absent, and ... (b) Brainstem functions are absent.
2. IRREVERSIBILITY is recognized when – (a) The cause of coma is established and is sufficient to account for the loss of brain functions, and ... (b) The possibility of recovery of any brain functions is excluded, and ... (c) The cessation of all brain functions persists for an appropriate period of observation and/or trial of therapy.

'In the absence of confirmatory tests, a period of observation of at least twelve hours is recommended... .' 'Confirmation of clinical findings by EEG is desirable when objective documentation is needed... .' 'Without complicating conditions, absent cerebral blood flow as measured by these tests (intracranial angiography or radioisotope cerebral angiography), in conjunction with the clinical determination of cessation of all brain functions for at least six hours, is diagnostic of death.'

Guidelines for the Determination of Brain Death in Children[14] was published in the USA in 1987.

These guidelines set criteria for objective confirmatory tests for children under the age of one year, and an observation period of at least twelve hours in older children.

In 1977 the Australian Law Reform Commission published *Report No. 7 on Human Tissue Transplants.*[15] The draft legislation was adopted by most Australian States and Territories. Death was defined as either: 'irreversible cessation of circulation of blood in the body of the person, or irreversible cessation of all function of the brain of the person'.

Vital points include: 'two medical practitioners of appropriate status must certify that they have each carried out a clinical examination and that they can confirm that *at the time of that examination, irreversible cessation of all function of the person's brain had, in the opinion of the medical practitioner, already occurred.*' It is essential to note that this requirement for certification, and for the two practitioners, exists only in the context of confirmation of brain death for the purposes of removal of tissue for medical purposes including transplantation. It does not apply when organ donation is not proposed.

Furthermore, no laws dictate any means or method by which the decision regarding brain death is to be reached.

> *Statement and Guidelines on Brain Death and Organ Donation – 1993*. A report of the Australian and New Zealand Intensive Care Society (ANZICS) Working Party on Brain Death and Organ Donation.[3]
>
> This 1993 publication was the first recent attempt at a review of practice. Practice in Australia and New Zealand has been based largely upon the Royal Colleges Statement. The Working Party recognized that variations in clinical standards and practice existed. It was recognized that the diagnosis of brainstem death was not consistent with Australian laws. The need for clarification of tests of confirmation was required. Repetition of testing and periods of observation were reviewed.
>
> The working party acknowledged that the same practices should be applied to all brain dead patients, irrespective of the decision to retrieve organs. It was agreed, after considerable argument, that repetition of testing was logical. An appropriate period of observation was reinforced, consistent with acceptance of death as a process, and an acceptance that complete loss of tissue may occur over a period of time.
>
> The statement included: 'It is recommended that whenever possible appropriate imaging techniques should have been performed to make the diagnosis of structural brain injury.' 'Blood-gas analysis must be available...' and 'The $PaCO_2$ must rise to at least 60 mmHg... .' "A fall in arterial blood pH to 7.30 will confirm that the respiratory stimulus is adequate in that situation.'[16] 'Objective demonstration of absence of (intracranial) blood flow is required when the preconditions for clinical confirmation can not be met.'

> In 1994 the American Academy of Neurology published: *Practice Parameters for Determining Brain Death in Adults*.[5]
>
> Diagnostic criteria were reinforced. Descriptions of clinical testing were more precise. It included: 'A repeat clinical evaluation six hours later is recommended, but this interval is arbitrary. A confirmatory test is not mandatory but is desirable in patients in whom specific components of clinical testing cannot be reliably performed or evaluated.'

The definitions

BRAINSTEM DEATH

'Brainstem death', not 'brain death', is the accepted and recognized term in the United Kingdom where there is no legal definition of brain death. The term 'brainstem death' was specifically chosen and approved to define the fundamental role and function of the brainstem and the unarguable consequences of loss of that function. 'Brainstem death' requires only that evidence of death of the brainstem be demonstrated, whereas 'brain death' implies that there is an expectation that there is sufficient clinical evidence to demonstrate both brainstem and cerebral death. Pallis wrote 'Although brainstem death may very occasionally occur as a primary event it is, in the vast majority of cases, *the infratentorial repercussion of catastrophic supratentorial disturbances.*'[17]

The definition of brainstem death and the means used to confirm it, when used in its narrowest context, does allow for the diagnosis of death and the removal of support in the rare situation of isolated brainstem cellular necrosis when the cerebral hemispheres remain intact. This can occur with primary brainstem infarction due to vertebrobasilar artery occlusion, cerebellar haemorrhage or after primary brainstem trauma. In these instances blood flow above the tentorium would be largely intact. It will also allow for a situation in which loss of supratentorial tissue is not complete. Such might occur if the 'closed box' cranium is not complete as in massive 'eggshell' fractures, unfused sutures in children or following surgical decompression.[18] It could occur in encephalitis or meningitis in which loss of tissue is diffuse but patchy. In any of these instances supratentorial pressure transmitted to the brainstem has at some point been sufficient to occlude blood supply to the infratentorium, with brainstem cell loss, but may not have been sufficient to occlude all the vessels above.

Accepted practice in the United Kingdom is that brainstem death can confidently be confirmed through observation of the unconscious person during care, knowledge of the causal injury, exclusion of confounding factors such as drugs and metabolic disturbances, and the use of defined tests of brainstem function including respiratory centre function and cranial nerve reflexes and responses. By definition, any investigations directed at demonstrating absence of supratentorial blood supply or function are unnecessary.

With the use of this definition medical and social acceptance of brainstem death is based upon the ultimate reality and implication of loss of the brainstem, and is not hindered or restricted by a formal legal definition, or confused and obfuscated by terminological argument. By limiting the definition, loss of brainstem function is emphasized and recognized as the essential component, irrespective of the possible survival of brain cells above the tentorium. In its simplest form this definition recognizes that persistent supratentorial function is immaterial when brainstem failure with consequent cessation of respiration and circulation would terminate all somatic bodily function. The reality and practicality of the definition also recognizes the impossibility of confirming now or in the foreseeable future that 'every brain cell' has died, and allows for acceptance of evidence of continuing anterior pituitary function as a persistent but pointless 'function' which nevertheless cannot alter the final outcome.

BRAIN DEATH (OR 'WHOLE' BRAIN DEATH)

'Brain death', as opposed to 'brainstem death', is the term used as the formal definition in the majority of countries. This definition is usually understood to mean that irretrievable cell failure has occurred both above and below the tentorium. In most of these countries a legal definition of death exists, including that of brain death. Specific legal requirements and conditions may also exist when removal of organs and tissues is proposed after brain death is confirmed. The definitions all vary to some degree and extent, but almost all include the use of two contentious words – 'all' and 'function'. While these words do not alter the reality of the pathophysiology, or the implications of the event, they do allow for confusion and doubt to exist when argument is continued about their use. The theoretical possibility of minimal though quite inadequate intracranial blood supply and evidence of persistence of anterior pituitary function are arguments which can be misused to diminish the strength and standing of both the legal definition and social acceptance of brain death.

It is essential to note that although the definition implies that supratentorial brain tissue has also begun the process of dissolution at the time, it is by no means universal practice to include objective tests in the diagnosis of brain death which will assist to prove that fact. The loss of supratentorial function is usually interpreted through knowledge of the causal injury, through clinical observation of the process of deterioration, and through knowledge of the pathological process causing brain death.

The concerns and issues referred to below refer largely to the term 'brain death' as defined above – an all-inclusive loss of intracranial tissue. 'Brainstem death', by definition, is not affected by most of these issues.

Brain death – concerns and issues

A number of controversial issues continue to confound an acceptance of brain death as a fact of medical life. This chapter cannot concern itself directly with the arguments relating to the philosophical acceptance of brain death as the equivalent of 'cardiac death', or with the morality of the cessation of artificial ventilation which therefore follows that acceptance. The reader is referred to other texts.[17,19-26]

The chapter cannot, however, be complete without reference to areas of doubt and concern which relate to an understanding of the mechanisms of loss of brain tissue, and the diagnosis and confirmation of brain death. These issues are fundamental to societal acceptance of brain death.

CONFUSION OF BRAIN DEATH WITH 'THE VEGETATIVE STATE'

The severe brain injury suffered by Karen Anne Quinlan[27] led to considerable confusion for many years as lay publications did not clearly differentiate her vegetative state from that of brain death. The so-called 'vegetative state' is one in which there is no evidence of consciousness but all or most brainstem functions continue. Breathing continues, blood pressure is well maintained and cranial nerve reflexes remain intact. Cerebral tissue has been extensively damaged, or the pathways which allow cerebral processes to be transmitted to all the nerves of the body are irretrievably interrupted. The condition is clearly not brain death and must not be confused with such. The 'locked-in syndrome' is a variation in which consciousness is maintained but the lower brain injury has severed most of the connections to the peripheral nervous system within the lower brain itself. Some isolated movement and sensation may persist in parts of the limbs as well as within the head and neck. Breathing always continues and communication with the person, however limited, can still be maintained.

INEVITABILITY OF CARDIAC STANDSTILL

Earlier descriptions of brain death stated that cardiac standstill was inevitable within days of the final event.[28] That evidence was used as moral justification for the inclusion of the word 'death': the irretrievable loss of brain function could be equated with the familiar 'asystolic' death because the two events bore a close temporal relationship. That moral argument can, however, now be contested through clear and convincing evidence of considerable prolongation of an adequate and stable circulation though the use of additional measures such as vasopressin with catecholamine infusions,[29-32] and with hormone replacement therapy.[33,34] The fundamental pathophysiological basis of brain death is not challenged by this ability to prolong cardiac output. The moral basis for the terminology and the decisions regarding organ retrieval and removal of ventilation *may* be challenged. That challenge is only possible if prolongation of somatic survival through the provision of extensive and extraordinary support is allowed to obscure the reality that loss of brainstem function is the end of that individual's existence, no matter how long support is continued.

Whether artificial support is ventilation only or ventilation with adrenaline and vasopressin is ultimately irrelevant. Without either, familiar 'death' will occur. With the use of both the practice is merely prolongation of the process of somatic death and a means of delaying any final acceptance of an end to life.

Removal of useless and pointless support remains the only objective of the diagnosis of brain death. Prolongation of support for the purposes of organ retrieval, if that were the stated desire of the person before death, is not pointless, and is thus entirely consistent. Support for the purpose of enhancing the survival of an unborn baby in a brain dead mother is not pointless. A delay for the good of surviving family may also not be pointless, if that were to aid in their own personal survival. Recognition of the indignity to the memory of the brain-dead person that is entailed by delay is the final responsibility of all those involved.

SOMATIC AND SPINAL REFLEX RESPONSES FOLLOWING BRAIN DEATH

Brain death, however defined, is loss of intracranial tissue only, simply because intracranial pressure exceeds the ability of systemic pressure to perfuse the skull contents. Since 1976 no definition has ever referred to loss of spinal cord tissue within the context of brain death. Separation of the spinal cord from the brain is an inescapable pathophysiological event, not an 'ad hoc' division.[23] The spinal cord is part of the central nervous system but it is not intracranial. Philosophical or religious desire to place the soul within the spinal cord cannot be contested, any more than that it resides within the liver or large bowel.

Spinal reflexes occurring as a response to the final hypoxia of cardiac arrest, 'the Lazarus sign,'[35] remain an extremely unpleasant experience for those attending the body at the time. This activity occurs only when systemic circulation has effectively ceased and is the result of disinhibition of spinal motor neurones through loss of brainstem control. These movements are not seen in cardiac arrest without prior loss of brainstem tissue.

Spinal reflexes persist after brain death. Pallis[17] has described this in detail, emphasizing retention of spinal reflexes in the decapitated cat[36] and rises in blood pressure following peripheral stimulation in the same model.[37] Dysautonomic responses and motor reflex responses are well recognized events during surgery in patients with spinal cord transection – spinal regional anaesthesia being the recommended method of anaesthesia in these patients. These responses can as logically be expected in surgery following brain death. Rises in circulating catecholamines have been shown to occur with hypercarbia in brain death apnoea testing.[38] The relationship between spinal cord reflexes and adrenal medullary secretion of catecholamines in brain-dead patients during organ retrieval has also been supported.[39] There is no basis for an interpretation of these physical events as being related to intracranial experience of sensation. The use of muscle relaxants and antihypertensive agents during surgery is no different to their similar use in the quadriplegic person with spinal cord transection.

SURVIVAL OF ALL OR SUBSTANTIAL SUPRATENTORIAL FUNCTION IN THE ABSENCE OF BRAINSTEM FUNCTION AND BRAINSTEM BLOOD FLOW

Clinical instances where this might occur are well recognized and readily diagnosed, provided that the conditions demanded for confirmation of brain death are clearly and strictly observed.

In traumatic injury direct brainstem injury may occur. Sudden death due to brain injury in motor vehicle trauma may occur through the hangman's fracture of the brainstem.[17] Primary vertebrobasilar artery occlusion will lead to infarction of the brainstem and cerebellar haemorrhage will cause brainstem compression locally with no effect above the supratentorium.[40-44] In these instances blood flow above the tentorium would be largely intact. All the requirements for the diagnosis (loss of consciousness and absent brainstem function) are satisfied **except the fundamental requirement for a causal brain injury appropriate to the diagnosis of brain death**. All major guidelines require that the diagnosis not be entertained unless there is clear proof of the underlying brain injury. The standards required for the care of the severely brain injured patient in modern medicine now demand diagnostic investigations such as computerized axial tomography (CAT) or magnetic resonance imaging (MRI), radionuclide studies or radiocontrast angiography, in all situations of spontaneous haemorrhage and trauma. In these instances the absence of supratentorial pathology, or the clear persistence of cerebral hemispheric blood flow in the absence of brainstem perfusion, is easily identified. For example, a patient presenting with a history of the onset over a few hours of slurring of speech, ataxia and loss of co-ordination followed by coma can be confidently diagnosed to have a primary brainstem/cerebellar infarction with minimal supratentorial pathology. Loss of consciousness and absent brainstem function can then be confidently attributed to primary brainstem infarction. CAT scan or MRI scanning would almost certainly be performed as part of clinical management in any event and neither would show a gross cause for the clinical state. That together with the history would necessitate cerebral angiography which would demonstrate blood flow to the cerebral hemispheres but absent flow to the brainstem. Under these circumstances, in any country or state where brain death is defined legally as loss of all brain function the diagnosis of brain death in its strictest sense could not be made. It is essential to note that this state does not preclude the removal of useless support. It simply cannot be referred to as brain death in the legal sense. In contrast, in the United Kingdom, or wherever the concept of brainstem death is accepted, the diagnosis of brainstem death can be made with equal confidence.

THE POSSIBILITY OF SURVIVAL OF VIABLE ISOLATED CELLS OR ISLANDS OF CELLS

For those who would demand absolute proof of loss of every intracranial neuronal cell there is and will be no answer. At this time, and for the foreseeable future, the possibility of persistent survival of isolated viable cells can neither be proven nor disproved. There is no method of examination of the human brain that can reliably prove that not one neuronal cell remains functional, while the brain remains within the body. Radiocontrast angiography, radionuclide imaging, positron emission (PET) scanning or even assessment of active phosphorous by nuclear magnetic resonance do not have the discrimination required to identify even a few cells as functioning cells.

Histological examination of the brain, outside the body, is the only possible absolute evidence. Studies performed on brains following brain death have not always shown uniform necrosis of tissue. This is consistent with the time elapsed between diagnosis and cessation of circulation – brain death as a process – the longer the time elapsed the more likely actual necrosis can be identified. Inadequate information correlating pathological studies and the sequence and timing of clinical events have not provided absolute scientific confidence in these studies. Black's analysis[45] of histological studies confirms the uniformity of evidence of tissue death – in all those with minimal changes, cardiac standstill occurred very soon after brain death diagnosis. In most observations the recognition of time elapsed was noted and is usually consistent. Exceptions as described (patchy tissue loss for example) are not accounted for in these studies. Absence of inflammatory changes has been suggested as required for the histological diagnosis of brain death. As described previously sudden and overwhelming occlusion of blood supply will be logically associated with an absence of inflammation. The more common event of gradual loss of tissue, and gradual and progressive posterior fossa vessel occlusion will entail sufficient flow over a long enough period to allow inflammatory changes to occur. The absence or presence of these changes cannot be used as evidence for or against the viability of the standards used to confirm brain death.

Objective evidence of a severe cerebral injury, with a clinical course consistent with the evolution of brain death, resulting in brain death confirmed by a rigorous process, taken together with knowledge and understanding of the pathophysiology and histopathology is all that can ever reassure the observer of the fact of death of most of the brain. The absence of blood flow demonstrated by radiocontrast angiography may be further evidence of brain death but even that will never be able to satisfy those who demand proof of death of 'all' of the brain.

CONTINUING EVIDENCE OF HYPOTHALAMIC AND PITUITARY FUNCTION DESPITE EVIDENCE OF ABSENT INTRACRANIAL BLOOD FLOW

Brain death had always been associated with diabetes insipidus although some studies showed that it did not always manifest itself as such.[46-48] In 1987 Novitzky[49] published evidence to show that the administration of anterior pituitary hormone replacement therapy would benefit the potential organ donor – thus increasing the likelihood of successful donation and possibly improving the function of donated organs.[49-54] These studies were supported by others.[55] Subsequent publications[46,48,56-58] provided contradictory evidence which were based upon evidence of continued function of the anterior pituitary after brain death. These studies concluded that it is only the 'sick euthyroid state' that occurs following brain death, and as such, hormone therapy is not indicated. That evidence is now, however, often quoted to question the validity of clinical brain death criteria as proving 'that all brain function has ceased'. It has also been used to further the case for 'upper brain death' as a more valid point of non-return.[23-25]

The most significant impact has been to question the methods used to demonstrate brain death. Evidence of continued anterior pituitary function has been improperly equated with substantial survival of neuronal tissue above the tentorium. Reasonable concern relates to the validity of present methods to confirm loss of tissue both above and below the tentorium. This concern has led to a renewed belief that cerebral angiography is essential to show loss of all brain function. Other more philosophical concerns are that if pituitary function is shown to continue the whole basis of brain death can be argued to be faulty.

The clinician faced with the terrible reality of brain death has found it disruptive and irksome to be faced with argument that persistent anterior pituitary function automatically implies that 'all function of the entire brain' has not ceased. The arguments cannot be ignored, and must be discussed for completeness.

Persistence of anterior pituitary function has been shown to continue even in the absence of any discernible blood flow using cerebral angiography and radionuclide scanning.[57,58] Angiography, therefore, is not a solution to the problem. It does, however point to the probable explanation for an otherwise unexplainable phenomenon. Retained pituitary function has been demonstrated in every patient diagnosed as brain dead in a number of studies, and for quite prolonged periods following confirmation of brain death.[48,57-59] With even the most conservative approach it would be impossible to conceive that every patient in those series had retained substantial supratentorial integrity and function.

Explanation must exist for this anomaly. The first may be found in the embryological origin of the anterior pituitary as the only intracranial tissue derived from ectoderm and of glandular structure. Such tissue may withstand loss of blood supply for much longer periods than neuronal tissue. In addition the rich blood supply with its unusual vascular system which permits bidirectional flow between pituitary and hypothalamus may also contribute to persistent blood flow.

Secondly, examination of the arterial supply to the pituitary in the dog clearly shows an extracranial (extradural) origin of the inferior hypophyseal artery within the cavernous sinus. It is not only physically extracranial but its response to noradrenaline clearly differentiates it from that of the superior hypophyseal arteries arising intracranially.[60] Ample evidence exists of arterial continuity between dural and inferior hypophyseal vessels and the neural lobe, median eminence of the hypothalamus and neurohypophysis.[61-63] The human pituitary is partially supplied by hypophyseal vessels originating from the internal carotid arteries as they pass through the cavernous sinus and before they pierce the dura to create the circle of Willis (circulosis arteriosus cerebri).[64,65] At this point the vessels are almost certainly protected from an elevated intracranial pressure. Although further vessels supply the hypothalamus and pituitary from more distal parts of the circulosis arteriosus the unique bidirectional flow of blood is most likely responsible for the continued minimal arterial supply to these regions. A recent autopsy study of 84 brain-dead patients showed extensive necrosis in all brains with the notable exception of survival of some hypothalamic cells especially those of the supraoptic nucleus in at least half of the brains examined.[66]

Finally, the pituitary gland lies within the sella turcica, a small recess within the sphenoid bone. The stalk connecting the hypothalamus with the pituitary passes through a tight opening of the diaphragm sella, a shelf of dura stretched between the clinoid processes. The diaphragm sella isolates the gland from the brain and cerebrospinal fluid (CSF). The diaphragm protects the pituitary from the suprasellar arachnoid cistern containing CSF. When the diaphragm is incompetent or absent it is associated with the 'empty sella' syndrome in which the pituitary remains as a flattened structure. It is therefore possible that the diaphragm sella further protects the pituitary from raised intracranial pressure. That diabetes insipidus remains an almost invariable feature of brain death is entirely consistent as it is the hypothalamus that produces this hormone before transfer via the stalk to the posterior pituitary. It is of further interest that diurnal fluctuations have been found to be absent in some studies,[59] and that all other studies show that levels of hormones are low – the sick euthyroid state. Further evidence of changes in hormone levels that relate directly to brain death continue to accumulate. Repeated studies have shown an immediate reduction in hormone levels following the experimental induction of brain death in animal experimentation.[67-69] These changes cannot be likened to the 'sick euthyroid' state which usually follows severe head injury.[70] The animals were previously well. Persistent minimal dural and hypophyseal flow may only enable transfer of remaining hormone from hypothalamus and pituitary gland.

One final statement can be made with regard to the philosophical misuse of this remnant intracranial function. It should be emphasized that through the ectodermal origin of the adenohypophysis, its glandular structure and blood supply, the anterior pituitary and adjacent tissue could as easily be recognized as not being part of that brain which all would consider is a complex structure of interconnecting nerve cells providing initiative, thought and spontaneous mental activity in addition to basic functions of action, sensation and co-ordination. In addition survival of one part of the brain which is due only to a unique extracranial vascular supply still does nothing to alter the fundamental fact of that final event in which massive rise in intracranial pressure finally deprives the whole of the rest of the brain of its blood supply and causes tissue death of all of the rest of that brain.

There is now support for the argument that laws framed before persistent hypothalamo-pituitary 'function' in brain death was recognized, are technically incorrect. None of this new evidence can contest the reality of extinction of the whole of the rest of the brain – the fundamental and conscientious intention of those who framed the laws and guidelines.

PERSISTENCE OF MORE EXTENSIVE AREAS OF FUNCTION AS EVIDENCED BY ELECTROENCEPHALOGRAPHIC (EEG) AND EVOKED POTENTIAL SIGNALS

EEG as a means of confirming brain death was being abandoned as long ago as 1968.[71] The techniques are basically unreliable in the confirmation of brain death. An absence of electrical signals can occur in deep coma due to drug intoxication and as such the EEG cannot be used as confirmation of loss of cerebral brain function in the same way as blood flow methods. The EEG is unable to prove or disprove the existence of lower brain function as the signals recorded are primarily cortical in origin. The existence in the intensive care unit of complex medical equipment and monitoring devices provides considerable difficulty in interpretation of any low voltage signals of the kind that might be expected from a brain without function, or minimal function if the tool is being used to exclude that possibility. Reports have been published of persistent EEG activity despite repeated clinical examinations which have shown absent brainstem activity.[72] Such activity may be due to the nature of the injury. A disease process which is slowly progressive, especially if there is variable vascular involvement, such as a meningitis or encephalitis, may produce a variation of the condition described previously: that is of death of the brainstem but with patchy death of cerebral tissue. Arguments based upon the persistence of EEG signals in such a case might perhaps have been better directed at the possible inappropriateness of the underlying condition which might not have satisfied the requirement for a pathological condition leading to brain death. In addition it should be emphasized that the clinical clues which lead to discovery of persisting activity are the same clues which protect the patient from an incorrect diagnosis. There was sufficient reason for the clinicians involved to question the validity of the clinical signs and to be guided by them.

Somatic (SEP) and brainstem auditory (BAER) evoked potential examinations have also been reported to have shown transmission when brainstem function is clinically absent. An abundance of literature exists but the reports are sufficiently confusing to have deterred the routine use of these techniques in the confirmation of brain death. The nature of the injury may again be responsible as described above. Patchy initial loss of tissue, hypothermia or the presence of any sedative drugs or anaesthetic agents may all cause conflicting results. Improper association of the first clinical signs of brain death with persistent responses may also contribute to the conflicting results. Evoked potentials are of course conducted by axonal transmission, but are also conducted directly through tissue (late responses) as in any electrically conductive medium (whether tissue is functional or not). Such late responses must be clearly differentiated from functioning axonal transmission and may be a source of confusion. Both left and right vestibulo-ocular reflexes must be assessed. If these reflexes are truly absent there can be no possibility of normal neuronal transmission via the eighth nerve, and no true auditory evoked responses are possible. Failure to perform both examinations adequately may be a confounding cause for persistent BAER responses.

OBJECTIVE ASSESSMENTS OF INTRACRANIAL BLOOD FLOW

Tests of blood flow have been widely used as routine confirmatory tests, being almost obligatory in some European countries. They have been espoused as the only true 'gold standard' of confirmation of brain death. Those with doubts about existing practice believe that such studies would be essential to resolve their anxieties. Counter-arguments have been that transfer to the imaging department with the attendant consequences of changes of ventilation and loss of vital measurements entails perhaps the final risk which will ensure the very event feared, or that the introduction of hypertonic radiocontrast media will complete the injury. In any event these radiocontrast flow studies are fundamentally flawed by the very nature of infarction of tissue. In death arterial vessels empty, venous vessels fill. At post-mortem examinations arteries can be injected with contrast medium or with plastics to provide exact models of arterial blood supply. Studies using high-powered pumps to inject radiocontrast medium can with sufficient force show flow into cerebral vessels – the pressure needs only to be above that of intracranial pressure at the time.[73] Radionuclide techniques had not previously demonstrated brainstem blood supply with any certainty so could not be argued to be an equivalent gold standard. Possibly the most reliable method for showing global loss of intracranial blood supply may be the use of Tc-99m hexamethyl-propylene-amine oxide (HMPAO) or similar agents which actually cross the blood–brain barrier in viable tissue. As these agents are picked up and held for some time by viable brain tissue they are theoretically more precise indicators of actual cell viability.[74-77] They are also more reliable for demonstrating absence of posterior fossa tissue blood flow and viability.

None of these tests have the discrimination to prove zero blood flow. **They remain only methods of confirming a clinical expectation.** Absence of discernible flow is a means of satisfaction in situations of doubt or suspicion. They would certainly be reliable in situations of uncertain intoxication – flow or no flow being absolute evidence. They would help to provide evidence of persistent supratentorial blood flow. They cannot be regarded as absolute independent tests. The study by Riischede[78] first showed absence of intracranial vessel filling in brain death. A subsequent study then examined whether absence of filling was due to spasm or intravascular thrombosis.[79] It was found to be neither, simply occlusion by high intracranial pressure. The report included two patients in whom angiography performed by unilateral carotid puncture immediately following spontaneous haemorrhage with intraventricular blood showed absent blood flow initially but some flow on angiography later.[79] In one instance contralateral carotid puncture the next day showed good flow. This might only mean that Circle of Willis circulation was sufficient to maintain sufficient cerebral blood flow from the other side in spite of unilateral carotid compression (Figure 6.4). The patient survived to leave hospital. In another case immediate ventricular puncture and relief of high intracranial pressure led to good intracranial vessel filling but the patient subsequently became brain dead. Return of filling did not mean that the test was faulty, only the conditions under which it was performed. These reports confirm that the tests are not absolute discriminators – only tests used to confirm a substantiated clinical expectation. One study reported 30 patients in whom barbiturate coma led to the use of angiography as a primary diagnostic tool.[80] Not surprisingly 18 patients showed continued flow. In another more logical study, once peak intracranial pressure exceeded mean arterial pressure and clinical signs of brain death preceded angiography,[81] intracranial circulation could no longer be demonstrated. Given sufficient time following brain death with dissolution of tissue and loss of that tissue through fractures or via the foramen magnum, repeat angiography may show filling of major intracranial arteries.[82,83] Autopsy studies after brain death have shown herniation of tissue through the foramen magnum and into the spinal canal following brain death.[84] The patency of those arteries allows filling, not flow. No return of function occurs and the basis of brain death remains that of a clinical expectation supported by whatever tests are believed to be the most helpful in specific situations.

EXCLUSIONS, SAFEGUARDS AND COMMON SENSE

In any of the unusual situations referred to previously there will always be clues within the history, clinical examination and investigations such as angiography performed to

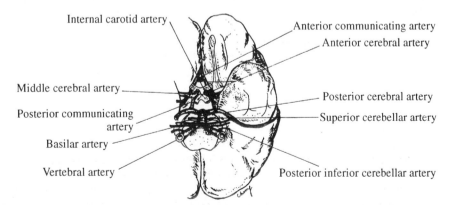

Figure 6.4 The Circle of Willis. Blood supply for the basilar artery or either internal carotid artery can flow through the connecting posterior and anterior communicating arteries to supply blood to all parts of the brain.

elucidate the injury. An absence of a supratentorial (cerebral) lesion on CAT or MRI scan, or absent brainstem flow with normal cerebral blood flow, will provide safeguards to an automatic acceptance of absent brainstem function as evidence of 'whole' brain death. A process which is known to cause a gradual injury, especially with patchy small vessel thrombosis or in which the process may be so slow as never to allow intracranial pressure to achieve the occlusion pressures necessary for global extinction, should be expected to fail the fundamental requirement for a brain injury *sufficient to precede the condition of brain death*. The same may apply in infants with open fissures and in traumatic injury in which the skull is fractured extensively, or in whom decompression via those fractures or via the nose and ears is continuous. A CAT or MRI scan showing an absence of sulci and compressed ventricular system will reassure – open sulci and open ventricles will not.

Competent clinical examination and an adequate knowledge of the fundamental requirements will be sufficient to identify those injured persons in whom the expected process may be varied. Maintenance of body temperature and blood pressure will alert the observer to an unusual situation in which persistence of patchy areas of function may be likely. In any such situation strict attention to the rules and the *expectation of those rules* must prohibit the diagnosis of brain death without objective confirmation by some other method such as cerebral angiography. If the objective method chosen reveals blood flow or activity then the diagnosis can with equal confidence be ruled out.

In a study of 140 brain-dead patients,[85] when all those rules were strictly observed, confirmation by doppler flow studies was uniformly consistent with clinical and angiographic studies. In that study temperature fall, blood-pressure instability and an absence of heart rate response to 2.4 mg intravenous atropine were included requirements for the diagnosis of brain death. The use of atropine as a screening test has been both recommended[86] and contested[87,88] but no other published studies have consistently used that technique as an essential requirement for brain-death diagnosis and confirmation. As long ago as 1984 atropine in a dose of 0.04 mg/kg was part of the demanding criteria used by the Presbyterian-University Hospital in Pittsburgh.[16] Rejection of the use of atropine as a test of brainstem function appears to be based only upon two published studies in which atropine did not completely ablate a heart rate response in patients who otherwise satisfied criteria of brain death. Our experience[85] has been that an absence of a response is a very final event. When radiocontrast studies show even traces of flow there is often still a heart rate response of 5 to 10 beats per minute. When no response occurs our study showed no flow by any of the study methods used. (Atropine must not be infused through any intravenous route already carrying inotropic drugs). Those previous studies may have failed to recognize the death of brain tissue as a process, one that means progressive loss of tissue over a period of time. The test is not at fault, only the timing of the test. There remains no doubt that in those studies brain death was occurring and that the tests of brainstem function accurately predicted brain death. **As in other studies which question the validity of clinical testing it is the tendency to equate the first loss of brainstem reflexes as the moment at which all other means of assessment are then to be judged.** With the simple criteria used in our study, doubt about persistent function will not be a concern.

The most elementary guide to practice is the injunction that whenever there exists any doubt as to the primary cause of brain injury, or where that brain injury is by its nature one that may lead to patchy brain necrosis, or where any of the clinical tests may be unreliable, objective imaging tests are essential if there is any desire for compliance with the strictest demands of laws or societal expectations.

The ANZICS guidelines[3] list as indications for objective testing:

- No clear cause for coma
- Possible metabolic or drug effect
- Cranial nerves cannot be adequately tested
- Cervical vertebral or cord injury
- Cardiovascular instability precluding apnoea test.

The American Academy of Neurology *Practice Parameters for Determining Brain Death in Adults* (1994)[5] recommends confirmatory tests under the following conditions:

- Severe facial trauma.
- Pre-existing pupillary abnormalities.
- Toxic levels of any sedative drugs, aminoglycosides, tricyclic antidepressants, anticholinergics, antiepileptic drugs, chemotherapeutic agents, or neuromuscular blocking agents.
- Sleep apnoea or severe pulmonary disease resulting in chronic retention of carbon dioxide.

To this list can be added any situation in which body temperature does not fall (cooling permitted), blood pressure remains unexpectedly stable, heart rate increases with atropine, or there is any suspicion that the requirement that a brain injury *sufficient to precede the condition of brain death* has not been satisfied, as described above.

Ultimately it is the responsibility of the caring physicians and the institution in which they work to provide guidelines for practice and to satisfy themselves and their society that the expectations of their society are being met.

CAUTION

This chapter attempts to advocate that clinicians be aware of clinical states which provide exceptions to the expected rules. It cautions those who question the validity of the clinical diagnosis, that case reports based upon failure to anticipate and to recognize exceptional circumstances or, that death of brain tissue is a process occurring over a period of time, are not only fallacious but also potentially destructive and disruptive.

In one early classical study [89] brain death had been diagnosed in two patients with fast activity on EEG. It was no surprise that autopsy of these two cases (9 and 13 of the series) showed only brainstem destruction. A more recent example[90] concluded that the clinical basis of brain death diagnosis could be questioned because angiography showed normal cerebral blood flow following a brainstem injury. The failure was in fact one to satisfy the fundamental requirement – that is of a brain injury which can be expected to lead to brain death. Reference[23] to evidence of absent EEG activity in the presence of a normal radionuclide angiogram[91] failed to note that the child involved was retarded with cerebral atrophy.

The subject of intracranial death of tissue and the clinical criteria required for diagnosis are matters demanding scientific objectivity and clarity. Proper and responsible use of observations and conclusions is essential to further inquiry into this complex subject.

Conclusion

Brain death is a state in which the brain has usually been deprived of its blood supply in a progressive fashion. Cessation of function first, and autolysis of cells next, will occur progressively. Brain death, as with death of all tissues, is not a moment in time but a process. It occurs over a period of many hours. Given time, temperature will fall, blood pressure will fall, diabetes insipidus will occur. Under these conditions the irretrievable necrosis of brain tissue, and the inevitable final cessation of function of the brain will be sure.

Brain death is not just a pathological event which has become available to all to argue or abuse in order to satisfy philosophical or theological argument. It is a real event for those caring for patients and next-of-kin in the intensive care unit. The history of brain death has been written admirably and exhaustively by Pallis.[17] The story has been that of patients maintained through gradual death of tissues, the onset of multiple nosocomial infections and decubitus ulcerations. Their eyes were unseeing, glazed, without tension, fixed and usually widely dilated. There was no response to usually painful mouth and tracheal care and no cough to clear secretions. Somatic 'death' occurred slowly, inexorably, and intractably, to the extreme distress of family, friends, nurses and doctors. For most the pointlessness was obvious. Cardiac standstill was known to be inevitable, but seldom soon enough to spare all involved from the days of misery and grief. There was no question for the individuals involved of philosophical interpretations of death. The possibility of surviving islands of neurones, persistent pituitary–hypothalamic function or the ability of a powered pump to fill large arteries with radiocontrast medium would mean nothing to the people involved, or for the dignity of the dying or dead person. It was for these real and terrible reasons that the terms 'coma depasse',[92] 'brain death', and 'brainstem death', were first coined.

Whatever the merits of the use of the word 'death' it is for those faced with this terrible condition, created only through our best attempts to care for the brain injured, that the reality of the pathophysiology needs to be reinforced. It is for them that the inevitability of somatic 'death' needs to be defended. It is for them that philosophical argument be directed responsibly. It is also for them that the standards of care, diagnosis and confirmation be continually reinforced, and that the demands of society in this area be met.

References

1. Anonymous. Diagnosis of brain death. Statement issued by the honorary secretary of the Conference of Medical Royal Colleges and their Faculties in the United Kingdom on 11 October 1976. *Br Med J* 1976, **2**: 1187–1188.
2. Anonymous. Guidelines for the Determination of Death: Report of the Medical Consultants on the Diagnosis of Death to the President's Commission for the Study of Ethical Problems in Medicine and Biomedical and Behavioural Research. *JAMA* 1981, **246**: 2184–2186.
3. Australian and New Zealand Intensive Care Society. *Statement and Guidelines On Brain Death and Organ Donation.* ANZICS, Melbourne, 1993.
4. Dobb GJ, Weekes JW. Clinical Confirmation of Brain Death. *Anaesth Intens Care* 1995, **23**: 37–43.
5. Practice parameters for determining brain death in adults (summary statement). The Quality Standards Subcommittee of the American Academy of Neurology [comment]. *Neurology* 1995, **45**: 1012–1014.

6. Fearnside MR, Cook RJ, McDougall P, McNeil RJ. The Westmead Head Injury Project outcome in severe head injury. A comparative analysis of pre-hospital, clinical and CT variables. *Br J Neurosurg* 1993, **7**(3): 261–279.
7. Miller J, Butterworth J, Gudeman S. Further Experience in the Management of Severe Head Injury. *J Neurosurg* 1981, **54**: 289–299.
8. Bowers SA, Marshall LF. Outcome in 200 Consecutive Cases of Severe Head Injury treated in San Deigo County: a Prospective Analysis. *Neurosurgery* 1980, **6**(3): 237–242.
9. Anonymous. National Organ Donor Registry, Adelaide, 1995.
10. Zurynski YA, Dorsch NWC, Pearson I. Incidence and Effects of Increased Cerebral Blood Flow Velocity after Severe Head Injury: Transcranial Doppler Study. I.: Prediction of Post-Traumatic Vasospasm and Hyperemia. *J Neurol Sci* 1995, **134**: 33–40.
11. Abrasko R, Zurynski Y, Dorsch NWC. The Importance of Traumatic Subarachnoid Haemorrhage. *J Clin Neurosc* 1996, **3**: 21–25.
12. Gilder SSB. Declaration of Sydney – A Statement of Death: Twenty-second World Medical Assembly. *Br Med J* 1968, **3**: 493–494.
13. Anonymous. Diagnosis of Death. Memorandum issued by the honorary secretary of the Conference of Medical Royal Colleges and their Faculties in the United Kingdom on 15 January 1979. *Br Med J* 1979, **i**: 3320.
14. Anonymous. Guidelines for the determination of brain death in children. Task Force for the Determination of Brain Death in Children. *Arch Neurol* 1987, **44**: 587–588.
15. The Law Reform Commission. *Report No. 7: Human Tissue Transplants.* Australian Government Publishing Service, Canberra, 1977.
16. Anonymous. Presbyterian-University Hospital Policy Manual: Certification of Brain Death and Management of Referred Brain Dead Organ Donors. Presbyterian University Hospital, Pittsburgh, 1984.
17. Pallis C. Brainstem Death. In *Handbook of Clinical Neurology*, Vol. 13(57), 13th edn, R Braakman (Ed.), Elsevier Science Publishers BV, Amsterdam, 1990.
18. Alvarez LA, Lipton RB, Hirschfeld A, Salamon O, Lantos G. Brain death determination by angiography in the setting of a skull defect. *Arch Neurol* 1988, **45**: 225–227.
19. Pallis C. Whole-brain death reconsidered – physiological facts and philosophy. *J Med Ethics* 1983, **9**: 32–37.
20. Pallis C. Death. In *Encyclopaedia Britannica*, 1030–1042, 1986.
21. Browne A. Whole-brain death reconsidered. *J Med Ethics* 1983, **9**: 28–31, 44.
22. Shann F. A Personal Comment: Whole Brain Death versus Cortical Death. *Anaesth Intensive Care* 1995, **23**: 14–15.
23. Truog RD, Fackler JC. Rethinking brain death. [Review]. *Crit Care Med* 1992, **20**: 1705–1713.
24. Veatch RM. The impending collapse of the whole-brain definition of death [published erratum appears in Hastings Cent Rep 1993 Nov–Dec; 23(6):4]. [Review]. *Hastings Cent Rep* 1993, **23**: 18–24.
25. McCullagh P. *Brain Dead, Brain Absent, Brain Donors.* Wiley, Chichester, New York, 1993.
26. Jorns KP. Theological theses on the ethics of organ transplantation and on a law concerning them. *Forensic Sci Int* 1994, **69**: 279–283.
27. Kennedy IM. The Karen Quinlan case: problems and proposals. *Journal of Medical Ethics* 1976, **2**(1): 3–7.
28. Black PM. Brain death (first of two parts). [Review]. *N Engl J Med* 1978, **299**: 338–344.
29. Yoshioka T, Sugimoto H, Uenishi M, et al. Prolonged hemodynamic maintenance by the combined administration of vasopressin and epinephrine in brain death: a clinical study. *Neurosurgery* 1986, **18**: 565–567.
30. Manaka D, Okamoto R, Yokoyama T, et al. Maintenance of liver graft viability in the state of brain death. Synergistic effects of vasopressin and epinephrine on hepatic energy metabolism in brain-dead dogs. *Transplantation* 1992, **53**: 545–550.
31. Kinoshita Y, Yahata K, Okamoto K, Yoshioka T, Sugimoto T. [Organ preservation with the

combination of vasopressin and catecholamine in brain dead donors]. [Japanese]. *Nippon Geka Gakkai Zasshi* 1991, **92**: 771–774.
32. Pennefather SH, Bullock RE, Mantle D, Dark JH. Use of low dose arginine vasopressin to support brain-dead organ donors. *Transplantation* 1995, **59**: 58–62.
33. Washida M, Okamoto R, Manaka D, et al. Beneficial effect of combined 3,5,3'-triiodothyronine and vasopressin administration of hepatic energy status and systemic hemodynamics after brain death. *Transplantation* 1992, **54**: 44–49.
34. Taniguchi S, Kitamura S, Kawachi K, Doi Y, Aoyama N. Effects of hormonal supplements on the maintenance of cardiac function in potential donor patients after cerebral death. *Eur J Cardiothorac Surg* 1992, **6**: 96–101, discussion.
35. Ropper AH. Unusual spontaneous movements in brain-dead patients. *Neurology* 1984, **34**: 1089–1092.
36. Miller FR, Waud RA. Viscero-motor reflexes. *Am J Physiol* 1925, **73**: 329–340.
37. Downman CBB. Reflexes elicited by visceral stimulation in the acute spinal animal. *J Physiol* 1946, **105**: 80–94.
38. Ebata T, Watanabe Y, Amaha K, Hosaka Y, Takagi S. Haemodynamic changes during the apnoea test for diagnosis of brain death. *Can J Anaesth* 1991, **38**: 436–440.
39. Gramm HJ, Zimmermann J, Meinhold H, Dennhardt R, Voigt K. Hemodynamic responses to noxious stimuli in brain-dead organ donors. [Review]. *Intensive Care Med* 1992, **18**: 493–495.
40. Ferbert A, Buchner H, Ringelstein EB, Hacke W. Brain death from infratentorial lesions: clinical neurophysiological and transcranial Doppler ultrasound findings. [Review]. *Neurosurg Rev* 1989, **12**: 340–347.
41. Ogata J, Imakita M, Yutani C, Miyamoto S, Kikuchi H. Primary brainstem death: a clinico-pathological study. *J Neurol Neurosurg Psychiatry* 1988, **51**: 646–650.
42. Darby J, Yonas H, Brenner RP. Brainstem death with persistent EEG activity: evaluation by xenon-enhanced computed tomography. *Crit Care Med* 1987, **15**: 519–521.
43. Ferbert A, Buchner H, Ringelstein EB, Hacke W. Isolated brain-stem death. Case report with demonstration of preserved visual evoked potentials (VEPs). *Electroencephalogr Clin Neurophysiol* 1986, **65**: 157–160.
44. Rodin E, Tahir S, Austin D, Andaya L. Brainstem death. *Clin Electroencephalogr* 1985, **16**: 63–71.
45. Black PM. Brain death (second of two parts). *N Engl J Med* 1978, **299**: 393–401.
46. Gramm HJ, Meinhold H, Bickel U, et al. Acute endocrine failure after brain death? *Transplantation* 1992, **54**: 851–857.
47. Hohenegger M, Vermes M, Mauritz W, Redl G, Sporn P, Eiselsberg P. Serum vasopressin (AVP) levels in polyuric brain-dead organ donors. *Eur Arch Psychiatry Neurol Sci* 1990, **239**: 267–269.
48. Howlett TA, Keogh AM, Perry L, Touzel R, Rees LH. Anterior and posterior pituitary function in brain-stem-dead donors. A possible role for hormonal replacement therapy. *Transplantation* 1989, **47**: 828–834.
49. Novitzky D, Cooper DK, Reichart B. Hemodynamic and metabolic responses to hormonal therapy in brain-dead potential organ donors. *Transplantation* 1987, **43**: 852–854.
50. Wicomb WN, Novitzky D, Cooper DK. Effects of hormonal therapy on subsequent organ (kidney) storage in the experimental animal. *Transplant Proc* 1988, **20**: 55–58.
51. Cooper DK, Novitzky D, Wicomb WN. Hormonal therapy in the brain-dead experimental animal. *Transplant Proc* 1988, **20**: 51–54.
52. Novitzky D, Wicomb WN, Cooper DK, Tjaalgard MA. Improved cardiac function following hormonal therapy in brain-dead pigs: relevance to organ donation. *Cryobiology* 1987, **24**: 1–10.
53. Wicomb WN, Cooper DK, Novitzky D. Impairment of renal slice function following brain death, with reversibility of injury by hormonal therapy. *Transplantation* 1986, **41**: 29–33.
54. Novitzky D, Cooper DK, Morrell D, Isaacs S. Change from aerobic to anaerobic metabolism after brain death, and reversal following triiodothyronine therapy. *Transplantation* 1988, **45**: 32–36.

55. Jeevanandam V, Todd B, Regillo T, Hellman S, Eldridge C, McClurken J. Reversal of donor myocardial dysfunction by triiodothyronine replacement therapy. *J Heart Lung Transplant* 1994, **13**: 681–687; discussion.
56. Keogh AM, Howlett TA, Perry L, Rees LH. Pituitary function in brain-stem dead organ donors: a prospective survey. *Transplant Proc* 1988, **20**: 729–730.
57. Schrader H, Krogness K, Aakvaag A, Sortland O, Purvis K. Changes of pituitary hormones in brain death. *Acta Neurochir (Wien)* 1980, **52**: 239–248.
58. Powner DJ, Hendrich A, Lagler RG, Ng RH, Madden RL. Hormonal changes in brain dead patients [see comments]. *Crit Care Med* 1990, **18**: 702–708.
59. Hall GM, Mashiter K, Lumley J, Robson JG. Hypothalamic-pituitary function in the 'brain-dead' patient [letter]. *Lancet* 1980, **2**: 1259.
60. Hanley DF, Wilson DA, Conway MA, Traystman RJ, Bevan JA, Brayden JE. Neural mechanisms regulating neurohypophysial resistance arteries. *Am J Physiol* 1992, **263**: H1605–H1615.
61. Bergland RM, Davis SL, Page RB. Pituitary secretes to brain. Experiments in sheep. *Lancet* 1977, **2**: 276–278.
62. Page RB, Bergland RM. The neurohypophyseal capillary bed. I. Anatomy and arterial supply. *Am J Anat* 1977, **148**: 345–357.
63. Page RB, Munger BL, Bergland RM. Scanning microscopy of pituitary vascular casts. *Am J Anat* 1976, **146**: 273–301.
64. Shlomo Melmed. *The Pituitary*. Blackwell Science, Oxford, 1995, 1–20.
65. Tindall GT. *Disorders of the Pituitary*. CV Mosby, St Louis, 1986, 9–16.
66. Ikuta F, Takeda S. [Neuropathology required from 'brain death']. [Review] [Japanese]. *Rinsho Shinkeigaku* 1993, **33**: 1334–1336.
67. Nishioka H, Ito H, Ikeda Y, Koike S, Ito Y. [Hypothalamic–adenohypophysial function of corticotropin releasing hormone (CRH)–ACTH.] *No To Shinkei* 1995, **47**: 1159–1163.
68. Sebening C, Hagl C, Szabo G, *et al.* Cardiocirculatory effects of acutely increased intracranial pressure and subsequent brain death. *Eur J Cardiothorac Surg* 1995, **9**: 360–372.
69. Bittner HB, Kendall SW, Chen EP, Van-Trigt P. Endocrine changes and metabolic responses in a validated canine brain death model. *J Crit Care* 1995, **10**: 56–63.
70. Chiolero MD, Lemarchand T, Schutz Y, de Tribolet N. Plasma Pituitary Hormone Levels in Severe Trauma with or without Head Injury. *J Trauma* 1988, **28**(9): 1368–1374.
71. A Definition of Irreversible Coma. Report of the Ad Hoc Committee of the Harvard Medical School to Examine the Definition of Brain Death. *JAMA* 1968, **205**: 337–340.
72. Fackler JC, Rogers MC. Is brain death really cessation of all intracranial function? *J Pediatr* 1987, **110**: 84–86.
73. Rosenklint A, Jorgensen PH. Evaluation of Angiographic Methods in the Diagnosis of Brain Death. Correlation with Local and Systemic Arterial Pressure and Intracranial Pressure. *Neuroradiology* 1974, **7**: 215–219.
74. Monsein LH. The Imaging of Brain Death. *Anaesth Intensive Care* 1995, **23**: 44–50.
75. Bonetti MG, Ciritella P, Valle G, Perrone E. 99mTc HM-PAO brain perfusion SPECT in brain death. *Neuroradiology* 1995, **37**: 365–369.
76. Schober O, Galaske R, Heyer R. Determination of brain death with 123I-IMP and 99mTc-HM-PAO. *Neurosurg Rev* 1987, **10**: 19–22.
77. Galaske RG, Schober O, Heyer R. 99mTc-HM-PAO and 123I-amphetamine cerebral scintigraphy: a new, non invasive method. *Eur J Nucl Med* 1988, **14**: 446–452.
78. Riishede J, Ethelberg S. Angiographic changes in Sudden and Severe Herniation of the Brainstem through Tentorial Incisura. *Arch Neurol Psychiatry* 1953, **70**: 399–409.
79. Pribram HFW. Angiographic Appearances in Acute Intracranial Hypertension. *Neurology* 1961, **11**: 10–21.
80. Aruga T, Ono K, Kawahara N, et al. Some practical problems of clinical diagnosis of brain death. [Japanese]. *No To Shinkei* 1983, **35**: 777–785.
81. Kouji H, Kawahara N, Mii K, Takakura K. Analysis of angiographical findings in patients with severe brain disorders progressing to brain death. [Japanese]. *No To Shinkei* 1989, **41**: 727–735.

82. Lucking CH, Gerstenbrand F. The Clinical Picture of Brain Death after Serious Brain Injuries. *Electroencephalogr Clin Neurophysiol* 1971, **30**: 272.
83. Schroder R. Later Changes in Brain Death. *Acta Neuropathol (Berl)* 1983, **62**: 15–23.
84. Herrick MK, Agamanolis DP. Displacement of cerebellar tissue into spinal canal. A component of the respirator brain syndrome. *Arch Pathol* 1975, **99**: 565–571.
85. Zurynski Y, Dorsch N, Pearson I, Choong R. Transcranial Doppler ultrasound in brain death: experience in 140 patients. *Neurol Res* 1991, **13**: 248–252.
86. Drory Y, Ouaknine G, Kosary IZ, Kellermann JJ. Electrocardiographic findings in brain death, description and presumed mechanism. *Chest* 1975, **67**: 425–432.
87. Ouaknine GE, Mercier C. Value of the atropine test in the confirmation of brain death. [French]. *Union Med Can* 1985, **114**: 76–80.
88. Siemens P, Hilger HH, Frowein RA. Heart rate variability and the reaction of heart rate to atropine in brain dead patients. *Neurosurg Rev* 1989, **12**: 282–284.
89. Mohandas A, Chou SN. Brain death. A clinical and pathological study. *J Neurosurg* 1971, **35**: 211–218.
90. Kaukinen S, Makela K, Hakkinen VK, Martikainen K. Significance of electrical brain activity in brain-stem death. *Intensive Care Med* 1995, **21**: 76–78.
91. Blend MJ, Pavel DG, Hughes JR, Tan WS, Lansky LL, Toffol GJ. Normal cerebral radionuclide angiogram in a child with electrocerebral silence. *Neuropediatrics* 1986, **17**: 168–170.
92. Mollaret P, Goulon M. Le coma depasse (memoire preliminaire). *Rev Neurol* 1959, **101**: 3–15.

CHAPTER 7

Legislation on organ and tissue donation

BERNARD M. DICKENS, SEV S. FLUSS, ARIEL R. KING

Introduction 95
Key issues in organ transplantation legislation 96
The current configuration of organ transplantation legislation 107
References 119

Introduction

Human organ transplantation commenced with a series of experimental studies conducted at the beginning of the twentieth century. In 1912, Alexis Carrel was awarded the Nobel Prize for Medicine or Physiology, for his pioneering work, and this was followed by major clinical and scientific advances that have been described in numerous publications. Surgical transplantation of human organs from deceased, as well as living, donors to ill and dying patients began after the Second World War. Over the past 30 years, organ transplantation has become a worldwide practice and has saved many thousands of lives. It has also improved the quality of life of countless other persons. Continuous improvements in medical technology, particularly in relation to tissue 'rejection', have brought about expansion of the practice and an increase in the demand for organs. A constant feature of organ transplantation since its commencement has been the shortage of available organs. Many experts have stressed the importance of legislation in this field, including Gerson,[1] who comments that:

> Ultimately the potential for organ transplantation will depend not only on advanced medical technology, but also on progress in the legal technology of organ donation.

In this chapter, we focus on current patterns in the legislation governing transplantation of 'organs', a term which we shall interpret to exclude blood and reproductive tissues such as sperm, ova, or pre-embryos. There have, of course, been a number of earlier studies on this subject, for example that published by the World Health

Organization[2] and the study published by the Australian Law Reform Commission.[3] More recently, a major survey describing the legislative situation in Latin America was published by Fuenzalida-Puelma.[4] Because of obvious constraints, this chapter cannot be comprehensive, and a detailed global survey of legislation, including guidelines, on organ and tissue transplantation would require a sustained effort on the part of a team of major dimensions. One valuable resource for such a study was recently published by Martinus Nijhoff, with the co-operation of WHO.[5] A substantial amount of new information is available as a result of EUROTOLD, a European Commission-funded European Multicentre Study on the Transplantation of Organs from Living Donors: Ethical and Legal Dimensions (based at the University of Leicester).

After an overview of some of the key current issues in organ transplantation legislation, we provide a brief outline of some recent developments at the international and national levels, focusing particularly on legislation in a number of countries, particularly France. Developments in these countries are, of course, not necessarily representative of the global situation, but yield helpful insights into the way in which national governments are responding to key issues (notably those identified below). We also describe recent developments in efforts to combat commerce in human organs, a topic that has attracted a significant amount of international and national attention in recent years, with WHO and the Council of Europe playing a major role and certain non-governmental organizations (notably the (international) Transplantation Society) likewise providing important guidance on legal and ethical issues.

Key issues in organ transplantation legislation

Legal systems differ among countries. Some have systems based on judge-made law, reflecting a region's customs and prioritization of values. This may be supplemented, consolidated or amended by politically-inspired legislation. These systems are sometimes described as customary or common law systems. In contrast, others have defined all prevailing rights and duties in a single code, legislated at the highest level of authority, so that any claimed right or duty must be derived from language in the code. Judges will interpret what the code means, but are not themselves an original source of the law. These systems are usually described as civil law systems. The term 'civil' is also used within legal systems to contrast criminal law, concerned with punishment of wrongdoers, with non-criminal or civil law concerned with the duty to give compensation that binds those who, even innocently, have violated others' rights. Systems of religious law also apply in many countries. These are based on authoritative interpretations of sacred texts, and may be the basis of customary or codified legal systems or exist in parallel forms governing matters of religious significance. In religious law systems, sacred texts are usually interpreted to prevail over other expressions of law in the case of conflict.

Not every country has legislation specifically addressing the taking of organs or tissues from living persons, from persons after death, or from either. Issues of death, respect for and disposal of dead bodies, and preservation of the physical integrity of bodies of the living are, however, almost invariably of spiritual and religious concern. Accordingly, any legislation affecting donations of body materials by living persons and how bodies of deceased persons may be treated must be seen as influenced by significant cultural, religious and social practices of peoples. It must be determined in each case whether legislation, standing alone or in a code, expresses or is intended to change such practices.

Few if any countries have legislation that deals exhaustively or comprehensively with organ or tissue acquisition from bodies of living or deceased persons and transplantations into bodies of recipients. Practitioners involved in either removal of materials from living or dead bodies or implantation into recipients, and those involved in preservation, storage, transmission and preparation of transplantable materials, must clearly be familiar with their governing legislation. The legislation in itself will not tell them all that they need to know, however, about the legal environment in which they act. Such issues as standards of mental competency to donate, objections by family members of a person now dead who previously consented to posthumous removal of transplantable body materials prior to burial or cremation, whether a dying patient can be given invasive treatment designed to maximize utility of body materials after death, the legal status of and control over preserved materials and, for instance, what information must be given to a prospective organ recipient, are unlikely to be addressed in detail in legislation. Some instances of legislation may indeed be very detailed on specific issues. Usually, however, those who read their governing legislation will find little more than generalized direction, and will have to find answers to key questions from general areas of law, often by interpretation and interpolation.

Legislation designed to prevent and criminalize commercial initiatives in the acquisition and supply of organs has become more widespread. Offers to buy organs, from living or deceased persons, and offers to sell, are legally prohibited or restricted by legislation in a rising number of countries. Similarly, medical and other healthcare practitioners who knowingly involve themselves in organ removals or implantations in which commerce is involved increasingly face legislated penalties. Legislation has not responded in specific details, however, to early awareness that transplantation was most successful where genetic compatibility was present between the source and the recipient of transplanted materials, nor to the more recent development of immunosuppressive drugs that make close genetic matching less necessary for a successful outcome.

LIVING DONORS

Legislation regarding donations from living persons may define a distinction between donations of their organs and of their tissues. In the United Kingdom, for instance, 'organ' means 'any part of a human body consisting of a structured arrangement of tissues which, if wholly removed, cannot be replicated by the body'. In Ontario, the legislation does not define 'organ' but 'tissue' in law 'includes an organ but does not include any skin, bone, blood, blood constituent or other tissue that is replaceable by natural process of repair'. Since, for instance, bone marrow removed for transplantation is naturally replicated or replaced, its removal is not governed by the provisions on organ removal under this legislation. However, in the United States of America, the National Organ Transplantation Act specifically defines a human organ to include bone marrow. The same legislation specifies that 'organ' means a human liver. Recent studies show that in cases of liver segment donation, the liver quickly resumes its previous size in the donor's body, indicating that a segment of the organ is not itself an organ but may be only tissue, where tissue is defined as material replaced in the body. These illustrations show with what care and precision legislation governing removal of material from living donors for transplantation into others must be read and understood.

Legislation directed to the same general purposes, in countries with similar cultures, may have different effects.

In the case of non-regenerative materials, removal of vital organs cannot be undertaken, even with informed and free consent, if removal would terminate or endanger the donor's life. Removal of a twin organ such as a kidney may be allowed by legislation, however, as may removal of minor segments of single organs such as livers and the small bowel when the donor's bodily functioning is not significantly impaired following recovery from removal surgery.

A danger associated with living organ and tissue donation is that donors act for unacceptable reasons, or without adequately free and informed consent. The most obviously unacceptable reason is hope of monetary or comparable gain. The incentive to make money is normally a legally acceptable reason to do what otherwise a person would not do, such as work for another or give away ownership of property. It is widely accepted, however, that a commercial market in transplantable human body materials should be prevented or discouraged, and that legislation should prohibit traffic in body materials. There is a margin of legal uncertainty concerning whether gift-exchanges are distinguishable in principle from commercial transfers of goods, services or body materials in exchange for money, or comparable rewards or advantages. A contrast may be proposed between personal reciprocity between acquaintances in the former and impersonal bargaining between strangers in the latter. It is usually presumed that a person would make a living donation to a stranger only for money. Following experiences of individuals coming to London to be paid as organ donors, the United Kingdom enacted legislation that permits living persons' organ donations only to genetically-linked recipients, with power to make exceptions, for instance for marriage partners. Legal concerns with living related donors and unrelated donors may be based on genetic relations, marital relations, or both.

Legislation is often emphatic to preclude parents or other legal guardians of children from 'volunteering' them as organ donors. Legislation is not uncommonly explicitly prohibitive, although in jurisdictions that recognize the autonomy of mature minors, competent minors may be deemed adult for the purposes of legislative interpretation, and able to donate on the same terms as adults. Some legislation prohibits minors from donating organs, but defines 'organs' to exclude bone marrow. Parents may thus consent to their children's donation, almost invariably to another child of the family who suffers from leukaemia or a similar condition treatable by bone marrow transplantation. Since courts are inclined to find that parental powers exist only to equip parents to discharge their duties,[6] and that their authority is only to provide therapeutic services, some benefit for the donor child must be present to justify the child being a donor. Donation towards a sibling's survival may achieve this justification, because it is in the donor child's interests that a sibling should survive, but donations to strangers or others with whom no significant relationship exists will not.

It is also usually explicit in legislation that adults can consent to be donors of transplantable materials only when they are mentally competent. Mentally incompetent adults are legally comparable to children, in the same way that mature or mentally competent minors are comparable to adults. Competency tends not to be defined in legislation on transplantation, but the general principles of legal systems usually relate standards of competency to particular functions. Standards of required competency are lower for minimally invasive procedures, and higher for more invasive procedures. For instance, competency to accept mild transitory discomfort is adequate for blood donation but intellectual comprehension of long-term implications and ability to withstand the physical and emotional stress of surgery is required, for instance, for kidney donation. Bone marrow donation is uncomfortable when given, but has few long-term effects. As for minors, mentally incompetent persons should not be donors unless they have some relationship with the recipient that is beneficial or significant for them.

The status of anencephalic newborns tends not to be addressed in legislation. These children have a survival from birth measured only in days, but they could be a valuable multiple source of organs on which the lives of other children may depend. Anencephalic children have no upper brain, but they are not brain dead (see Chapter 6) because they have functioning brainstems. If their death follows its natural course, their organs will have degenerated too much by the time they are recoverable to be transplantable. Techniques exist such as artificial ventilation to preserve the quality of their organs for transplantation, but the use of such techniques is not beneficial for them. It is not clear, therefore, that their parents have legal authority to consent to use of these techniques, because the children have no significant relationships with potential recipients. Nevertheless, protocols have been developed among transplantation specialists according to which dying persons can be treated in non-beneficial ways that maximize posthumous use of organs, provided that such ways are not harmful or distressing to them. It is likely but not yet established that courts would accept the legality of procedures that comply with these protocols. These children are born dying, and are at the point of transition from living to cadaveric donation.

At the same point of transition, and potentially of greater significance in transplantation practice, is the matter of fetal tissue transplantation. Fetal tissues include placentas available following natural childbirth. Mothers may legally consent to the therapeutic, research or industrial use of these tissues, and are usually deemed voluntarily to have abandoned any legal powers of control at the time of birth unless they positively assert them, perhaps according to cultural practices. There may be an emerging claim under general legal principles, however, that, when women presume incineration, burial or other such disposal of placental tissues, any other use should be dependent on their explicit consent.[7] In the event of spontaneous miscarriage or stillbirth, mothers can usually consent to recovery of transplantable fetal organs or tissues on analogy with principles applicable to cadavers, which permit family members to donate materials (see below). The legal distinction between miscarriage and stillbirth usually relates to legislation on recording vital statistics and burial, incineration or other sanitary disposal of materials, rather than legislation on transplantation.

The same set of rules should apply to recovery of transplantable tissues that result from induced abortion, which in principle are materials the mother may make available as a live donor. However, with the growing potential to transplant fetal neural cells into patients suffering from neurodegenerative disorders, such as Parkinson's disease and perhaps Huntington's and Alzheimer's disease, and the incentive that utility of these tissues may give women to resolve difficult dilemmas of continuation or termination of pregnancy in favour of the latter, some countries are considering legislative controls. There is a related fear that women's care may be distorted to enhance prospects of recovery of fetal tissues, and that women may be pressured or improperly induced to donate them. In the USA, state legislation based on the Uniform Anatomical Gift Act, which addresses only postmortem donations may apply to fetal tissue gifts. At present few jurisdictions have transplantation legislation that specifically governs tissues that result from induced abortion, but practitioners of fetal tissue recovery and implantation must stay alert to legislative developments, and to judicial declarations or interpretations in this area.

Removal of transplantable organs and tissues from living donors fits into a general framework of law of which specific legislation or civil code provisions are only a part.

Issues of authority to practise medicine, capacity of facilities to undertake procedures, and, for instance, of payment for intermediate services such as preservation and transportation of materials, have to be addressed through more general reference to provisions of legal systems.[8] The same applies to determination of legal criteria of the difference between living donors and cadaveric sources of materials. Legislation on transplantation may address the matter not by reference to tests of life or death, but by establishing procedures for application of non-specified medical criteria. For instance, in Turkey, Law No. 2238 of 29 May 1979 on organ and tissue removal, provides in section 11 that:

> For the purposes of this Law, a committee of four physicians, comprising a cardiologist, a neurologist, a neurosurgeon, and a specialist in anaesthesiology and resuscitation, shall unanimously determine that death has occurred by natural means, in accordance with contemporary medical knowledge and procedures.

Persons not so determined to be dead are living persons, whose consent to removal of materials is required. 'Contemporary medical knowledge and procedures' are legally established by reference to the opinions or testimony of expert medical authorities. The unusually strict Turkish requirement, to involve several medical specialists in determining death, limits where death can be identified, and therefore where organs and tissues can be recovered. Only leading medical facilities in larger cities may have this capacity. How courts interpret such language as 'by natural means' is also relevant. The materials from bodies of elderly persons who suffer 'natural' death may be less transplantable than those of younger, healthier people killed in accidents, for instance from head injuries due to road traffic accidents.

POST-MORTEM DONATION

Countries and sometimes individual state, provincial or other jurisdictions within countries have their own distinctive legal approaches to the definition of death and to recovery of organs and tissues for transplantation. In the case of cadavers, legal distinctions between organs and tissues, turning on the body's capacity for spontaneous regeneration, are of no concern. Similarly, criminal law offences of wounding those from whose living bodies materials are removed for transplantation without their competent informed and voluntary consent are inapplicable, but most if not all countries' laws condemn the deliberate and perhaps the negligent infliction of indignity on the body of a deceased person. How dead bodies may be treated often reflects religious convictions regarding afterlife and resurrection, which may be common to a country or legally protected according to individual personal or family beliefs. Dead bodies and their separated parts have not generally had the legal status of property, although organs from living donors may have this status,[9] but not for purposes of commerce.

Where legislation expressly allows post-mortem recovery of transplantable materials, it tends to fall into one of two general legislative patterns, namely 'express consent' or 'opting-in/contracting-in' legislation, and 'presumed consent' or 'opting-out/contracting-out' or implied consent legislation. The former are more traditional, and remain common for instance in Anglo-Saxon influenced jurisdictions. The latter are becoming more prevalent in especially Western European countries.[8] They are frequently favoured by transplantation advocates because they may facilitate recovery of transplantable organs and also present the enterprise of cadaveric organ recovery and transplantation as a benefit society may legitimately promote.

Express consent legislation permits individuals before death, or family members after death, to donate cadaveric materials, as an expression of altruism and autonomy. It may suggest, however, that donation is an exceptional act, and that the norm is that bodies are buried intact. Presumed consent legislation treats cadaveric materials as a public asset, but permits individuals who object to their own or deceased family members' materials being removed to prohibit recovery. The right to withhold permission preserves personal autonomy, but it may require individuals who object to recovery of material to identify themselves to health or other public authorities, to place their personal religious or other convictions on record, and to declare that they decline to contribute an asset to the public.

Express consent laws may be reinforced by legal provisions that require health personnel or facilities to ask eligible individuals whether they are willing to do so. These are often described as 'required request' laws (see Chapter 17).

Express consent laws

Early legislation that permitted donation of materials for transplantation after death was limited to recovery of corneas, the first human body part to become transplanted on a significant scale. Legislation now commonly empowers individuals to approve in advance post-mortem acquisition of any body materials, and also permits specified family members of a deceased person to consent to recovery, provided that the deceased when living had not expressed disapproval. Further, when patients apparently close to death have not consented to materials being recovered post mortem for transplantation but have not objected to this prospect, and physicians or other attendants cannot request their consent because they are unconscious or not able to form or express a choice, physicians or others may discuss recovery of materials with their relatives before death. Any consent that appropriate family members give prior to death, to posthumous removal of material, can be acted on as promptly as possible when death occurs, to maximize utility of recovered materials.

A legal concern is how clearly a person must express his consent to posthumous donation, and how that consent should be recorded. The common practice to attach a general and brief donation card to another significant document, particularly a vehicle driver's licence, may tend to exclude those ineligible to hold such a document. If ineligibility is due to immature age this may be acceptable, since young people may not have developed mature awareness of implications of donation for others, such as family members. In many countries and cultures, however, this form of consent to donation may also exclude women where it is not usual for women to hold relevant documents. Where organizations have been formed to promote post-mortem donation, they may produce separate donation forms or cards that are easily available and give assistance in applying any legislated or other provisions for the recording of such donation.

A comparable legal concern in posthumous organ donation is recording refusal of consent. Legislation usually empowers surviving family members to approve removal of materials for transplantation from bodies of deceased relatives who left no evidence that they did not wish their bodies to be used in this way. Evidence would exist in a prohibitive religious faith, in burial instructions incompatible with organ removal, and, of course, in clearly communicated disapproval of donation. A communication to a family member or a health or other service provider who proves not to be available when death unexpectedly occurs, and a request for posthumous donation made to another available person ineligible to consent, may fail to achieve its purpose. Accordingly, while express consent legislation requires donors to take the initiative to declare donation in some suitable way, it also requires people who do not want their organs to be removed after death on consent of their family members to demonstrate their refusal in a way that reasonably anticipates the usually unforeseeable circumstances of their deaths.

Legislation often declares that a family member eligible to consent commits no offence if he or she consents in good faith to posthumous recovery of materials from a deceased person who had in life expressed an objection of which the consenting family member is unaware. Similarly, a hospital or other facility involved in removing or transplanting materials on such consent incurs no legal liability, provided that its relevant officers and staff had no independent evidence of the deceased person's objection. A procedure should be in place in a facility that recovers organs after death to ensure that any patients' objections are adequately recorded. Mechanical access to medical or health information shared among health facilities within an area will contribute to the effectiveness of patients' declarations both to consent and to decline to consent to posthumous donation.

Legislation frequently provides that a person's competent declaration of post-mortem donation is legally sufficient to empower appropriate authorities to recover materials for transplantation. This is not an invariable rule, and for instance in Denmark, Italy, Sweden and Norway removal of materials is unlawful on the deceased person's earlier consent if relatives object to it for reasons of their own.[8] Additional consent of family members is usually not required, however, since normally they have no legal power of veto. Despite this legal reassurance, however, there is considerable evidence that many health care personnel will not take materials on deceased persons' declarations when family members object.[10]

Accommodation of family members' objections may be a dysfunction of a legislated rule that is sound in itself. Legislation usually provides that determination of a potential donor's death and surgical removal of materials shall be performed by physicians who have no commitment to or involvement with a prospective recipient. The provision is designed to preclude any conflict of interest in diagnosis of death and in management of the patient in the process and aftermath of death. However, personnel involved with such potential donors tend to remain engaged with their family members following death, and committed to family members' successful management of bereavement and grief. When surgery on the deceased to recover materials would cause distress to family members and obstruct their recovery from mourning, hospital personnel may consider it preferable not to undertake the surgery. 'Wastage' of recoverable materials may be considered necessary for surviving family members' well-being.

Legislation that renders a patient's declaration the full legal authority to recover materials after death, in fact, remains only permissive; it does not oblige facilities or personnel eligible to acquire suitable materials to do so. Medical personnel who decline offered materials have no duties of care in negligence law and no fiduciary duties towards potential recipients, because of the legislated disconnection between personnel who manage a potential donor's terminal care and determination of death, and patients eligible to receive donated materials.

The legislation that implements the express consent system permits prospective donors, whether individuals in anticipation of their own deaths or their family members afterwards, to set conditions or limitations to donation. They may donate some identified organs but not others, donate to a specific facility as opposed to others, or for treatment of patients affected by a particular disease. It can be difficult to determine, however, the extent to which a donor's specification can be accommodated by the general legal context of donation. A legal challenge concerns whether a donation violates legislation on non-discrimination that is binding on private individuals or public health facilities. A donor who limits recipients on grounds of their race, religion, origin or other characteristic unrelated to sickness or need may compel a physician or a health facility administrator to confront and resolve a complex legal question, speedily and without warning. A possible response is to decline donations that discriminate among recipients. This may be appealing, but may also result in unnecessarily early deaths of qualified patients while not protecting the interests or well-being of patients who are not qualified.

The legal requirement that, in order to be effective, a donation must be made without coercion or undue inducement may appear more relevant to living than post-mortem donations. However, family members of a deceased person may be subject to both coercion and over-inducement to consent to posthumous recovery of transplantable materials. Pressure may be exerted by health professionals, officers of transplantation facilities and, for instance, family members of likely recipients.

Legislation on 'required request' may suggest that health personnel and facilities are required to obtain a high proportion of consents. Inducements may come in the form of non-monetary advantages offered to donors, such as access for themselves or others to health services otherwise inaccessible to them. A family member's consent to donation based on appreciation of a facility's services or a healthcare provider's dedication has not been deliberately induced, but the distinction between an adequately autonomous donation and a consent a donor was conditioned or manipulated to provide, may be narrow. Since a physician or facility that has recovered materials on the basis of a consent that was improperly procured is liable to suffer in a number of legal and disciplinary proceedings, they have strong self-protective incentives to prevent both coercion and over-inducement of family members to make donations.

Presumed consent laws

A number of countries, predominantly in Western Europe, have enacted legislation that authorizes removal of materials from cadavers unless in life the deceased had expressed an objection, and thereby opted out of automatic liability to posthumous removal of bodily materials for transplantation (see Chapter 17). In France, for example, Law No. 94-654 of 29 July 1994 on the donation and use of elements and products of the human body, medically assisted procreation, and prenatal diagnosis, includes detailed provisions (introduced into the Public Health Code) that are based on this approach as shown in Articles L. 671-7 and L. 671-8 and reprinted in Table 7.1.

That law repealed a 1949 law on corneal grafting as well as a 1976 law on organ removal, and it is understood that decrees of specific details will implement the above-mentioned provisions as well as other provisions in the law. Pending the issuance of such decrees it is understood that Decree No. 78-501 of 31 March 1978 remains in force. That decree prescribes that persons who wish to prohibit organ removal after their death may formulate the prohibition in any manner, and a hospital licensed to remove organs from dead bodies may record a prohibition at any time in a special register. If a person is incapable of expressing views, any indication found on his or her person, among his or her personal effects or elsewhere, that suggests that he or she would prohibit the removal of organs following death must be entered in the register. Anyone who can testify that a hospitalized person is opposed to organ removal after death must enter a substantiated statement to this effect in the register. Physicians responsible for recovery of cadaveric organs must, if not already aware of a prohibition, ensure that there is no prohibition in the register and ensure that the deceased was not a minor nor

Table 7.1 French law on donation, Law 94-654, 29 July 1994

- **Article L. 671-7.** Organs may only be removed from a dead person for therapeutic or scientific purposes and after death has been determined in accordance with the conditions laid down by a decree made after consulting the Conseil d'État.

    Such removal may be carried out provided that the person concerned has not expressed his refusal of removal during his lifetime.

    Such refusal may be expressed by an indication of the person's wishes in a computerized national register established for this purpose. It may be withdrawn at any time. The conditions governing the operation and management of the register shall be determined by a decree made after consulting the Conseil d'État.

    If the physician has no direct knowledge of the wishes of the deceased, he must endeavour to obtain the testimony of his family.

- **Article 671-8.** If the deceased person was a minor or had reached the age of majority and was the subject of legal protection measures, removal for the purposes of donation may only be performed if each person exercising parental authority or the person's legal representative gives his express consent in writing.

mentally incompetent. If the deceased was a minor or mentally incapable, organs may be removed only on written authorization of the deceased's legal representative, entered in the register. In this regard, there must be express consent by the representative before or possibly following the deceased person's death.

Details of the central concept expressed in the French legislation differ among countries. In Belgium, for instance, the law of 13 June 1986 on the removal and transplantation of organs provides in section 10(1) that:

> Organs and tissues for transplantation, and for the preparation of therapeutic substances... may be removed from the body of any person recorded in the Register of the Population or any person recorded for more than six months in the Aliens Register, unless it is established that an objection to such a removal has been expressed.
>
> It shall be a requirement, in the case of persons other than those mentioned above, that they have explicitly expressed their consent to the removal.

Capable persons aged 18 years and over may object, but so also may a younger person who is capable of objection or, during his or her lifetime, a close relative living with the youth may record the youth's opposition to posthumous removal of organs and/or tissues. Legal representatives, guardians or closest relatives of mentally incapable persons may express objection during such persons' lifetimes. Objections to posthumous removal of materials made by capable persons or on their behalf must be signed and dated, and transmitted to the Data Processing and Information Centre of the Ministry of Public Health and Family Affairs.

It may be questionable under civil liberties and privacy legislation whether persons who object to posthumous organ recovery, perhaps on religious grounds, should be required to record this fact in governmental registers. Giving such information to hospital authorities is less objectionable, since knowledge of this nature permits hospitals to treat patients and their dead bodies according to patients' religious requirements, for instance concerning dietary practices and notifications and procedures on death.

In Singapore, the Human Organ Transplant Act of 1987 is unusual in combining a general presumed consent system with an express consent system for the Muslim community. The law prohibits trading in any human organ and in blood, but its provisions otherwise apply only to kidneys, other legislation applying to removal of other materials from deceased persons. Section 5(1) permits a designated officer of a government hospital or specially approved private hospital to authorize 'the removal of any organ [that is, a kidney] from the body of a person who had died in the hospital for the purpose of the transplantation of the organ'. Section 5(2) provides that authority shall not be given, however, unless death was caused by accident or injuries caused by accident, and if the person now deceased had registered an objection with the state's Director of Medical Services. If the person was aged under 21 years or was not of sound mind, authorization can be given only with consent of a parent or guardian. No authorization is possible if the deceased person was above 60 years of age. Section 5(2)(g) also prohibits authorization of kidney removal from the body of a deceased person 'who is a Muslim'.

Actual objection and presumed objection under section 5(2)(g) to posthumous kidney recovery are relevant to section 12 of the law, which governs recipients of organs. This provides that:

> ... in the selection of a proposed recipient of any organ removed pursuant to section 5 – (a) a person who has not registered any objection with the Director [to posthumous donation] ... shall have priority over a person who has registered such objection (s. 12(1)).

Further,

> a person referred to in section 5(2)(g) [that is, 'a Muslim'] shall have priority over a person who has registered such objection only if he has made a gift of his organ, to take effect upon his death, under section 3 of the Medical (Therapy, Education and Research) Act (s. 12(2)(a)).

The legality of this accommodation of Muslims may be assessed in the context of the local community and culture, and also of the international human rights conventions to which Singapore is a party. If members of a religious faith are subject to religious precepts against posthumous surrender of organs, it may be objectionable for them to have to record dissent from doctrines of their religious faith to gain equal priority with others in eligibility to receive kidneys, particularly if religious non-conformity renders them liable to religious sanctions or retaliation.

Accordingly, the legal problems and concerns that are raised by legislation implementing the express consent system of cadaveric recovery of transplantable materials are not obviated by adoption of the presumed consent system. The problems are different in character, but not necessarily different in degree of complexity or the challenge they pose to the achievement of justice.

Confidentiality

Living persons who are not paid will rarely donate significant materials from their bodies to strangers, but naturally replaceable body materials such as blood and bone marrow may be donated to unknown recipients. In the case of cadaveric donations, whether by persons in anticipation of their own deaths or by donors of materials from deceased relatives, donations are frequently required by legislation to be mutually anonymous. Families of deceased donors are unable directly to assert any moral or other claims against recipients, and recipients cannot directly discharge any moral or other obligations they may feel they owe donors or their families. Donation becomes impersonal, and serves society. Many countries have legislated provisions to ensure confidentiality that prevent linkage of particular donors with particular recipients.

Legislation mandating preservation of confidentiality usually has suitably guarded exceptions through which information on donation or receipt of materials can be given to courts of law, coroners' courts, health professional licensing and disciplinary tribunals, other administrative tribunals, police officers and, for instance, medical researchers. Hospitals or distribution services are usually expected to be able to link recipients to donated materials, for instance for therapeutic follow-up and warning if conditions such as infections are subsequently discovered in donors that may affect recipients. Where registers are kept such as of donation consents or conditions, access to them will be carefully guarded and protected through legal sanctions. Entry on such a register does not establish, of course, that donation actually occurred.

Mechanical data-handling has progressed to facilitate reciprocal tracing between donors of materials and recipients, which creates a need for legal protection of confidentiality against improper linkage. Absolute confidentiality such as through legally compelled destruction of data is undesirable, however, because post-transplantation knowledge acquired about the human source of material may be essential to share with a recipient or a recipient's physician. For instance, realization that a deceased person suffered from an infection transmissible through transplantation should lead to a recipient being warned, offered appropriate tests and treated as medically indicated. Expanding knowledge of genetic characteristics and predispositions may similarly show advantages of being able to link sources of transplanted materials with recipients. At the same time, however, such knowledge also shows with what care confidentiality must be protected, since knowledge of a source's genetic status may disclose knowledge for instance of a deceased person's living family members such as children, brothers, sisters, and parents.[11]

Transplantation tourism

Restrictive legislation, for instance against paid donation and purchase of materials or against donations by living unrelated minors, may cause intended commercial sellers and purchasers or minors, to go to jurisdictions where prohibitions do not exist or are unenforced. Practitioners within a country may face import restrictions limiting acquisition of materials procured from other countries, unless, for instance, they can show certification of acquisition according to appropriate legal standards. Because of international protection of rights to travel, however, prospective recipients may travel to countries where they can acquire materials for implantation by means or sources prohibited in their own countries, and those willing to give materials from their bodies that they are prohibited to donate in their own countries may travel elsewhere where such prohibitions do not apply.

Countries that follow the common law tradition tend to apply territorial principles to the scope of their legislation. It binds only those who act within their territorial jurisdiction, so their permanent residents may go elsewhere to undertake acts that are prohibited within their territory of residence without liability to sanction on return. In contrast, countries in the civil law tradition, like those that apply religious laws, consider nationals to be bound by their laws wherever they go. Even if their conduct is lawful in the territory in which it is performed, they may be punished on return to their countries if their conduct abroad is domestically prohibited.

Nationals of countries that apply territorial principles of jurisdiction have incentives to travel elsewhere, where laws permit transplantation practices that their national legislation prohibits. Acting on their own initiative or through brokers, they may find locations where donors and recipients of transplantable materials may meet. If this is recognized to constitute an international outrage, concerted international action may result in this being considered an offence of universal jurisdiction. Such offences are punishable against any person by any country in which a suspect is lawfully present. An historical instance of a universal crime is piracy, but modern enemies against human kind (*hostis humani generis*) include war criminals and perpetrators of genocide.

It is a matter of judgment whether illegal traffickers in transplantable body materials, from living or cadaveric sources, rank in this order. If not, an international multilateral human rights treaty may be negotiated that, on due ratification and legal effect, would provide widespread jurisdiction against violators. Ratifying states would be expected to give effect to the treaty in their domestic legal systems, and states applying territorial principles of jurisdiction might have to consider offences under the treaty as having extra-territorial effect. At present, few if any offences beyond treason, murder, paedophilia and bigamy are generally considered to be extra-territorial by states that apply territorial jurisdiction. The scope, significance and international offensiveness of transplantation tourism might have to be more clearly determined, however, before an initiative of this nature becomes practicable.

The current configuration of organ transplantation legislation

According to information available to the World Health Organization (WHO), the countries listed in Table 7.2 currently have some form of legislation in the field of organ transplantation (countries are grouped in this table according to the WHO regional classification system). In a number of countries, listed in Table 7.3, legislation is currently pending, while in certain countries, listed in Table 7.4, available information suggests that transplantation continues to be governed by non-statutory measures. Examples of legislative approaches in France, Portugal, Russia and the United States are considered below.

FRANCE

Reference has already been made to Law No. 94-654 of 29 July 1994, which deals with organ transplantation as well as a number of other topics often subsumed under the subject of 'bioethics'. The Law introduces a new Title I in Book VI (Donation and use of elements and products of the human body) of the Public Health Code. That Title, 'General principles applicable to the donation and use of elements and products of the human body', establishes the fundamental principle of consent. It is laid down that the donor's prior consent is essential for the removal of elements of the human body and the collection of its products; such consent may be revoked at any time. No payment may be granted to any person who submits to the removal of elements from his body or the collection of products therefrom. A donor may not know the identity of the recipient, nor the recipient that of the donor. There is an exemption in the case of certain products of the human body (no doubt including hair, skin, etc.) and these are to be listed in due course by a decree.

The provisions concerning organ removal from living persons merit reproduction and are outlined in Table 7.5.

Table 7.2 Jurisdictions known to the World Health Organization Headquarters to have legislation on organ transplantation (by WHO region)

European region

Austria	Luxembourg
Belgium	Norway
Bulgaria	Portugal
Croatia	Romania
Czech Republic	Russian Federation
Denmark	Slovakia
Finland	Spain
France	Sweden
Germany (new Länder only)	Switzerland (17 cantons)
Greece	Turkey
Hungary	United Kingdom
Italy (= certain regions)	The former Yugoslavia (status of legislation unclear)

Region of the Americas

Argentina	El Salvador
Bolivia	Guatemala
Brazil	Honduras
Canada (all Provinces and Territories)	Mexico
Chile	Panama
Colombia	Paraguay
Costa Rica	Peru
Cuba	USA (Federal + all States and District of Columbia)
Dominican Republic	
Ecuador	Venezuela

Western Pacific region

Australia (all States and Territories)	Singapore
Japan	Tonga
Malaysia	Viet Nam
Philippines	

Eastern Mediterranean region

Cyprus	Libyan Arab Jamahiriya
Iraq	Syrian Arab Republic
Kuwait	Tunisia
Lebanon	

African region

Algeria	South Africa
Malawi	Zimbabwe
Niger	

South-East Asia region

India (Union + Bombay, Karnataka, Kerala, Maharashtra, Tamil Nadu, Union Territory of Delhi)	Indonesia
	Sri Lanka

Table 7.3 Countries/jurisdictions in which new organ transplantation legislation is under consideration/pending/in preparation

Bulgaria
Czech Republic
Germany
Hong Kong
Israel
Japan
Netherlands
Poland
Portugal
Switzerland (Federal)

Table 7.4 Selected countries/jurisdictions in which commerce in human organs is governed by non-statutory measures

Egypt
Germany (except five new Länder)
Hong Kong
Ireland
Israel
Liechtenstein
Malta
Saudi Arabia
Switzerland (at national level)

Other divisions in Chapter I of Book IV deal with the criteria for the authorization of establishments engaging in organ removal for transplantation purposes, and there is an interesting provision to the effect that practitioners who perform organ transplantations as part of their regular activities may receive no specific remuneration for the actual transplantation procedure as such.

The new French legislation is also of particular interest in that the provisions dealing with 'solid' organs are separate from those dealing with tissues, cells, and products of human origin. Such provisions are contained separately in Chapter II. That chapter includes a number of divisions, including Divisions 2 and 3, reproduced in Table 7.6 because of their likely interest in countries other than France.

Finally, a second law was promulgated on 29 July 1994, namely Law No. 94-653 on respect for the human body. That law introduces new provisions in a Chapter II (Respect for the human body) of Title I (Civil rights) of Book I of the Civil Code, affirming that '[t]he Law shall assure the primacy of the person, prohibit all affronts to the dignity of the person, and guarantee respect for the human being from the commencement of life' (see Table 7.7).

Table 7.5

Division 2 (Removal of organs from a living person) of Chapter I (Organs) of Title III (Organs, tissues, cells, and products of the human body) of Book VI of the Public Health Code, France

- **Article L. 671-3.** The removal of organs from a living person who is making a donation thereof may be performed only in the direct therapeutic interest of the recipient. The recipient must be the father or mother, son or daughter, or brother or sister of the donor, except in the case of removal of bone marrow for a transplant.

    In emergency cases, a spouse may act as a donor.

    The donor, informed in advance of the risks to which he is exposed and the possible consequences of removal, must express his consent before the Presiding Judge of the *tribunal de grande instance*, or the judge designated by him. In an emergency, consent shall be obtained, by any appropriate manner, by the Public Prosecutor. Such consent may be withdrawn without formality and at any time.

- **Article L. 671-4.** No procedure to remove organs for the purpose of donation may be performed on a living minor or living person who has reached the age of majority, who is the subject of legal protection measures.

- **Article L. 671-5.** By way of derogation from the provisions of Article L. 671-4, bone marrow may be removed from a minor for the benefit of his brother or sister.

    Such removal may only be carried out subject to the consent of each person exercising parental authority or the minor's legal representative. Consent shall be expressed before the Presiding Judge of the *tribunal de grande instance*, or the judge designated by him.

    In an emergency, consent shall be obtained, by any appropriate manner, by the Public Prosecutor.

    The authorization to carry out removal shall be granted by a committee of experts, which shall ensure that the minor has been informed of the intended removal, so that he may express his wishes if he is capable of doing so.

    Removal shall not be carried out if the minor refuses.

- **Article L. 671-6.** The committee of experts referred to in Article L. 671-5 shall be composed of three members appointed for three years by order of the Minister responsible for Health. It shall consist of two physicians, including one paediatrician, and a person who is not a member of the medical professions.

    The committee shall reach a decision in accordance with the general principles and rules laid down in Title I of this Book. It shall evaluate the medical grounds for the operation, the risks that such an operation is likely to entail, and its foreseeable physical and psychological consequences.

    The committee of experts shall not indicate its reasons in the event of its refusal to grant authorization.

PORTUGAL

Another recent law in a European country is the Portuguese Law of 22 April 1993 on the removal and transplantation of human organs and tissues. This law contains a number of general provisions, as well as particularly interesting sections dealing with removal of organs from living persons and from cadavers, respectively (see Table 7.8).

Table 7.6

Division 2 *Removal of tissues and cells and collection of products* *of the human body for the purpose of donation*

• **Article L. 672-4.** The removal of tissues or cells or the collection of products of the human body from a living person may be performed only for therapeutic and scientific purposes.

• **Article L. 672-5.** No removal of tissues or cells and no collection of products of the human body may be performed on a living minor or a living person who has reached the age of majority, who is the subject of legal protection measures.

• **Article L. 672-6.** A decree made after consulting the Conseil d'État shall determine the medical situations and conditions where the removal of tissues and cells and the collection of products of the human body from a deceased person are authorized.

Such removal may only be performed for therapeutic or scientific purposes and under the conditions laid down in Articles L. 671-7, L. 671-8, and L. 671-9.

Division 3 *Authorization of establishments and entities performing the* *removal of tissues or cells of the human body* *for the purpose of donation*

• **Article L. 672-7.** Tissues and cells of the human body for the purpose of donation may only be removed in health establishments authorized for this purpose by the administrative authority.

The authorization shall be issued for a period of five years. It shall be renewable.

• **Article L. 672-8.** Practitioners who remove tissues as part of this activity may receive no remuneration for the act of removal.

• **Article L. 672-9.** The technical, health, and medical requirements necessary to ensure that procedures are conducted in accordance with the general principles laid down in Title I of this Book, and which must be fulfilled by health establishments in order to be authorized to perform the removal of tissues, shall be determined by a decree made after consulting the Conseil d'État.

Section 10 is an almost paradigmatic example of a statute based on the presumed consent principle. Section 11 of the law provides for the establishment of a computerized National Register of Non-Donors (RENNDA), its purpose being to record the names of all persons who have informed the Ministry of their wish not to be donors. On 25 September 1994, a Decree-Law was issued which regulates the organization of RENNDA. It deals with, *inter alia*, registration with RENNDA, the purpose of the computerized database, the procedures for the collection of data, data protection, and the issuance of non-donor cards.

RUSSIAN FEDERATION

In the Russian Federation, a law of 22 December 1992 on the transplantation of human organs and/or tissues filled what appeared to be an existing void in this area in the

Table 7.7

- **Article L. 16-1.** Every person shall have the right to respect for his body.
  The human body shall be inviolate.
  The human body, its elements, and its products may not be the subject of a right of inheritance.

- **Article L. 16-2.** A judge may prescribe any measures that are appropriate to prevent or cease illicit affronts to the human body or any illicit dealings [*agissements*] concerning elements or products thereof.

- **Article L. 16-3.** The integrity of the human body may be prejudiced only in the event of therapeutic necessity for the person concerned.
  The prior consent of the person concerned must be obtained except in cases where his condition necessitates a therapeutic intervention to which he is not in a position to consent.

- **Article L. 16-5.** Any conventions made to confer an inheritable value on the human body or its elements or products shall be null and void.

- **Article L. 16-6.** No payment may be made to a person who participates in experimentation upon his body, in the removal of elements from his body, or in the collection of products therefrom.

Table 7.8 Sections of the Portuguese law on organ donation of 1993

6. (1) Without prejudice to the provisions of subsection 2 of this Section, only regenerative substances may be removed from living persons.
   (2) The donation of non-regenerative organs or substances shall be permitted provided that the donor and recipient are related up to the third degree.
   (3) Any donation of non-regenerative substances by minors or incompetent persons shall be prohibited.
   (4) Donations shall under no circumstances be permitted if there is a strong probability of serious and permanent impairment of the physical integrity and health of the donor.

10. (1) All Portuguese citizens, stateless persons, and foreign nationals residing in Portugal shall be considered as potential post-mortem donors, provided that they have not informed the Ministry of Health of their wish not to be donors.
    (2) In cases where unavailability for donation is restricted to certain organs or tissues or to certain purposes, the restrictions concerned must be expressly indicated in the appropriate records and card.
    (3) In the case of minors or incompetent persons, unavailability for donation shall be expressed, for the purposes of registration, by the legal representatives concerned; it may also be expressed by minors who are capable of understanding and able to express their wishes.

former USSR. Because of its particular interest, this Law (which came into force on 1 May 1993) is reproduced in Appendix 1 on page 116.

Also on 22 December 1992, the then Supreme Soviet of the Russian Federation adopted a Resolution assigning responsibility to the Government to undertake the following tasks: formulate State programmes for the training of specialists in transplantology and resuscitation and other medical and technical personnel to undertake operations for the transplantation of human organs and/or tissues; draw up a list of health establishments at which transplantations may be performed; draw up a list of State health establishments responsible for the collection and preparation of organs and/or tissues from living and cadaveric donors, and carry out strict supervision of their activities; establish independent regional co-ordinating centres for the registration and allocation of donor organs; and examine the question of the creation of a fund for assistance to recipients and to (genetically) related donors in connection with operations for the transplantation of human organs and/or tissues.

UNITED STATES OF AMERICA

It is not possible within the confines of this chapter to do justice to rapidly evolving developments in the United States. However, reference should be made to an interim rule issued by the Food and Drug Administration on 8 December 1993. The summary reads as follows:

> The Food and Drug Administration (FDA) is issuing an interim rule to require certain infectious disease testing, donor screening, and record keeping to help prevent the transmission of AIDS and hepatitis through human tissue used in transplantation. The regulations are effective upon publication. FDA is taking this action in response to growing concerns that some human tissue products are being offered for transplantation use without even the minimum donor testing and screening needed to protect recipients against human immunodeficiency virus (HIV) infection and hepatitis infection. The new regulations require all facilities engaged in procurement, processing, storage, or distribution of human tissues intended for transplant to ensure that minimum required infectious disease testing has been performed and that records documenting such testing for each tissue are available for inspection by FDA. The regulations also provide authority for the agency to conduct inspection of such facilities and to detain, recall, or destroy tissue for which appropriate documentation is not available.

Reference should also be made to detailed regulations (published in the *Federal Register* on 8 September 1994) on the functions of and standards for organ procurement organizations. Among other provisions, these require such organizations to ensure that tests are performed on prospective organ donors to prevent the acquisition of organs that are infected with HIV.

OTHER INTERNATIONAL DEVELOPMENTS

There have been numerous international developments in recent years, focusing particularly on commerce in human organs. Table 7.9 lists some of these responses. This is an area in which WHO has been particularly active, as illustrated in Table 7.10. The Guiding Principles developed by WHO appear in Table 7.11.

Table 7.9 Selected international responses to commerce in human organs, 1970–1995

- Transplantation Society (1970 Statement)
- Council of Europe (May 1978 Resolution)
- Council of the European Society for Organ Transplantation (October/November 1983 Policy position)
- Transplantation Society (September 1985 Guidelines)
- 37th World Medical Assembly (October 1985 Statement)
- Annual General Meeting of the European Dialysis and Transplantation Association and the European Renal Association (July 1986 Statement)
- Council of Arab Ministers of Health (March 1987 Draft Law)
- 39th World Medical Assembly (October 1987 Declaration)
- Conference of European Health Ministers (Council of Europe) (Final Text of November 1987)
- XIV International Congress on Penal Law (October 1989 Resolution)
- First Joint Meeting of the European Society for Organ Transplantation and the European Renal Association (December 1990 Resolution)
- World Health Assembly (May 1991 Resolution)
- European Parliament (Discussion on Schwartzenberg Report, and adoption of Resolution, September 1993)
- First International Conference of Medical Parliamentarians (February 1994 Declaration)
- Steering Committee on Bioethics of the Council of Europe (Draft Protocol on Organ Transplantation to the Draft Bioethics Convention)
- Transplantation Society (September 1994 Guidelines)

Table 7.10 Principal stages in the development of WHO's guiding principles on human organ transplantation

Date	Event	Participants
May 1986	39th World Health Assembly discusses organ transplantation	Member States
May 1987	40th World Health Assembly adopts resolution on 'Development of guiding principles for human organ transplants'	Member States
May 1989	42nd World Health Assembly adopts resolution on 'Preventing the purchase and sale of human organs'	Member States
May 1990	WHO convenes Informal Consultation on Organ Transplantation (Geneva)	20 experts + WHO Secretariat
October 1990	WHO convenes 2nd Informal Consultation on Organ Transplantation (Geneva)	5 experts + WHO Secretariat
January 1991	87th session of the WHO Executive Board recommends endorsement of Draft Guiding Principles on Human Organ Transplantation	Members of the Executive Board
May 1991	44th World Health Assembly endorses Guiding Principles on Human Organ Transplantation	Member States
September 1991	WHO publishes *Human Organ Transplantation: A Report on Developments Under the Auspices of WHO (1987–1991)*	–

Table 7.11 Guiding principles on human organ transplantation (WHO, Geneva, 1991)

Guiding Principle 1

Organs may be removed from the bodies of deceased persons for the purpose of transplantation if:

(a) any consents required by law are obtained; and
(b) there is no reason to believe that the deceased person objected to such removal, in the absence of any formal consent given during the person's lifetime.

Guiding Principle 2

Physicians determining that the death of a potential donor has occurred should not be directly involved in organ removal from the donor and subsequent transplantation procedures, or be responsible for the care of potential recipients of such organs.

Guiding Principle 3

Organs for transplantation should be removed preferably from the bodies of deceased persons. However, adult living persons may donate organs, but in general such donors should be genetically related to the recipients. Exceptions may be made in the case of transplantation of bone marrow and other acceptable regenerative tissues.

An organ may be removed from the body of an adult living donor for the purpose of transplantation if the donor gives free consent. The donor should be free of any undue influence and pressure and sufficiently informed to be able to understand and weigh the risks, benefits and consequences of consent.

Guiding Principle 4

No organ should be removed from the body of a living minor for the purpose of transplantation. Exceptions may be made under national law in the case of regenerative tissues.

Guiding Principle 5

The human body and its parts cannot be the subject of commercial transactions. Accordingly, giving or receiving payment (including any other compensation or reward) for organs should be prohibited.

Guiding Principle 6

Advertising the need for or availability of organs, with a view to offering or seeking payment, should be prohibited.

Guiding Principle 7

It should be prohibited for physicians and other health professionals to engage in organ transplantation procedures if they have reason to believe that the organs concerned have been the subject of commercial transactions.

Guiding Principle 8

It should be prohibited for any person or facility involved in organ transplantation procedures to receive any payment that exceeds a justifiable fee for the services rendered.

Guiding Principle 9

In the light of the principles of distributive justice and equity, donated organs should be made available to patients on the basis of medical need and not on the basis of financial or other considerations.

APPENDIX (SUMMARY OF RUSSIAN LAW AS PUBLISHED IN INTERNATIONAL DIGEST OF HEALTH LEGISLATION 1993, VOL. 44, NO. 2, PAGES 239–242)

RUSSIAN FEDERATION. Law of 22 December 1992 of the Russian Federation on the transplantation of human organs and/or tissues. Dated 22 December 1992. (7 pp.) RF 93.12

The Preamble to this Law reads as follows:

This Law establishes the conditions and procedures for the transplantation of human organs and/or tissues, based on the contemporary achievements of science and medical practice and also taking into account the recommendations of the World Health Organization.

Transplantation of human organs and/or tissues constitutes a means for saving the lives and restoring the health of citizens, and must be undertaken on the basis of compliance with the legislation of the Russian Federation and human rights, in conformity with the humanitarian principles proclaimed by the international community.

The following are the provisions of the Law:

CHAPTER I

General provisions

Conditions and procedures for the transplantation of human organs and/or tissues

1. The transplantation of organs and/or tissues from a living donor or a cadaver may be carried out only in cases where other medical techniques cannot guarantee the saving of the life of the patient (recipient) or the restoration of his health.

The removal of organs and/or tissues from a living donor shall be permissible only in cases where, according to the findings of a committee [concilium] of medical specialists, his health will suffer no significant damage.

The transplantation of organs and/or tissues shall be permitted solely with the consent of the living donor and, as a rule, with the consent of the recipient.

Human organs and/or tissues may not be bought or sold. The buying and selling of human organs and/or tissues, as well as the advertising of such activities, shall entail criminal liability in accordance with the legislation of the Russian Federation.

Operations to transplant organs and/or tissues in recipients shall be carried out on the basis of medical indications, in accordance with the general rules governing the conduct of surgical operations.

List of human organs and/or tissues that may be transplanted

2. Transplantation procedures may be conducted involving the heart, lungs, kidneys, liver, bone marrow, and other organs or tissues included on a list established by the Ministry of Health of the Russian Federation, in conjunction with the Russian Academy of Medical Sciences.

The provisions of this law shall not apply to organs, parts of organs, and tissues that are associated with human reproductive processes including tissues or components associated with reproduction viz. ovules, sperm, ovaries, testicles, or embryos, nor to blood or blood constituents.

Restrictions on the selection of living donors

3. No organ and/or tissue may be removed for transplantation purposes from a living donor under 18 years of age (other than in the case of bone-marrow grafts) or from living donors who have been duly recognized as incompetent.

No organ and/or tissue may be removed if it is established that it belongs to a person suffering from a disease that constitutes a danger to the recipient's life or health. No organ and/or tissue may be removed for transplantation purposes from persons who are dependent upon the recipient, either because of their functions or in any other manner.

Any person who coerces a living donor to consent that organs and/or tissues be removed from him shall be liable to criminal proceedings in accordance with the legislation of the Russian Federation.

Health establishments engaging in the collection, preservation, and transplantation of human tissues and/or organs

4. Human organs and/or tissues may be collected and preserved only in State health establishments.

The transplantation of human organs and/or tissues shall be authorized within specialized health establishments.

A list of health establishments engaging in the collection, preservation, and transplantation of human organs and/or tissues, and the regulations concerning their activities, shall be drawn up by the Ministry of Health of the Russian Federation, in conjunction with the Russian Academy of Medical Sciences.

Medical report as to the necessity for the transplantation of human organs and/or tissues

5. A medical report as to the necessity for the transplantation of human organs and/or tissues shall be drawn up by a committee of physicians of the health establishment concerned; this group shall include the attending physician, a surgeon, an anaesthesiologist, and, as appropriate, physicians in other specialties, on the basis of an Instruction to be issued by the Ministry of Health of the Russian Federation.

Consent of the recipient to the transplantation of human organs and/or tissues

6. The transplantation of human organs and/or tissues shall be carried out with the written consent of the recipient. In addition, the recipient must be warned of the possible complications that the envisaged intervention may present for his health. In the event that the recipient is less than 18 years of age, or if he has been duly recognized as incompetent, transplantation may be carried out only with the written consent of his parents or of his legal representative.

Organs and/or tissues may be transplanted in a recipient without his consent or that of his parents or his legal representative in exceptional cases, where a delay in carrying out the appropriate intervention would threaten the recipient's life and it is impossible to obtain consent.

Validity of international agreements

7. In the event that an international agreement to which the Russian Federation is a party lays down rules different from those laid down in this Law, the rules contained in the international agreement shall prevail.

CHAPTER II

Removal of organs and/or tissues from a cadaver for purposes of transplantation

Presumed consent to organ and/or tissue removal

8. The removal of organs and/or tissues from a cadaver shall not be authorized if the health establishment is aware, at the time of removal, of the fact that, during his lifetime, the deceased person or his close relatives or legal representative expressed their opposition to the removal of organs and/or tissues after his death, with a view to transplantation in a recipient.

Determination of the time of death

9. Organs and/or tissues may be removed from a cadaver for transplantation purposes provided that death has been certified in an irrefutable manner by a committee of medical specialists.

The declaration of death shall be based on a determination of the irreversible cessation of the functions of the whole brain (brain death), established in accordance with procedures laid down by the Ministry of Health of the Russian Federation.

No transplantation specialist and no member of the team working for the organ donation service and remunerated by the latter service may participate in the diagnosis of death in cases where the use of the deceased person as a donor is envisaged.

Authorization for organ and/or tissue removal from a cadaver

10. Organs and/or tissues may be removed from a cadaver only with the authorization of the chief physician of the health establishment and in compliance with the requirements laid down by this Law.

In the event where a forensic-medical expert appraisal is required, authorization to remove organs and/or tissues from a cadaver must likewise be obtained from the expert in forensic medicine, with appropriate notification thereof to the Public Prosecutor.

CHAPTER III

Removal of organs and/or tissues from a living donor for transplantation purposes

Conditions for the removal of organs and/or tissues from a living donor

11. The removal of organs and/or tissues from a living donor for transplantation to a recipient shall be authorized subject to compliance with the following conditions:

— the donor must have been warned of any complications that could affect his health as the result of an operation for the removal of organs and/or tissues;

— the donor must have given his free and informed consent in writing to the removal of his organs and/or tissues; and

— the donor must have undergone a comprehensive medical examination in respect of which a report must have been drawn up by a committee of medical specialists as to the possibility of removing his organs and/or tissues for transplantation purposes.

The removal of organs from a living donor shall be authorized if there exists a genetic relationship with the recipient, other than in cases of bone-marrow grafts.

Rights of the donor

12. Any donor who has expressed his consent to the transplantation of his organs and/or tissues shall be entitled:

— to seek, from the health establishment, complete information on the possible effects upon his health of the operation for the removal of his organs and/or tissues; and

— to receive free treatment, including treatment with medicaments, in the health establishment and associated with the operation he has undergone.

Restrictions on the transplantation of organs and/or tissues from living donors

13. A paired organ, a part of an organ, or a tissue whose absence does not entail any irreversible damage to health may be removed from a living donor for transplantation purposes.

CHAPTER IV

Liability of health establishments and their staff

Liability for divulging information concerning donors and recipients

14. It shall be prohibited for physicians and other staff members of health establishments to divulge information concerning donors and recipients.

The divulgence of such information shall entail liability in accordance with the legislation of the Russian Federation.

Prohibition on the sale of human organs and/or tissues

15. Health establishments authorized to engage in operations for the collection and preparation of organs and/or tissues from cadavers shall be prohibited from selling them.

This law shall not apply to preparations and transplant materials for the preparation of which tissue components are used.

Liability of health establishments

16. In the event that the health of a donor or recipient is damaged as a result of failure to comply with the conditions and procedures for the removal of organs and/or tissues, or the conditions and procedures transplantation provided for by this Law, the health establishment shall incur material liability towards the above-mentioned persons in accordance with the procedures laid down by the legislation of the Russian Federation.

References

1. Gerson WN. Refining the law of organ donation: lessons from the French law of presumed consent. *Journal of International Law and Politics* 1987, **19**: 1013–1032.
2. World Health Organization. *Use of human tissues and organs for therapeutic purposes: a survey of existing legislation.* WHO, Geneva, 1969.
3. Australian Law Reform Commission. *Human tissue transplants.* Australian Government Publishing Service, Canberra, 1977.
4. Fuenzalida-Puelma HL. Organ transplantation: the Latin American legislative response. *Bulletin of the Pan American Health Organization* 1990, **24**: 424–445.
5. World Health Organization. *Legislative responses to organ transplantation.* Martinus Nijhoff, Dordrecht, The Netherlands and Boston, 1994.
6. Dickens BM. The modern function and limits of parental rights. *Law Quarterly Review* 1981, **97**: 462–485.
7. Canadian Royal Commission on New Reproductive Technologies. *Proceed with care.* Government Services, Ottawa, 1994, page 967 et seq.
8. Giesen D. *International medical malpractice law: a comparative study of civil liability arising from medical care.* JCB Mohr (Paul Siebeck), Tübingen, and Martinus Nijhoff, Dordrecht, 1988, pages 607–623.
9. Dickens BM. Living tissue and organ donors and property law: more on *Moore. Journal of Contemporary Health Law and Policy* 1992, **8**: 73–93.
10. Mason JK, McCall Smith RA. *Law and medical ethics* (4th edn). Butterworths, London, 1994, page 304.
11. American Association for the Advancement of Science. *The genome, ethics and the law: issues in genetic testing.* AAAS Publication No. 92-115, Washington, DC, 1992.

Note: Except where clearly indicated, any views expressed are not necessarily those of the World Health Organization.

CHAPTER 8

Transmission of disease by organ transplantation

SVETLOZAR N. NATOV, BRIAN J.G. PEREIRA

Introduction   120
Transmission of bacterial infections   120
Transmission of viruses   122
Transmission of parasites   136
Transmission of malignancy by the allograft   138
Conclusions and summary   140
References   141

Introduction

Despite significant advances in organ transplantation, the transmission of infection by tissue and organ donation continues to play an important role in the post-transplant morbidity and mortality. Consequently, policies that deal with testing of organ donors for potentially transmissible infectious agents, and the rational use of organs from infected donors remain a subject of debate. This chapter will focus on transmission of viral infections by organ transplantation and will only briefly address transmission of bacterial and protozoal infections, and malignancies. Every effort has been made to check donation protocols and tests for transmission of disease, however, it is still possible that errors have been missed. Furthermore, dosage schedules are constantly being revised and new side effects recognized, new diseases are discovered and tests for transmissible diseases are constantly being improved. These variations in disease frequency in different populations mean the local considerations must always be made with respect to the particular tests required to screen donors to prevent transmission of disease.

Transmission of bacterial infections

Exogenous (iatrogenic) contamination of the graft can occur during harvesting, preservation, handling, or at the moment of transplantation. Its incidence is associated with the complexity of the organ procurement process.[1] Iatrogenic contamination usually involves bacteria and rarely viruses and/or fungi.[1-5] Strict adherence to aseptic procedures should obviate this hazard.[2]

Bacterial infection transmitted by the allografts is rarely of donor origin. Nevertheless, when a graft from an infected donor is transplanted, it exposes the recipient to a large inoculum of donor pathogens.[6] Transmission of bacterial infections usually originates from latent or subclinical infection in the donor.[5] Occasionally, the infectious focus may be located within the tissues or organs procured for donation.[1,4,7] However, the risk of infection in the transplant recipient is a result of the interaction of the exposure to the pathogen transmitted by the contaminated allograft and the recipient's net state of immunosuppression.[1,8-11]

The exact vector of transmission within the allograft remains unidentified for most infectious agents. However, as many of them are typically bloodborne, it is likely that they are carried by residual blood cells retained in the allograft rather than by its native cells. Therefore, the method of allograft processing and preservation, for example continuous pulsatile perfusion of the kidney as compared to simple cold storage in ice slush, may be crucial for decreasing the infectious load and even eliminating or inactivating the infectious agent, thus reducing the risk of infection transmission. In proof of this hypothesis, Gottesdiener summarized data from ten studies on the incidence of bacterial contamination of donor kidneys.[6] He observed a vast variance (2.1% to 23.4%) in the incidence of bacterial contamination and postulated that this large range is, at least in part, related to the difference in the methods of preservation and the use of various antibiotics for preservation fluids and for prophylaxis after transplantation. However, the difference in protocols for bacterial surveillance developed in these studies also accounts for the differences in the ability to accurately identify bacterial contamination of the allograft.[6]

Obtaining cultures from the blood or suspected sites of infection (urine, sputum, wound) is the microbiologic method commonly used for screening potential donors for the presence of transmissible pathogens. Transplantation of organs from donors with positive blood cultures (bacterial or fungal) usually results in increased rates of graft loss and mortality in the recipient.[2] In addition, transplantation of organs from donors with localized bacterial or fungal infections, particularly when the harvested allograft itself contains the infectious focus, may be a potential hazard that can have severe consequences.[8,9] McCoy and colleagues[8] reported that 14 out of 81 cadaver renal allografts (17%) had evidence of bacterial contamination before transplantation. Among the infected allografts, gram-negative organisms were cultured in one third of the cases (five out of 14 kidneys) and accounted for one recipient death and the loss of two other allografts. In another instance,[9] *Staphylococcus aureus* septicemia was observed in both renal recipients of a single donor. Complications of infection required transplant nephrectomy in each case and resulted in death in one of the recipients. Although it was not known whether the donor had *Staphylococcus aureus* sepsis at the time of organ procurement, the occurrence of septicemia with the same pathogen in both recipients of a single donor indicates that the donor was most likely the common source of infection. In a series of 200 transplant kidneys, the incidence of contamination (based on a positive culture from the preservation fluid) was 19.5%.[1] Major complications, such as severe local infections and a ruptured mycotic aneurysm, led to transplant nephrectomy in four out of 39 recipients of infected kidneys. Despite the fact that the actual 1-year survival of contaminated grafts (70%) did not differ statistically from that of the 'sterile' kidneys (69%), the authors advocate that all efforts should be made to prevent contamination of the graft and thus to preclude severe complications. Nevertheless, several studies have reported excellent results following inadvertent use of infected organs.[1,12] In a series of 232 cadaver renal transplants, 15 patients received kidneys from donors with positive bacterial cultures from blood, urine, wound, or flush solution. The combination of perioperative broad spectrum prophylactic antibiotics and specific treatment directed to the isolated organism resulted in the absence of any instance of post-transplant infection with the donor organism.[5]

Based on these observations, many organ procurement organizations do not accept donors with positive blood cultures and do not use lungs from donors with positive sputum cultures or kidneys from donors with positive urine cultures. However, organs remote from the site of infection may be considered for transplantation.[13] Measures to minimize serious sequelae from the use of contaminated grafts should include antibiotics during donor nephrectomy,[1] culture of preservation fluid,[1] reporting of positive cultures of donor blood or urine and preservation fluid to the recipient centre,[1] alerting the centres that receive other organs from the same donor if a pathogen is cultured from the preservation fluid of one of the organs,[1] continuation of antibiotic coverage until culture results of the preservation fluids are known,[1,5] and use of specific treatment for positive donor cultures.[5] However, if positive donor cultures are reported to the transplant centre later in the postoperative period and the patient is clinically well, it may not be necessary to institute such therapy.[5]

Transmission of viruses

CYTOMEGALOVIRUS (CMV)

Human cytomegalovirus is a beta herpes virus of the herpesvirus family[14] with a linear double-stranded DNA genome of 230 kilobase pairs.[15] The complete nucleotide sequence of the virus has been recently described.[16] Antigenic variations exist and can be detected by cross-neutralization tests, nucleic acid hybridization techniques and restriction endonuclease cleavage of DNA.[17,18] No two epidemiologically unrelated isolates of human CMV have been found to be completely identical.[18] This has made possible the identification of several different strains.[18,19] It is not known whether they differ in virulence. However, these strains share at least 80% of genetic information[18] and therefore, the variability of the human CMV is not great enough to provide sufficient evidence for the designation of distinct types.[19]

Distribution

Human CMV is ubiquitous and humans are the only known reservoir. Once acquired, the virus establishes latency and remains silent indefinitely. Infected individuals excrete the virus in saliva, urine, tears, breast milk, sperm and vaginal secretions and can transmit infection to their close contacts.[20] However, casual contact has not been shown to transmit CMV infection.[21] The cellular receptor for human CMV has not yet been identified with certainty, although class I MHC proteins and a putative receptor of around 30 kDa on fibroblasts seem to be good candidates.[22-24] The cellular sites at which the virus persists in individuals with latent CMV infection are also uncertain. However, blood cells, epithelial cells and capillary endothelial cells can probably harbour the virus.[25] A recent study has found evidence of CMV infection in practically all tissues in normal CMV-seropositive subjects who underwent autopsies after fatal accidents, thus indicating that any cell can be latently infected with CMV.[26] Furthermore, Hendrix et al.[27] have reported that CMV DNA can be detected by PCR in almost all major arteries in CMV-seropositive individuals.

Tests

In otherwise healthy individuals, CMV infection is clinically inapparent and consequently remains undetected unless laboratory tests are performed. Diagnosis is established by isolation of the virus[28] or by demonstrating its presence by molecular or serologic techniques.[29] Routine culture of the virus is a highly sensitive test,[28] but it is time consuming and therefore impractical for screening potential donors. Rapid virus assays for detection of CMV early antigens have been developed based on the use of monoclonal antibodies and indirect immunofluorescence.[30-33] Molecular hybridization techniques employing labelled viral DNA probes to detect the presence of CMV genome are also currently available.[34-37] Although most of the DNA-hybridization assays can be performed within 24 hours, they are inconvenient for widespread application in clinical practice and their sensitivity and specificity need to be further assessed.[36,38]

A typical feature of CMV infection, particularly in the immunosuppressed patient, is the persistence of the virus in the host despite the production of serum antibodies.[39-42] The presence of antibody to the virus appears to be a good marker of CMV infection. Consequently, serological testing has become the method of choice for early and rapid detection of human CMV infection. Numerous serologic techniques have been applied to the diagnosis of human CMV, however, the most reliable and commonly used ones are complement fixation, indirect hemagglutination, indirect immunofluorescence, immunoenzymatic techniques and latex agglutination.[43-45] Most of these assays have shown sensitivity and specificity ranging from 93% to 100%.[45] Although, radioimmunoassay was initially considered to have highest sensitivity,[44] others have more recently found that the latex agglutination test is the most sensitive and also the easiest to perform.[45] Most serologic techniques can be performed within 24 hours and are therefore suitable for donor screening.[43,46]

Modes of transmission

Transmission of CMV infection is the result of the interaction between viral and host factors. In transplantation, the viral factor of major importance appears to be the source and quantity of the virus transmitted to the host, as no difference in virulence and predisposition to dissemination have been reported between the different strains of CMV. It is unclear which cell transmits the virus from the donor – infected cells of the allograft itself or blood cells retained in the allograft after processing. Attempts to isolate or identify the virus in kidneys from donors with latent infection have been unsuccessful[47-50] in all but one[51] studies. On the contrary, although not found free in plasma, the virus is present in white blood cells of blood donors with cytomegalovirus infection and is transmitted by blood transfusion via the leukocyte fraction.[52] However, transfusion is unlikely to be the mode of transmission of CMV in the setting of transplantation. The risk of CMV transmission by an unit of blood has been estimated to be 2.7% or less,[53-56] and numerous studies have failed to establish any correlation between blood transfusion and transmission of the virus unless more than 20 units of blood are transfused.[53,57-60] Furthermore, animal studies have shown that the virus is much less effectively transmitted by blood transfusion than by the allograft.[61] Even if transfusion could have played some role in the past,[58,59] the use of CMV-screened blood[62,63] or leukocyte-depletion filters[64] at present has decreased its importance as a risk factor.[62,65]

Host factors have an important role in determining the susceptibility to CMV infection and the consequent morbidity. Among these factors, pre-existing immunity to CMV seems to be of particular importance in the control of virus replication and spread.[66] Studies in renal transplantation have found that seroconversion occurs in 36% to 95% of the seronegative patients who received kidneys from seropositive cadaver or living-related donors[60,65,67-78] compared to only in 0% to 30% of the seronegative recipients of kidneys from seronegative cadaver donors.[60,65,67,73,76-78] Furthermore, only seronegative recipients of organs from seropositive donors, and not seronegative recipients from seronegative donors, were reported to shed CMV post-transplantation.[65,75] However, a high incidence of post-transplantation CMV infection has also been observed in seropositive renal allograft recipients, irrespective of the donor CMV status. This emphasizes the role of endogenous latent CMV infection in the recipient. Summarizing 16 studies with a total of 1276 renal transplant recipients, Ho concluded that CMV infection developed in 84% of the seropositive renal allograft recipients, regardless of the donor CMV status, while the incidence of CMV infection in seronegative kidney recipients was 53% and almost always related to the use of renal allografts from seropositive donors.[79] Moreover, in some studies the rate of CMV-infection in seropositive kidney recipients was as high as 90% to 100%, which essentially implies that almost all seropositive patients reactivate their own latent virus after transplantation, due to pharmacologic immunosuppression.[79]

Clinical CMV syndromes in transplant recipients

Three clearly distinguishable epidemiologic patterns of CMV infection have been observed in transplant recipients:

1. *Primary infection.* Occurs in transplant recipients with no previous exposure to CMV, e.g. seronegative at the time of transplantation, who receive an organ from a seropositive donor. These patients are identified by the appearance of antibody to CMV and/or by positive cultures.
2. *Reinfection.* Occurs when seropositive transplant recipients receive an allograft from a seropositive donor and shed the virus of donor origin.
3. *Reactivation infection.* Occurs in seropositive individuals who reactivate their endogenous latent virus after transplantation regardless of the serologic status of the donor. Reinfection and reactivation are commonly referred to as *secondary infection* and are identified by a four-fold increase in the titre of antibody to CMV.

The presence of identical CMV strains either in both donor and recipient(s) or in each of the recipients of a single cadaver donor strongly suggests that the grafted kidney is the vector of CMV transmission and the donor is the source of CMV infection.[70,74,75,80-82]

Seropositive kidney recipients may develop secondary infection – reinfection or reactivation, depending on the serologic status of the donor. Post-transplantation CMV infection has been observed in 69% of the seropositive patients who received allografts from seropositive donors compared to only 27% of the seropositive recipients of kidneys from seronegative donors.[75] Reinfection probably accounted for the difference between the groups. Thus, in seropositive kidney recipients, reinfection appears to be almost twice as frequent as reactivation. Restriction enzyme analysis for strain typing of virus isolates from recipient pairs has shown that reinfection with donor strain occurs in 86% of the seropositive recipients of kidneys from seropositive donors.[75] Therefore, at least in the setting of immunosuppression, prior immunity to CMV does not protect from transmission of infection by transplantation of organs from a donor with CMV infection. Further, reinfection with the donor strain may in some way suppress reactivation of the recipient latent virus, as the majority of cases with secondary infection are due to reinfection with an exogenous virus rather than to reactivation of the endogenous virus.

Cytomegalovirus disease is defined as a symptomatic CMV infection characterized by fever, leukopenia, and often organ-specific symptoms. The overall incidence of CMV disease in renal allograft recipients is 22–28%.[73,75] However, a disease rate as high as 57% has been reported among those who excrete the virus post-transplantation.[75] CMV disease is significantly more common among recipients with primary CMV infection (61–91%) than among those with secondary infection (23–42%).[73,75,77] Primary infection is also associated with more severe disease and significantly longer duration of fever and leukopenia as compared to patients with disease due to reactivation.[73] Pre-transplant immunity is likely to offer some protection against the donor virus and to a certain degree may prevent its dissemination.[73] Among patients with secondary infection, reinfection is associated with more severe disease than reactivation.[73] Furthermore, in recipients with pre-transplant CMV infection, Grundy et al.[75] observed post-transplantation CMV disease only among those who received organs from seropositive donors (39%) and not among those who had transplants from seronegative donors. In contrast, Smiley et al.[73] observed that the incidence of CMV disease in seropositive recipients of kidneys from seropositive donors was similar to this in seropositive kidney recipients of seronegative donors (24% vs 20%). However, while reactivation was a mild disease with no major complications, reinfection was associated with clinically more severe disease.[73]

Transplant policies

The allocation of organs based on the CMV antibody status of the donor and recipient is a matter of debate. Smiley et al.[73] have estimated that a total ban on the use of organs from seropositive donors would decrease the incidence of CMV disease from 37.5% to 10%, and the severity of disease would also be markedly reduced. However, with the current shortage of organs, such a policy is untenable since seropositive donors account for at least 42% of kidneys.[73] Some groups 'reserve' kidneys from seronegative donors exclusively for seronegative recipients precluding transplantation of kidneys from seropositive donors into seronegative recipients.[83] Compared to a historical group in which the serologic status of donor and recipient was not taken into consideration for organ matching, this policy has resulted in a significant decrease in morbidity (from 10.6% to 1.7%), mortality (from 3.7% to 0%) and graft loss (from 2.5% to 0%) due to lack of primary infection. Despite the imposed constraints, neither decrease in the rate of transplantation of seronegative recipients nor prolongation of their waiting time on the transplant list have been observed because of acceptance of a greater HLA-A, -B, -DR mismatch for the listed seronegative patients. Furthermore, seronegative recipients constituted only 26% of the tested recipients whereas 36% of the tested donors were seronegative. Clearly, this practice is beneficial for seronegative recipients. However, based on the findings that seropositive recipients of organs from seropositive donors have significantly more severe disease than seropositive recipients from seronegative donors,[73] it is disadvantageous for seropositive patients to receive kidneys exclusively from seropositive donors. No strict guidelines have been adopted as of now, and CMV disease is currently considered an acceptable risk in the transplant population (Table 8.1).

Strategies to reduce CMV disease in organ transplant recipients

Several strategies have been employed to reduce the severity of CMV disease in organ transplant recipients including CMV vaccine, CMV hyperimmune globulin (CMVIG), acyclovir, and non-specific globulin.[84,85] In seronegative kidney recipients from seropositive donors, the incidence and onset of CMV infection was not altered by CMVIG, but the occurrence of CMV disease decreased from 60% to 21%, when CMVIG was started within 72 hours of transplantation and continued for 16 weeks post-transplantation.[86] Furthermore, CMVIG prophylaxis has been associated with reduced rate of serious and life-threatening opportunistic infections, which frequently complicate CMV disease and a 40% reduction in viremia.[86] CMVIG prophylaxis has also decreased serious CMV-associated disease from 54% to 15% in kidney transplant recipients at risk for primary infection, who received anti-rejection therapy.[86]

Combining the results of two liver transplant studies – one random-assignment trial and one open-label trial, Snydman et al.[87] have demonstrated that the use of CMVIG resulted in a significant decrease (from 89% to 57%) in the rate of CMV infection in the donor-positive/recipient-negative group as well as in a 50% reduction in severe CMV-associated disease. It was also observed that CMVIG prophylaxis among patients who receive OKT3, regardless of their serologic status, significantly reduced severe CMV-associated disease.[88]

In a study in renal allograft recipients, acyclovir administered pre-operatively and for 3 months thereafter has been found to prevent primary CMV infection and disease in CMV-seronegative recipients.[89] However, these observations have not been confirmed by others.[90] Similarly, several studies have shown that two to four weeks of ganciclovir prophylaxis does not prevent primary CMV disease in heart, liver, lung and heart/lung

Table 8.1 Virus transmission by organ transplantation

Agent	Tests	Transplantation from positive donor*
Cytomegalovirus	Serology for antibody	Yes
Herpes simplex virus	Clinical examination of donor	Yes
Human immunodeficiency virus	Clinical history, examination Serology for antibody	Absolute contraindication
Hepatitis B virus	Clinical history, examination HepBsAg test	Contraindicated ?Endemic areas
Hepatitis C virus	Clinical history, examination Serology for antibody	Contraindicated ? Positive recipients
Creutzfeldt–Jakob disease	Clinical history	Absolute contraindication

*Please see text for details

transplant recipients,[91-94] though longer treatment periods are under review and that non-specific immune globulin was of no benefit in preventing CMV disease in the high-risk patient.[95,96] In conclusion, reviewing the current strategies for prevention of CMV disease in the high-risk transplant recipients, Snydman reports that CMVIG has been the only agent demonstrated to be effective as prophylaxis against the most severe forms of CMV disease.[88]

HERPES SIMPLEX VIRUS (HSV)

Transmission

Herpes simplex virus is another member of the herpesvirus family. Like the other herpesviruses, it is also an ubiquitous pathogen that can persist in a latent state in adults. Although reactivation of HSV infection is common in the immunocompromised host, including transplant recipients, a number of prospective studies have failed to identify any renal or cardiac allograft recipients who have acquired HSV through transplantation.[49,58,62,97-99] These findings are compatible with the fact that the neuron is the only cell which can harbour HSV in a latent form.[100-102] However, the allograft contains neurons and nerve endings, and hence transmission of HSV by organ transplantation, although rare, is possible. Dummer et al.[103] described disseminated infection with HSV type 2 in two seronegative patients who received kidneys from a single seropositive cadaver donor. DNA restriction endonuclease analysis of the patients' isolates demonstrated that the strains were identical and could be of common origin. The authors speculated that the infection originated from the donor's urogenital tract and was transmitted by the kidney transplants.

Goodman et al.[104] reported two possible instances of transmission of HSV infection by transplantation of organs from two different seropositive donors. A HSV-seronegative patient who received a pancreatic allograft from one of the donors, seroconverted after transplantation, developed disseminated HSV infection and died despite the transplant pancreatectomy. Immunoperoxidase staining of the pancreatic allograft for HSV antigens was strongly positive. Likewise, a seronegative cardiac allograft recipient from a second donor also developed antibody to HSV and disseminated disease after transplantation. Both these cases provide strong circumstantial evidence favouring transmission of HSV by organ transplantation. However, two other recipients of organs from the implicated donors (a kidney from the one and a liver from the other) were HSV-seropositive pre-transplant by complement fixation and did not develop disease. It may be speculated that their pre-existing immunity to HSV protected them from reinfection. Therefore, although it is likely that HSV is transmitted by organ transplantation, the risk of post-transplant HSV infection may not be uniform and could be influenced by putative viral and host factors.

Transplant policies

Based on the available data, it is recommended that potential donors should be examined for herpetic lesions, HSV cultures from suspicious lesions should be performed and concurrent testing for anti-HSV antibody should be done.[103,104] Neutralizing antibody has been traditionally considered the 'gold standard' for detecting HSV infection.[104] The presence of HSV infection in a potential donor is not a contraindication for donation, because of the rarity of this mode of transmission.[103,104]

Nonetheless, if a seronegative patient receives a graft from a seropositive donor, the potential for HSV transmission should be considered. Careful clinical observation of the recipient is mandatory with respect to signs of herpetic disease or unexplained illness suggestive of HSV. Early initiation of antiviral therapy with intravenous acyclovir in such cases has been helpful in modifying the course of the disseminated HSV infection and preventing serious consequences.[103,104] Therefore, it is recommended that antiviral therapy should be started immediately whenever HSV infection is suspected, even in the absence of positive cultures from the usual mucosal sources.[103,104]

HUMAN IMMUNODEFICIENCY VIRUS (HIV)

Transmission

Human immunodeficiency virus belongs to the lentivirus subgroup of retroviruses. Transmission of HIV by kidney,[105–107] liver,[105,107,108] heart,[107,109,110] pancreas,[110] bone,[107,111] and possibly skin[112] transplantation has been documented. Most cases had occurred before the implementation of donor screening for HIV-antibody in 1985. However, even after 1985, transmission of HIV infection from seronegative donors has been reported in isolated instances[107,113] due to limitations of HIV-antibody testing. In one case,[114] massive blood transfusion in the donor led to a false-negative result for HIV-antibody, because of hemodilution. The blood sample was collected after the donor had received multiple blood transfusions during an attempted resuscitation. Other serum samples, obtained on the day of admission, before the blood transfusions, and two days after the transfusions, were retrospectively found to be positive for HIV-antibody.

Despite a true negative result for HIV-antibody, a donor may still be infectious and, therefore, able to transmit HIV.[107] This occurs if testing is performed in the 'window' period of several weeks to six months between the acquisition of the infection and the appearance of HIV-antibody. Simonds et al.[107] have documented transmission of HIV infection from a multiple organ donor who was HIV-1-antibody negative at the time of organ and tissue procurement and had no known risk factors for HIV-infection. Four solid organs and 54 other tissues were harvested and 52 of them were distributed. Subsequent testing of donor's frozen spleen cells for HIV-1 by culture and the polymerase chain reaction (PCR) established the presence of HIV-1 in the donor. Forty-eight recipients could be identified, 41 were tested for HIV-1 antibody, and a total of seven of them were found to be infected with HIV-1. HIV was transmitted to all recipients of solid organs and unprocessed fresh-frozen bone. The recipients of tissues that were relatively avascular and/or processed by lyophilization, ethanol extraction, gamma irradiation, or mechanical removal of blood and hematogenous elements were not infected. It is likely that some of the physical and chemical agents used for processing and preservation of the procured tissues may have inactivated or eliminated the virus. Alternatively, it may be that avascular tissues did not contain the virus.

Finally, a donor could acquire HIV infection during the interval between initial testing and actual organ donation. Quarto et al.[113] have reported a case of transmission of HIV infection from a living related kidney donor with a history of homosexual behaviour. The donor was seronegative at the time of initial evaluation, eight months prior to transplantation, and his high-risk behaviour was unknown at that time. The donor was not retested immediately before donation, and was subsequently found to have seroconverted during the interval between initial evaluation and actual organ donation.

Transplant policies

Precise donor screening is essential for preventing HIV transmission through organ and tissue transplantation and is mandated by law in many countries.[115] The current recommendations for donor screening[115] include:

1. Screening for behaviours and risk factors that are associated with acquisition of HIV infection.
2. Physical examination for signs and symptoms related to HIV infection.
3. Laboratory screening for antibody to HIV.

The testing algorithm for HIV-antibody assays[115] requires an initial EIA test and, if reactive, a retest on the same specimen. Because of the time constraints in some situations, the sample should be set in triplicate in the initial EIA. A repeatedly reactive result (positive screening test) is defined as reactivity above the test cut-off in two or more of the three assays. In order to avoid false-negative results, HIV-antibody testing should be performed on a blood sample obtained before administration of any transfusions or infusions and collected as close to the time of organ/tissue procurement as possible.[115] If a pre-transfusion/infusion sample was, however, not obtained or has been exhausted, post-transfusion/infusion specimens may be considered for testing only after having been assessed for evidence of dilution.[115] The use of new assays, such as the HIV p24-antigen and PCR, should be considered as they may identify the rare donors who are HIV-infected, and yet antibody-negative. However, studies on the utility of these tests are limited,[116,117] and currently no definitive recommendation can be made.[115]

Although some methods of tissue processing seem to prevent HIV transmission, more definitive studies are needed to ensure that these inactivating or sterilizing agents render the tissue safe for transplantation without impairing its functional and structural integrity.[115] No technique for inactivating the virus in whole organs is currently available.[115] Until more is known in this area, it is advisable to use processing techniques for bone and bone fragment that have been shown to limit the risk of HIV transmission and carefully evacuate all marrow components from whole bone whenever feasible.[115]

HEPATITIS B VIRUS

Hepatitis B virus (HBV) is a DNA virus, member of the hepadnavirus family. Currently, the most important mode of transmission is sexual contact.[118,119] The parenteral route is also common – the risk of acquiring the HBV from an HBV-contaminated needle is as high as 78%.[120] Horizontal transmission among household contacts of HBV carriers has also been observed.[121,122] However, transmission via casual contact or food has never been documented.

Three antigen–antibody systems are closely associated with hepatitis B virus infection.[123] Hepatitis B surface antigen (HBsAg) is a protein of the viral surface coat. Several subtypes have been identified. The presence of HBsAg in the serum provides evidence of HBV infection and implies infectivity of the blood. The presence of antibody to HBsAg (anti-HBs) indicates past infection with HBV or immune response from HBV vaccine. It may also represent passive antibody from hepatitis B immune globulin (HBIG). Hepatitis B e antigen (HBeAg) is found only in HBsAg positive individuals. It correlates with HBV replication and therefore indicates higher infectivity of serum. The presence of antibody to HBeAg (anti-HBe) in serum of HBsAg carrier suggests lower titre of HBV and lesser infectivity. Hepatitis B core antigen (HBcAg) is associated with the viral inner core. IgM class antibody to HBcAg (anti-HBc) indicates recent infection with HBV, while IgG class anti-HBc is a marker of past infection with HBV at some undefined time.

Transmission by transplantation

Hepatitis B is transmitted by organ transplantation. The risk of transmission depends on the serological status of both donor and recipient. Wolf et al.[124] reported virological outcomes among five HBsAg negative/anti-HBs negative recipients of kidneys from three HBsAg positive cadaver donors, two of whom were also HBeAg positive. The single recipient from the first HBsAg positive/HBeAg positive donor had anti-HBc antibody prior to transplantation and did not test positive for HBsAg in the post-transplant period. Both recipients of organs from the second HBsAg positive/HBeAg positive donor tested positive for HBsAg after transplantation and the HBV subtypes in the recipients were identical to that in the donor. The third donor was HBsAg positive/HBeAg negative, and neither of his two recipients tested positive for HBsAg after transplantation. However, one of the patients developed anti-HBs, and the antibody production was presumably triggered by exposure to the donor's HBs antigen. This observation clearly demonstrates that transmission of HBV infection occurs and is more likely with the use of allografts from HBsAg positive donors who are also HBeAg positive. Indeed, HBe antigenemia is a marker of active viral replication and therefore related to high infectivity.[125,126]

The kidney tissue itself is unlikely to carry HBV. The residual blood retained in the transplant is generally believed to be the vehicle of transmission. If this is true, certain techniques for handling and preservation of the kidney allografts after harvesting may have impact on the risk of the allograft infectivity by modulating the infectious burden transmitted to the recipients. In support of this possibility, is a case report by Lutwick et al.[127] wherein two kidneys were harvested from a HBsAg positive/HBeAg positive donor in whom serological results were not available at the time of organ procurement. Both kidneys were transplanted into HBsAg negative/anti-HBs negative recipients. One of the kidneys was packed in ice and transplanted six hours later. After transplantation, the recipient developed asymptomatic HBs-antigenemia. HBsAg subtyping revealed that recipient and donor had both the same subtype. The other kidney was mechanically perfused for 40 hours prior to transplantation. The recipient of this kidney subsequently tested positive for anti-HBs but never developed HBs-antigenemia. Of note, the perfusate from the kidney was also found to be positive for HBsAg. As a certain amount of infectious load is probably needed to ensure viral transmission, the continuous mechanical perfusion may have cleared some of the virus, thus reducing the infectious burden to an extent insufficient to establish infection in the recipient. Nevertheless, the production of anti-HBs in the post-transplant period indicates that the recipient was exposed to HBsAg, most probably from the donor, as no other source of infection could be identified.

Transplantation policies

Although the vast majority of organ procurement organizations do not accept kidneys from HBsAg positive donors, such a moratorium has been recently questioned. Based on currently available data it seems that only HBeAg positive carriers of HBsAg transmit hepatitis B infection to their recipients.[124,127] On the other hand, in areas endemic for hepatitis B such as Hong Kong and Saudi Arabia, the HBsAg carrier rate is so high that it is difficult to reject all HBsAg-positive donors (10% of all potential donors are HBsAg positive), especially when they are HLA-matched living relatives.[128–130] At the same time, the prevalence of naturally occurring immunity to hepatitis B virus infection among the adult population in these areas is as high as 40% to 50%, which makes it easy to find immune patients who could potentially receive allografts from HBsAg positive donors with no or low risk of acquiring new infection with the donor strain.[130,131] Therefore, the potential exists for the use of kidneys from HBsAg positive/HBeAg negative donors in HBsAg negative patients who are immune to hepatitis B virus. Chan and Chang[128] reported four HBsAg negative recipients of kidneys from living related HBsAg positive/HBeAg negative donors. All but one of the recipients had serologic evidence of exposure to HBV prior to transplantation. Hyperimmune gammaglobulin was given to all patients at the time of transplantation. None of the recipients tested positive for HBsAg or manifested any evidence of hepatic dysfunction after transplantation. In a similar study, Al-Khader et al.[129] reported post-transplantation outcomes in three HBsAg negative recipients of renal allografts from HBsAg positive/HBeAg negative donors (two cadaver and one living related). Post-transplantation, all three patients remained negative for HBsAg and had normal liver function. Hyperimmune gammaglobulin and a booster dose of recombinant hepatitis B vaccine (to boost the pre-existing immunity) were simultaneously administered to all patients at the time of transplantation. In another study,[132] ten out of eleven recipients of kidneys from HBsAg positive donors were HBsAg negative with pre-existing immunity due to a previous contact with the virus or active immunization with subsequent adequate antibody response. Only one recipient was HBsAg positive at the time of transplantation. All ten HBsAg negative recipients remained HBsAg negative after transplantation and no case of active HBV disease was observed, including in the HBsAg positive patient. The allograft apparently did not transmit HBV to any of the recipients.

In conclusion, HBV can be transmitted via organs donated from HBsAg positive donors who are simultaneously positive for HBeAg. However, kidneys from HBsAg donors who are negative for HBeAg may be considered for transplantation in HBsAg negative recipients with pre-existing immunity to naturally acquired hepatitis B, or patients who manifest an adequate antibody response to hepatitis B vaccination. In both these groups peri-operative hyperimmune globulin is recommended.

Since HBsAg carriage can be associated with delta agent superinfection,[133] simultaneous transmission of delta agent with HBsAg can occur. Lloveras et al.[134] reported transmission of delta agent to two healthy HBsAg carriers by kidneys procured from a single HBsAg positive donor with known heroin addiction. One of the recipients developed severe hepatitis. Liver histology was consistent with severe acute hepatitis. Aminoperoxidase staining identified the presence of delta agent. Thirteen months later, anti-delta antibodies were present in the blood. The other recipient developed fulminant hepatitis 14 weeks post-transplantation and died within 48 hours. Liver histology was not reported. The authors recommend that routine serologic screening for delta infection should be performed whenever organs from HBsAg positive donors are to be used for transplantation into HBsAg positive recipients, particularly if risks factors, such as a history of drug addiction, are present in the donors.

HEPATITIS C VIRUS

Hepatitis C virus (HCV), the principal cause of parenterally transmitted non-A, non-B hepatitis, is a single stranded RNA virus belonging to the flaviviridae family.[135] The

40–60 nm virus has a lipid envelope and a viral genome of about 9,400 nucleotides.[135] The N-terminus encodes the basic nucleocapsid (C) followed by two glycoprotein domains, the envelope (E1) and second envelope/non-structural (E2/NS1) regions. Downstream to this region are the non-structural genes NS2, NS3, NS4 and NS5 respectively. Several different strains of the virus have been identified.[135,136] In general, the C, NS2 and NS4 regions are highly conserved among the different strains of the virus, and tests that employ antigens from these regions have a high sensitivity. The E1, E2/NS1 and NS5 regions are highly variable and tests directed towards these antigens have been used to identify different strains of the virus.

Tests

The first serological test for HCV was an enzyme-linked immunosorbent assay (ELISA) which detects a non-neutralizing antibody (anti-HCV) to a recombinant antigen (c100) from the non-structural region of the HCV genome (ELISA1).[137] The limitations in the sensitivity and specificity of this test[138–141] led to the development of more sensitive and specific tests which detect antibodies to multiple HCV antigens from the non-structural and core regions of the HCV genome. Currently, the more widely used screening test is a second generation ELISA (ELISA2) which detects antibodies to the c100, c200 and c22 antigens. Although the second generation anti-HCV tests are more sensitive than their predecessors, they too detect antibody to the virus, not the virus itself. Thus, they could be negative despite ongoing infection especially in the early phase of infection, and in immunosuppressed individuals.[141–143] Conversely, a positive anti-HCV test does not necessarily imply presence of infectious virus. It could also represent current infection, convalescent antibody or a false positive reaction to non-HCV antigens. In contrast, the reverse transcriptase polymerase chain reaction (PCR) can detect the presence of HCV RNA in serum shortly after acquiring HCV infection, prior to the appearance of anti-HCV, and in individuals who do not manifest an antibody response detected by the currently available tests.[141,144–146] However, the reliability of this test too is limited by false positive and false negative results. Imperfect handling and/or storage of blood samples can lead to degradation of serum HCV RNA and failure to detect ongoing infection,[147] and false positive results can occur in the presence of even minor contamination.[148] Nonetheless detection of serum HCV RNA by PCR has been used as the 'standard' to detect HCV infection.

The branched-chain DNA (bDNA) assay is a new quantitative assay for HCV RNA in which the signal/probe rather than the viral nucleic acid is amplified.[149,150] The lower limit of detection of the bDNA assay is 350,000 molecules/ml[145] compared to 2000 molecules/ml for PCR.[150] Although the bDNA assay is less sensitive than PCR, it is simple, automated, reproducible and could exclude false positive results due to contamination.[150] The development of these tests has opened avenues to study the transmission of HCV infection by organ transplantation, and the role of hepatitis C in post-transplantation liver disease.

Transmission by organ transplantation

Shortly after the introduction of the first generation anti-HCV tests, the New England Organ Bank initiated studies to evaluate the risk of transmission of HCV infection by anti-HCV positive cadaver organ donors. Stored sera from 716 consecutive cadaver organ donors between 1986–1990 (prior to the availability of anti-HCV tests) were screened for anti-HCV using the ELISA1, and 13 (1.8%) anti-HCV positive donors were identified.[142,151] Of the twenty-nine recipients of organs from these anti-HCV positive donors, fourteen (48%) developed post-transplantation non-A, non-B hepatitis, a prevalence that was seven to eight-fold higher than that among recipients of untested donors.[152,153] Post-transplantation, 67% of these recipients tested positive for anti-HCV, and 96% tested positive for HCV RNA by PCR. Among HCV RNA negative recipients of organs from HCV RNA positive donors, 100% tested positive for HCV RNA after transplantation. These observations unequivocally demonstrated the transmission of HCV by organ transplantation. Similar studies have been undertaken by other organ procurement organizations/transplant centres. Overall, among recipients of organs from anti-HCV positive donors, 35% (range 15%–100% develop post-transplant liver disease, 52% (range 32% to 100%) test positive for anti-HCV after transplantation, and 74% (range 57% to 96%) test positive for HCV RNA by PCR.[142,151,154–161]

Based on high prevalence of HCV infection among recipients of organs from anti-HCV positive donors, several organ procurement organizations adopted a policy restricting the use of anti-HCV positive donors to life-saving transplants (heart, liver or lung). Similar policies have been recommended by the US Public Health Service Inter-Agency Guidelines.[162] However, as shown above, there is a wide range in the prevalence of liver disease and HCV infection after transplantation of organs from anti-HCV positive donors. This is similar to blood donation wherein transfusion of anti-HCV positive blood products does not uniformly lead to the transmission of HCV infection.[138,140,163,164] Consequently, several authors have argued against a moratorium on renal transplantation from organ donors with a positive anti-HCV test until more information is available regarding the prevalence of viremia among cadaver organ donors, and the performance of anti-HCV tests in detecting donors at risk of transmitting the virus.[165]

The differences in the rate of transmission of HCV infection by anti-HCV positive donors reported by different centres, however could be due to several factors. First, clinical or laboratory evidence of liver disease, and testing for anti-HCV among organ transplant recipients significantly underestimates the transmission of HCV. Hence, failure to test recipients for HCV RNA at some centres could erroneously suggest a low rate of transmission of HCV infection. Second, the risk of transmission of HCV infection by anti-HCV positive cadaver organ donors could be related to the prevalence of HCV RNA among these donors. In principle, a lower prevalence of HCV RNA among anti-HCV positive cadaver organ donors at some centres could explain the lower rate of transmission of HCV by anti-HCV positive donors, reported by these centres.[154,161,166] The latter possibility is discussed further in the next section.

Prevalence of markers of HCV infection among cadaver organ donors

In a US national collaborative study sera from 3078 cadaver organ donors were tested for anti-HCV at the individual organ procurement organizations by ELISA1. The prevalence of a positive ELISA1 was 5.1% with a range from 1.5% to 16.7%. Sera from these ELISA1 positive donors and 100 randomly selected ELISA1 negative cadaver organ donors were retrieved for testing for anti-HCV by ELISA2, and HCV RNA by PCR. Based on this testing, the extrapolated prevalence of ELISA2 and HCV RNA among cadaver organ donors was 4.2% and 2.4% respectively. The prevalence of anti-HCV among cadaver organ donors was several-fold higher than the 0.6% prevalence among healthy blood donors in the US. The higher prevalence among cadaver organ donors could reflect a higher prevalence of risk factors associated with the spread of viral infections, such as unsuspected intravenous drug use or sexual promiscuity. Indeed, anti-HCV positive cadaver organ donors were more likely to be male, with a history of alcohol or drug abuse, elevated blood alcohol levels, a positive toxic screen, hepatitis B core antibody and antibody to CMV.[167] These characteristics are consistent with known epidemiologic features of populations exposed to parenterally transmitted viruses.

Screening and confirmatory testing strategies for HCV infection among cadaver organ donors

In the US national collaborative study of HCV infection in cadaver organ donors, none of the anti-HCV negative organ donors tested positive for HCV RNA, indicating a test sensitivity of the ELISA2 of 100%.[167] Consequently, the negative predictive value was 100%, and transmission of HCV infection by anti-HCV negative donors would be extremely unlikely. Therefore, a policy of screening organ donors with ELISA2, and discarding those that test positive for anti-HCV could virtually eliminate transmission of HCV infection. The specificity of ELISA2 was also high (98.1%). However, because of the low prevalence of infection, the positive predictive value of ELISA2 was only 55%. Consequently, this policy would lead to waste of organs from 45% of donors with a positive test, or 1.8% of all donors.

One strategy to minimize waste would be to identify clinical or laboratory characteristics that differentiate anti-HCV positive donors with and without HCV RNA. Such a strategy could avoid wasting organs from the 1.8% of donors who are anti-HCV positive but HCV RNA negative. Some investigators have suggested that anti-HCV positive donors without history of drug abuse or homosexual lifestyle, absence of anti-HBs or anti-HBc, and normal serum alanine aminotransferase levels are at low risk of transmitting disease and could hence be used for transplantation.[161,166] However, in the US national collaborative study, these characteristics did not distinguish anti-HCV positive donors with and without serum HCV RNA, suggesting that patients with positive tests for anti-HCV but negative for HCV RNA, indeed had HCV infection in the past rather than being false positive. These data indicate that risk factors for HCV infection cannot differentiate between anti-HCV positive donors with and without serum HCV RNA. Thus, there are no 'low-risk' anti-HCV positive cadaver organ donors.

Another strategy to minimize waste would be to develop confirmatory tests for use in donors with a positive screening test. However, in the US national collaborative study, the RIBA2, which has been suggested as a confirmatory test in blood donors, was not specific enough to distinguish ELISA2 positive organ donors with and without serum HCV RNA. Hence, newer confirmatory tests with an even greater specificity need to be developed in order to reduce organ waste.

Transplantation of kidneys from anti-HCV positive donors into recipients with pre-transplantation HCV infection

If anti-HCV positive but HCV RNA negative donors could be identified and utilized, 2.4% of cadaver organ donors that test positive for serum HCV RNA by PCR would still remain unsuitable for transplantation of non-life saving organs. Hence, several authors have suggested the use of kidneys from anti-HCV positive donors in recipients with pre-existing HCV infection.[168,169] This practice could potentially eliminate both the need for confirmatory testing in donors with anti-HCV and discarding of organs from anti-HCV positive donors. However, studies in a post-transfusion hepatitis C model in chimpanzees have demonstrated that previous infection with HCV did not protect from reinfection with a different strain or even the same strain of the virus.[136] In fact, repeat exposure did not protect against reappearance of viremia, biochemical or histological evidence of liver disease. Similar data are not available from clinical studies in humans. In a preliminary report of a prospective study from Spain, there were no differences in the post-transplantation prevalence of liver disease, graft or patient survival between anti-HCV positive renal transplant recipients who received organs from anti-HCV positive vs negative donors.[169] These data suggest that superinfection may not have serious clinical consequences in humans. However, at the current time, there are insufficient data to fully evaluate this strategy. Until the completion of clinical trials to evaluate such strategies, we consider transplantation of organs from anti-HCV positive donors into anti-HCV positive recipients to be an experimental treatment.

CREUTZFELDT–JAKOB DISEASE

Creutzfeldt–Jakob disease (CJD) is a progressive, inevitably fatal, slow virus encephalopathy, which occurs in mid-life and affects both sexes. It is characterized by very long incubation period, protracted prodromal period of vague symptoms, dementia, and relentlessly progressive course. The disease has a worldwide yearly incidence of about one case per million population in countries where it has been vigorously sought.[170–172] The natural mechanism of spread and the reservoir of the CJD virus are unknown.

Transmission by transplantation

Animal experimental studies have demonstrated that corneal epithelia from hamsters and guineapigs, as well as suspension of brain from guineapig, were infectious and able to transmit mink encephalopathy (a slow virus animal model for Creutzfeldt–Jakob disease) and the Creutzfeldt–Jakob agent.[173–175] However, the infection acquired in this way appears to be a less fulminant form of experimental Creutzfeldt–Jakob disease.[174] Duffy et al.[176] reported a fatal case of CJD presumably acquired by a corneal transplant from a donor with initially unrecognized infection. Transmission by cadaveric dural matter graft has been also well documented.[177] Several cases of Creutzfeldt–Jakob disease have been observed following administration of purified growth hormone.[178–181] The latter was prepared from human pituitary glands collected at autopsy from patients free of systemic infections but not necessarily of degenerative neurologic diseases.[178–180] It is possible that pituitary glands, or fragments of hypothalamus adjacent to them, from patients with unrecognized Creutzfeldt–Jakob disease contaminated batches of purified growth hormone. Thus, the disease may have been transmitted by human growth hormone administration.

Transplantation policies

Awareness of Creutzfeldt–Jakob disease is important in selecting potential cadaveric donors for tissue and solid organ donation as the virus has been found widely distributed in the cerebrospinal fluid, cornea, liver, kidney and lung of animals and humans with the infection.[181–183] Therefore, patients with Creutzfeldt–Jakob disease, dying of the disease itself or of any other cause, are unacceptable for organ donation. Because of the years-long incubation period and the vague prodromal clinical presentation, it becomes extremely difficult and even impossible to identify the infection in a potential cadaver donor. That is why, any adult who eventually dies with the clinical picture of unclear neurologic disorder and/or with rapidly progressive dementia in the absence of a space-occupying intracranial lesion should be suspected of having Creutzfeldt–Jakob disease.

Transmission of parasites

Transmission of parasites by the allograft occurs less frequently. As an example, we will review transmission of *Toxoplasma gondii*, which can cause a serious disease.

TOXOPLASMA GONDII (T. GONDII)

In the immunologically competent subject, acquired toxoplasmosis is usually an asymptomatic or mild disease, sometimes recognizable only by serological tests. In the absence of serological diagnosis, latent infection in a potential donor will commonly remain unrevealed. The prevalence of antitoxoplasma antibodies indicates that 17%–50% of the adult population in the United States may have been infected with *Toxoplasma gondii* at some point of time during life.[184] In the immunosuppressed patient, toxoplasmosis frequently causes fulminant, disseminated and often fatal disease.[185] *T. gondii* is typically transmitted by the oral route. However, transmission by blood products has been documented in seronegative recipients of leukocyte transfusions from donors with serologic evidence of recently acquired toxoplasma infection.[186]

Transmission by transplantation

Transmission of *T. gondii* by organ transplantation has been reported. Reynolds et al.[187] reported a case of disseminated toxoplasmosis following renal transplantation in an allograft recipient of a living related donor. The recipient was seronegative for toxoplasma antibody before and immediately after transplantation. However, within one month post-transplantation, he seroconverted and developed a fatal disease. At autopsy, *T. gondii* was present in the brain, myocardium, skeletal muscle, liver, ureter, testis, parathyroid, and the native kidneys, but could not be found in the transplanted kidney. Although the serological status of the donor was not known, the absence of antibody in the recipient's pre-transplant sera and the appearance and rise in antibody titre after transplantation, suggest that the patient could have been infected with *T. gondii* at, or about the time of transplantation. In another report,[188] acute toxoplasmosis was observed in a pair of renal allograft recipients from a single donor who was retrospectively found to have evidence of *T. gondii* infection. One recipient died, and the autopsy revealed parasites in the myocardium, lungs, liver and bone marrow. However, *T. gondii* could not be identified in the transplanted kidney. The recipient's pre-transplant status was unknown, but ante-mortem serum, collected 25 days after transplantation, was found to have high titre of IgM antibody to *T. gondii*, thus providing evidence of a new infection which coincided with transplantation. The other recipient tested negative for *T. gondii* antibody at two weeks after renal transplantation, but was found to be positive at eight weeks and a blood specimen collected at this time grew *T. gondii*. During this period the patient was constantly hospitalized and there was no apparent exposure to a possible source of toxoplasma infection other than the transplanted kidney. Transplant nephrectomy was performed, and the patient was treated with intravenous trimethoprim-sulfamethoxazole, followed by oral pyrimethamine and sulfadiazine. The failure of the pathologic examination to reveal *T. gondii* in the transplanted kidneys of these patients does not preclude transmission by the renal allografts as thorough microscopic study was not performed on either of the transplants. In addition, the vehicle of transmission may have been donor leukocytes entrapped in the allografts.

Probable transmission of *T. gondii* by cardiac transplantation has been reported by Ryning et al.[189] They observed occurrence of toxoplasmosis in two cardiac allograft recipients who were seronegative for toxoplasma antibodies before transplantation. Both of them received organs from donors with serologic evidence of recently acquired toxoplasma infection. Oral transmission and transmission by blood transfusions were excluded. Consequently, it was believed that the donor hearts were the vector of transmission. Both heart recipients developed severe infection with *T. gondii*, and one of them died. The donor of this recipient was a multiorgan donor and his kidneys were transplanted into two patients. Eight weeks after transplantation, neither kidney recipient had clinical or serological evidence of acute toxoplasmosis. This case illustrates that toxoplasmosis was transmitted by the heart, but not by the kidney allografts. Hence, it appears that cardiac allograft recipients are at a higher risk of acquiring toxoplasma infection than recipients of other solid organs. This is consistent with data that during parasitemia the myocardium is preferably invaded by *T. gondii*.[190]

Clinical features

In primary infection, symptoms usually occur concomitantly with seroconversion[189] or shortly after transplantation.[187,188,191] In contrast, in one study[192] clinical evidence of disease in cardiac transplant recipients did not occur until five to ten months after seroconversion emphasizing the variability between the time of seroconversion to Toxoplasma and the onset of clinical disease. Pre-existing immunity to *T. gondii* seems to make reactivation or reinfection unlikely regardless of the current immunosuppressive state. Luft et al.[192] reported that of 19 patients with antibody to Toxoplasma before transplantation, ten had a significant increase in antibody titre after receiving a cardiac transplant either from a seronegative donor or a donor with unknown serologic status. However, none of them developed clinical illness that could be related to toxoplasma infection. The other nine recipients had no significant rise in antibody titre and no clinical signs of disease. In another study,[191] only one out of 39 seropositive recipients developed mild clinical disease that could be attributed to infection with *T. gondii*. The serologic status of the donors in this group is not known. These reports emphasize the differences in the clinical presentations of newly acquired infection from an exogenous source and reactivation of latent endogenous infection. They also indicate that pre-existing immunity might be protective against developing infection with the donor parasite – reinfection. However, in contrast to the above data, parasitemia with *T. gondii* and occurrence of severe clinical disease as a result of reactivation of latent infection has been reported in bone marrow transplant recipients.[193]

Pyrimethamine prophylaxis has been used to protect seronegative recipients of organs from seropositive donors. Hakim et al.[191] reported that none of the seven seronegative patients who receive hearts from seropositive donors developed primary infection with *T. gondii* when they were given oral pyrimethamine as a single daily dose of 25 mg for six weeks. In contrast, clinical disease developed in four and death occurred in two out of seven seronegative cardiac recipients of seropositive donors who did not receive prophylaxis with pyrimethamine. However, prophylaxis in recipients with pre-existing infection is probably not necessary as none of ten patients with serological evidence of reactivation, who were briefly treated (less than one week) or untreated, developed serious disease due to toxoplasmosis.

In conclusion, because of the high risk of severe disease caused by *T. gondii* in seronegative patients receiving an allograft from a seropositive donor, it is advisable that both donors and recipients be routinely screened for toxoplasma antibody. Potential donors with evidence of acute toxoplasma infection are unacceptable for donation. Unfortunately, decisions based on serological findings are often precluded by the short time period between organ harvesting and transplantation which is not enough to perform serologic testing. In case a previously seronegative patient receives an allograft from a donor who is retrospectively found to be seropositive, the recipient should be followed closely for evidence of seroconversion to *T. gondii*. However, the appearance of toxoplasma antibody does not mean disease, and the time interval between seroconversion and clinical disease may be quite long. It seems reasonable to administer pyrimethamine prophylaxis as a single oral dose of 25 mg daily for six weeks to all seronegative recipients of seropositive donors and treat all patients who develop infection with *T. gondii* with oral pyrimethamine 25 mg twice daily and spiramycine (a macrolide antibiotic) 1 g twice daily immediately on seroconversion.[194] In addition, it is recommended to give initially sulphadiazine 1 g four times daily intravenously and replace it later with sulphatriad (a mixture of three sulphonamides: sulphadiazine 185 mg, sulphamerazine 130 mg, and sulphathiazole 185 mg) three times daily.[194]

Transmission of malignancy by the allograft

Inadvertent transmission of malignancy by tissue and organ transplantation has been observed wherein the tumour that developed in the recipient was histologically identical to the neoplasm in the donor. Donor origin malignant cells are antigenically different from the host, and therefore, are submitted to immunologic attack and elimination when transferred to a healthy subject. However, the compromised immunity in the transplant recipient may impair an appropriate response. Hence, potential donors with disseminated cancers are generally unsuitable, except for procurement of the cornea, while those with localized malignancy were initially accepted for organ and tissue donation.[195]

Transmission of malignancy by organ transplantation has been observed, albeit occasionally. In one instance,[196] epidermoid carcinoma occurred in a kidney harvested from a donor with bronchogenic carcinoma with cerebral metastases. Since histopathology of the recipient tumour was indistinguishable from that of the donor, the neoplasm was considered transmitted by the allograft. Discontinuation of immunosuppression, in association with irradiation and transplant nephrectomy resulted in complete remission. However, rejection of the transplanted tumour cells does not always follow cessation of immunosuppressive therapy. McPhaul and McIntosh[197] reported the occurrence of anaplastic carcinoma in a kidney recipient from a cadaveric donor with squamous-cell carcinoma of the pyriform sinus. Discontinuation of immunosuppression did not prevent the rapidly fatal course of the malignancy.

The Cincinnati Transplant Tumor Registry[198] includes 113 recipients of renal, cardiac, pancreatic and hepatic allografts, procured from living and cadaver donors who had cancer at the time of, within five years before, or within 18 months after organ procurement. Malignancies in the recipients, which were histologically identical to those in the corresponding donor, were identified as transplanted tumours. Fifty-two recipients (46%) developed transplanted malignancies through various periods of time ranging from days up to 38 months after transplantation. Tumours were either confined to the allograft, in some cases invading the surrounding structures, or had distant metastases. Immunosuppressive therapy was discontinued in 13 patients which, in association with reduction of the tumour burden by graft removal, resulted in complete disappearance of disseminated malignancy in five recipients. However, cessation of immunosuppression and concurrent additional interventions (graft nephrectomy, chemotherapy, immunotherapy and local radiotherapy) had no effect in the other eight, and resulted in an early fatal outcome.

The findings of the Cincinnati Transplant Tumor Registry emphasize that it is mandatory to screen potential donors for possible neoplasm and exclude those with cancers. However, donors with low-grade skin cancers and primary brain tumours are usually considered acceptable. In renal transplantation, careful gross examination of the harvested kidneys is necessary. A biopsy must be performed whenever a suspicious nodule is present. If a cancer is diagnosed, the neoplasm needs to be widely excised. If large excision is feasible, this kidney can be safely transplanted as proven by the lack of tumour recurrence in all seven patients in this registry. However, such recipients must be carefully observed for signs of malignancy. A grafted kidney from a cadaver donor, whose autopsy reveals a previously unsuspected but widely spread cancer, should be removed. If, however, left in place, the patient needs careful and frequent evaluation. If a transplant cancer subsequently develops, the allografted kidney should be removed and immunosuppressive therapy discontinued immediately. In case the tumour regresses and ultimately undergoes complete remission, further renal transplantation is possible but should be delayed until the patient has been free of cancer for at least one year.

The Cincinnati Transplant Tumor Registry data, however, does not include donors with primary neoplasms confined to the brain. As we already noted, due to the extremely low incidence of systemic spread from primary central nervous system (CNS) neoplasms, donors with these malignancies are generally acceptable. However, the safety of this practice and the suitability of such donors seems to be questionable.[199-201] Extraneural spread of primary CNS neoplasms can occur in as many as 2.3% of cases.[202] Likewise, the overall incidence of transmission of primary CNS malignancies by the allograft has been recently estimated at 3.0%.[201] The histopathology of the malignant primary CNS tumour has been long considered the most important predictor of extraneural spread. Glioblastoma multiforme and medulloblastoma are the two cell types that most readily spread outside the CNS. Lefrancois et al.[199] have reported transmission of malignancy from a single donor with medulloblastoma to three organ recipients (heart, pancreas and kidneys). Transmission of glioblastoma multiforme in both kidney recipients of a single donor has been also observed.[201] In contrast, no transmission of malignancy has occurred from a cadaver renal donor with malignant astrocytoma.[203] The cell type and the grade of malignancy of the tumour could potentially determine the risk of transmission by transplantation. However, the more aggressive the malignancy, the more frequently it will require interventions such as craniotomy and ventriculosystemic shunts, which themselves are associated with high risk of extraneural spread. Therefore, it is difficult to ascertain the real significance of histology alone.[201,203] The history of craniotomy is now believed to be the only identified risk factor for the extraneural spread of a primary CNS malignancy. This has certainly an impact on organ procurement policies as usually most of the potential donors with primary CNS malignancy have undergone craniotomy by the time of donation and therefore are at a higher risk to transmit tumour cells by the harvested organs. However, in the report of Delhey et al.[203] tumour transmission did not occur despite the presence of a previous craniotomy in the donor.

Ventriculosystemic shunts have been incriminated as a possible route for metastatic spread of CNS malignancies. However, large studies have reported that approximately 90% of the cases with extraneural spread occur in the absence of a shunt and that the pattern, timing, and incidence of spread is generally the same with or without a shunt.[204,205] Although the shunt does not appear to alter the incidence of extraneural dissemination of the primary CNS malignancy, its absence in no way provides security against a possible spread.[201,205] In many cases of transplanted malignancy from a donor with a primary CNS tumour, a shunt was absent,[201] while at the same time donors with shunts have donated organs with no subsequent transmission of malignancy to the recipients.[203] Therefore, a shunt by itself does not appear to be a reliable predictor of transmission of malignancy by organ donation. Nevertheless, some authors consider its presence as a contraindication to donation.[199]

It has been suspected that the low incidence of extraneural spread of primary CNS tumours could be due, at least in part, to the generally short survival of these patients. Prolonged duration of the disease and significant meningeal involvement obviously increase the risk of metastatic spread of the malignancy thus increasing the risk of transmission of the malignancy by organ transplantation.[200]

Transplantation policies

In almost every instance of malignancy transplanted via the allograft from a donor with primary CNS malignancy there has been more than one risk factor in the donor – high grade malignancy, ventriculosystemic shunt, craniotomy, long duration of the disease, and extensive infiltration of the surrounding brain.[199–201] Consequently, it appears appropriate that donors with CNS malignancy and multiple risk factors for extraneural spread should be excluded.

Conclusions and summary

With the increasing waiting lists for organ transplantation, the need to include marginal donors is constantly growing. However, the risk of using organs from donors with infections or malignancies has to be weighed against the benefit of transplantation on patient survival. Guidelines are constantly changing reflecting new data. We recommend a careful clinical history and examination of the donor and absolute exclusion of donors who:

1. Test positive for HIV.
2. Have positive blood cultures (excluding skin contamination) within the past seven days.
3. Have had intravenous drug abuse within the past two years.
4. Have had malignancy other than primary intracranial tumour or basal cell or localized squamous cell carcinoma of skin within the past five years.

In primary brain tumour with multiple risk factors, only eye donation is acceptable.

In cases where donors are seropositive for HBV and HCV, we would suggest that the decision to reject or accept an organ should be based on an appropriately informed discussion between the potential recipient and the physician.

ACKNOWLEDGMENT

The authors wish to acknowledge the contribution of Andrew S. Levey, MD in the formulation of several of the policies advocated in this chapter. The authors also acknowledge the support from the New England Organ Bank, Newton, MA, Dialysis Clinics Inc., Nashville, TN, American Gastroenterology Association/Smith Kline Beecham Clinical Research Award (to Dr Pereira), American Society of Transplant Physicians/Ortho Grant-in-Aid Award in Transplantation (to Dr Pereira), a National Kidney Foundation Young Investigator Award (to Dr Pereira), and the Nephrology Clinical Research Fellowship of New England Medical Center and St Elizabeth's Hospital, Boston (to Dr Natov).

References

1. Bijnen AB, Weimar W, Dik P, Oberop H, Jeekel J. The hazard of transplanting contaminated kidneys. *Transplant Proc* 1984, **16**: 27–28.
2. Slapak M. The immediate care of potential donors for cadaveric organ transplantation. *Anaesthesia* 1978, **33**: 700–709.
3. Spees EK Jr, Light JA, Oakes DD, Reinmuth B. Experiences with cadaver renal allograft contamination before transplantation. *Br J Surg* 1982, **69**: 482–485.
4. Harrington JC, Bradley JW, Zalneraitis, Cho SI. Relevance of urine cultures in the evaluation of potential cadaver kidney donors. *Transplant Proc* 1984, **16**: 29–30.
5. Odenheimer DB, Matas AJ, Tellis VA, Quinn T, Glicklich D, Soberman R, Veith F. Donor cultures reported positive after transplantation: A clinical dilemma. *Transplant Proc* 1986, **18**: 465–466.
6. Gottesdiener KM. Transplanted infections: donor-to-host transmission with the allograft. *Ann Intern Med* 1989, **110**: 1001–1016.
7. Eastlund T. Infectious disease transmission through tissue transplantation: Reducing the risk through donor selection. *J Transplant Coordination* 1991, **1**: 23–30.
8. McCoy GC, Loening S, Braun WE, Magnusson MO, Banowsky LH, McHenry MC. The fate of cadaver renal allografts contaminated before transplantation. *Transplantation* 1975, **20**: 467–472.
9. Doig RL, Boyd PJR, Eykyn S. Staphylococcus aureus transmitted in transplanted kidneys. *Lancet* 1975, **ii**: 243–245.
10. Rubin RH, Tolkoff-Rubin NE. Infection: The new problems. *Transplant Proc* 1989, **21**: 1440–1445.
11. Tolkoff-Rubin NE, Rubin RH. The interaction of immunosuppression with infection in the organ transplant recipient. *Transplant Proc* 1994, **26** (Suppl 1): 16–19.
12. Anderson CB, Haid SD, Hruska KA, Etheredge EA. Significance of microbial contamination of stored cadaver kidneys. *Arch Surg* 1978, **113**: 269–271.
13. Soifer BE, Gelb AW. The multiple organ donor: Identification and management. *Ann Intern Med* 1989, **110**: 814–823.
14. Mathews REF. Classification and nomenclature of viruses. Third report of the International Committee on Taxonomy of Viruses. *Intervirology* 1979, **12**: 132–296.
15. DeMarchi JM, Blankenship ML, Brown GD, Kaplan AS. Size and complexity of human cytomegalovirus DNA. *Virology* 1978, **89**: 643–646.
16. Chee MS, Bankier AT, Beck S, Bohni R, Brown CM, Cerny R, Horsnell T, Hutchison CA III, Kouzarides T, Martignetti JA, Preddie E, Satchwell SC, Tomlinson O, Weston KM, Barrel BG. Analysis of the protein-coding content of the sequence of human cytomegalovirus strain AD 169. *Curr Topic Microbiol Immunol* 1990, **154**: 126–169.
17. Weller TH, Hanshaw JB, Scott DE. Serologic differentiation of viruses responsible for cytomegalic inclusion disease. *Virology* 1960, **12**: 130–132.

18. Huang ES, Kilpatrick BA, Huang YT, Pagano JS. Detection of human cytomegalovirus and analysis of strain variation. *Yale J Biol Med* 1976, **49**: 29–43.
19. Zablotney SL, Wentworth BB, Alexander ER. Antigenic relatedness of 17 strains of human cytomegalovirus. *Am J Epidemiol* 1978, **107**: 336–343.
20. Ho M. Epidemiology of cytomegalovirus infection in man. In: Ho M (Ed.). *Cytomegalovirus: biology and infection*. Plenium Medical Book Company, New York, 1991, pages 155–187.
21. Wenzel RP, McCormick DP, Davis JA, Davies JA, Berling C, Beam WE Jr. Cytomegalovirus infection: A seroepidemiologic study of a recruit population. *Am J Epidemiol* 1973, **97**: 410–414.
22. Grundy JE, McKeating JA, Ward PJ, Sanderson AR, Griffiths PD. Beta 2 microglobulin enhances the infectivity of cytomegalovirus and when bound to the virus enables class I HLA molecules to be used as a virus receptor. *J Gen Virol* 1987, **68**: 793–803.
23. Taylor HP, Cooper NR. Human cytomegalovirus binding to fibroblasts is receptor mediated. *J Virol* 1989, **63**: 3991–3998.
24. Adlish JD, Lahijani RS, St Jeor SC. Identification of a putative cell receptor for human cytomegalovirus. *Virology* 1990, **176**: 337–345.
25. Wiley CA, Nelson JA. Role of human immunodeficiency virus and cytomegalovirus in AIDS encephalitis. *Am J Pathol* 1988, **133**: 73–81.
26. Toorkay CB, Carrigan DR. Immunohistochemical detection of an immediate early antigen of human cytomegalovirus in normal tissue. *J Infect Dis* 1989, **160**: 741–751.
27. Hendrix MG, Daemmer M, Bruggeman CA. Cytomegalovirus nucleic acid distribution within the human vascular tree. *Am J Pathol* 1991, **138**: 563–567.
28. Ho M. Virological diagnosis and infections in cells and tissues. In Ho M (Ed.). *Cytomegalovirus: biology and infection*. Plenium Medical Book Company, New York, 1991, pages 75–99.
29. Ho M. Serologic tests for cytomegalovirus infection. In Ho M (Ed.) *Cytomegalovirus: biology and infection*. Plenium Medical Book Company, New York, 1991, pages 101–126.
30. Shuster EA, Beneke JS, Tegtmeier GF, Pearson GR, Gleaves CA, Wold AD, Smith TF. Monoclonal antibody for rapid laboratory detection of cytomegalovirus infections: Characterization and diagnostic application. *Mayo Clinic Proc* 1985, **60**: 577–585.
31. Lui SF, Sweny P, Grundy JE, Blaxill A, Griffiths PD, Moorhead JF, Fernando ON. The clinical application of a rapid diagnostic test for the detection of cytomegalovirus infection in renal transplant recipients. *Transplant Proc* 1987, **19**: 2129–2130.
32. Paya CV, Wold AD, Smith TF. Detection of cytomegalovirus infections in specimens other than urine by the shell viral assay and conventional tube cultures. *J Clin Microbiol* 1987, **25**: 755–757.
33. Schirm J, Timmerije W, van der Bij W, The TH, Wilterdink JB, Tegzess AM. Rapid detection of infectious cytomegalovirus in blood with the aid of monoclonal antibodies. *J Med Virol* 1987, **23**: 31–40.
34. Agut H, Deny P, Rousseau E, Garbarg-Chenon A, Beliveau C, Chabolle F, Chouard H, Bricout F, Nicolas JC. DNA hybridization for detection of human cytomegalovirus in bronchoalveolar lavage and pharynx biopsy. *Ann Inst Pasteur Virol* 1988, **139**: 101–111.
35. Chou S, Roark L, Merigan TC. DNA of 21 clinical cytomegalovirus strains detected by hybridization to cloned DNA fragments of laboratory strain Ad-169. *J Med Virol* 1984, **14**: 263–268.
36. Hilborne LH, Nieberg RK, Cheng L, Lewin KJ. Direct *in situ* hybridization for rapid detection of cytomegalovirus in bronchoalveolar lavage. *Am J Clin Pathol* 1987, **87**: 766–769.
37. Masih AS, Linder J, Shaw BW Jr, Wood RP, Donovan JP, White R, Markin RS. Rapid identification of cytomegalovirus in liver allograft biopsies by *in situ* hybridization. *Am J Surg Pathol* 1988, **12**: 362–374.
38. Chou S, Merigan TC. Rapid detection and quantitation of human cytomegalovirus in urine through DNA hybridization. *N Engl J Med* 1983, **308**: 921–925.

39. Stulburg CS, Zuelzer WW, Page RH, Taylor PE, Brough. Cytomegalovirus infection with reference to isolations from lymph nodes and blood. *Proc Soc Exp Biol Med* 1966, **123**: 976–982.
40. Lang DJ, Hanshaw JB. Cytomegalovirus infection and the postperfusion syndrome. Recognition of primary infections in four patients. *N Engl J Med* 1969, **280**: 1145–1149.
41. Tamura T, Chiba S, Chiba Y, Nakao T. Virus excretion and neutralizing antibody response in saliva in human cytomegalovirus infection. *Infect Immun* 1980, **29**: 842–845.
42. Waner JL, Hopkins DR, Weller TH, Alfred EN. Cervical excretion of cytomegalovirus: Correlation with secretory and humeral antibody. *J Infect Dis* 1977, **136**: 805–809.
43. Horodniceanu F, Michelson S. Assessment of human cytomegalovirus antibody detection techniques. *Arch Virol* 1980, **64**: 287–301.
44. Booth JC, Hannington G, Bakir TM, Stem H, Kango H, Griffiths PD, Heath RB. Comparison of enzyme-liked immunosorbent assay, radioimmunoassay, complement fixation, anticomplement immunofluorescence and passive haemagglutination techniques for detecting cytomegalovirus IgG antibody. *J Clin Pathol* 1982, **35**: 1345–1348.
45. McHugh TM, Casavant CH, Wilbert JC, Stites DP. Comparison of six methods for the detection of antibody to cytomegalovirus. *J Clin Microbiol* 1985, **22**: 1014–1019.
46. Colimon R, Michelson S. Human cytomegalovirus: pathology, diagnosis, treatment. *Adv Nephrol* 1990, **19**: 333–356.
47. Nelson JM, Smith KO. Examination of adult primary kidney cell cultures for adventitious viral agents. *Proc Soc Exp Biol Med* 1971, **137**: 1253–1257.
48. Lopez C, Simmons RL, Mauer SM, Najarian JS, Good RA, Gentry S. Association of renal allograft rejection with viral infections. *Am J Med* 1974, **56**: 280–289.
49. Naraqi S, Jackson GG, Jonasson O, Yamashiroya HM. Prospective study of prevalence, incidence, and source of herpesvirus infections in patients with renal allografts. *J Infect Dis* 1977, **136**: 531–540.
50. Balfour HH Jr, Slade MS, Kalis JM, Howard RJ, Simmons RL, Najarian JS. Viral infections in renal transplant donors and their recipients: A prospective study. *Surgery* 1977, **81**: 487–492.
51. Orsi EV, Howard JL, Baturay N, Ende N, Ribot S, Eslami H. High incidence of virus isolation from donor and recipient tissues associated with renal transplantation. *Nature* 1978, **272**: 372–373.
52. Winston DJ, Ho WG, Howell CL, Miller MJ, Mickey R, Martin, WJ, Lin C-H, Gale, RP. Cytomegalovirus infections associated with leukocyte transfusions. *Ann Intern Med* 1980, **93**: 671–675.
53. Prince AM, Szmuness W, Millian SJ, David DS. A serological study of cytomegalovirus infections associated with blood transfusions. *N Engl J Med* 1971, **284**: 1125–1131.
54. Armstrong JA, Tarr GC, Youngblood LA, Dowling JN, Saslow AR, Lucas JP, Ho M. Cytomegalovirus infection in children undergoing open-heart surgery. *Yale J Biol Med* 1976, **49**: 83–91.
55. Wilhelm JA, Matter L, Schopfer K. The risk of transmitting cytomegalovirus to patients receiving blood transfusions. *J Infect Dis* 1986, **154**: 169–171.
56. Preiksaitis JK, Brown L, McKenzie M. The risk of cytomegalovirus infection in seronegative transfusion recipients not receiving exogenous immunosuppression. *J Infect Dis* 1988, **157**: 523–529.
57. Kääriäinen L, Paloheimo JA, Klemola E, Makela T, Koivuniemi A. Cytomegalogvirus mononucleosis: Isolation of the virus and demonstration of subclinical infections after fresh blood transfusion in connection with open-heart surgery. *Ann Med Exp Fenn* 1966, **44**: 297–301.
58. Pien FD, Smith TF, Anderson CF, Webel ML, Taswell HF. Herpesviruses in renal transplant patients. *Transplantation* 1973, **16**: 489–495.
59. Spencer ES. Cytomegalovirus antibody in uremic patients prior to renal transplantation. *Scand J Infect Dis* 1974, **6**: 1–4.
60. Ho M, Suwansirikul S, Dowling JN, Youngblood LA, Armstrong JA. The transplanted kidney as a source of cytomegalovirus infection. *N Engl J Med* 1975, **293**: 1109–1112.

61. Hamilton JD, Seaworth BJ. Transmission of latent cytomegalovirus in a murine kidney tissue transplantation model. *Transplantation* 1985, **39**: 290–296.
62. Preiksaitis JK, Rosno S, Grumet C, Merigan TC. Infections due to herpesviruses in cardiac transplant recipients: role of the donor heart and immunosuppressive therapy. *J Infect Dis* 1983, **147**: 974–981.
63. Tegtmeier GE. The use of cytomegalovirus-screened blood in neonates. *Transfusion* 1988, **28**: 201–203.
64. Eisenfeld L, Silver H, McLaughlin J, Klevjer-Anderson P, Mayo D, Anderson J, Herson V, Krause P, Savidakis J, Lazar A, Rosenkrantz T, Pisciotto P. Prevention of transfusion-associated cytomegalovirus infection in neonatal patients by the removal of white cells from blood. *Transfusion* 1992, **32**: 205–209.
65. Chou S, Norman DJ. The influence of donor factors other than serologic status on transmission of cytomegalovirus to transplant recipients. *Transplantation* 1988, **46**: 89–93.
66. Ho M. Advances in understanding cytomegalovirus infection after transplantation. *Transplant Proc* 1994, **26** (Suppl 1): 7–11.
67. Betts RF, Freeman RB, Douglas RG Jr, Talley TE, Rudell B. Transmission of cytomegalovirus infection with renal allograft. *Kidney Int* 1975, **8**: 387–394.
68. Fryd DS, Peterson PK, Ferguson RM, Simsons RL, Balfour HH, Najarian JS. Cytomegalovirus as a risk factor in renal transplantation. *Transplantation* 1980, **30**: 436–439.
69. Peterson PK, Balfour HH, Marker SC, Fryd DS, Howard RJ, Simmons RL. Cytomegalovirus disease in renal allograft recipients: A prospective study of the clinical features, risk factors and impact on renal transplantation. *Medicine* 1980, **59**: 283–300.
70. Glenn J. Cytomegalovirus infections following renal transplantation. *Rev Infect Dis* 1981, **3**: 1151–1178.
71. Johnson PC, Lewis RM, Golden DL, Oefinger PE, Van Buren CT, Kerman RH, Kahan BD. The impact of cytomegalovirus infection on seronegative recipients of seropositive donor kidneys versus seropositive recipients treated with cyclosporine-prednisone immunosuppression. *Transplantation* 1988, **45**: 116–121.
72. Kurtz JB, Thompson JF, Ting A, Pinto A, Morris PJ. The problem of cytomegalovirus infection in renal allograft recipients. *Quart J Med* 1984, **211**: 341–349.
73. Smiley ML, Wlodaver CG, Grossman RA, Barker CF, Perloff LJ, Tustin NB, Starr SE, Plotkin SA, Friedman HM. The role of pretransplant immunity in protection from cytomegalovirus disease following renal transplantation. *Transplantation* 1985, **40**: 157–161.
74. Weir MR, Hemy ML, Blackmore M, Smith J, First MR, Irwin B, Shen S, Genemans G, Alexander JW, Corry RJ, Nghiem DD, Ferguson RM, Kittur D, Shield CF III, Sommer BG, Williams GM. Incidence and morbidity of cytomegalovirus disease associated with a seronegative recipient receiving seropositive donor-specific transfusion and living-related donor transplantation. *Transplantation* 1988, **45**: 111–116.
75. Grundy JE, Super M, Sweny P, Moorhead J, Lui SF, Berry NJ, Fernando ON, Griffiths PD. Symptomatic cytomegalovirus infection in seropositive kidney recipients: Reinfection with donor virus rather than reactivation of recipient virus. *Lancet* 1988, **ii**: 132–135.
76. Boyce NW, Hayes K, Gee D, Holdsworth SR, Thomson NM, Scott D, Atkins RC. Cytomegalovirus infection complicating renal transplantation and its relationship to acute transplant glomerulopathy. *Transplantation* 1988, **45**: 706–709.
77. Ludwin D, White N, Tsai S, Chernesky M, Achong M, Smith EKM. Result of prospective matching for cytomegalovirus status in renal transplant recipients. *Transplant Proc* 1987, **19**: 3433–3434.
78. Waltzer WC, Arnold AN, Anaise D, Hurley S, Rainsbeck A, Egelandsdal B, Pullis C, Rapaport FT. Impact of cytomegalovirus infection and HLA matching on outcome of renal transplantation. *Transplant Proc* 1987, **19**: 4077–4080.
79. Ho M. Human cytomegalovirus infections in immunosuppressed patients. In Ho M (Ed.). *Cytomegalovirus: biology and infection*. Plenium Medical Book Company, New York, 1991, pages 249–301.

80. Wertheim P, Buurman C, Geelen J, Van der Noordaa J. Transmission of cytomegalovirus by renal allograft demonstrated by restriction enzyme analysis. *Lancet* 1983, **i**: 980–981.
81. Chou S. Acquisition of donor strains of cytomegalovirus by renal transplant recipients. *N Engl J Med* 1986, **314**: 1418–1423.
82. Grundy JE, Super M, Lui S, Sweny P, Griffiths PD. The source of cytomegalovirus infection in seropositive renal allograft recipients is frequently the donor kidney. *Transplant Proc* 1987, **19**: 2126–2128.
83. Ackermann JR, LeFor WM, Weinstein S, Kahana L, Shires DL, Tardif G, Baxter J. Four-year experience with exclusive use of cytomegalovirus antibody (CMV-Ab)-negative donors for CMV-Ab-negative kidney recipients. *Transplant Proc* 1988, **20** (Suppl 1): 469–471.
84. Davis C. The prevention of cytomegalovirus disease in renal transplantation. *Am J Kidney Dis* 1990, **16**: 175–188.
85. Tsevat J, Snydman DR, Pauker SG, Durand-Zaleski I, Werner BG, Levey AS. Which renal transplant patients should receive cytomegalovirus immune globulin? A cost-effectiveness analysis. *Kidney Int* 1991, **52**: 259–65.
86. Snydman DR, Werner BG, Heinze-Lacey B, Berardi VP, Tilney NL, Kirkman RL, Milford EL, Cho SI, Bush HL, Levey AS, Strom TB, Carpemter CB, Levey RH, Harmon WE, Zimmerman CE, Shapiro ME, Steinman T, LoGerfo F, Idelson B, Schröter GPJ, Levin MJ, McIver J, Leszczynski J, Grady GF. Use of cytomegalovirus immune globulin to prevent cytomegalovirus disease in renal-transplant recipients. *N Engl J Med* 1987, **317**: 1049–1054.
87. Snydman DR, Werner BG, Dougherty NN, Griffith J, Rohrer RH, Freeman R, Jenkins R, Lewis WD, O'Rourke E, the Boston Center for liver transplantation CMVIG-study group. A further analysis of the use of cytomegalovirus immune globulin in orthotopic liver transplant patients at risk for primary infection. *Transplant Proc* 1994, **26** (Suppl 1): 23–27.
88. Snydman DR. Cytomegalovirus prophylaxis strategies in high-risk transplantation. *Tranplant Proc* 1994, **26** (Suppl 1): 20–22.
89. Balfour HH, Chace BA, Stapleton JT, Simmons RL, Fryd DS. A randomized, placebo-controlled trial of oral acyclovir for the prevention of cytomegalovirus disease in recipients of renal allografts. *N Engl J Med* 1989, **320**: 1381–1387.
90. Wong T, Toupance O, Chanard J. Acyclovir to prevent cytomegalovirus infection after renal transplantation. *Ann Intern Med* 1991, **115**: 68.
91. Singh N, Yu VL, Mieles L, Wagener MM, Miner RC, Gayowski T. High-dose acyclovir compared with short-course preemptive ganciclovir therapy to prevent cytomegalovirus disease in liver transplant recipients. A randomized trial. *Ann Intern Med* 1994, **120**: 375–381.
92. Merigan TC, Renlund DG, Keay S, Bristow MR, Starnes V, O'Connell JB, Resta S, Dunn D, Gamberg P, Ratkovec RM, Richenbacher WE, Millar RC, DuMond C, DeAmond B, Sullivan V, Cheney T, Buhles W, Stinson EB. A controlled trial of acyclovir to prevent cytomegalovirus disease after heart transplantation. *N Engl J Med* 1992, **326**: 1182–1186.
93. Bailey TC, Trulock EP, Ettinger NA, Storch GA, Cooper JD, Powdery WG. Failure of prophylactic ganciclovir to prevent cytomegalovirus disease in recipients of lung transplants. *J Infect Dis* 1992, **165**: 548–552.
94. Martin M, Mañez R, Linden P, Estores D, Torre-Cisneros J, Kusne S, Ondick L, Ptachoinski R, Irish W, Kisor D, Felser I, Rinaldo C, Stieber A, Fung J, Ho M, Simmons R, Starzl T. A prospective randomized trial comparing sequential ganciclovir-high dose acyclovir to high dose acyclovir for prevention of cytomegalovirus disease in adult liver transplant recipients. *Transplantation* 1994, **58**: 779–785.
95. Kasiske BL, Heim-Duthoy KL, Tortorice KL, Ney AL, Odland MD, Rao KV. Polyvalent immune globulin and cytomegalovirus infection after renal transplantation. *Arch Intern Med* 1989, **149**: 2733–2736.
96. Bailey TC, Ettinger NA, Storch GA, Trulock EP, Hanto DW, Dunagan WC, Jendrisak MD, McCullough CS, Kenzora JL, Powderly WG. Failure of high dose oral acyclovir with

or without immune globulin to prevent primary cytomegalovirus disease in recipients of solid organ transplants. *Am J Med* 1993, **95**: 273–278.
97. Korsager B, Spencer ES, Mordhorst CH, Anderson HK. Herpesvirus hominis infections in renal transplant recipients. *Scan J Infect Dis* 1975, **7**: 11–19.
98. Rand KH, Rasmussen LE, Polland RB, Arvin A, Merigan TC. Cellular immunity and herpesvirus infections in cardiac-transplant recipients. *N Engl J Med* 1977, **296**: 1372–1377.
99. Pass RF, Long WK, Whitley RJ, Soong S-J, Diethelm AG, Reynolds DW, Alford CA Jr. Productive infection with cytomegalovirus and herpes simplex virus in renal transplant recipients: Role of source of kidney. *J Infect Dis* 1978, **137**: 556–563.
100. Bastian FO, Rabson AS, Yee CL, Tralka TS. Herpesvirus hominis: isolation from human trigeminal ganglion. *Science* 1972, **178**: 306–307.
101. Baringer JR. Recovery of herpes simplex virus from human sacral ganglion. *N Engl J Med* 1974, **291**: 828–830.
102. Warren KG, Brown SM, Wroblewska Z, Gilden D, Koprowski H, Subak-Sharpe J. Isolation of latent herpes simplex virus from the superior cervical and vagus ganglions of human beings. *N Engl J Med* 1978, **298**: 1068–1069.
103. Dummer JS, Armstrong J, Somers J, Kusne S, Carpenter BJ, Rosenthal JT, Ho M. Transmission of infection with herpes simplex virus by renal transplantation. *J Infect Dis* 1987, **155**: 202–206.
104. Goodman JL. Possible transmission of herpes simplex virus by organ transplantation. *Transplantation* 1989, **47**: 609–613.
105. CDC. Human immunodeficiency virus infection transmitted from an organ donor screened for HIV antibody – North Carolina. *MMWR* 1987, **36**: 306–308.
106. Kumar P, Pearson IE, Martin DH, Leech SH, Buisseret PD, Bezbak HC, Gonzales FM, Royer JR, Streicher HZ, Saxinger WC. Transmission of human immunodeficiency virus by transplantation of a renal allograft, with development of the acquired immunodeficiency syndrome. *Ann Intern Med* 1987, **106**: 244–245.
107. Simonds RJ, Holmberg SD, Hurtwitz RL, Coleman TR, Bottenfield S, Conley LJ, Kohlenberg SH, Castro KG, Dahan BA, Schable CA, Rayfield MA, Rogers MF. Transmission of human immunodeficiency virus type 1 from a seronegative organ and tissue donor. *N Engl J Med* 1992, **326**: 726–732.
108. Samuel D, Castaing D, Adam R,. Saliba F, Chamaret S, Misset JL, Montagnier L, Bismuth H. Fatal acute HIV infection with aplastic anaemia, transmitted by liver graft. *Lancet* 1988, **i**: 1221–1222.
109. Dummer JS, Erb S, Breinig MK, Ho M, Rinaldo CR Jr, Gupta P, Ragni MV, Tzakis A, Makowka L, Van Thiel D, Starzl TE. Infection with human immunodeficiency virus in the Pittsburgh transplant population: A study of 583 donors and 1043 recipients, 1981–1986. *Transplantation* 1989, **47**: 134–140.
110. Erice A, Rhame FS, Heussner RC, Dunn DL, Balfour HH Jr. Human immmunodeficiency virus infection in patients with solid-organ transplants: report of five cases and reviews. *Rev Infect Dis* 1991, **13**: 537–547.
111. CDC. Transmission of HIV through bone transplantation: case report and public health recommendations. *MMWR* 1988, **37**: 597–599.
112. Clarke JA. HIV transmission and skin grafts. *Lancet* 1987, **i**: 983.
113. Quarto M, Germinario C, Fontana A, Barbuti S. HIV transmission through kidney transplantation from a living related donor. *N Engl J Med* 1989, **320**: 1754.
114. Bowen A, Lobel S, Caruana R, Leffell MS, House MA, Rissing JP, Humphries AL, Transmission of human immunodeficiency virus (HIV) by transplantation: Clinical aspects and time course analysis of viral antigenemia and antibody production. *Ann Intern Med* 1988, **108**: 46–48.
115. CDC. Guidelines for preventing transmission of human immunodeficiency virus through transplantation of human tissue and organs. *MMWR* 1994, **43**: 1–17.
116. Pepose JS, Buerger DG, Paul DA, Quinn TC, Darragh TM, Donegan E. New developments in serologic screening of corneal donors for HIV-1 and hepatitis B virus infections. *Ophthalmology* 1992, **99**: 879–888.

117. Callaway T, McCreedy B, Pruett T. Polymerase chain reaction for HIV screening of tissue donors [Abstract #02638 (S44)]. *XIV International Congress of the Transplantation Society, Paris, France.* August 16–21, 1992.
118. Rosenblum L, Darrow W, Witte J, Cohen J, French J, Gill PS, Potterat J, Sikes K, Reich R, Hadler S. Sexual practices in the transmission of hepatitis B virus and prevalence of hepatitis Delta virus infection in female prostitutes in the United States. *JAMA* 1992, **267**: 2477–2481.
119. Kingsley LA, Rinaldo CR, Lyter DW, Valdiserri RO, Belle SH, Ho M. Sexual transmission efficiency of hepatitis B virus and HIV among homosexual men. *JAMA* 1990, **264**: 230–234.
120. Alter HJ, Seeff LB, Kaplan PM, McAuliffe VJ, Wright EC, Gerin JL, Purcell RH, Holland PV, Zimmerman HJ. Type B hepatitis: the infectivity of blood positive for e antigen and DNA polymerase after accidental needlestick exposure. *N Engl J Med* 1976, **295**: 909–913.
121. Tabor E, Bayley AC, Cairns L, Gerety RL. Horizontal transmission of hepatitis B virus among children and adults in five rural villages in Zambia. *J Med Virol* 1985, **15**: 113–120.
122. Botha JF, Ritchie MJJ, Dusheiko GM, Mouton HWK, Kew MC. Hepatitis B virus carrier state in black children in Ovamboland: Role of perinatal and horizontal infection. *Lancet* 1984, **i**: 1210–1212.
123. Locarnini SA, Gust ID. *Hepadnaviridae*: Hepatitis B virus and the delta virus. In Balow A, Hausler WJ Jr, Lennette EH (Ed.). *Laboratory diagnosis of infectious diseases. Principles and practice.* Springer-Verlag, New York, 1988, **2**: 750–796.
124. Wolf JL, Perkins HA, Schreeder MT, Vincenti F. The transplanted kidney as a source of hepatitis B infection. *Ann Intern Med* 1979, **91**: 412–413.
125. Shikata T, Karasawa T, Abe K, Uzawa T, Suzuki H, Oda T, Imai M, Mayumi M, Moritsugu. Hepatitis B e antigen and infectivity of hepatitis B virus. *J Infect Dis* 1977, **136**: 571–576.
126. Fairley CK, Mijch A, Gust ID, Nichilson S, Dimitrakakis M, Lucas CR. The increased risk of fatal liver disease in renal transplant patients who are hepatitis B e antigen and/or HBV DNA positive. *Transplantation* 1991, **52**: 497–500.
127. Lutwick LI, Sywassink JM, Corry RJ, Shorey JW. The transmission of hepatitis B by renal transplantation. *Clin Nephrol* 1983, **19**: 317–319.
128. Chan MK, Chang WK. Renal transplantation from HBsAg positive donors to HBsAg negative recipients. *Br Med J* 1988, **297**: 522–523.
129. Al-Khader AA, Dhar JM, Al-Sulaiman M, Al-Hasani MK. Renal transplantation from HBsAg positive donors to HBsAg negative recipients. *Br Med J* 1988, **297**: 854.
130. Chan PCK, Lok ASF, Cheng IKP, Chan MK. The impact of donor and recipient hepatitis B surface antigen status on liver disease and survival in renal transplant recipients. *Transplantation* 1992, **53**: 128–131.
131. Talukder MA, Gilmore R, Bacchus RA. Prevalence of hepatitis B surface antigen among male Saudi Arabians. *J Infect Dis* 1982, **146**: 446.
132. Bedrosian J, Akposso K, Metivier F, Moal MC, Pruna A, Idatte JM. Kidney transplantation with HBsAg+ donors. *Transplant Proc* 1993, **25**: 1481–1482.
133. Rizzetto M, Canese MG, Aricò S, Crinelli O, Trepo C, Bonino F, Verrne G. Immunofluorescence detection of new antigen-antibody system (δ/anti-δ) associated to hepatitis B virus in liver and in serum of HBsAg carriers. *Gut* 1977, **18**: 997–1003.
134. Lloveras J, Monteis J, Sanchez-Tapias JM, Bruguera M, Masramon J, Aubia J, Orfila A, Cuevas X. Delta agent transmission through renal transplantation with severe hepatitis induction in two HBsAg healthy carriers. *Transplant Proc* 1986, **18**: 467–468.
135. Houghton M, Weiner A, Han J, Kuo G, Choo QL. Molecular biology of the hepatitis C viruses: Implications for diagnosis, development and control of viral disease. *Hepatology* 1991, **14**: 381–388.
136. Farci P, Alter HJ, Govindarajan S, Wong DC, Engle R, Lesniewski RR, Mushahwar IK, Desai SM, Miller RH, Ogata N, Purcell RH. Lack of protective immunity against reinfection with hepatitis C virus. *Science* 1992, **258**: 135–140.

137. Kuo G, Choo QL, Alter HJ, Gitnick GL, Redeker AG, Purcell RH, Miyamura T, Dienstag JL, Alter MJ, Stevens CE, Teghmeier GE, Bonino F, Colombo M, Lee W-S, Kuo C, Berger K, Shuster JR, Overby LR, Bradley DW, Houghton M. An assay for circulating antibodies to a major etiologic virus of human non-A, non-B hepatitis. *Science* 1989, **244**: 362–364.
138. Alter HJ, Purcell RH, Shih JW, Melpolder JC, Houghton M, Choo Q-L. Detection of antibody to hepatitis C virus in prospectively followed transfusion recipients with acute and chronic non-A, non-B hepatitis. *N Engl J Med* 1989, **321**: 1494–1500.
139. Farci P, Alter HJ, Wong D, Miller RH, Shih JW, Jett B, Purcell RH. A long-term study of hepatitis C virus replication in non-A, non-B hepatitis. *N Engl J Med* 1991, **325**: 98–104.
140. Esteban JI, Gonzales A, Hernandez JM, Viladomiu L, Sanchez C, Lopez-Talavera JC, Lucea D, Martin-Vega C, Vidal X, Esteban R, Guardia J. Evaluation of antibodies to hepatitis C virus in a study of transfusion-associated hepatitis. *N Engl J Med* 1990, **323**: 1107–1112.
141. Weiner AJ, Kuo G, Bradley DW, Bonino F, Saracco G, Lee C, Rosenblatt J, Choo QL, Houghton M. Detection of hepatitis C viral sequences in non-A, non-B hepatitis. *Lancet* 1990, **335**: 1–3.
142. Pereira BJG, Milford EL, Kirkman RL, Quan S, Sayre KR, Johnson PJ, Wilber JS, Levey AS. Prevalence of HCV RNA in hepatitis C antibody positive cadaver organ donors and their recipients. *N Engl J Med* 1992, **327**: 910–915.
143. Sugitani M, Inchauspe G, Shindo M, Prince AM. Sensitivity of serological assays to identify blood donors with hepatitis C viremia. *Lancet* 1992, **339**: 1018–1019.
144. Simmonds P, Zhang LQ, Watson HG, Rebus S, Ferguson ED, Balfe P, Leadbetter GH, Yap PL, Peutherer JF, Ludlam CA. Hepatitis C quantification and sequencing in blood products, hemophiliacs, and drug users. *Lancet* 1990, **336**: 1469–1471.
145. Lau JYN, Davis GL, Orito E, Qian KP, Mizokami M. Significance of antibody to the host cellular gene derived epitope GOR in chronic hepatitis C virus infection. *J Hepatol* 1993, **17**: 253–257.
146. Ulrich PP, Romeo JM, Lane PK, Kelly I, Danial LJ, Vyas GN. Detection, semi-quantitation, and genetic variation in hepatitis C virus sequences amplified from the plasma of blood donors with elevated alanine aminotransferase. *J Clin Invest* 1990, **86**: 1609–1614.
147. Busch MP, Wilber JC, Johnson PJ, Tobler L, Evans CS. Impact of specimen handling and storage on detection of hepatitis C virus RNA. *Transfusion* 1992, **32**: 420–425.
148. Kwok S, Higuchi R. Avoiding false negatives with PCR. *Nature* 1989, **339**: 237–238.
149. Urdea MS, Horn T, Fultz TJ et al. Branched DNA amplication multimers for the sensitive, direct detection of human hepatitis viruses (Nucl Acids Res Symp Ser no. 24). Oxford University Press, Oxford, 1991, pages 197–200.
150. Lau JYN, Davis GL, Kniffen J, Qian KP, Urdea MS, Chan CS, Mizokami M, Neuwald PD, Wilber JC. Significance of serum hepatitis C virus RNA levels in chronic hepatitis C. *Lancet* 1993, **341**: 1501–1504.
151. Pereira BJG, Milford EL, Kirkman RL, Levey AS. Transmission of hepatitis C virus by organ transplantation. *N Engl J Med* 1991, **325**: 454–460.
152. LaQuaglia MP, Tolkoff-Rubin NE, Dienstag JL, Cosimi AB, Herrin JT, Kelly M, Rubin RH. Impact of hepatitis on renal transplantation. *Transplantation* 1981, **32**: 504–507.
153. Weir MR, Kirkman RL, Strom TB, Tilney NL. Liver disease in recipients of long-surviving renal allografts. *Kidney Int* 1985, **28**: 839–844.
154. Roth D, Fernandez JA, Babischkin S, De Mattos A, Buck BE, Quan S, Olson L, Burke GW, Nery JR, Esquenazi V, Schiff ER, Miller J. Detection of hepatitis C infection among cadaver organ donors: evidence for low transmission of disease. *Ann Intern Med* 1992, **117**: 470–475.
155. Huang CC, Lai MK, Lin MW, Pao CC, Fang JT, Yao DS. Transmission of hepatitis C virus by renal transplantation. *Transplant Proc* 1993, **25**: 1474–1475.
156. Otero J, Rodriguez M, Escudero D, Gomez E, Aguado S, de Ona M. Kidney transplants with positive anti-hepatitis C virus donors. *Transplantation* 1990, **50**: 1086–1087.

157. Vincenti F, Weber P, Kuo G, Forsell J, Hunt S, Melzer J, Salvatierra O, Jr., Stempel C. Hepatitis C virus in cadaver organ donors: Prevalence and risk of transmission to transplant recipients. *Transplant Proc* 1991 **23**: 2651–2652.
158. Gomez E, Aguado S, Gago E, Martinez A, Cimadevilla R, Melon S, de Ona M, Alvarez-Grande. A study of renal transplants obtained from anti-HCV positive donors. *Transplant Proc* 1991, **23**: 2654–2655.
159. Triolo G, Squiccimarro G, Baldi M, Messina M, Salomone M, Torazza MC, Pratico L, Bonino F, Amoroso A, Segoloni GP, Vercellone A. Antibodies to hepatitis C virus in kidney transplantation. *Nephron* 1992, **61**: 276–277.
160. LeFor WM, Wright CE, Shires DL, Kahana L, Spoto E, Ackermann JR. A preliminary outcome evaluation of the impact of HCV-AB studied in 521 cadaver vascular organ donors over a 6 year period. *X Annual Meeting of the American Society of Transplant Physicians, Chicago*, May 28–29, 1991, P-2-65.
161. Tesi RJ, Waller K, Morgan CJ, Delaney S, Elkhammas EA, Henry ML, Ferguson RM. Transmission of hepatitis C by kidney transplantation – the risks. *Transplantation* 1994, **57**: 826–831.
162. CDC. Public health service inter-agency guidelines for screening donors of blood, plasma, organs, tissues and semen for evidence of hepatitis B and hepatitis C. *MMWR* 1991, **40**: 1–17.
163. Van der Poel CL, Lelie PN, Choo QL, Reesink HW, Leentvaar-Kuypers A, Kuo G. Anti-hepatitis C antibodies and non-A, non-B post-transfusion hepatitis in the Netherlands. *Lancet* 1989, **ii**, 297–298.
164. Van Der Poel CL, Cuypers HTM, Reesink HW, Weiner AJ, Quan S, diNello R, van Boven JJP, Winkel I, Mulder-Folkerts D, Exel-Oehlers PJ, Schaasberg W, Leentvaar-Kuypers A, Polito A, Houghton M, Lelie PN. Confirmation of hepatitis C virus infection by new four-antigen recombinant immunoblot assay. *Lancet* 1991, **337**: 317–319.
165. Diethelm AG, Roth D, Ferguson RM, Schiff ER, Hardy MA, Starzl TE, Miller J, Van Thiel DH, Najarian JS. Transmission of HCV by organ transplantation. *N Engl J Med* 1992, **326**: 410–411.
166. Mendez R, Aswad S, Bogaard T, Khetan U, Asai P, Martinez A, Flores N, Mendez RG. Donor hepatitis C antibody virus testing in renal transplantation. *Transplant Proc* 1993, **25**: 1487–1490.
167. Pereira BJG, Wright TL, Schmid CH, Bryan CF, Cheung RC, Cooper ES, Hsu H, Heyn-Lamb R, Light JA, Norman DJ, Van Thiel DH, Werner BG, Wright CE, Levey AS. Screening and confirmation testing of cadaver organ donors for hepatitis C virus infection – A U.S. national collaborative study. *Kidney Int* 1994, **46**: 886–892.
168. Pirsch JD, Belzer FO. Transmission of HCV by organ transplantation. *N Engl J Med* 1992, **326**: 412.
169. Morales JM, Campistol JM, Andres A, Fuertes A, Ercilla G, Rodicio JL, Pereira BJG. Transplantation of kidneys from HCV-infected donors into recipients with pre-transplantation HCV infection. *J Am Soc Nephrol* 1993, **4**: 950.
170. Masters CL, Harris JO, Gajdusek DC, Gibbs CJ Jr, Bernoulli C, Asher DM. Creutzfeldt–Jakob disease: Patterns of worldwide occurrence and the significance of familial and sporadic clustering. *Ann Neurol* 1979, **5**: 177–188.
171. Brown P, Cathala F. Creutzfeldt–Jakob disease in France: 1. Retrospective study of the Paris area during the ten-year period 1968–1977. *Ann Neurol* 1979, **5**: 189–192.
172. Brown P. An epidemic critique of Creutzfeldt–Jakob disease. *Epidemiol Rev* 1980, **2**: 113–115.
173. Marsh RF, Hanson PR. Transmissible mink encephalopathy: infectivity of corneal epithelium. *Science* 1975, **187**: 656.
174. Manuelidis EE, Angelo JN, Gorgacz EJ, Kim JH, Manuelidis L. Experimental Creutzfeldt–Jakob disease transmitted via the eye with infected cornea. *N Engl J Med* 1977, **296**: 1334–1336.
175. Manuelidis EE, Angelo JN, Gorgacz EJ, Manuelidis L. Transmission of Creutzfeldt–Jakob disease to syrian hamster. *Lancet* 1977, **i**: 479.

176. Duffy P, Wolf J, Collins G, DeVoe AG, Streeten B, Cowen D. Possible person-to-person transmission of Creutzfeldt–Jakob disease. *N Engl J Med* 1974, **290**: 692–693.
177. CDC. Fatal degenerative neurologic disease in patients who received pituitary-derived human growth hormone. *MMWR* 1985, **34**: 359–360, 365–366.
178. Powell-Jackson J, Weller RO, Kennedy P, Preece MA, Whitcombe EM, Newsom-Davis J. Creutzfeldt–Jakob disease after administration of human growth hormone. *Lancet* 1985, **ii**: 244–246.
179. Koch TK, Berg BO, De Armond SJ, Gravina RF. Creutzfeldt–Jakob disease in a young adult with idiopathic hypopituitarism: Possible relation to the administration of cadaveric human growth hormone. *N Engl J Med* 1985, **313**: 731–733.
180. Gibbs CJ Jr, Joy A, Heffner R, Franco M, Miyazaki M, Asher DM, Parisi JE, Brown PW, Gajdusek DC. Clinical and pathological features and laboratory confirmation of Creutzfeldt–Jakob disease in a recipient of pituitary-derived human growth hormone. *N Engl J Med* 1985, **313**: 734–738.
181. Brown P, Gajdusek DC, Gibbs CJ Jr, Asher DM. Potential epidemic of Creutzfeldt–Jakob disease from human growth hormone therapy. *N Engl J Med* 1985, **313**: 728–731.
182. Asher DM, Gibbs CJ Jr, Gajdusek DC. Pathogenesis of subacute spongiform encephalopathies. *Ann Clin Lab Sci* 1976, **6**: 84–103.
183. Gajdusek DC, Gibbs CJ Jr, Asher DM, Brown P, Diwan A, Hoffman P, Nemo G, Rohwer R, White L. Precautions in medical care of, and in handling materials from, patients with transmissible virus dementia (Creutzfeldt–Jakob disease). *N Engl J Med* 1977, **297**: 1253–1258.
184. Feldman HA, Miller LT. Serological study of toxoplasmosis prevalence. *Am J Hyg* 1956, **64**: 320–335.
185. Ruskin J, Remington JS. Toxoplasmosis in the compromised host. *Ann Intern Med* 1976, **84**: 193–199.
186. Siegle SE, Lunde MN, Gelderman AH, Halterman RH, Brown JA, Levine AS, Graw RG Jr. Transmission of toxoplasmosis by leukocyte transfusion. *Blood* 1971, **37**: 388–394.
187. Reynolds ES, Walls KW, Pfeiffer RI. Generalized toxoplasmosis following renal transplantation. *Arch Intern Med* 1966, **118**: 401–405.
188. Mason JC, Ordelheide KS, Grames GM, Tharsher TV, Harris RD, Bui RHD, Mackett MCT. Toxoplasmosis in two renal transplant recipients from a single donor. *Transplantation* 1987, **44**: 588–591.
189. Ryning FW, McLeod R, Maddox JC, Hunt S, Remington JS. Probable transmission of *Toxoplasma gondii* by organ transplantation. *Ann Intern Med* 1979, **90**: 47–49.
190. Theologides A, Kennedy BJ. Toxoplasmic myocarditis and pericarditis (editorial). *Am J Med* 1969, **47**: 169–174.
191. Hakim M, Esmore D, Wallwork J, English TAH. Toxoplasmosis in cardiac transplantation. *Br Med J* 1986, **292**: 1108.
192. Luft BJ, Naot Y, Araujo FG, Stinson EB, Remington JS. Primary and reactivated toxoplasma infection in patients with cardiac transplants. *Ann Intern Med* 1983, **99**: 27–31.
193. Shepp DH, Hackman RC, Conley FK, Anderson JB, Meyers JD. *Toxoplasma gondii* reactivation identified by detection of parasitemia in tissue culture. *Ann Intern Med* 1985, **103**: 218–221.
194. Hakim M, Wreghitt TG, English TAH, Stovin PGI, Cory-Pearce R, Wallwork J. Significance of donor transmitted disease in cardiac transplantation. *J Heart Transplant* 1985, **4**: 302–306.
195. Couch NP, Curran WJ, Moore FD. The use of cadaver tissues in transplantation. *N Engl J Med* 1964, **271**: 691–695.
196. Wilson RE, Hager EB, Hampers CL, Corson JM, Merrill JP, Murray JE. Immunologic rejection of human cancer transplanted with a renal allograft. *N Engl J Med* 1968, **278**: 479–483.
197. McPhaul JJ Jr, McIntosh DA. Tissue transplantation still vexes. *N Engl J Med* 1965, **272**: 105.
198. Penn I. Transmission of cancer with donor organs. *Transplant Proc* 1988, **20**: 739–740.

199. Lefrancois N, Touraine JL, Cantarovich D, Cantarovich F, Faure JL, Dubernard JM, Dureau G, Colpart JJ, Bouvier R, Traeger J. Transmission of medulloblastoma from cadaver donor to three organ transplant recipients. *Transplant Proc* 1987, **19**: 2242.
200. Morse JH, Turcotte JG, Merion RM, Campbell DA Jr, Burtch GD, Lucey MR. Development of a malignant tumor in a liver transplant graft procured from a donor with a cerebral neoplasm. *Transplantation* 1990, **50**: 875–877.
201. Colquhoun SD, Robert ME, Shaked A, Rosenthal JT, Millis JM, Farmer DG, Jurim O, Busuttil RW. Transmission of CNS malignancy by organ transplantation. *Transplantation* 1994, **57**: 970–974.
202. Campbell AN, Chan HSL, Becker LE, Daneman A, Park TS, Hoffman HJ. Extracranial metastases in childhood primary intracranial tumors. A report of 21 cases and review of the literature. *Cancer* 1984, **53**: 974–981.
203. Delhey K, Lewis R, Dunn J, Berry J, Gray R, Buren CV, Kahan B. Absence of tumor transmission from a cadaveric renal donor with malignant astrocytoma and a ventriculoperitoneal shunt – two-year recipient follow-up and review of the literature. *Transplantation* 1991, **52**: 737–738.
204. Kleinman GM, Hochberg FH, Richardson EP. Systemic metastases from medulloblastoma. Report of two cases and review of the literature. *Cancer* 1981, **48**: 2296–2309.
205. Berger MS, Baumeister B, Geyer JR, Milstein J, Kanev PM, LeRoux PD. The risks of metastases from shunting in children with primary central nervous system tumors. *J Neurosurg* 1991, **74**: 872–877.

CHAPTER 9

Organ recovery from cadaveric donors

MARK DEIERHOI

Introduction 152
Donor evaluation 152
Donor management 156
Organ retrieval 157
References 161

Introduction

Cadaveric organ recovery until recently simply involved bilateral nephrectomy and the removal of a few lymph nodes and the donor spleen for histocompatibility testing. With the expansion of extrarenal solid organ transplantation, organ recovery has become a complex process involving multiple co-ordinators, teams of surgeons, and timing of activities between multiple transplant centres. The recovery process may be considered now to consist of several distinct but overlapping phases, including donor evaluation, resuscitation and management, and the actual surgical procedure of organ retrieval. This chapter will summarize the recovery process from the perspective of these separate phases, with particular emphasis on practical points which the surgeon applies to the management of an individual donor.

Donor evaluation

The process of organ recovery from a cadaveric donor begins with evaluation of the suitability of the donor. This includes an assessment of donor suitability from a general standpoint as well as the consideration of organ specific selection criteria. In practical terms, the general suitability of a donor relates to the risk of transmission of infectious diseases, the risk of transmission of cancer and certain aspects of the donor's overall condition. The risk of transmission of infectious diseases is covered in Chapter 8 and we will concentrate here on transmission of cancer and general criteria for donor suitability.

The evaluation of a potential cadaveric donor beings with a thorough history. It cannot be stressed strongly enough how important it is to obtain as much information as possible regarding the donor's medical history. This is one of the most critical responsibilities of the donor retrieval co-ordinator. Every source of information which is available should be utilized. This includes the hospital record, nurses and physicians caring for the patient and particularly family members. For young, fit trauma patients whose first exposure to the healthcare system may be their injury, family information regarding social factors such as cigarette smoking and alcohol or drug use is extremely important. Family members may also have information regarding other possible risk factors such as prior heart disease, history of hyperlipidaemia, or hypertension. Pertinent factors in the history and hospital course which should be elicited from nurses and physicians caring for the patient include mechanism of injury, history of hypotensive episodes, and resuscitation measures.

Cancer is generally a contraindication to organ donation.[1,2] Specific exceptions include non-melanotic skin cancer and low grade non-metastasizing brain tumours such as Grade 2 astrocytomas. More aggressive central nervous system tumours have been documented to be transferred with solid organ transplants. In addition, the presence of a ventriculoperitoneal shunt or a recent craniotomy preclude organ donation. The risk of a previously treated cancer is unknown. A malignancy treated more than five years prior to donation with no evidence of extra-organ spread and no recurrence can perhaps be disregarded. This is particularly true for cardiac transplantation, since few tumours other than melanoma and renal cell carcinoma metastasize to the heart.

There are other situations in which a systemic disease may preclude organ donation altogether. Diabetes with established secondary complications is one example. These patients are likely to have vascular disease in all transplantable organs. Patients with diffuse atherosclerosis may be unsuitable as donors. In most cases, however, patients with a systemic disease should be evaluated using organ specific criteria, as one or more organs may, in fact, be suitable for recovery. Patients with a history of cardiac arrest, significant episodes of hypotension, or who require high doses of pressor drugs should also be evaluated for each organ individually.

Once the general suitability of a donor is established, attention should be turned to assessment of individual organ function. This includes initial screening studies, more definitive evaluation, and finally assessment of the organs anatomically in the operating room. Important organ-specific studies are listed in Table 9.1.

CARDIAC DONATION

Routine examinations in the evaluation of cardiac donors should include an EKG and some centres seek an echocardiogram. CPK isoenzymes are useful if a myocardial contusion from trauma is suspected. Cardiac catheterization with coronary angiography is increasingly being performed to assess marginal donors because of the urgent need for cardiac donors. Most centres perform coronary angiography routinely in donors over the age of 45 because of the increased risk of coronary disease in this population.[3] Hemodynamic monitoring may be useful in cardiac evaluation, particularly the use of Swan–Ganz monitoring of cardiac output.[4] Exclusion criteria for recovery of the donor heart include global dyskinesia or wall motion abnormalities on echocardiography, valvular heart disease or coronary disease. In the situation where suitability of a donor is

Table 9.1 Organ specific tests for donor assessment

1. Heart
    ECG
    Echocardiogram
    Cardiac enzymes
    Cardiac catheterization
2. Lung
    Chest film
    Arterial blood gas
    Ventilatory parameters
    O_2 challenge
    Bronchoscopy
3. Liver
    Liver enzymes – AST, ALT, GGT, alkaline phosphatase
    Prothrombin time
    Liver biopsy
4. Kidney
    Creatinine, BUN
    Urine output
    2 hr creatinine clearance
    Renal biopsy
5. Pancreas
    Blood glucose
    Serum amylase, lipase
    Glycosylated haemoglobin

uncertain, visual inspection in the operating room is often important in making a final decision.

LUNG DONATION

Evaluation of a potential lung donor routinely includes arterial blood gases, a chest X-ray film and some type of O_2 challenge. Respiratory parameters, including peak ventilatory pressures are extremely helpful, particularly in patients with a history of smoking or other respiratory diseases. Bronchoscopy is critical in evaluation of the pulmonary donor and may also be a management tool as well, allowing for clearing of secretions with resulting improvement in pulmonary function. Findings which exclude a potential lung donor include pulmonary contusion by chest film and pulmonary oedema which does not clear with the volume restriction and diuresis. High peak pulmonary pressures and a poor response to O_2 challenge are also indications for declining lung recovery. In regards to O_2 challenge, most centres use a cutoff of $PO_2 > 100$ mm Hg on 40% inspired O_2 or a $PO_2 > 300$ mm Hg on 100% inspired O_2.[3] Bronchoscopy findings which preclude donation generally include significant purulent discharge from the lungs or any endobronchial abnormality. The usual age range for pulmonary donors is up to age 45. Smoking history is not an absolute contraindication to a lung recovery, provided all measurable parameters are normal. In general, chronic pulmonary diseases such as chronic obstructive pulmonary disease and sarcoidosis will preclude lung recovery. Asthma requiring multiple medications is usually a contraindication as well.

LIVER DONATION

Functional studies are less important in evaluation of a potential liver donor. Liver enzymes, including AST and ALT, should be obtained as well as GGT, alkaline phosphatase, bilirubin and prothrombin time. The single most critical study is a liver biopsy performed at the time of organ retrieval. Elevated liver enzymes, particularly if they rise over time or remain high, indicate significant ischemia and preclude recovery of the liver. An elevated coagulation time coupled with enzyme abnormalities, provides further evidence of ischemia. An isolated, elevated prothrombin time, however, may often be seen in donors as a result of brain injury and disseminated intravascular coagulation and by itself is not necessarily an exclusion criteria. Critical factors on liver biopsy include lymphocytic infiltrates indicative of hepatitis and fatty infiltration. A significant degree of diffuse fatty infiltration correlates with primary non-function of the liver. A history of alcohol use is not a critical problem if direct observation of the liver is normal and there is no evidence of chronic disease on biopsy. Similarly, there is a broad range for age acceptability of potential liver donors. Age up to 65 is certainly acceptable if anatomic and functional parameters are all acceptable.

RENAL DONATION

Renal evaluations should include serum creatinine and BUN or urea and measurement of urine output. Here again a biopsy may be extremely helpful, particularly if there is any abnormality in renal function or doubt about the donor's suitability. Since it is known that renal mass and creatinine clearance decline with age, a one hour creatinine clearance may be useful, particularly in small, older female donors. In one respect, evaluation of a renal donor is somewhat more difficult than evaluation of the extra-renal organs. Because a patient with delayed graft function may be supported by dialysis, the need for immediate function of a kidney is not nearly as critical as it is for more life-saving organs. Since it is known that particularly young donors have considerable functional reserve and delayed function is less critical, abnormal parameters for renal function may be accepted in these donors. For example therefore, an elevated creatinine, particularly if it declines during resuscitation may not preclude renal recovery. Age criteria again are less important for renal donors and kidneys from donors up to, or occasionally over, the age of 70 have been used successfully. There is, however, a clear-cut decline in long-term graft survival for donors over age 60[6] and in this age group it would seem that more restrictive parameters should be utilized. Sclerosis of more than 30% of glomeruli on renal biopsy should be an indication for excluding a donor.

PANCREAS DONATION

Assessment of the pancreas prior to retrieval is more problematic than that of other solid organs. In the situation of a severe head injury, functional parameters are extremely non-specific. Usual measurements include blood glucose, amylase and lipase. Glucose is frequently elevated in head injury patients or those receiving large volumes of fluid as a result of diabetes insipidus. Amylase may be elevated as a result of abdominal trauma and is not important if there is no evidence of direct trauma to the pancreas. If there is no history of diabetes, an elevated blood sugar may be disregarded.[8] If there is some question because of difficulty obtaining a satisfactory history, a glycosolated haemoglobin level can be obtained. This will only be elevated if long-standing hyperglycemia is present. An age cutoff is important in pancreas transplantation as it has been recognized that there is a significant decline in islet cell mass above the age of 45.

SMALL BOWEL DONATION

Small bowel transplantation has only recently become feasible and is still a rare occurrence. There is currently no specific criteria for selection of small bowel donors. Donors under 40 kg should probably not be used because of the small size of the vessels for anastamosis. Small bowel retrieval precludes pancreas retrieval due to the necessity of taking the entire superior mesenteric artery and the superior mesenteric vein to the level of the splenic vein.

Donor management

Resuscitation and management of a potential organ donor should begin at the same time as evaluation. Once the general suitability of the donor has been determined, efforts should be made to stabilize the donor, correct significant electrolyte abnormalities, and optimize organ function. Resuscitation should proceed simultaneously with organ specific evaluation.

Resuscitation should commence with an assessment of the donor's haemodynamic status. Important laboratory studies include electrolytes, haemoglobin and haematocrit, glucose and arterial blood gases. To evaluate haemodynamic status, vital signs should be obtained as well as central venous pressure. Central venous pressure (CVP) monitoring is critical for the resuscitation of a multiorgan donor to optimize fluid volume loading. A central venous line should be placed at this time if one is not present. An arterial line is also helpful for monitoring blood pressure and obtaining arterial blood gases.[9]

In the usual situation, donors have sustained serious head trauma and treatment has been directed at protecting the brain. The patients are therefore volume depleted and quite often hypernatremic because of volume restriction. In addition they are often hyperglycemic because of the injury or administration of high-dose corticosteroids. They are quite likely to have diabetes insipidus with a massive diuresis. This is compounded by any degree of hyperglycemia present. If the patient has an open head injury or multiple fractures, a significant fall in haematocrit is likely to be seen as well.

Initial resuscitation measures should be directed at optimizing haemodynamic status. In the typical donor, resuscitation may be achieved with a glucose-containing low salt solution such as 5% dextrose to provide volume expansion and reduce hypernatremia. If significant hyperglycemia is seen, initial resuscitation may be achieved with colloid solution such as albumin or hetastarch. Blood glucose consistently elevated should be treated with intravenous insulin, taking care to avoid hypokalemia.

Once initial corrections have been made, if the patient has persistent massive diuresis indicative of diabetes insipidus, a pitressin infusion should be commenced. The maximum dose should be 5 mg/hr recognizing that high doses can result in significant splanchnic vasoconstriction. Acceptable parameters for adequate organ function include a urine output of 1–2 μg/kg/hr, CVP 4–10 cm H_2O, and a systolic blood pressure of 90–140 mm Hg. Pressor agents should be used for hypotension only after adequate intravascular volume has been established. Dopamine in doses up to 15 μg/kg/hr should be utilized initially. Use of dobutamine and alpha vasoconstrictive agents such as levophed and norepinephrine should be limited to low doses and only for refractory hypotension with evidence of peripheral vasodilatation. When resuscitating a multiorgan donor, particularly if cardiac retrieval is contemplated, persistent hypotension, despite apparently adequate hydration should be investigated with insertion of a Swan–Ganz catheter. Although the majority of donors can be managed adequately without this form of monitoring, optimization of cardiac function may require measurement of left-sided filling pressures, systemic vascular resistance, and cardiac output. In addition, a Swan–Ganz catheter can provide important information about left or right ventricular dysfunction that would preclude heart retrieval.

Other adjustments to optimize haemodynamic status may be required as well. For a patient clearly volume overloaded with evidence of pulmonary oedema, the use of frusemide may be employed along with volume restriction. In some centres mannitol is used to establish a diuresis and assist in volume expansion. However, mannitol use should be restricted in patients who are clearly volume overloaded or those who already have large urine outputs from diabetes insipidus or hyperglycemia. Hypertension with systolic blood pressures >180 mm Hg should be treated with a rapid acting antihypertensive agents such as nitroprusside.

Once satisfactory haemodynamic status has been achieved, efforts should be made to move the donor rapidly to the operating room. In the operating room oxygenation should be maintained and the patient should be given volume replacement to replace the urine output. Pressors which have been administered should be weaned as much as possible during the retrieval, ideally only low-dose dopamine should be administered at the time of aortic cross-clamping and organ perfusion.

Organ retrieval

Multi-organ retrieval requires a co-ordinated approach with several surgeons working simultaneously. There often may be teams present for individual heart, lung and abdominal organ retrieval (Figure 9.1). Although intra-abdominal organ retrieval is now often performed by a single surgeon with organs that are shipped between centres, most cardiac and lung transplant teams continue to prefer to retrieve their own organs. With improvements in preservation time for heart and lung, it is hoped this will change.

Nevertheless, retrieval still requires co-operation and the first priority is to organize the conduct of the operation. Retrieval surgeons should discuss their preferences in recovery procedures and determine a protocol prior to beginning surgery, so as to minimize confusion and anticipate possible problems and conflicts.

Multiorgan retrieval begins with a long mid-line incision from the sternal notch to the symphisis pubis. Cardiac, pulmonary and intra-abdominal organ dissection may be carried on simultaneously. The general sequence for initial dissection is becoming more standardized and is well described in several textbooks on surgical techniques.[4,7]

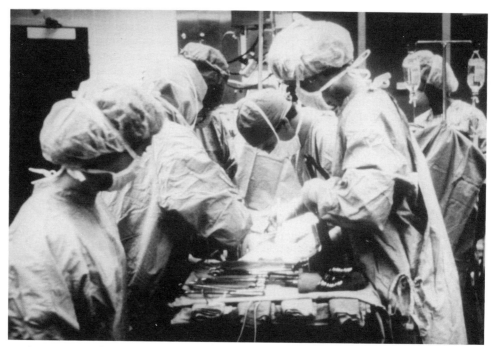

Figure 9.1 Multiple organ donation involves large teams of individuals with different skills.

Dissection of the heart is straightforward and involves examination for ventricular function and wall motion abnormalities. The heart is inspected as well for evidence of coronary disease. The aorta and superior vena cava are then dissected and the vena cava encircled. For lung retrieval both pleural spaces should be widely opened and the lung inspected for masses or firmness or evidence of diffuse disease. The bulk of the lung dissection is generally performed following perfusion and cross-clamping and at this point minimal dissection is required.

Intra-abdominal organ dissection begins with control of the great vessels, identification of the ureters, and dissection of the portal hepatic structures. On entering the abdomen, the liver is examined for evidence of chronic disease and possible trauma. The structures in the porta hepatis are palpated looking for aberrant hepatic arterial anatomy which occurs 25% of the time. The left lobe of the liver is mobilized by dividing the left triangular ligament and the gastrohepatic ligament is inspected for a replaced left hepatic artery. The colon and small bowel are then mobilized, the aorta and vena cava identified and isolated and the ureters identified and isolated as well. If pancreas retrieval is to be undertaken, the pancreas should be inspected at this time but minimal dissection should be performed. No dissection is required of the small bowel prior to the *in situ* flush and cooling.

At this point the abdominal surgeon will choose between *en bloc* and sequential organ retrieval. *En bloc* retrieval has become popular in a number of centres for all retrievals. The chief advantages of the *en bloc* technique are rapidity of retrieval and lack of dissection trauma until the organs have been flushed and cooled.[10] The principal disadvantage is occasional difficulty performing the subsequent organ dissection, particularly in an obese donor. This may lead to increased bleeding after liver and pancreas transplant in certain cases. *En bloc* recovery is extremely valuable in the unstable donor. In the situation where hypotension or cardiac arrest occurs, all the abdominal organs may be salvaged by *en bloc* recovery. In addition, combined retrieval of the pancreas and liver is facilitated by *en bloc* dissection if there is a replaced right hepatic artery, which may be found in up to 13% of cadaveric donors.[11]

Precooling dissection is the same for both techniques. The portal structures are identified. The common bile duct is divided and the gall bladder flushed. The hepatic artery is traced to the coeliac trunk and the gastroduodenal artery ligated. The portal vein is identified and may be encircled. The duodenum and head of the pancreas are dissected off the inferior vena cava. Preparations are then made for cannulation and infusion of preservation solutions.

The sequence for flushing and organ removal has become well standardized. Co-ordinators should have all preservation solutions available and identified and should communicate with the receiving centres regarding timing. This is especially critical if the recipient dissection is anticipated to be difficult, such as in a retransplant. Teams should be prepared to delay the cross-clamping and organ retrieval for preparation of difficult recipients. Arrangements should be made for transportation at this time. The surgical teams should agree on the site for decompression. Most prefer division of the inferior vena cava in the chest with drainage into the right pleural space. The most common alternate technique is to insert a large cannula in the IVC at the bifurcation and decompress into a reservoir off the operating room table. Issues related to organ separation at the time of removal should be discussed between the surgeons. Critical issues include the site for division of the pulmonary arteries and veins in separate heart and lung retrieval, and the site for division of the inferior vena cava for retrieval of the heart and liver. In addition, the allocation of iliac vessels between the liver and pancreas should be decided.

Prior to cross-clamping, blood should be obtained for cultures and histocompatibility testing. If organs are being shared, blood should be included for serologic studies at the receiving institutions. Any necessary drugs should be administered at this time. This includes heparin, vasodilators, and in the case of lung retrieval, prostaglandins.

The next step is to insert cannulas for *in situ* flushing and core cooling. The heart is perfused through a cannula inserted into the proximal aorta and the lung flushed via a cannula in the left main pulmonary artery. For abdominal organ perfusion, a cannula is inserted in the distal aorta at its bifurcation. The site for liver perfusion depends on the status of the pancreas and small bowel. If netierh are to be used, then a cannula may be placed in a branch of the superior mesenteric vein in the root of the small bowel mesentery. If the pancreas or small bowel is to be retrieved, some surgeons prefer to divide the portal vein at the time of cross-clamping and perfuse the liver through a cannula held manually in the portal vein.

After cross-clamping, organ removal is commenced. The order is standardized and begins with the heart followed by the lungs and subsequently the intra-abdominal organs. The procedure, particularly for the lungs and the liver and pancreas may be quite involved. The liver is removed first of the abdominal organs, followed by the pancreas and finally the kidneys. Co-ordinators should be certain that ample space and necessary instruments are available for any subsequent back table dissections required.

Mention should be made about the differences in the technique for organ retrieval from non-heartbeating donors. In this situation, the donor is prepared for retrieval with the groins exposed. Femoral arterial and venous cannulas are placed and heparin and a vasodilator such as phentolamine are administered. Ventilatory support is withdrawn and organ retrieval is commenced after cardiac arrest has occurred. The abdominal organs are perfused through the cannulae and then retrieved *en bloc* and reperfused with preservation solution during the back table dissection before organ separation. Organ recovery from non-heartbeating donors has been used with considerable success for kidneys, liver, and pancreas.[12]

Several important tasks remain following removal of the organs. Lymph nodes should be recovered for histocompatibility testing. Liver biopsies and, if indicated, renal biopsies should be obtained and all specimens and blood samples to be sent with each organ should be clearly labelled and attached to the containers for the organs.

Expanding waiting lists will continue to exert pressure to retrieve all possible organs from every donor. Multiorgan retrieval has become routine for most organ procurement organizations and the procedures for donor management and surgical removal have become standardized. Teamwork and communication are critical in achieving maximum utilization of all donors, as are experienced co-ordinators (Figure 9.2) and surgeons. Further increase in organ utilization will depend on better assessment of organ suitability for transplantation and better preservation techniques.

Figure 9.2 The co-ordinator leaves the operating room.

References

1. Penn I. Malignancy in transplanted organs. *Transplant Int* 1993, **6**: 1–3.
2. Penn I. Donor transmitted disease: Cancer. *Transplant Proc* 1991, **23** (5): 2629–2631.
3. Allen M. Donor management. In *Thoracic Transplantation*, Shumway SJ, Shumway N (Eds), Blackwell Science Publishers, Cambridge, Massachusetts, 1995, pages 84–99.
4. Flye MW. Multiple cadaveric organ recovery. In *Atlas of Organ Transplantation*, Flye MW (Ed.), W.B. Saunders Company, Philadelphia, PA, 1995, pages 47–78.
5. Shumway SJ. Operative techniques in thoracic organ procurement. In *Thoracic Transplantation*, Shumway SJ, Shumway NE (Ed.), Cambridge, Massachusetts, 1995, pages 187–194.
6. Cecka JM, Terasaki PI. The UNOS Scientific Renal Transplant Registry. In *Clinical Transplants*, Terasaki PI (Ed.) 1994, pages 1–16.
7. Potter CDO, Wheeldon DR, Wallwork J. Functional assessment and management of heart donors: a rationale for characterization and a guide to therapy. *J Heart Lung Transplant* 1995, **14**: 59–65.
8. Gores PF GK. Donor hyperglycemia as a minor risk factor and immunologic variables as major risk factors for pancreas allograft loss in a multivariate analysis of a single institution's experience. *Ann Surg* 1992, **215**: 217–230.
9. Soifer BE, Gelb AW. The multiple organ donor: identification and management. *Ann Int Med* 1989, **110**: 814–823.
10. Nakazato PCW. Total abdominal evisceration: an en bloc technique for abdominal organ harvesting. *Surgery* 1992, **111**: 37–46.
11. Shaffer DLW. Combined liver and whole pancreas procurement in donors with a replaced right hepatic artery. *Surg Gynecol Obstet* 1992, **175**: 204–207.
12. D'Alessandro AM, Hoffmann RM, Knechtle SJ, Eckhoff DE, Love RB, Kalayoglu M, Sollinger HW, Belzer FO. Successful extrarenal transplantation from non-heartbeating donors. *Transplant* 1995, **59**: 977–982.

CHAPTER 10

The living organ donor

RICHARD D.M. ALLEN, STEPHEN V. LYNCH, RUSSELL W. STRONG

Introduction   162
Ethical issues   163
Living kidney donation   169
Living partial liver donation   186
Other living organ donation   192
References   194

Introduction

Donation of a solid organ by a living person was born of necessity and remains a necessity. It is an established and unquestionably viable therapeutic option for treatment of end-stage kidney and, more recently, liver disease. The role of living donation of lung, small bowel and pancreas is less clear.

The viability of living organ donation is based on three essential and inclusive prerequisites:

1. The chance of success of the subsequent transplant procedure must be good, providing both a better quality of life and survival for the recipient than other therapeutic options realistically available to the recipient.
2. The risk involved in organ donation must be low and acceptable to the donor, recipient and the responsible clinician.
3. Living organ donation must be voluntary and from a fully informed donor.

Provision of the first two prerequisites should be straightforward for an established transplant unit. However, the third and most important, is more complex and problematic, for it is here that the potential exists for outside pressures on the donor and the attitude and value judgements of the healthcare professional to have an effect on donor and recipient decisions. If there is controversy over living organ donation, it is with this third prerequisite.[1,2]

The role and timing of organ donation will vary from community to community and from time to time. Before the introduction of effective and safe immunosuppression living kidney donation from an identical twin was the only realistic option for meaningful graft survival. For example, in a report in 1964 of 87 non-twin living donor grafts, only one graft was surviving beyond two years.[3] Cadaver renal transplantation became a reality with the introduction of azathioprine and steroids in 1961, but the improved one-year graft survival in the order 20% favouring living related grafts, dictated preference for living over cadaver donors. In the 1990s, long after the introduction of cyclosporine and other newer immunosuppressants, graft survival is no longer an overriding consideration compared to the availability of cadaver organs. In communities without brain-death legislation, or with cultural and religious restrictions, living organ donation is the only realistic life-saving option. For this reason, the medical ethics associated with living organ donation must be attuned to the practicalities of each community's needs.[4]

In this chapter, we discuss the benefits and risks of living kidney and liver donation and outline the management practices necessary to minimize the risk to the donor. However, before embarking on the more straightforward issues of living organ donation, it is important that the healthcare professional be aware of the ethical dilemmas, raised both by their profession and the community, before they offer or deny their patients the therapeutic option of living organ donation. They should be able to discuss the option of living organ donation in a dispassionate and unbiased manner. The passion and the bias may help the well-informed donor and recipient but the role of the healthcare professional is to inform and not to persuade or dissuade a potential donor and recipient.

Ethical issues

The first successful renal transplant took place between monozygotic 'identical' twin brothers in 1954 and in an era without effective immunosuppression.[5,6] The decision to embark on the procedure was based on the prior success of skin grafts. The renal transplant functioned for eight years, failing because of recurrent renal disease. Since then, it is estimated that more than 300,000 kidney transplants have been performed worldwide, of which about 70,000 have come from living donors. Weight of numbers alone might therefore suggest that there are few, if any, ethical dilemmas with the role of living related renal transplantation. However, closer inspection of the figures suggest otherwise with considerable variation seen between and within countries, and even between clinicians within a given transplant centre. The country to country variation may be explained by the availability of cadaver organs (Table 10.1).

The well-documented replacement renal therapy programmes in the USA demonstrate a progressive increase in the numbers of renal transplant procedures being performed over the last ten years and with a comparatively constant ratio of living to cadaveric donors (Figure 10.1). However, a study examining the practice patterns of the use of living donors in the 231 United Network of Organ Sharing (UNOS) approved transplant centres in the United States found that only 31% of responding centres performed such procedures in 1992.[7] This was despite widespread acceptance of the procedure by 92% of responding centres. A similar survey of European centres reported that 22% considered the use of living donors to be ethically unacceptable and another 15% of centres had abandoned the practice because of adverse effects on the donors.[8]

The implication of the mixed attitudes to living kidney donation is that considerations apart from the availability of cadaver organs are at play. These considerations almost certainly reflect reluctance to subject an otherwise fit and well person to a major surgical procedure that is in no way advantageous to the donor's physical health.

Table 10.1 Renal transplant activity in Europe and Australia in 1995 (pmp = patients per million population)

Country	Cadaveric kidney transplants		Living kidney transplants	
	Number	pmp	Number	pmp
Australia	346	19.4	79	4.4
Austria	293	39.1	11	1.5
Belgium	331	32.1	19	1.8
Denmark	114	21.9	40	7.7
Finland	164	32.8	2	0.4
Germany	2128	26	83	1
Greece	42	4.2	85	8.5
Hungary	278	27	3	0.3
Norway	123	29.3	66	15.7
Portugal	368	36.8	2	0.2
Spain	1765	46	35	0.9
Sweden	203	24.2	79	9.4
Switzerland	158	22.6	41	5.8
UK/Ireland	1765	28.9	127	2.1

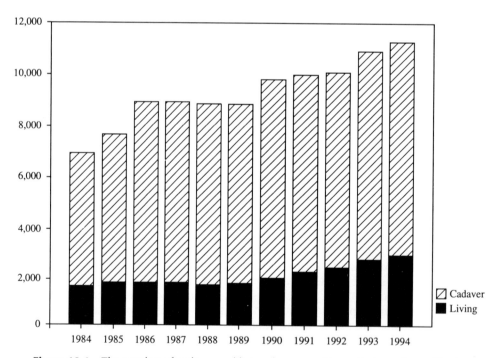

Figure 10.1 The number of cadaver and living donor renal transplants performed in the United States of America between 1984 and 1994.

For every surgical procedure it is crucial to weigh up the risks of the procedure and compare them with the benefits. It is the informed donor who accepts the risk and gives consent, not the surgeon who performs the nephrectomy. The approximate risks to the donor (which will be discussed in detail later in the chapter) are a short-term morbidity of 20% and mortality, of 0.03%.[7,9-11] The long-term risks of developing renal failure are less well documented but appear to be no greater than for the normal population. The benefit to the donor is clearly not physical, but should at least be emotional. An emotional bond can exist in any relationship whether it be between siblings, parent and child, other relatives, spouses and even close friends. It can be argued that the potential living donor should not be denied the opportunity of ameliorating the effects of either a chronic disease in the case of renal failure, or death in the case of liver failure, for a family member, spouse or friend with whom there is a strong emotional tie. Determining the strength of the relationship between donor and recipient and extent of the potential emotional benefit is not always easy. Coersion from family members, cultural expectations, guilt and material gain may unfortunately also lead individuals to consent. Hence there is merit in the suggestion that the donor evaluation be performed by a physician separate from both the recipient's physician and the recipient, thus diminishing the bias on the part of the former towards their patient or transplant programme.

There is little argument over the benefit of living donor renal transplantation for the recipient, both in the short and long term (Table 10.2). The convenience of a planned elective renal transplant procedure should not be under-estimated, particularly for children and adolescents in whom delays in development, growth and educational milestones are important.[12] It may be no less important for the family wage earner or a parent responsible for the care of young children since dialysis is an imperfect substitute for a functioning kidney, with the associated lifestyle restrictions, progressive vascular and bone disease and inferior patient survival.

The advantage of living donor renal transplantation in terms of graft and patient survival is clear. International and national registry data and that from large individual centres, all show an advantage.[13-16] The advantage was greater in the era before cyclosporine and was the major impetus for living related renal transplantation at that time. Nevertheless, it persists today, even for unrelated HLA mismatched living donor transplantation (Figure 10.2). The Minneapolis group have reported a 63% ten-year living related kidney graft survival with cyclosporine compared to 40% for cadaver grafts.[15] Corresponding patient survival was 76% and 60% respectively. The advantage for living donor kidney grafts is probably not confined to HLA matching since HLA matched cadaver transplants have an inferior graft survival to living related zero, one or two haplotype matched transplants as well as kidneys from living unrelated donors.[14,17,18] This suggests factors apart from HLA matching are important, including shorter renal ischaemia times, healthier kidneys and optimal preparation of the recipient for an elective transplant procedure.

If living related renal transplantation provides better graft and patient survival compared to cadaver transplantation, there would be an expectation that recipients would have less rejection, require less immunosuppression and perhaps have less immunosuppression-related complications. Anecdotal evidence supports this impression, but documentation is difficult to find. It can nevertheless be seen in the ease of steroid withdrawal compared to cadaver kidney recipients[19] and cyclosporine withdrawal for living related transplant recipients.[20]

Would there then be a continuing role for living renal transplantation if there were sufficient cadaver kidneys available such that waiting lists for renal transplantation did

Table 10.2 Advantage of living related renal transplantation

Short term
1. Avoidance of dialysis
2. Elective procedure to suit needs of the recipient
3. Less rejection
4. Less immunosuppression

Long term
1. Better graft survival
2. Better patient survival
3. Less long term immunosuppression related complications

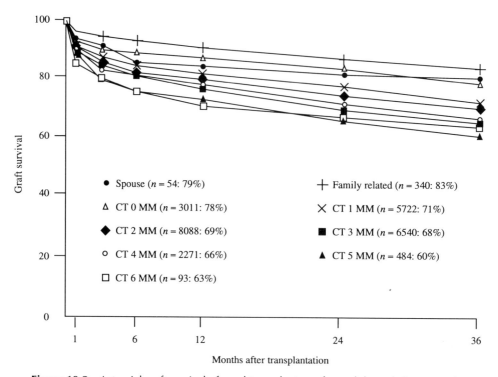

Figure 10.2 Actuarial graft survival of renal transplants performed through Eurotransplant demonstrating the results of spouse and family related donor kidneys compared to cadaveric transplants (CT) with 0 to 6 HLA mismatches at HLA-A,-B,-DR (n = number initially at risk, % = 36-month survival) (reproduced with permission).

not exist? Enviable cadaver organ donor rates in countries such as Spain, Portugal, Finland and Austria have certainly impacted on the role of living donor transplantation (Table 10.1). The answer is therefore probably yes, even though clinical outcome for the recipient may be less satisfactory. However, the question remains hypothetical whilst cultural, religious, political, financial and legal factors inhibit attempts to increase cadaver organ donor rates.[21] The reality is represented by the 75% increase in the UNOS waiting list for cadaver renal transplants in the five-year period to 1993.[22] Twenty-three percent of renal transplants performed in the USA in 1993 were from living donors compared with 18.7% in 1988. In many developing countries, cadaver donor transplantation is virtually non-existent.[21] Hence, for the foreseeable future, there will be a continuing need for living donor transplantation worldwide.

Time is an important ingredient for successful living renal transplantation. The decision in favour of pursuing living renal transplantation is usually made early and is taken for emotional and practical reasons and is generally not changed unless medical contraindications exist. Nevertheless, the transplant procedure should be not be rushed into, for there must be time for the potential donor or recipient to consider and change their decision. If the potential living donor kidney recipient elects to be placed initially on a cadaver waiting list, it may be a manifestation of uncertainty. The recipient or the clinician caring for the recipient may perceive that there is a degree of unwillingness on the part of the potential donor to donate. Alternatively, despite the willingness of the potential donor, the recipient may consider it an unfair imposition for the potential donor or for their family. The ability for the decision making within the family to be disrupted by a pending living donor transplant procedure should not be underestimated.[23] In a study of 536 donors, 14% experienced direct pressure not to donate and 23% reported financial hardship as a result of organ donation.[24]

The luxury of time does not as yet exist for patients with terminal liver failure for the alternative is death rather than dialysis for such individuals. In liver transplant programmes, the challenge of the shortage of size-matched whole liver grafts for pediatric patients was met by the innovative introduction of reduced-size grafts from adult cadaver donors. As experience increased, the results proved to be equivalent to those achieved with whole liver grafts and even superior when applied to small infants. However, the reduction in waiting list mortality was offset by an increase in the number of adults awaiting transplantation. The redistribution, rather than expansion of the resource, has been challenged to be of questionable value if accompanied by an increase in adult waiting list mortality.

The use of liver partition to produce two grafts, with the right hemi-liver for an adult and the left segmental graft for a child, appeared to overcome the obvious disadvantage of discarding a potentially usable second allograft. The early results were less than satisfactory. Poor donor organ selection and failure to recognize anatomic variation negated the possibility of providing two suitable grafts in many cases and necessitated compromise. The conversion of an elective patient to an urgent retransplantation candidate because of technical failure risked the patient's life and devalued these so called 'split-liver' transplants as a method to increase the organ supply. The procedure was considered unjustified in elective patients, unless the donor anatomy permitted division into two grafts without compromise.[25] The increasing discrepancy between the supply and demand for donor livers has seen a gradual increase in the use of the method, particularly in Europe, where the results achieved are now equivalent to those reported for whole liver and reduced-size grafts by the European Transplant Registry.[26]

The concept of partial liver transplantation from a living donor evolved from the experiences with liver resection for a variety of disease processes, together with the experience in transplantation of reduced-size and split-liver grafts from cadaveric donors. The performance of a safe partial hepatectomy with maintenance of hepatic function and avoidance of injury to both the donor and to the allograft is mandatory. The risk to the donor is that associated with a major operation in the form of perioperative complications and long-term sequelae. As for living kidney donation, the only benefit to the donor is psychological. The risk for the recipient is comparable to that experienced with the use of a reduced-size graft as obtained from liver partition for split-liver transplantation. The perceived benefits for the recipient when transplanted electively are the uniform good quality of the graft, and reduction of debilitating complications and mortality before transplantation, all of which severely compromise the chance of success after transplantation.[27] The originally proposed theoretical immunologic advantage, has not yet been established.

Should an elective living related liver transplant (LRLT) be performed in preference to a cadaver donor transplant? There is no single answer to this question. Circumstances vary from country to country. The basis of any answer must be that there is some measurable advantage for the recipient or the use of a living donor is not justified. Although the results of LRLT reported so far are excellent, there are no controlled studies comparing one with the other and no data, therefore, to show a measurable advantage. There may be a measurable advantage for liver transplant programmes overall in countries where heart beating cadaveric donation is established. The increased supply of donor organs resulting from the introduction of LRLT not only reduces the waiting list mortality but allows other patients to receive a cadaver donor organ at an earlier period than they would have otherwise achieved and prior to severe deterioration, with the potential for improved post-transplant survival. However, proposing such a argument that LRLT gives an overall benefit to liver transplant programmes is ethically unacceptable.

There may be powerful motivation of love and altruism by parents toward their child and this is commendable, but transplant surgeons must not allow the donor to be inflicted by undue hazards. Although the risk for the donor in LRLT is as yet unknown, there has been one donor death reported and morbidity of 3% to 17% in reported series.[28,29] It would be expected to be at least commensurate with that experienced in living related kidney transplantation. In the early report from Kyoto, 41% of donors had complaints such as fatigue, wound pain, gastritis and duodenal ulcer.[30] The disturbing number of psychosocial problems experienced by families in the early series from Chicago reported by Whitington and colleagues adds credence to the notion that there are significant risks and concerns for the welfare of the donor and the family.[27]

It appears from recent publications that there has been an increasing use of LRLT procedures in Europe and the United States.[29,31,32] The early reservation, scepticism and even condemnation appears to have been replaced by cautious optimism and even enthusiasm. Success is a major factor in the acceptance of new therapeutic endeavour and human need can cause a change in attitude and value.[33] The ethical issues, however, have not changed even though medical acceptance or bias has. The pressures on the development and viability of individual transplant programmes have resulted in increased acceptance of candidates for transplantation despite recurrent disease which leaves little prospect of survival beyond a limited, finite period. Is there a fair selection of candidates, given the scarce resource? Attention should thus be focused as much on recipient selection as increasing the donor pool.

Is LRLT an option in an urgent situation? The overwhelming desire of parents to save their child can blind them to the potential dangers which they may, at that time, consider irrelevant. The element of urgency removes time for reflection and reconsideration, both basic steps in living donor renal programmes. The transplant team must not allow heroic parents to participate in an unworthy enterprise where the chance of survival of the recipient is poor. 'The desperation of illness never justifies the infliction of a hopeless remedy.'[34] If there is no alternative, and analysis indicates that the prognosis for the recipient's recovery is reasonably good and harm to the donor is likely to be minimal, it seems reasonable to agree to parents' strong motivation to participate.

The 'statistics of probability' are used as an aid in providing a solution to a humanistic problem.[34]

Although there have been more than 750 LRLTs performed worldwide, more than half have been performed in countries where organ transplantation from brain-dead donors is not practised or almost non-existent. With no alternative, the community acceptance of living organ donation is high, and declining to be a donor for a loved one in such an environment would be an exception. Transplant surgeons in these countries have embraced the concept with prodigious fervour, but, it may be a retrograde step for the progress of transplantation in such communities. Previous concerted efforts to introduce cadaver organ transplantation have dissipated somewhat, to the disadvantage of other patients in need of kidneys, heart and adult liver transplants. Notwithstanding these comments, the rights of the individual to donate or not donate part of their liver through self-determination, without coercion and armed with all the facts, should be respected.

Living kidney donation

The mechanics of living kidney donation are designed to minimize risk to the donor. The decision on the part of the donor to consent to the procedure of donor nephrectomy is inherently conditional on the knowledge that it will be performed by a surgeon experienced in the procedure and in an environment where post-operative management is optimal. Equally, there is the understanding that the donated kidney will be transplanted in circumstances that will maximize the success of the procedure.

Established renal transplant units will have a protocol in place for the formal assessment for living kidney donors. This often involves a transplant co-ordinator to facilitate the process by ensuring its completeness and minimizing the inconvenience for the potential donor. The steps taken between identification of the potential donor and the donor nephrectomy procedure are outlined in Figure 10.3 and have a suggested time frame of four to eight weeks. As the process involves a healthy donor and an otherwise healthy recipient, there is little advantage in shortening this period. It is usually much longer. The imminent need for dialysis should not be seen as an indication to take short cuts in this process. Indeed, reducing the level of recipient uraemia by temporary haemodialysis probably enhances the safety of the subsequent transplant procedure. For the donor, there is a need to plan for the time required to undertake investigations to assess suitability for kidney donation as well as the five to ten day stay in hospital and a further four to eight weeks of convalescence.

RECIPIENT CONSIDERATIONS

The first prerequisite for living renal donation is that the subsequent transplant procedure should have a reasonable chance of success. The recipient should represent a good surgical risk and have an acceptable life expectancy if the subsequent transplant procedure proves to be successful. In theory, clinicians should not differentiate between acceptable recipient risk for living and for cadaver transplantation, as both sources of kidneys are precious. In practice, however, acceptable risk for a cadaver transplantation in terms of cardiovascular risk and life expectancy seem to be comparatively less stringent than for living donor renal transplantation.

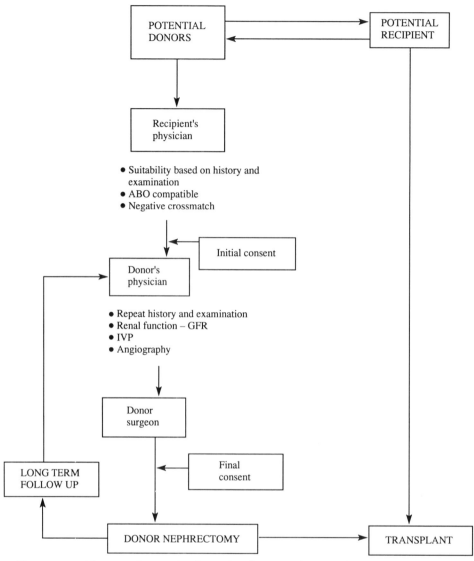

Figure 10.3 The steps that need to be undertaken in evaluation of a potential living donor renal transplant.

Recurrent primary renal disease can also affect the success of the subsequent transplant procedure. For example, presence of anti-glomerular basement membrane antibody is an absolute contraindication. In patients with focal sclerosing glomerulonephritis whose progression to end-stage renal failure was rapid, there is a 70% chance of developing recurrent renal failure after transplantation. In such circumstances it would be difficult to recommend living kidney donation. Recurrence of some other glomerulonephritidies is also common but less likely to lead to graft failure.[35,36] There is no evidence to suggest the incidence of recurrence of primary renal disease is greater in a living related renal transplant.

Long-term survival in living donor renal allograft recipients with hepatitis B infection has been reported in India.[37] Six years after follow up, the incidence of chronic hepatitis in these patients was 75%, but there was no difference in patient and graft survival between hepatitis B antigen positive and negative groups. The study considered living related renal transplantation justified, thus highlighting the need to evaluate the relative contraindications to living kidney donation on the basis of individual need. If dialysis facilities are not available and cadaver renal transplantation is not a practical option, the use of a living donor kidney may be the only realistic treatment option for the recipient.

THE POTENTIAL DONOR

The majority of potential kidney donors will volunteer either to the potential recipient or the recipient nephrologist. Care must be taken on the part of the nephrologist raising the issue of potential kidney donors when discussing the treatment options for renal replacement therapy. If not volunteered, it is reasonable for the nephrologist and other healthcare workers to discuss the possibility of living renal transplantation with the potential recipient but not directly with family members. It should be for the potential recipient to discuss the issue with his or her family. The direct approach of nephrologist to a family member generally provides an unrewarding outcome. Merely raising the issue of living renal donation directly with the recipients' family instigates powerful psychological processes beyond the potential donor's voluntary control and may not leave room for refusal without psychological cost.[38] Acceptable but unwilling donors who refuse to donate should be given an opportunity for counselling or perhaps be provided with a medical excuse.[39]

An active policy of recruiting living donors exists at the Huddinge Hospital in Stockholm, with 36% of their renal transplants coming from living donors.[40] This policy however, has necessitated a 63% non-acceptance rate with the most common reasons being hypertension and renal disease in parents, unwillingness in siblings and interestingly, immunological incompatibility in spouses. In the large living donor programme in Minneapolis in the seven-year period from 1985, 8.4% of 1054 potential donors already screened by exclusion criteria, were found to have unknown health problems.[41] This was considered an unforseen benefit of living donor medical evaluation.

DONOR EXCLUSION CRITERIA

Exclusion criteria, both absolute and relative, are listed in Table 10.3. If any of the absolute exclusion criteria are present, then no further discussion should be had with that potential donor apart from a detailed and empathetic explanation of unsuitability on medical grounds. Because of the inherent risks associated with living renal donation, it is essential that the potential donor has reached the legal age to provide informed consent. Almost all countries will also have legal criteria for living organ donation which must be observed (see Chapter 7). Use of living donors under the age of 18 is usually illegal under any circumstances, but has been contested in court and has been accepted in circumstances such as identical twins. The donor should not be intellectually impaired or have a psychiatric illness that might impair their decision making. Perhaps the most important and difficult problem is to ensure that a covert commercial transaction does not take place in relation to the donation. The upper age for living renal donation should be flexible since several retrospective studies have demonstrated the safety for donors over the age of 60 provided all other criteria are met.[42,43]

Table 10.3 Exclusion criteria for living renal donation

Absolute	Relative
Legal criteria	Age > 60 years
Coercion	Obesity
Age < 18 years	Female of childbearing age
Infectious disease (HIV, hepatitis B, hepatitis C)	Family history of diabetes
Malignancy	CMV positivity
Renal disease	History of deep vein thrombosis or emboli
Hypertension	Cigarette smoking
Diabetes mellitus	
Cardiovascular disease	
Respiratory disease	
Pregnancy	
Coagulopathy	
Psychiatric illness	
Intellectual impairment	
Systemic illness with potential to develop renal disease	
Drug abuse	
Recreational drug abuse	

The potential donor with cardiovascular or respiratory disease that would increase the risk of major surgery must not be considered. Generally, the donor would be considered a normal healthy person or a Class 1 patient as assessed by the American Society of Anaesthiologists' classification of physical status.[45] This would place the risk of anaesthesia-related death at about 1 in 200,000 procedures.[46] Well-controlled medical problems such as asthma would need to be evaluated on an individual basis. Diabetes mellitus and other systemic diseases that might later lead to renal disease are self explanatory exclusion criteria. The family history of diabetes is more problematic but a donor who is at least ten years older than the age at which the recipient developed diabetes can be considered, provided a three-hour oral glucose tolerance test is normal.[47]

Many renal diseases can be inherited in forms that cause renal failure in the recipient but may not be overt in the younger sibling. If the recipient has reflux nephropathy, care is taken to ensure that the potential donor does not have the same problem. Adult polycystic kidney disease, Alport's syndrome, diabetes, and haemoglobinopathy should alert the physician to undertake specific investigations.[48] Family member donors for recipients with either hereditary nephritis or adult polycystic kidney disease, apart from having negative investigations for those problems, should be greater than 25 years of age to exclude occult disease.

As most of the reported deaths following living kidney donation result from pulmonary emboli, a history of deep vein thrombosis or emboli is considered by many to be an absolute contraindication. However, the circumstances of the development of the deep vein thrombosis should be considered as the risk can be minimized with the aggressive use of prophylactic measures. The decision can also be influenced by the presence of other risk factors including obesity.

DONOR–RECIPIENT MATCHING

If all absolute criteria are absent, ABO blood grouping and tissue typing may then be performed to ensure compatibility before embarking on more detailed assessment of suitability for living renal donation. Tissue typing is best performed twice, initially to define the HLA match and to confirm that both the T- and B-cell crossmatches are negative. In some patients with autoreactive antibodies, a positive crossmatch will occur even if the donor is HLA identical. Most tissue-typing laboratories are now able to provide further analysis of the antibody causing such positive crossmatches, but unless there is good evidence that only autoantibodies are present, a living related transplant should not proceed. The second crossmatch is performed within one week of transplantation. Screening of recipient family members by tissue typing before even a cursory discussion of the implications of living kidney donation should be avoided for the best match may well be an unwilling donor. After the decision to donate has been made it is then possible to use knowledge of the HLA matching to provide advice on the likely outcome (Figure 10.4).

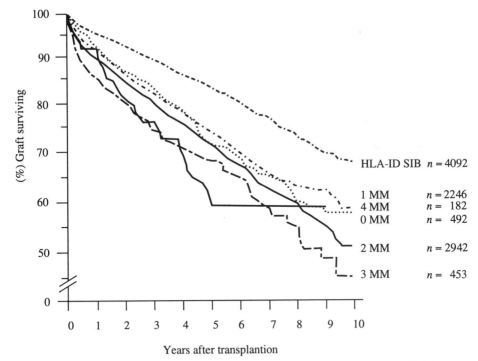

Figure 10.4 The results of living related renal transplants reported to the Collaborative Transplant Study, and demonstrating the effect of different levels of HLA-B,-DR mismatching. HLA identical sibling transplants are compared with 1 haplotype mismatches who had 0,1 or 2 HLA-B,-DR mismatches and 2 haplotype mismatches with 3 or 4 HLA-B,-DR mismatches (reproduced with permission).

DONOR MEDICAL SUITABILITY

It is common practice for potential donors to be assessed for both medical suitability and bona fide intention by an independent physician, usually a nephrologist with understanding of the processes involved in the treatment of renal failure by transplantation. Some institutions also include a psychiatric assessment on a routine basis to verify the voluntary nature of the decision making of the donor. Others, however, believe that a sensitive and informative approach by a renal physician and a transplant surgeon provides an acceptable alternative. Either way, it is important that the assessment be performed in privacy with only the initial discussion about organ donation and transplantation involving the donor and recipient together. All subsequent parts of the assessment should be independent of the recipient. The involvement of a second physician increases this degree of independence. The second physician repeats the history and medical examination, seeks proof of age, makes an independent assessment of the potential donor's level of intellectual capacity and psychiatric status and provides an unbiased view on the expectations and risks of living renal donation. Initial blood and urine investigations are undertaken before the more invasive investigations necessary to assess the level of donor renal function and anatomical suitability for donation (Table 10.4). The same physician can be involved in the follow up after kidney donation.

Table 10.4 Initial screening of ABO blood group compatible living donor

1. **History**
   Relationship to recipient
   Age
   General medical history
   Medications, allergies, social history – occupation, sport
   HIV/hepatitis risk factors
   Alcohol, smoking, drug abuse

2. **Examination**
   Full routine examination, particularly respiratory and cardiovascular, BP on at least two occasions

3. **Initial investigation**
   Repeat ABO blood grouping
   Tissue typing, T and B cell cross-match
   Blood – urea, creatinine, liver function tests, Ca, PO_4
   – fasting glucose, oral lipids
   – full blood count
   – clotting indices
   – viral studies, HIV, CMV, hepatitis B and C
   – VDRL
   Urine – urinalysis
   – microscopy and culture of urine
   – 24-hour urinary protein
   ECG
   CXR
   Stressed thallium scan (± coronary angiography) (if donor >50 years, renal ultrasound)

ASSESSMENT OF RENAL FUNCTION

The long-term risk of deterioration of function of the remaining kidney after donor nephrectomy as a result of hyperfiltration injury remains a possible concern.[49] A number of papers have been published with a remarkable consistency of reported results.[50,51] In summary, the glomerular filtration rate rises within one week of donation by half to two-thirds of the pre-operative level. The reserve capacity of the kidney to increase its glomerular filtration rate (GFR) is largely dependent on the number of functioning nephrons and can be influenced by age and size of the kidney. Younger donors have a greater potential for functional increase, as do male donors compared to female donors.[52] A minimum level of GFR as assessed by clearance of creatinine or ^{99m}Tc-DTPA has not been clearly defined. Nevertheless, most nephrologists would not recommend a living kidney donor procedure if the corrected GFR was less than 90 ml/min.

Most transplant centres would preferentially remove the kidney with the lower GFR unless there was a significant anatomical advantage offered by the other kidney. In such circumstances, it may be necessary to differentiate a normal variant from a diseased kidney. This can be achieved by a comparison of renal function before and after protein loading. No change in differential GFR as determined by ^{99m}Tc-DTPA scanning is supportive evidence for bilateral normal parenchymal renal function.[53]

ANATOMICAL CONSIDERATIONS

The most obvious consideration is to ensure the potential donor has two kidneys and that they are of normal size. Routine ultrasound examination and an intravenous pyelogram will detect the presence of a horseshoe, hypoplastic, aplastic or otherwise diseased kidney. The presence of double ureters in the absence of a history of urinary tract infection is not a contraindication to donation either for the recipient or the donor. All other factors being equal, the authors would leave the kidney with the double ureters with the recipient and remove the kidney with the single ureter for transplantation (Figure 10.5). Fortunately, this decision need not be made frequently for the incidence of double ureters is 1–2%. A micturating cystogram is necessary when evaluating a family member of a potential recipient with reflux nephropathy.

Although surgeons prefer to transplant a left-sided donor kidney, virtually any kidney of normal size and function can be used provided that the risk taken because of anatomical abnormalities is explained and accepted by the donor and recipient. Preference for a left-sided kidney is based on the longer and often stronger renal vein on that side. The right-sided renal vein, apart from being two centimetres shorter, is often thin walled and more likely to be multiple in number.[54] Vascular contrast imaging of the renal veins is unrewarding. The popularity of the left kidney can be gauged from a single centre Korean study in which 385 of 450 living donor kidneys transplanted in a three-year period were left sided.[54]

The last and most invasive investigation is angiography of the kidneys (Figure 10.6). It is performed as an outpatient procedure in search of a kidney with a single artery. The preferred technique is by intra-arterial digital subtraction angiography in which computer processing is used with small doses of contrast. Comparatively small intra-arterial catheters minimize the incidence of groin haematoma to only a few percent. Selective renal angiography is best avoided because of the risk of renal artery damage.[55] Flush aortography with the liberal use of posteroanterior and oblique projections

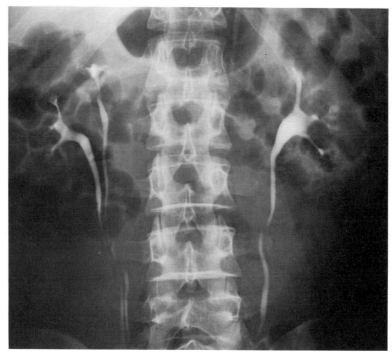

Figure 10.5 Intravenous pyelography demonstrating a single ureter leading from the potential donor's left kidney (on right side of figure), but two ureters leading from the right kidney (on left side of figure).

provides an accuracy rate of 96%.[56] About two-thirds of potential donors will have single vessels to both kidneys and 7% will have multiple vessels to both kidneys.[57] Not surprisingly, the most experience in transplanting kidneys of multiple vessels is from transplant centres in countries in which the major source of kidneys is from live donors.[58,59]

Recipient factors can also dictate which kidney is donated. In infants, a right-sided kidney may be preferred as it provides for orthotopic placement in the right side of the peritoneal cavity with the renal artery passing behind the inferior vena cava.[60] In adults, many surgeons by convention prefer to use a right-sided kidney if a left-sided transplant procedure is to be performed and vice versa.

INFORMED CONSENT

Much of the ethical debate over living organ donation revolves around whether or not informed consent is possible. Appropriate information is provided for the potential donor throughout the evaluation period. At the initial interview, the rationale for recommending living kidney donation is explained and is based on the shortage for available cadaveric kidneys and improved graft survival. As the evaluation proceeds, further information is fed back to the donor on their individual risk from undergoing the donor nephrectomy procedure. The common scenario is a potential donor who has received information from the recipient's physician, their own physician involved in

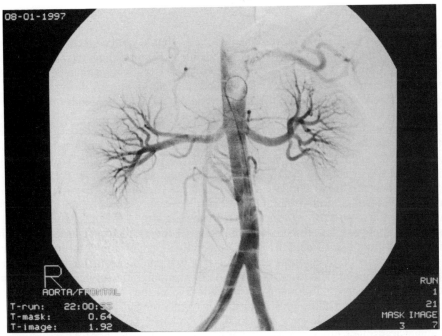

Figure 10.6 Angiography is required to delineate the anatomy of the blood vessels supplying the kidneys. The injection catheter can be seen lying within the aorta from which the single left and right renal arteries can be seen to branch to the kidneys.

evaluation for living kidney donation and the surgeon performing the procedure. It is important that these three clinicians, and any others involved in the donor evaluation, are consistent in their views and are convinced of the voluntary nature of the kidney donation. The donor should sign a written declaration of free and comprehending consent to donation, stipulating the voluntary nature of the donation, freedom from obligation of recipient, especially by financial favour, and freedom from risk of transmissible diseases. A minimum cooling-off period is advisable and is mandatory by law in some countries. Equally, the physician responsible for the evaluation of the donor must certify in writing that the donor has received an adequate explanation of the nature and effect of living kidney donation and that the consent was freely given.

Dealing with the donor who has added risk is more problematic. Public opinion and attitudes outside the transplanting centres would suggest it is the donor who should make the final decision in circumstances of added donor risk.[61] The role of the surgeon and the provision of informed consent by the donor is different from other surgical procedures. Short-term morbidity and mortality following donor nephrectomy relates directly to the surgical procedure, the risks of which are quoted by the nephrectomy surgeon. Description of the risks associated with long-term effect of donating a kidney and the expectation of living a full life expectancy with one kidney, however, is more in the domain of the physician undertaking the donor medical assessment. The added risk of transplanting a kidney with multiple vessels is the responsibility of the transplanting surgeon.

DONOR NEPHRECTOMY

Responsibility for performing a donor nephrectomy should not be taken lightly because, more than any other surgical procedure, there is the expectation that the patient will live and survive normally. The technique is not taught by text, but rather by assisting an experienced nephrectomy surgeon. Published descriptions and illustrations of the surgical procedure are uniformly disappointing, although a recommended exception has been authored by Flye and colleagues.[62]

The principles of donor nephrectomy are to obtain a viable kidney with a minimal morbidity to the donor. Adequate exposure and careful handling of tissues minimizes vascular spasm and preserves the perihilar and periureteral fat to ensure adequate vascularity of the urinary collecting system. Maximum safe lengths of the renal vessels are obtained and the ureter is divided as it crosses the bifurcation of the common iliac artery. The donor is admitted the evening prior to surgery to a hospital ward in close proximity to the recipient. A simple bowel preparation is given, particularly if the transperitoneal approach is to be used. The donor is reviewed by the anaesthetist and a description of the mode of delivery of post-operative analgesia is described in detail to the donor, especially if epidural anaesthesia or patient controlled administration of analgesia is prescribed. Blood is taken for ABO grouping and serum stored. The need for transfusion during or following the procedure is in the vicinity of 2%. As well as intravenous overnight fluids, oral fluid intake is encouraged up to two hours prior to surgery. Morning scheduling of the donor nephrectomy is preferable, allowing the subsequent transplant procedure to be completed by mid-afternoon. After induction of muscle relaxant general anaesthesia, prophylactic antibiotics are given and a urinary catheter inserted. Three commonly used surgical techniques exist. Two involve an extraperitoneal approach to the kidney whilst the third is intraperitoneal. Although large numbers of procedures have been performed, prospective studies assessing the benefits of one approach over the other are uncommon. The best technique is the one that the surgeon feels most comfortable with.

As the donor nephrectomy procedure is most commonly performed by urologists, it is not surprising that the posterior extraperitoneal approach is favoured by most.[63] The patient is placed in oblique lateral cubitus position on the operating table with the table flexed to open up the space between the lower costal margin and the iliac crest. The 11th and 12th ribs are marked with the level of the planned incision having as much to do with surgeon's preference as the position of the kidney. It can be made at the level of the 11th rib, its interspace, the 12th rib or subcostally (Figure 10.7). The incision extends from the paraspinal muscles posteriorly to the edge of the rectus muscles anteriorly. The muscle layers of the abdominal wall are divided and if necessary, the portion of the 12th or 11th rib resected subperiosteally and pleura reflected superiorly. The peritoneum is reflected anteriorly. Gerota's fascia, the fascial envelope in which the kidney is situated, is opened longitudinally and the kidney is reflected posteriorly. Mobilization of the upper pole of the kidney is more easily achieved at this stage because of the ability to use supporting tissues at the lower pole for downwards retraction of the kidney. Superiorly, the fibrous tissue attachment of the adrenal vein to the kidney is divided, leaving the adrenal gland *in situ*. The gonadal vein is divided as it crosses the ureter about 3 cm below the lower pole of the kidney. The ureter is dissected distally together with periureteral tissue to the level of the birfurcation of the iliac vessels. Laterally, the ureteral dissection is carried up to the point where the periureteral adipose tissue merges with the perirenal fat at the lower pole of the kidney, ensuring preservation of the blood supply to the ureter from the renal artery. Medially, the

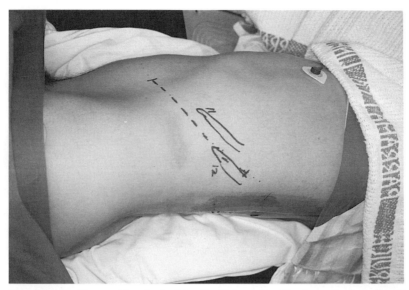

Figure 10.7 The position for incision for kidney donation is marked on the skin with the broken line, while the position of the 11th and 12th ribs are also shown drawn upon the skin.

ureteric dissection includes the gonadal vein from the point where it is divided to its entrance into the renal vein on the left side or the vena cava on the right side. The kidney is reflected posteriorly to permit dissection of the renal vein and then anteriorly for dissection of the artery. On the left side, a posterior lumbar vein is almost invariably encountered entering the left renal vein either adjacent to or with the gonadal vein, as well as the adrenal vein draining into the superior aspect of the renal vein.

The ureter is divided and adequate diuresis ensured with intravenous fluids in preference to use of frusemide or mannitol. There is no universal agreement regarding the advisability of systemic heparinization prior to dividing the renal vessels although a large retrospective review has demonstrated no difference in subsequent transplant renal function.[64]

Perhaps the most critical part of the operation is the point where the renal artery and then vein are clamped. Maximum length of vessels is sought but not at the expense of placing the donor at risk. Both are cross-clamped, the artery with a right-angled forcep and the vein with a Satinsky clamp (Figure 10.8). The kidney is passed to a separate surgeon prior to ligation of the renal artery and closure of the renal vein stump using a running 6/0 prolene suture (Figure 10.9). The operating table is flattened and the wound is closed in layers. If rib resection has been undertaken, routine chest X-rays are obtained in the recovery room to exclude pneumothorax.

The anterior extraperitoneal approach for donor nephrectomy has advantages of providing excellent exposure of the renal vasculature and less likelihood of a pneumothorax, and should be considered when the patient has multiple renal arteries or a skeletal deformity. The theoretical disadvantages of this approach are increased difficulty in the obese patient and less access to the distal ureter to obtain ureteral length.[65] The donor is placed on a flat operating table in an 10° oblique position using rolled blankets placed under the shoulder and hip. A curvilinear subcostal incision is

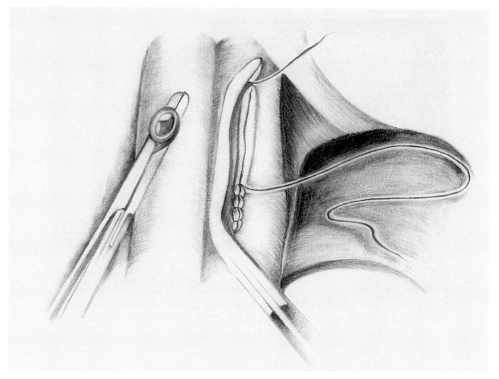

Figure 10.8 The renal artery and renal vein of the donor are shown cross-clamped, the artery with a right-angled forcep, the vein with a Satinsky clamp, while the vein is being sutured securely.

made from the midline and extended laterally at least 3 cm below the costal margin to the tip of the 12th rib. The most difficult part of the procedure is development of the retroperitoneal space. If this cannot be achieved, the peritoneum is opened and an anterior intraperitoneal approach to the kidney is undertaken with mobilization of either the hepatic or splenic flexure of the colon as appropriate. Some surgeons prefer this transperitoneal approach because of the impression of a safer and more accurate dissection of the renal vessels and less manipulation of the kidney. Otherwise, the procedure is similar to that for the posterior extraperitoneal approach.

After removal of the kidney and on a back table, a cannula is placed in the renal artery and the donor kidney perfused with 200 to 400 ml of organ preservation fluid at 40 °C (Figure 10.10). The time between cross-clamping the donor renal artery and commencement of the perfusion of the kidney is defined as the warm ischaemia time and less than 5 minutes is acceptable. Two minutes is achievable. The rate of perfusion will be limited by the size of the cannula, with the bag of perfusion fluid kept about one metre above the level of the kidney. High pressure perfusion is not necessary and best avoided. Adequate perfusion is achieved when there is no longer evidence of blood coming out of the renal vein and the kidney is uniformly pale. The kidney is then either transferred directly to another operating theatre or placed in a sterile container for short-term storage. Subsequent primary renal transplant function can be expected in at least 95% of instances.

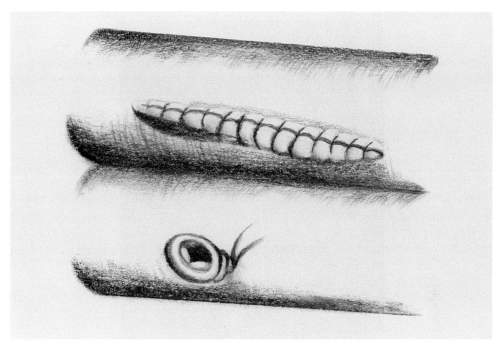

Figure 10.9 The renal artery and renal vein stumps are shown after ligation and suturing respectively.

Post-operative nursing of the donor nephrectomy patient is preferably undertaken in a ward away from the sights and sounds of the recipient's post-operative management and is the same as for conventional nephrectomy. Paralytic ileus is not uncommon. Nasogastric drainage and routine use of a urinary catheter is avoided because of their role in initiating respiratory and urinary tract infections respectively. Chest physiotherapy and subcutaneous heparin are given routinely until the patients are fully mobilized, usually five to seven days after surgery. With the assistance of a wheelchair, the donor should be able to visit the recipient a day or so after surgery. Depending upon the nature of the employment, most donors are able to return to their workplace three to six weeks after surgery, even though some wound discomfort may persist for several more months.

EARLY MORBIDITY AND MORTALITY

Published complication rates after donor nephrectomy vary considerably. Table 10.5 provides a summary of recent publications on large numbers of patients undergoing living donor nephrectomy.[16,63,66-72] Only two papers mention the presence of atelectasis and associated fever which realistically, probably occurs in up to a third of patients. Pneumothorax incidence based and the need for a chest tube varies from 1% to 7%, but the overall incidence is probably greater if the reported 39% incidence of inadvertent entry of the pleural cavity is representative.[68] The two studies involving the anterior approach reported no pneumothoraces.[67,61] Only one publication reported on the

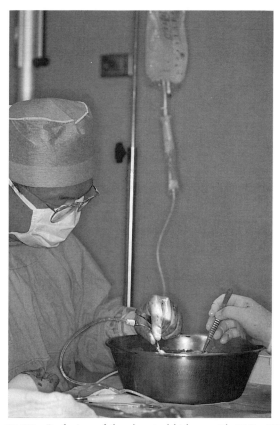

Figure 10.10 Perfusion of the donated kidney with 200–400 ml of cold (4 °C) perfusate using a hydrostatic pressure of one metre or less. The kidney is in the sterile bowl and connected to the bag of perfusate through a line, the tip of which is in the open renal artery.

incidence of abdominal wall nerve injury resulting in paraesthesia and numbness. The incidence of 6% is probably an underestimation and would probably be greater if objectively assessed. Surprisingly, incisional herniae are not reported, perhaps a credit to the surgeons or just overlooked. Overall, the reported minor complication rate related to surgery is around 20%, relating mostly to wound and urinary tract infection, the need for transfusion or pneumothorax. Major morbidity from the surgical procedure is in the vicinity of 1% to 2% and includes pulmonary embolus, the need for splenectomy and re-exploration for haemorrhage (Table 10.6).

Deaths have occurred after donor nephrectomy. None of the large series presented in Table 10.5 reported death following donor nephrectomy. However, a survey in the United States in which 75% of UNOS Transplant Centres replied, identified three deaths in 9,692 procedures between 1987 and 1992 (0.03% or 1 : 3,231 procedures).[7] A survey of the American Society of Transplant Surgeons (ASTS) documented five deaths in 19,368 living related donor nephrectomies between 1980 and 1991 (0.026% or 1 : 3,874).[10] In a survey published in 1987 of the 12 largest transplant centres in the USA, two deaths were identified with a mortality rate of 0.035%.[73] The ASTS

Table 10.5 Complication rates in reported series of living donor nephrectomies

Publication and reference no.	Patients	Surgical approach	Complications rate	Hospital stay	Minor chest	Major chest	Transfusion	Wound infection	Pneumothorax	UTI	Pulmonary	Death
Dunn 1986[71]	314	Anterior	7%	–	35%	0.6%	0.6%	3.5%	0	–	0.3%	0
Streem 1989[63]	115	Posterior	12%	6.4	–	–	3%	1%	5.2%	–	–	
Blohme 1992[67]	490	Anterior	14%	–	–	4.7%	1%	2.2%	–	4%	0.2%	
Ottelin 1995[68]	333	Posterior	17%	–	–	1%	4%	2%	1.5%	3%	0	
Duraj 1995[70]	413	Posterior	–	–	15%	–	4%	3%	3%	10%	1%	
Waples 1995[69]/ D'Alessandro 1995[16]	681/ 681	Posterior	17%	–	–	0.7%	0.3%	4%	13%	5%	0.3%	

Table 10.6 Summary of the incidence of early complications after donor nephrectomy

Atelectasis	15–30%
Paraesthesia/nerve injuries	6%
Urinary tract infection	5%
Pneumothorax (requiring chest tube)	5%
Wound infection	3%
Transfusion (>1 unit)	2%
Pneumonia	1%
Splenectomy	0.3%
Pulmonary embolus	0.3%
Death	0.03%

membership survey was also able to document a total of 17 deaths in USA and Canada and one in Europe, with seven due to pulmonary embolus.[10] This underscores the need for routine prophylaxis against the deep vein thrombosis after living donor nephrectomy.

LATE COMPLICATIONS

Living donor nephrectomy is followed by early compensatory changes made possible by the reserve function of the remaining kidney. The long-term concern is of possible development of proteinuria, hypertension and progressive renal failure. Twenty-four hour urinary protein excretion increases in all donors after nephrectomy and is more marked in men than women.[74,75] In a study of donors 10 years or more after nephrectomy, serum creatinine concentration was 20% higher in donors and creatinine clearance 20% lower than corresponding values in siblings.[75] Proteinuria was common, with 2 of 38 patients having excretion of more than 300 mg/24 hours, but there was no correlation between presence of hypertension, level of renal function or time since nephrectomy. Consistent findings have also been reported in donors more than 20 years after nephrectomy in a study by Najarian and colleagues in which kidney donors were compared with siblings as a control group.[10] The findings confirm the safety of living kidney donation in terms of preservation of the donor's renal function (Table 10.7) and are reflected in the results of a survey of 54 life insurance companies within the USA. All companies would offer life insurance to kidney donors with only one indicating an increase in premium.[76]

Even longer follow up at 45 years after nephrectomy was obtained by studying the records of United States army personnel who lost kidneys from trauma during the Second World War.[77] When compared to servicemen of the same age, the incidence of hypertension and mortality was similar. In patients with unilateral renal agenesis, however, death from failure of the solitary kidney due to focal glomerulosclerosis is common.[78,79] The presence of reduced nephron number early in life or perhaps malformations of the solitary kidney accompanying unilateral renal agenesis are thought to be responsible for focal glomerulosclerosis in these patients. This theory is supported by a study in patients with a solitary kidney in whom a partial nephrectomy was performed.[80]

Table 10.7 Long-term renal function >20 years in renal donors compared to their siblings (data from Najarian et al.[10])

	Renal donors $n = 57$	Siblings $n = 65$
Age	61	58
Serum creatinine	1.1 mg/dl	1.1 mg/dl
Creatinine clearance	82 ml/min	89 ml/min
Hypertension	32%	44%
Proteinuria (>150 mg/d)	23%	22%

The UNOS living donor survey demonstrated considerable variation in patient follow up protocols,[7] with indefinite follow up recommended by only 13% of transplanting centres. Irrespective of the inability to demonstrate an increased incidence of hypertension or progressive proteinuria when compared to the general population, it would seem reasonable to recommend at least yearly follow up for measurement of hypertension, proteinuria and assessment of renal function. There is, however, no documentation to support this recommendation, nor evidence to support a reduction in dietary protein intake.

LIFESTYLE ADJUSTMENT

Leisure activities that might involve trauma to the remaining kidney, such as horseriding for example, should be undertaken with care. Pregnancy does not seem to be contraindicated after donor nephrectomy and women considering future pregnancy should not be automatically eliminated as potential donors.[81,82]

Quality of life studies have suggested that kidney donation does not cause long-term adverse psychological sequelae.[83,84] The majority of donors, irrespective of outcome of the transplant, express positive feelings towards kidney donation. Nevertheless, psychiatric morbidity amongst organ donors including depression, and anxiety adjustment disorder, marital stress, financial hardship and suicide does exist, and may be more common among those donors whose recipients had unsuccessful outcomes.[39,75] As with other possible long-term sequelae of living donor nephrectomy, comparison needs to be made with the incidence of these problems in the general population.

NEW INITIATIVES

Whilst cadaver transplantation rates remain low or essentially non-existent in some countries, there are several challenges remaining for living kidney donor transplant programmes. The opportunities for patients to receive a renal transplant are limited by the need for an ABO blood group compatible kidney, although even this is being challenged in some centres.[85] The frustration of having a fully motivated but incompatible related donor has been overcome in Korea with the so called 'swap programme'. In a non-profit situation, living unrelated kidneys are shared between patients' families.[86] From the same transplant centre has also come an intriguing description of retroperitoneoscopic donor nephrectomy. Using a 5 cm incision and applying specifically designed retractors, the donor kidney, its vessels and ureter are extracted without dividing muscles. The post-operative wound was smaller, pain was reduced and convalescence was shorter.[87,88]

The use of kidneys from emotionally rather than blood related living kidney donors such as spouses, accounted for 4% of transplants in the USA in 1994. Initially championed by the Wisconsin group, this donor source is now being embraced more enthusiastically because of unexpectedly high rates of survival (Figure 10.2).[10,89] A measure of the increasing support for genetically unrelated living kidney donors, even if more in principal than in practice, is seen in recent surveys in USA and the Middle East.[90,91]

Living partial liver donation

The ultimate application of reduction hepatectomy in orthotopic liver transplantation has been the use of a living donor to provide the graft. This has the potential of greatly expanding the donor pool, but is intimately enmeshed in the ethical issues mentioned earlier.[92] In São Paulo, Raia and colleagues reported two cases of liver transplantation from live donors in 1989.[93] Both recipients unfortunately died. In Brisbane and a week after the second Brazilian transplant procedure, a 15-month-old Japanese boy received a segment 2 and 3 graft from his mother.[94] A year later, his initial graft was replaced with a cadaver organ, with graft failure secondary to a combination of viral hepatitis and chronic rejection. Seven years later, he is alive with normal liver function.

Within a few weeks, Nagasue and colleagues in Shimane, Japan transplanted another child, followed by Broelsch and colleagues in Chicago who commenced a clinical programme using living related donors.[95,95] To date, at least one living liver donor has died and morbidity and significant donor surgical misadventure have been reported.[28] The largest programme is based in Kyoto, and in Japan, more than 400 children have received grafts from living donors. One year graft and recipient survival approaching 90% is being attained in many units.

As with living related kidney donation, the safety of the donor is utmost in the planning and implementation of liver donation from a living donor. Also as with the renal model, the liver can be thought of as two separate organs. These two hemilivers are fused along the principle plane or Cantlie's line. The procurement of one hemiliver, or part thereof, unlike renal donation, involves the safe detachment of one from the other without injury to either. By necessity, this separation always involves the skill of surgeons experienced in liver resection and with a thorough appreciation of the anatomy of the organ and its frequent variations. The time frame for the completion of most of the vital steps to LRLT, which are very similar to those in renal donation, can be reduced to a few days in emergency cases. For elective patients and their families, several weeks are available in which to make and reconsider decisions. During this time, if a suitable cadaver organ was offered, most units would abandon the living related option and proceed by using the cadaver graft.

RECIPIENT CONSIDERATIONS

As most recipients of living donor liver grafts are children and the commonest indication for transplantation in this group of patients is biliary atresia, the question of disease recurrence is not an issue. The underlying pathology is cured totally and permanently by successful liver replacement. However, there are a number of potential recipients with afflictions which have resulted in end stage liver disease with a known predisposition for recurrence. In the case of some viral infections such as hepatitis B and C, this is really a persistence rather than recurrence of the virus and infection of the allograft can be considered inevitable in most cases. The notable exception to this situation is fulminant hepatitis B where clearance of the virus prior to transplantation occurs. The wisdom of proceeding with living donation in conditions known to have a high incidence of recurrence or persistence in the recipient must be closely scrutinized although promising protocols of newer antiviral agents are emerging.

That the procedure has a reasonable chance of success depends much more on recipient variables in liver transplantation than in its renal counterpart. The number and extent of previous attempts at surgical correction can greatly hamper eventual transplantation. A veritable minefield of difficult adhesions to which is added portal hypertension, coagulopathy, thrombocytopaenia, malnutrition and haemodynamic instability make many recipients daunting technical challenges. To this must be added all of the additional problems of the implant itself and immunological and infectious sequelae.

Paradoxically recipients who present in acute fulminant hepatic failure, often of unknown aetiology, offer less of a technical challenge as their peritoneal cavities are usually intact and there is no portal hypertension. If transplantation is timely, their recovery is rapid and complete. It is this group of patients, however, that pose the greatest ethical dilemma. Can informed and considered consent really be obtained in a crisis where a previously well child has over a period of a few days, become so ill that the only chance for recovery is liver transplantation? The workup of the donor in such instances can barely be performed before the recipient's rising intracranial pressure precludes further treatment. Yet, the chances of success are higher the faster that process is completed. The difficulties for all involved in such situations cannot be overstated.

THE POTENTIAL DONOR

Growing public awareness of the possibility of living donor liver transplantation in countries where cadaveric organ donation is well established has been borne out of spectacular media coverage of the first cases which were performed only a few years ago.[28,94,95,96] In most instances, it is the family that broach the topic rather than the clinicians. In countries where brain death is not legally accepted, living donation and liver transplantation are synonymous.

Issues relating to the age of the donor, intellectual impairment, pre-existing psychiatric illness, the absence of commercialism and the process of obtaining informed consent have been addressed earlier in this chapter in the context of renal donation and are virtually identical for living donor liver transplantation.[97,98] Co-morbid conditions in a potential donor are similarly assessed with respect to cardiorespiratory disease to ensure that the donor is fit for general anaesthesia and major surgery. Factors predisposing to thrombo-embolic complications such as obesity, the oral contraceptive pill, and smoking will usually disqualify a potential donor. Absolute contraindications include known hepatic disease, excessive alcohol intake, viral hepatitis, significant hepatic steatosis, bleeding dyscrasias and some inherited liver disorders. Aspirin and other medications known to inhibit platelet function are ceased to reduce the likelihood of excessive haemorrhage during resection. Prophylaxis against deep venous thrombosis relies on mechanical rather than pharmacological measures. A thorough history is taken and full physical examination performed. Blood is taken for full blood count, urea, creatinine and electrolytes, standard liver function tests, serology for hepatitis A, B and C and the herpes groups of viruses. Other specific tests are performed if familial or genetically transmitted disease is suspected.

The donor is HLA typed and blood grouped. Tissue typing and lymphocyte cross matching bears no practical influence on the decision to transplant, but, is performed for future analysis. Generally, liver transplantation will only proceed when blood group compatibility exists between donor and recipient, however a number of cases have been performed across major blood group barriers in countries where heart beating cadaveric donation is prohibited. The Kyoto group have a standard protocol for such cases including preoperative recipient plasmapheresis and immunoadsorption of anti-blood group antigen antibodies.[99] In small infants, exchange transfusion is performed. The Kyoto group also advocates elaborate lavage of the allograft after procurement and induction immunosuppression which includes OKT3 and FK506. They report similar results for their blood group incompatible cases as with identical and compatible recipients, and do not regard ABO incompatibility as a contraindication for living donor liver transplantation.

ANATOMICAL CONSIDERATIONS AND THE ASSESSMENT OF LIVER FUNCTION

There is considerable variability between living donor programmes in the assessment protocols of function of the portion of the liver partitioned for transplantation, its anatomical suitability and reconstructions necessary for implantation, and most importantly, whether the donor will be left with adequate function in the short term prior to regeneration. It is well known that host size determines liver size, so provided perioperative function is adequate, both the donor and recipient will attain full regenerated hepatic volume rapidly after transplantation under the influence of their own hepatotrophic factors.[100]

Imaging with volumetric assessment is performed initially by computerized axial tomography and ultrasound. The presence of suspicious space-occupying lesions, hemiliver atrophy or parenchymal disease will exclude donation. Magnetic resonance imaging (MRI) with in-phase and out-of-phase views is particularly useful in cases of suspected steatosis. Doppler ultrasound, MRI or angiography are used to define the anatomy of the hepatic arteries, their branches, and the portal and hepatic veins. Intraoperative ultrasound is also useful during partition to confirm pre-operative findings and to further define hepatic venous anatomy. Liver biopsy, endoscopic retrograde or percutaneous transhepatic cholangiography are more invasive and rarely, if ever, indicated unless performed for the personal well being of an excluded potential donor.

Cholangiography should be routinely performed during procurement of the graft via the cystic duct to define biliary anatomy. This is always important in left hepatectomy as segments of the right hemiliver occasionally drain into the left hepatic duct. Once recognized, the plane of dissection can be modified to accommodate this variation, hence pre-operative cholangiography is not indicated.

In most circumstances the donor will be a parent and the recipient a young child. Hence, the donor will be left with at least a hemiliver and the child will receive a graft more than adequate for its needs. The potential for technical complication is greater if the graft is too large for the child's abdomen. Occasionally, the donor and recipient pairs will be adult and an accurate estimation of liver volume is required.[101,102,103] At least 30% of the standard liver volume is required for safe initial function in the recipient.[104,105,106] It has been long appreciated that resection of up to 70% of hepatic mass is tolerated in most non-cirrhotic individuals.

In the absence of suspicious features in the history and examination, and accompanied by normal standard biochemical parameters and imaging, no further assessment of liver function is required. However, some units also include such tests as arterial ketone body ratio, indocyanine green, caffeine or lignocaine clearance in their assessment.[107] While these investigations may be a helpful adjunct in the workup of the donor, they probably more accurately reflect ongoing research interests of individual LRLT programmes.

DONOR HEPATECTOMY

Anaesthetic induction and positioning of the patient is performed as for a standard liver resection. Invasive monitoring includes radial arterial and central venous lines. An epidural catheter is placed after induction to provide post-operative analgesia. Blood for later autologous transfusion is removed and a cell saver device used to scavenge blood lost during resection.

A chevron or bilateral subcostal incision is favoured and a general laparotomy performed. The ligamentous attachments of the liver are divided and the hepatic veins and inferior vena cava exposed. The porta hepatis is dissected exposing the common bile duct. Intraoperative ultrasound is used to define the position of hepatic veins and confirm their anatomical configuration. In most cases, the allograft will consist of segments II and III, although whole left and right hemilivers have been transplanted.[108,109] The anatomy of the caudate lobe (segment I) precludes its use. In choosing the plane of dissection for the procurement of a segment II and III graft, it is crucial to appreciate the anatomy of segment IV which belongs to the left hemiliver. Otherwise, a segment of devascularized and/or undrained liver in the donor will be the result (Figure 10.11).

Detailed and elegant studies of the appropriate resection planes have been performed by Couinaud and Houssin.[110] Although LRLT programmes with extensive experience have regularly procured grafts by amputation of segments II and III through the umbilical fissure, Couinaud in the French literature has dubbed this as scandalous because of the risk of an infarcted segment IV in the donor or a compromised segment II and III graft.[111] The argument is identical to that of Emond and colleagues in the context of the Chicago group's considerable experience with split livers.[112] The choice of the principle plane is vindicated by plastic corrosion cast studies that illustrate that in some instances, dissection through or to the right of the umbilical fissure may complicate reconstruction in the recipient or leave a partially devascularized cut surface in an otherwise healthy donor.[113,114,115] For this reason some units routinely divide along the principle plane and after removal of the left hemiliver, perform an excision of segment IV on the back table and discard it.[93,116] The advantages of this approach are that single larger vessels and bile ducts are obtained on the graft side and the complete viability and biliary drainage of the remaining donor organ are guaranteed. The disadvantage is a longer parenchymal dissection.

Programmes which select a plane just to the right of the falciform ligament, either identify and preserve the individual elements to segments II, III and IV or deliberately sacrifice some or all of the segment IV structures.[31,117] While this approach undoubtedly facilitates the procurement of the graft, it may add to the complexity of the implantation as multiple arterial or biliary anastomoses may be necessary. Obvious duskiness of

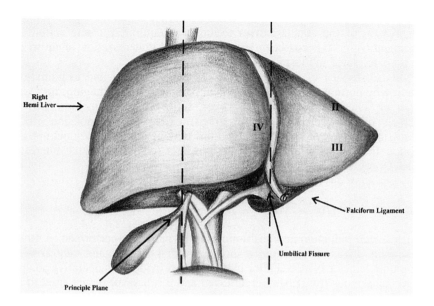

Figure 10.11 (a) and (b) The anatomy of the liver showing the principle plane, umbilical fissure and segments II, III and IV. Segments II and III usually provide the donated liver from living donors, but the anatomy of biliary drainage shown in (b) demonstrates that care must be taken in dissecting the segments to ensure adequate vascular supply and biliary drainage of both the remaining segment IV and the transplanted segments II and III.

segment IV occurred in 10 cases in the Kyoto series and committed the donor to a second resection of the devitalized area to avoid subsequent necrosis.[118]

If the principle plane is chosen, the gall bladder is mobilized from its fossa and an operative cholangiogram performed via the cystic duct to delineate biliary anatomy, after which cholecystectomy is completed. The left hepatic duct is divided at an appropriate level and the donor stump closed. Techniques for parenchymal transection vary. However, the use of some form of ultrasonic apparatus such as the Cavitron Ultrasound Surgical Aspirator (CUSA) facilitates the location of small vascular and biliary tributaries crossing the plane of dissection which can then be cauterized or individually ligated prior to division. Using this technique, the entire transection is possible without need to incur warm ischaemia from inflow occlusion. Blood loss is minimal and the need for heterologous transfusion is rare. At the conclusion of parenchymal division, the left hemiliver is attached only by the left hepatic vein, hepatic artery and portal vein (Figure 10.12). The middle hepatic vein is preserved with the remaining right hemiliver.

In situ perfusion of the left side can be performed, using a cold isotonic flush such as Hartmann's or Ringer's solution, by introducing a cannula into the left portal vein. The left hepatic vein is vented after isolating it from the suprahepatic cava. Standard preservation solutions such as University of Wisconsin (UW) solution, because of high potassium content, are not used for any *in situ* perfusion, but are introduced immediately after removal. Alternatively, an unperfused graft may be rapidly transferred to the back table and flushing initiated by an assistant.

The surgeon simultaneously suture ligates the origins of the divided vessels in the donor, checks haemostasis, inserts a closed suction drain and attends to abdominal closure. If bridging vascular grafts are anticipated, a length of long saphenous vein is

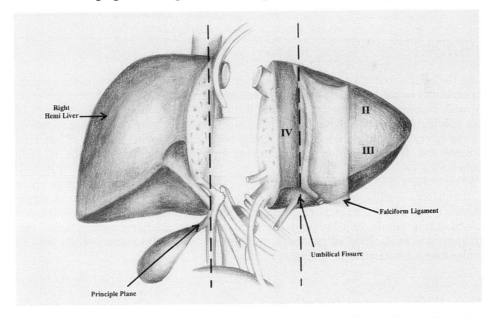

Figure 10.12 Division of the liver in the principle plane provides living hemiliver and split cadaver liver donation and does not place segment IV at risk of inadequate blood supply or biliary drainage.

removed from the donor's thigh. In female donors, the ovarian vein has been used.[119] Alternatively, banked human vascular grafts can be used if available.[120] In situations where the principle plane is chosen and a segments II and III graft is judged appropriate for implantation, further resection of segment IV is then performed *ex situ* with the graft submerged in UW solution using techniques similar to those used for the initial transection.

If a plane closer to the falciform ligament is selected, considerable back table reconstruction may be necessary to lengthen or combine vascular structures suitable for a single portal and a single arterial anastomosis in the recipient. More importantly, the retained segment IV in the donor must be checked for viability and bile leakage.

MORBIDITY AND MORTALITY

Although the literature relating to living donor liver transplantation is growing exponentially, there is a paucity of publications devoted to issues relating to donor morbidity.[121,122,123] A single death has been reported perioperatively as a result of pulmonary embolus. Sporadic instances of major biliary injury, splenic trauma, median nerve injury, subphrenic abscess, peptic ulcer, incisional herniae and minor wound complications have occurred.[31,109,120,124] Morbidity from a retained damaged segment IV is recognized as a bile leak or high post-operative transaminase levels.[108,117] The Kyoto group has reported biliary fistulae in 9 of 118 (8%) of donors where segment IV was left *in situ*.[108] In 42 cases at the same centre where it was included as part of the allograft, no biliary fistula were seen. This suggests that the most common donor complications could be virtually eradicated if the practice of attempting to retain segment IV in the donor is abandoned.

THE FUTURE

Living liver donor transplantation will continue to expand in regions where the concept of brain death is not legally or culturally recognized. Living donor and recipient pairs may well become a precious resource for testing and developing the 'holy grail' of donor specific unresponsiveness. In countries where heart beating donation is accepted, there are rarely insufficient donors to meet the needs of a growing number of potential recipients and living donation has a less major though significant place in such societies.

Furthermore, a renaissance is occurring with split liver transplantation, that is where two recipients, usually an adult and a child, receive partial grafts from a single cadaveric organ donor. For elective cases, graft and patient survival is approaching that obtained from whole and reduced sized livers. The predicted outcome of this development is eradication of paediatric waiting lists and a diminution of the need to consider living donation as an option.

Other living organ donation

Although undertaken infrequently and in a very limited number of centres, heart, lung, pancreas and small bowel can all be transplanted from living donors. The use of hearts from an unrelated living source is opportunistic and could obviously occur only if the donor were to receive a heart from a cadaver donor. Parental donation of a lobe of a lung or portion of small bowel is undertaken in desperate situations, whereas segmental pancreas living related donor transplantation was introduced to minimize graft loss from rejection.

HEART

The use of donor hearts from heart/lung recipients with primary pulmonary hypertension can potentially increase the number of heart transplants performed. Most of the published experience from the so-called 'domino procedure' comes from the United Kingdom and a report from the Papworth Hospital which claimed that the number of heart transplantations performed in their institution had increased by 19% as a result of the domino procedure.[125] Furthermore, there was no difference in one year graft survival when compared to cadaver hearts. The use of a conditioned hypertrophied heart is considered an advantage in heart recipients with increased pulmonary vascular resistance. However, the frequency with which the procedure is performed is limited by heart allocation priorities in different countries and more recently by the increasing success of single and double lung cadaver transplant procedures.

LUNG

Lung transplantation has become an established therapeutic option for patients with end-stage organ transplants and is in part limited by the shortage of organ donors.[126] Thus far, lobar lung transplantation only has been undertaken, although pneumonectomy from a live donor to a patient in need of a single lung transplant is technically feasible. The ethical issues related to doing no harm to the donor are either similar to an order of magnitude or greater than that of living hepatic donation.[127] It is difficult to assess the risk of elective pneumonectomy for a potential adult non-smoking healthy donor. Comparative data from older patients undergoing pneuonectomy for lung cancer is obviously not applicable. Nevertheless, even a mortality rate of 1% would be considered by many to be too high. Perhaps this is why the procedure is not widely practised and the reported experience is small.[128,129] Nevertheless, in urgent situations in the existing milieu of donor shortage, living related lobar lung transplantation is feasible and can provide a donor source in a limited time frame. For the donor, the major concern thus far would seem to be prolonged post-operative air leaks.

PANCREAS

Pancreas transplantation for the treatment of Type I diabetes mellitus was pioneered by, and has long been the domain of, the University of Minnesota.[130] Its success rate has lagged behind that of other solid organs, principally because of technical problems and rejection. In an attempt to increase the success rate by reducing rejection, live donors were accepted onto the programme in Minneapolis in 1979, before the widespread use of cyclosporine.[130] In the subsequent five-year period to 1983, 36 of 89 pancreas transplants performed at that centre were from living related donors. A variety of duct techniques were used, with preference in 1983 given to enteric drainage. The technical complication rate was higher for pancreas transplants from related donors but the overall patient and graft survival of 94% and 43% respectively at one year was better than for cadaver pancreases at 78% and 20% respectively. The best results were achieved in patients who previously had a kidney from the same donor and in those receiving cyclosporine. By 1994, 82 living donor distal segment pancreas transplants had been performed in Minneapolis without donor operative mortality and preservation of the spleen in 95%. When discussing prospective treatment options with potential donors and recipients, the centre emphasizes the increased risk of technical failure of 15%.[131] Over a sixteen-year period to 1994, 31 sequential living donor segmental pancreas and kidney transplants had been performed with a one-year pancreas survival of 82%. It is therefore not surprising that the centre embarked upon and was able to report the results of the first two successful simultaneous kidney and segmental pancreas transplants from living related donors.[131]

Prospective pancreas donor evaluation is done with care. They must be at least ten years older than the age of onset of diabetes in the recipient and no family member other than the recipient can be diabetic. Provided post-intravenous glucose stimulation first phase insulin levels are above the 30th percentile of normal range, donors have not become hyperglycaemic after hemipancreatectomy. Nevertheless, donors have decreased pancreatic alpha and beta cell function[132] and it has been suggested that donors maintain normoglycaemia after hemipancreatectomy by increasing glucose disposal.

SMALL BOWEL

Clinical small bowel transplantation is fraught with problems, in particular, graft versus host disease, rejection, opportunistic infection and post-transplant lymphoproliferative disorders secondary to the use of antilymphocyte preparations. Experience with living related donor small bowel transplantation is limited to a single report in the literature from Germany.[133] The graft from mother to child failed because of rejection twelve days after transplantation.

References

1. Rapaport FT. Living donor kidney transplantation. *Transplant Proc* 1987, **19**: 169.
2. Starzl TE. Living donors: Con. *Transplant Proc* 1987, **19**: 1974.
3. Woodruf MFA. Ethical problems in organ transplantation. *Br Med J* 1964, **1**: 1457.
4. Dosseter JB, Manickavel V. Ethics in organ donation: Contrasts in two cultures. *Transplant Proc* 1991, **23**: 2508–2511.
5. Merrill JP, Murray JE, Harrison JH et al. Successful homotransplantation of the human kidney between identical twins. *JAMA* 1956, **1960**: 277.
6. Murray JE. Nobel Prize Lecture: The first successful organ transplants in man. In *History of transplantation: Thirty-five recollections*. Terasaki PI (Ed.) UCLA Tissue Typing Laboratory, Los Angeles, pages 121–138.
7. Bir MJ, Ramos EL, Danovich GM et al. Evaluation of living renal donors – a current practice of UNOS transplant centers. *Transplantation* 1995, **60**: 322–327.
8. Bay WH, Hebert LA. The living donor in kidney transplantation. *Ann Int Med* 1987, **106**: 719–727.
9. Living related kidney donors (Editorial). *Lancet* 1982, **ii**: 696.

10. Najarian JS, Chavers BN, McHugh LE, Matas AJ. 20 years or more of follow up of living kidney donors. *Lancet* 1992, **340**: 807–810.
11. Jones J, Payne WD, Matas AJ. The living donor – risks, benefits, and related concerns. *Transplantation Reviews* 1993, **7**: 115–128.
12. Offner G, Hoyer F, Meyer B et al. Pre-emptive renal transplantation in children and adolescents. *Transpl Int* 1993, **6**: 125–128.
13. Smits JA, Persijn G, De Meester JM. Living unrelated renal transplantation: the new alternative? *Transpl Int* 1996, **9**: 252.
14. Cecka MJ, Terasaki PI. The UNOS Scientific Renal Transplant Registry. In *Clinical Transplants 1994*, Terasaki TI, Cecka JM (Eds), UCLA Tissue Typing Laboratory, Los Angeles, 1995, pages 1–18.
15. Matas AJ, Najarian JS. Kidney transplantation – the living donor. In *Clinical Transplants 1994*, Terasaki TI, Cecka JM (Eds), UCLA Tissue Typing Laboratory, Los Angeles, 1995, pages 362–363.
16. D'Allessandro AM, Sollinger HW, Kenchtle SJ et al. Living related and unrelated donors for kidney transplantation: A 28 year experience. *Ann Surg* 1995, **222**: 353–364.
17. Terasaki PI, Cecka JM, Gjertson DW, Takemoto S. High survival rates of kidney transplant from Starzl and living unrelated donors. *N Engl J Med* 1995, **333**: 333–336.
18. Jones JW, Gillingham KJ, Sutherland DER et al. Successful long term outcome with O haplotype matched living related kidney donors. *Transplantation* 1994, **57**: 512–515.
19. Stratta RJ, Armbrust MJ, Oh CS et al. Withdrawal of steroid immunosuppression in renal transplant recipients. *Transplantation* 1988, **45**: 1323–1328.
20. Kahn D, Ovnat A, Pontin AR et al. Long term results with elective cyclosporin withdrawal at 3 months after renal transplantation: Appropriate for living related transplants. *Transplantation* 1994, **58**: 1410–1412.
21. Chugh KS, Jha V. Differences in the care of ESRD patients well world wide: required the sources and future outlook. *Kid Int* 1995, (48 Suppl) **50**: S7–S13.
22. Feduska NJ, Cecka JM. Donor factors. In *Clinical Transplants 1994*, Terasaki TI, Cecka JM (Eds), UCLA Tissue Typing Laboratory, Los Angeles, 1995, pages 381–394.
23. Hilton BA, Starzomski RC. Family decision making about living related kidney donation. *ANNA Journal* 1994, **20**: 346–355.
24. Smith MD, Cappell DF, Province A et al. Living related kidney donors: A multicentre study of donor education, socioeconomic adjustment and rehabilitation. *Am J Kid Dis* 1986; **8**: 223.
25. Lynch SV, Strong RW, Ong TH, Pillay SP, Balderson GA. Reduced size liver transplantation in children. *Transplant Rev* 1992; **6**: 89–101.
26. de Ville de Goyet J. Split liver transplantation in Europe 1988–1993. *Transplantation* 1995, **59**: 1371–1376.
27. Whitington PF, Siegler M, Broelsch CE. In *Organ Replacement Therapy: Ethics, Justice, Commerce*, Land W, Dossetor JB (Eds), Springer-Verlag, Berlin, 1990, page 117.
28. Broelsch CE, Burdelski M, Rogiers X et al. Living donor for liver transplantation. *Hepatology* 1994, **20** (suppl 1): 49.
29. Sloof MJH. Reduced size liver transplantation, split liver transplantation and living related liver transplantation in rotation to the donor organ shortage. *Transplant Int* 1995, **8**: 65.
30. Morimoto T, Yamaoka Y, Tanaka K, Ozawa K. Quality of life among donors of liver transplants to relatives. *N Engl J Med* 1993, **329**: 363.
31. Jurim O, Shackleton CR, McDiarmid SV et al. Living donor liver transplantation at UCLA. *Am J Surg* 1995, **169**: 529–532.
32. Otte JB. Is it right to develop living related liver transplantation? Do reduced and split livers not suffice to cover the needs? *Transplant Int* 1995, **8**: 69.
33. Hunt R, Arras J. *Ethical Issues in Modern Medicine*, Mayfield, Palo Alto, CA, 1977, page 2.
34. Moore FD. Three ethical revolutions: ancient assumptions remodelled under pressure of transplantation. *Transplant Proc* 1988, **20**: 1061.
35. Ramos EL, Tischer CC. Recurrent diseases in the kidney transplant. *Am J Kid Dis* 1994, **24**: 142–154.

36. Cameron JS. Recurrent renal disease after renal transplantation. *Curr Opin in Nephrol Hypertens* 1994, **3**: 602–607.
37. Roy DM, Thomas PP, Dakshinamurthy KV et al. Long term survival in living related donor renal allograft recipients with hepaitits C infection. *Transplantation* 1994, **58**: 118–119.
38. Russell S, Jacob RG. Living related organ donation: The donor's dilemma. *Patient Educ Couns* 1993, **21**: 89–99.
39. Reither AM, Mahler E. Organ donation: Psychiatric, social and ethical considerations. *Psychosomatics* 1995, **36**: 336–343.
40. Ferhman-Ekholm I, Gabel H, Magnusson G. Reasons for not accepting living kidney donors. *Transplantation* 1996, **61**: 1264–1265.
41. Jones JW, Halldorson J, Elick B et al. Unrecognised health problems diagnosed during living donor evaluation: a potential benefit. *Transplant Proc* 1993, **25**: 3083–3084.
42. Kumar A, Kumar RZ, Srinadh DS et al. Should elderly donors be accepted in live related renal transplant programs? *Clin Transplantation* 1994, **8**: 523–526.
43. Lezaic V, Djukanovic L, Blagojevic-Lazic R et al. Living related kidney donors over 60 years old. *Transpl Int* 1996, **9**: 109–114.
44. Yoon YS, Bang BK, Jin DC et al. Factors influencing long term outcome of living donor kidney transplantation in the cyclosporin era. *Clinical Transplants* 1992, **6**: 257–266.
45. American Society of Anaesthesiologists Classification of Physical Status. *Anaesthesiology* 1963, **24**: 111.
46. Warden JC, Holland R. Anaesthesia mortality. *Anaes Intens Care* 1995, **23**: 255–256.
47. Barbosa J, King R, Goetz SC et al. Histocompatibility antigen (HLA) in families with juvenile insulin dependent diabetes mellitus. *J Clin Invest* 1977, **60**: 898.
48. Hannig VL, Erickson SM, Phillips JA. Utilisation and evaluation of living related donors for patients with adult polycystic kidney disease. *Am J Med Genet* 1992, **44**: 409–412.
49. Terasaki PI, Koyama H, Ceka JM et al. The hyperfiltration hypothesis in human renal transplantation. *Transplantation* 1994, **57**: 1450–1454.
50. Robitaille P, Mongeau JG, Lortie L et al. Long term follow up of patients who underwent unilateral nephrectomy in childhood. *Lancet* 1985, **1**: 1297–1299.
51. Ter Wee PM, Tegzess AM, Donker AJ. Reserve renal filtration capacity before and after kidney donation. *J Int Med* 1990, **228**: 393–399.
52. Sesso R, Whelton PK, Klag MJ. Effective age and gender on kidney function in transplant donors: A prospective study. *Clin Nephrol* 1993, **40**: 31–37.
53. Moxey-Mims MM, Venuto RC, Feld LG, Bock GH. A proposed method for the non-invasive evaluation of renal asymmetry in a living related donor candidate. *Clin Neph* 1997, **42**: 291–294.
54. Yang SC, Suh DH, Suh JS et al. Anatomical study of the left renal vein and draining veins as encountered during living donor nephrectomy. *Transpl Proc* 1992, **24**: 1333–1334.
55. Gleeson MJ, Brennan RP, McMullin JT. Renal angiography in potential living related renal donors in Iraq. *Clin Radiol* 1988, **39**: 625–627.
56. Flechner SM, Sandler CM, Houston GK et al. 100 living related kidney donor evaluations using digital subtraction angiography. *Transplantation* 1985, **40**: 675–678.
57. Weinstein SH, Navarre RJ, Loening SA, Corry RJ. Experiences with live donor nephrectomy. *J Urol* 1980, **124**: 321–323.
58. Rossi M, Alfani D, Berloco P et al. Bench surgery for multiple renal arteries in kidney transplantation from living donor. *Transplant Proc* 1991, **23**: 2328–2329.
59. Guerra EE, Didone EC, Zanotelli ML et al. Renal transplants with multiple arteries. *Transplant Proc* 1992, **24**: 1868.
60. Frawley J, Fairnsworth RH. Adult donor kidney transplantation in small children: A surgical technique. *Aust NZ J Surg* 1990, **60**: 911–912.
61. Spital A, Spital M. Living kidney donation: Attitudes outside the transplant center. *Arch Int Med* 1988, **148**: 1077–1080.
62. Flye MW, Anderson CB, Woodle ES et al. Living related renal transplantation. In *Atlas of Organ Transplantation*, Flye MW (Ed.), 1995, WB Saunders, Philadelphia, pages 127–134.

63. Streem SB, Novick AC, Steinmuller DR et al. Flank donor nephrectomy: Efficacy in the donor and recipient. *J Urol* 1989, **141**: 1099–1101.
64. Bentley FR, Amin M, Garrison RN et al. Value of systemic heparinisation during living donor nephrectomy. *Transplant Proc* 1990, **22**: 346–348.
65. Connor WT, van Buren CT, Floyd M et al. Anterior extraperitoneal donor nephrectomy. *J Urol* 1981, 443–447.
66. Cosimi BA. The donor and donor nephrectomy. In *Kidney Transplantation, Principles and Practice*, 4th edn, PJ Morris (Ed.), WB Saunders, 1994, pages 56–70.
67. Blohme I, Fehrmen I, Norden G. Living donor nephrectomy. *Scand J Urol Nephrol* 1992, **26**: 149–153.
68. Ottelin MC, Bueschen AJ, Lloyd LK et al. Review of 333 living donor nephrectomies. *South Med J* 1994, **87**: 61–64.
69. Waples MJ, Beizer FO, Uehling DT. Living donor nephrectomy: A 20 year experience. *Urol* 1995, **45**: 207–210.
70. Duraj F, Tyden G, Blom B. Living donor nephrectomy: How safe is it? *Transplant Proc* 1995, **27**: 803–804.
71. Dunn JF, Richie RE, MacDonell RC et al. Living related kidney donors: a 14 year experience. *Ann Surg* 1986, **203**: 637–643.
72. De Marko T, Amin M, Harty JI. Living donor nephrectomy: factors influencing morbidity. *J Urol* 1982, **127**: 1082–1083.
73. Francis DMA, Walker RJ, Becker GJ et al. Kidney transplantation from living related donors: a 19 year experience. *Med J Aust* 1993, **158**: 244–247.
74. Williams SL, Oler J, Jorkasky DK, Narkun-Burgess DM, Nolan CR, Norman JE et al. Forty-five year follow-up after uninephrectomy. *Kid Int* 1993, **43**: 1110–1115.
75. Liounis B, Roy LP, Thompson JF et al. The living, related kidney donor: A follow up study. *Med J Aust* 1988, **148**: 438–444.
76. Spital A. Life insurance for kidney donors – an update. *Transplantation* 1988, **45**: 819.
77. Narkun-Burgess DM, Nolan CR, Norman JE et al. Forty-five year follow-up after uninephrectomy. *Kid Int* 1993, **43**: 1110–1115.
78. Kiprov DD, Colvin RB, McCluskey RT. Focal and segmental glomerulosclerosis and proteinuria associated with unilateral renal agenesis. *Lab Invest* 1982, **46**: 275–281.
79. Thorner PS, Arbus GS, Celermajer DS et al. Focal segmental glomerulosclerosis and progressive renal failure associated with a unilateral kidney. *Paediatrics* 1984, **73**: 806–810.
80. Novick AC, Gephardet G, Guz B et al. Long term follow up after partial removal of a solitary kidney. *N Engl J Med* 1991, **325**: 1058.
81. Jones JW, Acton RD, Elick B et al. Pregnancy following kidney donation. *Transplant Proc* 1993, **25**: 3082.
82. Buszta C, Steinmuller DR, Novick AC et al. Pregnancy after donor nephrectomy. *Transplantation* 1985, **40**: 651–654.
83. Sharma VK, Enoch MD. Psychological sequelae of kidney donation. A 5 to 10 year follow up study. *Acta Psychiatry Scand* 1987, **75**: 264–267.
84. Gouge F, Moore J, Bremer BA et al. The quality of life of donors, potential donors and recipients of living related donor renal transplantation. *Transplant Proc* 1990, **22**: 2409–2413.
85. Tanabe K, Takahashi K, Sonda K et al. ABO incompatible living kidney donor transplantation: results and immunological aspects. *Transplant Proc* 1995, **27**: 1020–1023.
86. Park K, Kim YS. The use of kidneys from living related or unrelated donors: Pro. In *Clinical Transplants 1994*, Terasaki TI, Cecka JM (Eds), UCLA Tissue Typing Laboratory, Los Angeles, 1995, pages 374–375.
87. Yang SC, Lee DH, Rha KH et al. Retroperitoneoscopic living donor nephrectomy: Two cases. *Transplant Proc* 1994, **26**: 2409.
88. Ratner LE, Ciseck LJ, Moore RG et al. Laparoscopic live donor nephrectomy. *Transplantation* 1995, **60**: 1047–1049.

89. Teraski PI, Cecka JM, Gjertson BW et al. High survival rates of kidney transplants from spousal and living unrelated donors. *N Engl J Med* 1995, **333**: 333–336.
90. Spital A. Unrelated living kidney donors. An update on attitudes and use among US transplant centres. *Transplantation* 1994, **57**: 1722–1726.
91. The living non-related Renal Transplant Study Group. Physicians attitudes toward living non-related renal transplantation (LNRT). *Clin Transplantation* 1993, **7**: 289–295.
92. Singer PA, Siegler M, Whitington PF et al. Ethics of liver transplantation with living donors. *N Engl J Med* 1989, **321**: 620.
93. Raia S, Nery JR, Mies S. Liver transplantation from live donors. *Lancet* 1989, **2**: 497.
94. Strong RW, Lynch SV, Ong TH, Matsunami H, Koido Y, Balderson GA. Successful liver transplantation from a living donor to her son. *N Engl J Med* 1990, **322**: 1505–1507.
95. Nagasue N, Kohno H, Matsuo S et al. Segmental (partial) liver transplantation from a living donor. *Transplant Proc* 1992, **24**: 1958–1959.
96. Broelsch CE, Emond JC, Whitington PE et al. Application of reduced size liver transplants as split grafts, auxiliary orthotopic grafts and living related segmental transplants. *Ann Surg* 1990, **212**: 368.
97. Emond JC. Clinical application of living related liver transplantation. *Gastroenterol Clin North Am* 1993, **22**: 301–315.
98. Renz JF, Mudge CL, Heyman MB et al. Donor selection limits use of living related liver transplantation. *Hepatology* 1995, **22**: 1122–1126.
99. Ozawa K. 'Preoperative evaluation'. In *Living Related Donor Liver Transplantation*. Karger, Basel, 1994.
100. Kam I, Lynch S, Svanas G et al. Evidence that host size determines liver size: studies in dogs receiving orthotopic liver transplants. *Hepatology* 1987, **7**: 362–366.
101. Habib N, Tanaka K. Living related liver transplantation in adult recipients: a hypothesis. *Clin Transplant* 1995, **9**: 31–34.
102. Hashikura Y, Makuuchi M, Kawasaki S et al. Successful living related partial liver transplantation to an adult patient. *Lancet* 1994, **343**: 1233–1234.
103. Ichida T, Matsunami H, Kawasaki S et al. Living related donor liver transplantation from adult to adult for primary biliary cirrhosis. *Ann Intern Med* 1995, **122**: 275–276.
104. Yamaoka Y, Tanaka K, Ozawa K. Liver transplantation from living related donors. In *Clinical Transplants 1993*, Terasaki TI, Cecka JM (Eds), UCLA Tissue Typing Laboratory, Los Angeles, 1994, pages 179–184.
105. Urata K, Kawasaki S, Matsunami H et al. Calculation of child and adult standard liver volume for liver transplantation. *Hepatology* 1995, **21**: 1317–1321.
106. Kawasaki S, Makuuchi M, Matsunami H et al. Preoperative measurement of segmental liver volume of donors for living related liver transplantation. *Hepatology* 1993, **18**: 1115–1120.
107. Kawasaki S, Makuuchi M, Matsushita K *et al*. The arterial ketone body ratio in living related donor. *Transplantation* 1994, **58**: 1412–1414.
108. Ikai I, Marimoto Y, Yamamoto Y et al. Left lobectomy of the donor: operation for larger recipients in living related liver transplantation. *Transplant Proc* 1996, **28**: 56–58.
109. Bassas A, Malago M, Rogiers X et al. Living related liver transplantation in children. *Transplant Proc* 1996, **28**: 428–429.
110. Couinaud C, Houssin D. *Controlled partition of the liver for transplantation, anatomical limitations*. Published C. Couinaud, Paris, 1991 (ISBN 2-903 672-03-2).
111. Couinaud C. Un scandale: segment IV et transplantation due foie. *J Chir Paris* 1993, **130**: 443–446.
112. Emond JC, Whitington PF, Thistlewaite JR et al. Transplantation of two patients with one liver. *Ann Surg* 1990, **212**: 14.
113. Kazemier G, Hesselink EJ, Lange JF et al. Dividing the liver for the purpose of split grafting or living related grafting: A search for the best cutting plane. *Transplant Proc* 1991, **23**: 1545.
114. Kazemier G, Hesselink EJ, Terpstra OT. Hepatic anatomy. *Transplantation* 1990, **49**: 1029.

115. Czerniak A, Loton G, Hiss Y et al. The feasibility of in vivo resection of the left lobe of the liver and its use for transplantation. *Transplantation* 1989, **48**: 26.
116. Yeung CK, Ho JK, Lau WY et al. Institution of a pediatric liver transplantation program with living related orthotopic liver transplantation: initial experience in Hong Kong. *Transplant Proc* 1994, **26**: 2215–2217.
117. Boillot O, Voigho E, Dawahra M, Benchetrit S, Procheron J, Gille D. Surgical technique of left lateral hepatic lobectomy in a related living donor for pediatric transplantation. *Transplant Proc* 1995, **27**: 1708–1709.
118. Ozawa K. The operative process. In *Living Related Donor Liver Transplantation*. Karger, Basel, 1994, pages 85–152.
119. Tokunaga Y, Tanaka K, Yamaoka Y, Ozawa K. Portal vein graft in living related hepatic transplantation. *J Am Coll Surg* 1994, **178**: 297–299.
120. Wood RP, Kak SM, Ozaki CF et al. Development of the living related donor liver transplant programme at the Texas Medical Center: initial results and surgical complications. *Transplant Proc* 1993, **25**: 50–52.
121. Goldman LS. Liver transplantation using living donors. Preliminary donor psychiatric outcomes. *Psychosomatics* 1993, **34**: 235–240.
122. Shimahara Y, Awane M, Yamaoka Y et al. Safety and operative stress for donors in living related partial liver transplantation. *Transplant Proc* 1993, **25**: 1081–1083.
123. Yamaoka Y, Morimoto T, Inamoto T et al. Safety of the donor in living related liver transplantation, an analysis of 100 parental donors. *Transplantation* 1995, **59**: 224–226.
124. Emond JC, Heffron TG, Kork EO et al. Improved results of living related liver transplantation with routine application in a pediatric programme. *Transplantation* 1993, **55**: 835–840.
125. Oaks TE, Aravot D, Dennis C et al. Domino heart transplantation: The Papworth experience. *J Heart Lung Transplant* 1994, **13**: 433–437.
126. Davis RD, Pasqu MK. Pulmonary transplantation. *Ann Surg* 1995, **221**: 14–28.
127. Kramer MR, Sprung CL. Living related donation in lung transplantation. Ethical considerations. *Arch Int Med* 1995, **155**: 1734–1738.
128. Starnes VA, Barr ML, Cohen RG. Lobar transplantation. Indications, technique and outcome. *J Thoracic Cardiovasc Surg* 1994, **108**: 403–410.
129. Cohen RG, Barr ML, Schenkel FA et al. Living related donor lobectomy for bilateral lobar transplantation in patients with cystic fibrosis. *Ann Thoracic Surg* 1994, **57**: 1423–1427.
130. Sutherland DER, Goetz FC, Najarian JS. Pancreas transplants from related donors. *Transplantation* 1984, **38**: 625–633.
131. Gruessner RW, Sutherland DE. Simultaneous kidney and segmental pancreas transplants from living related donors – the first two successful places. *Transplantation* 1996, **61**: 1265–1268.
132. Seaquist ER, Robertson RP. The effects of hemi-pancreatectomy on pancreatic alpha and beta cell function in healthy human donors. *J Clin Invest* 1992, **89**: 1761–1766.
133. Delk E, Mengel W, Hamelmann H. Small bowel transplantation: Report of a clinical case. *Prog Paediatr Surg* 1988, **25**: 90–96.

CHAPTER 11

Organ and tissue preservation

VERNON C. MARSHALL

Historical background 200
Loss of viability during storage: ischaemic/hypoxic damage 202
Reperfusion damage after reimplantation 202
Effects of hypothermia 203
Principles of organ and tissue preservation 203
Extracorporeal organ storage 204
Individual components of solutions for continuous perfusion
 and single-pass flush 208
Importance of storage temperature 211
Preventing reperfusion damage 212
Applied strategies of organ preservation 212
Preservation of individual organs for clinical transplantation 215
Preservation of other tissues: non-viable scaffolding or viable graft? 220
The future 222
References 222

Historical background

Attempts to preserve human tissues and organs from putrefaction and decay after death began in antiquity. Embalming and mummifying techniques were brought to a considerable degree of technical skill by the Egyptian dynasties. Replacement of cadaveric blood by an appropriate embalming fluid was a harbinger of later flushing and perfusing organ preservation fluids. Modern modifications of such fluids can preserve function and form of a wide variety of organs and tissues, enhancing their subsequent transplantation as autograft, allograft or xenograft. In clinical practice the times required and provided for extracorporeal storage of differing tissues extend over a large range — minutes, days, weeks, months, or even years. Prolonged storage for the latter times has only been possible for individual cells or small aggregations of cells and tissues. This can now include frozen eggs and embryos. For vascularized organs and composite tissue blocks the maximum tolerated storage times possible are still numbered only in days.

Major pathways of progress can be summarized. The fundamental importance of *cold* in extracorporeal storage of all tissues and organs was recognized very early. Blood, the first widely transferred tissue allograft, illustrated well the important principle of cold preservation. Cooling to around 4 °C extended storage from a few hours out to two or three weeks. Hypothermia was also applied to simple surface cooling of organs, and was observed to be enhanced by flushing the contained blood from the organ by a chilled preserving solution. Search for optimal preserving solutions as modern elixirs of life-support continues today.

In the 1930s the pioneering experiments of the Nobel-prize-winning French surgeon Alexis Carrel, and the aviator technologist Charles Lindbergh, developed sophisticated perfusion techniques to procure and culture organs for transplantation. Ground rules were established for continuous perfusion of organs outside the body. Expert technology, perfect asepsis and controlled biological conditions were prerequisites. Failure occurred if any one of these aspects was deficient.[1]

The discovery of cryoprotectants in the late 1940s was another major stimulus to extracorporeal preservation. This made possible the storage of individual cells, of simple cell clusters and of tissues like blood with dispersed cells, under conditions approaching the requirements for optimum long-term preservation – freezing to subzero temperatures using agents such as glycerol or dimethyl sulphoxide, prolonged storage in liquid nitrogen and subsequent thawing for implantation.

Application of freezing to more complex vascularized organs has not found clinical application. The ability of polar land and sea creatures to function in temperatures close to freezing by using built-in antifreeze mechanisms, and the means by which hibernating animals can withstand the extremes of winter, are aspects of great interest. To date these mechanisms express species differences unable to be translated to clinical human practice. Unfortunately refrigeration of humans for long-term intergalactic space travel or to await a future millenium remains fictional rather than functional science.

In clinical practice requirements for preservation of different tissues and organs have been shown to have individual specificities, but several common denominators have aided search for a unified approach to preservation, applicable across all tissues and organs.

Living donors still contribute significantly to transplantation. Here the time of storage can often be made quite short. When transplantation is predominantly from the dead to the quick (as is the case with most forms of clinical transplantation apart from bone marrow) preservation times need to be extended (to 24 hours or more). The time gained is needed to finalize tissue typing and cross-matching, to exclude by careful autopsy the risk of transmitting disease from the cadaver donor, to mobilize chosen recipients and operating teams and to transport organs and tissues from procurement site to recipient hospital. This may require local, national, international or intercontinental surface or air transport.

The above aims and objectives have been largely fulfilled by modern organ procurement and preservation techniques. Other objectives have not – several stand out. We cannot adequately revivify cadaver organs, nor can we establish definitively their certain viability and effective function before transferring them to a new host. Initially impaired function rates of 5% or 10% of most cadaver organ grafts are still higher than we would wish. This precious extension of time given by preservation is also available for immunological manipulation of the cadaver donor organ (or the new host) to improve graft function and survival. Few of the promising experimental approaches so far investigated are as yet applicable to clinical practice. Some evidence suggests that ischaemic damage can potentiate immunological rejection. Ischaemia certainly releases a variety of cell to cell messengers which are relevant to inflammatory and immune responses; so preservation times should never be needlessly prolonged.

The sections that follow consider the effects of ischaemic and reperfusion injury, the development and application of preservation solutions as they concern the principles and practice of organ transplantation, the variations between requirements for individual organs and tissues; and the differences between preservation for transplantation of viable organs or tissues, and of non-viable grafts which can serve as useful human scaffolding.

Loss of viability during storage: ischaemic/hypoxic damage

Extracorporeal storage is accompanied by delayed, but ultimately inevitable deterioration of viability of the organ to be transplanted. Once blood flow to the organ ceases, rapid anoxic depletion of cellular energy sources occurs. Metabolism transfers from oxidative phosphorylation to the much less efficient anaerobic glycolysis. The latter generates only one twentieth of the high energy phosphate compounds normally provided by perfusion with oxygenated blood.

A critical effect of ischaemic anoxia is to inactivate progressively (by fuel depletion) the enzyme system controlling the cell membrane sodium pump (sodium/potassium adenosine triphosphatase: Na/K ATP-ase). Individual cells gradually but progressively gain sodium and water, and lose potassium, giving harmful cellular oedema and electrolyte imbalance of the cytosol. Mitochondrial respiration and calcium/magnesium ATP-ase are also inactivated; so calcium also enters cells and mitochondria, and magnesium is depleted. Anaerobic metabolism generates hydrogen ions, causing intracellular acidosis. Initially this may be autoregulatory and protective against further damage; but acidosis ultimately activates lysosomal enzymes contributing to cellular autolysis and death.

Parenchymal cells of transplantable organs and tissues show relatively little variation in their tolerance of ischaemia, although rapidly metabolizing tissues are least tolerant. Most can tolerate ischaemia for 30 to 60 minutes at body temperature without permanent damage. Most are irreversibly damaged if ischaemia exceeds 120 minutes.

Non-parenchymal cells, particularly those of the delicate vascular lining of the microcirculation, have been shown to be more sensitive to damage suffered during storage and after reimplantation; and have thus tended to be a major limiting factor in storage times. These cells include vascular endothelial cells, hepatic sinusoidal cells, and macrophages.

Reperfusion damage after reimplantation

Reperfusion injury upon revascularization is an additional hazard after prolonged ischaemia. Accumulation of metabolites generated under anaerobic conditions can set the stage for further injury when blood flow and oxygen supply are restored. Harmful compounds such as active free radicals of oxygen, which are normally produced only in minute quantities and are rapidly cleared by endogenous scavenging mechanisms, can lead to further cellular and microvascular injury. Additional strategies are required to deal with these problems.

Effects of hypothermia

Organ preservation techniques begin with cooling: this remains the most important method of diminishing the detrimental effects of ischaemic anoxia. Cooling diminishes metabolic activity and curtails oxygen demand of the preserved organ. However, metabolism still persists during hypothermia, even in ice at 0 °C.[2] Oxygen consumption decreases exponentially to reach 10% of normal at 10 °C, 5% at 5 °C, and 3% at 0 °C.[3] Cooling from 37 °C (body temperature) to around 0–5 °C (storage temperatures) extends the tolerance of most organs to ischaemia from between 1–2 hours to about 12 hours. Hypothermia alone is only a partial solution. Cold does not slow all biological functions uniformly. Several metabolic processes (normally concordant) become discordant when the organ is cooled. Transmembrane passive diffusion of ions is not appreciably affected by hypothermia. But active transport mechanisms, such as those governed by Na/K ATP-ase and Ca/Mg ATP-ase, are inhibited below 10 °C.[4] Hypothermia does not prevent but merely slows deleterious cell swelling during storage (Figure 11.1).

Principles of organ and tissue preservation

These are:
- Reduce by hypothermia the metabolic demand of the stored organ or tissue.
- Maintain viability of macro- and microvasculature, interstitium, parenchyma, and ducts using physicochemical, biological and pharmacological agencies.

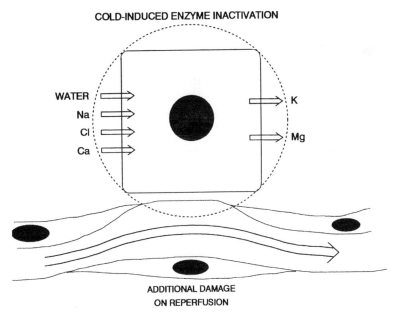

Figure 11.1 The hypothermically stored cell.

- Integrate strategies of organ procurement, storage, and reimplantation so that function in the recipient is immediate and continuous.

Extracorporeal organ storage

Two main techniques of cold storage have been used: machine perfusion or simple hypothermic storage.

MACHINE PERFUSION

This technique was most widely used initially. Continuous perfusion aims to simulate conditions *in vivo*. The organ can be continually supplied with oxygen and nutrients; waste products are continuously removed. However, the method is complex, costly and more liable to complication (by technical error or aseptic breach) than is simple static cold storage. Perfusion techniques still provide the maximum extension of successful storage of most organs. Carrel's experiments in the 1930s showed the way. Heart–lung machines applied to cardiac surgery demonstrated the feasibility of short-term extracorporeal machine support from the 1940s; and by the 1960s kidneys could be successfully cold-perfused for three days. The essential requirements were sterile non-thrombogenic and non-allergenic delivery tubing, an organ chamber, a pump and heat exchanger, a mechanism for oxygenation (membrane oxygenation or surface diffusion); appropriate monitoring of perfusate temperatures and pressures, and an effective perfusion fluid (Figure 11.2).

Advances in technology have meant that perfusion machines have become progressively smaller and more compact. Perfusion can be continuous or intermittent,

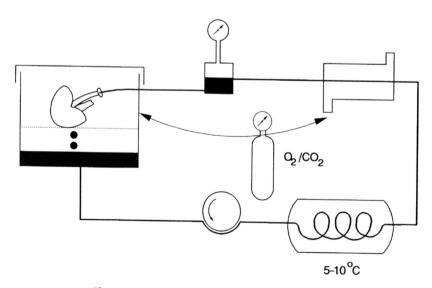

Figure 11.2 Continuous hypothermic perfusion.

and can if necessary be combined with simple hypothermic storage; the latter being more suitable for long distance transport.

Stable solutions were needed to enhance the rapidity and uniformity of cooling by continuous or single pass perfusion and to provide solutes which improved preservation of vasculature, parenchyma and interstitium. Blood was demonstrably not the ideal medium; cold blood adds rheological problems of greater viscosity and sluggish flow. Simple balanced electrolyte solutions mimicking the extracellular milieu, such as Ringer's-lactate solution or saline, were shown to be unsatisfactory organ-preserving solutions which exacerbated interstitial and cellular oedema. For continuous perfusion attention was then directed by Belzer and others at plasma-like solutions containing albumin, dextrans, starch or other colloids (combined with extracellular electrolyte content) to provide oncotic support and to prevent interstitial oedema.

Cold perfusion has shown greater capacity to revive and resuscitate organs previously damaged by warm ischaemia. This is its area of greatest potential; together with the prospect (as yet unrealized) of infusing immunosuppressants or agents capable of deleting antigen-presenting cells prior to implantation.

SIMPLE HYPOTHERMIC STORAGE

Currently, however, perfusion preservation is rarely used, because simple storage techniques are applicable more easily to all organs; and because the advantages of machine perfusion are incrementally small, and are most apparent at the extremity of the time limits of tolerated storage. These advantages are offset by the complexity, the cost, and the potential for organ damage which are inherent with perfusion.

Collins ushered in a new era of simple hypothermic storage by showing that a solution of crystalloid solutes alone, more resembling 'intracellular' than 'extracellular' composition, and quite free of colloid, could extend renal preservation reliably for 48 hours and more.[5] Subsequently solutions of widely different composition, based on the citrate anion (Marshall, Ross), aminoacids (Bretschneider) or simple sugar (sucrose) were shown to be equally or more effective than the original Collins' solution and the significance of 'intracellular' composition became less relevant.

Composition of equally effective flush solutions can differ significantly one from the other; but each has components aimed at preventing cellular swelling, and each contains a buffer. Various adjuvant solutes can further enhance preservation (Table 11.1).

COLLINS' SOLUTIONS

Prolonged kidney preservation by simple cold storage became possible after the development of Collins' 'intracellular' solutions.[5,6,7,8,9] However, initial function after 72-hour cold storage of kidneys was poor, and any added warm ischaemia markedly diminished the efficacy of these solutions. The early Collins' solutions (C2, C3, C4) had high concentrations of potassium, magnesium, phosphate, sulphate and glucose (120 millimolar). Precipitation of magnesium phosphate was a defect. Additions of procaine (C3, C4) and of phenoxybenzamine (C4) were found unnecessary, and in the case of procaine, harmful. Subsequently mannitol, magnesium, procaine and phenoxybenzamine were omitted from the very widely used EuroCollins' solution containing high concentrations of potassium (110 millimolar), phosphate (60 millimolar) and increased glucose (180 millimolar). The higher glucose concentrations

Table 11.1 Major components and modifications of flushing solutions for hypothermic storage

Solution	Major components	Modifications	Main organs
Collins'	Glucose, mannitol, phosphate, high K, low Na and Cl	EuroCollins' – mannitol omitted, sucrose for glucose improves efficiency	Kidney, liver, pancreas, small bowel, lung
Citrate	Magnesium and potassium citrate, mannitol, sulphate	Hypertonic or istonic equally effective	Kidney, liver, pancreas, small bowel
PBS	Sucrose, phosphate buffer	Sucrose 120 or 140 mmol	Kidney, liver, pancreas
Bretschneider – HTK	Histidine, tryptophane, alpha-keto glutarate, low Na and K	Further additives can enhance efficiency	Heart, kidney, liver
UW	HES, lactobionate, raffinose, phosphate, adenosine, allopurinol, glutathione, insulin, steroids, high K and low Na	Improved or unaltered by: removal HES, insulin, steroids, reversal Na/K ratio, sucrose for raffinose. Further additives can enhance efficiency	Gold standard for multiorgan storage. Liver, pancreas, kidney, heart, lung, small bowel
Sodium lactobionate Sucrose – SLS	Lactobionate, sucrose, reversed Na/K ratio, otherwise as for UW	Additives can enhance efficiency (chlorpromazine, PEG)	Preferred UW – modification for multiorgan storage?
Stanford, St Thomas's	ECF-based. High Na and low K, magnesium, buffer. Ca low or absent	Further additives can enhance efficiency	Heart and lung. Cardioplegia for open heart surgery

reduce the potential efficacy of EuroCollins' solution; and make it less effective in preservation of organs other than kidney. Glucose is slowly permeable across cell membranes, stimulates cellular anaerobic glycolysis and augments tissue acidosis. These potentially harmful effects become more apparent with prolonged storage times. Replacement of glucose by other less permeable and non-metabolizable solutes (sucrose, mannitol) improved the results of renal preservation by Collins' solution in rat[10] and dog kidneys.[11] A significant reduction in post-transplantation renal failure occurred when human kidneys were preserved in EuroCollins' solution with mannitol substituted for glucose.[12] Replacement of glucose by sucrose may be even more relevant in liver and pancreas grafts, whose cells are more permeable to glucose.

CITRATE SOLUTIONS

Citrate solution[13,14] (Marshall/Ross) was developed as an alternative to Collins' solution. The solution also has high concentrations of potassium and magnesium, but citrate replaces phosphate and mannitol replaces glucose. Citrate provides buffering capacity as well as interacting with magnesium to form impermeable, stable chelates. Isotonic and hypertonic solutions are equally effective; the sulphate ion is unimportant and can be replaced by other anions. Mannitol can be replaced by sucrose with equivalent results; replacement of mannitol with glucose was detrimental in kidney preservation. An acid pH (6.5) proved also deleterious. Solutions with levels of pH from 7.1 to 7.8 were equally effective.[15] Citrate solutions gave results comparable to Collins' solutions experimentally and clinically for kidney, liver and pancreas grafts.

SUCROSE SOLUTIONS

The demonstration that sucrose improved efficacy of preserving solutions[10] led to formulation of a simple isosmolar sodium phosphate-buffered sucrose solution (PBS). This very simple solution gave excellent preservation of rat, pig, dog and human kidneys.[16,17,18,19] Sucrose-based solutions, optimally at a sucrose concentration of around 140 millimolar, are effective also in preservation of other organs.

BRETSCHNEIDER'S HTK SOLUTION

This solution was initially developed for cardioplegia in open heart surgery, and was subsequently shown to be effective in experimental and clinical preservation of kidney and liver.[20] It contains relatively impermeable amino-acid and other solutes: the main component is histidine (200 millimolar, molecular weight 155 Daltons), with added mannitol, tryptophane (MW 146) and alpha-ketoglutaric acid (MW 204), with low electrolyte concentrations of sodium, potassium and magnesium. Histidine is an excellent buffer; and tryptophane, histidine and mannitol can act as free radical scavengers. The solution has given a low incidence of delayed function in kidney transplantation clinically; and was shown to be better than EuroCollins' solution in kidney transplantation in a prospective randomized clinical trial.[21]

Table 11.2 Flushing solutions for hypothermic storage: University of Wisconsin solution (UW)

Potassium (mmol/l)	135
Sodium (mmol/l)	35
Magnesium (mmol/l)	5
Lactobionate (mmol/l)	100
Phosphate (mmol/l)	25
Sulphate (mmol/l)	5
Raffinose (mmol/l)	30
Adenosine (mmol/l)	5
Allopurinol (mmol/l)	1
Glutathione (mmol/l)	3
Insulin (Units/l)	100
Dexamethasone (mg/l)	8
Hydroxyethyl starch (g/l)	50
Bactrim (ml/l)	0.5
Osmolality (mmol/kg)	320
pH (at room temperature)	7.4

UNIVERSITY OF WISCONSIN (UW) SOLUTION

This complex flushing solution was originally developed for transplantation of the pancreas.[22] Its components included impermeable solutes and buffering agents, colloid and electrolytes, with adenosine, allopurinol, glutathione, insulin and steroids as additional free radical scavengers, antioxidants, and hormones (Table 11.2). Subsequently, UW was effectively applied experimentally and clinically to liver, kidney, heart and small bowel preservation. The clinical success of UW solution and its variants in heart and lung, liver, pancreas, kidney, small bowel, and organ cluster grafts has made it the preferred solution and gold standard for multiple organ harvest in the 1990s.[20,23,24,25,26]

A multicentre trial in Europe comparing UW with EuroCollins' showed UW preservation gave more rapid recovery of renal function and less post-transplant dialysis,[27] other studies showed equivalent function of kidneys preserved by UW and Collins' solution in a randomized prospective clinical trial.[28] The UW solution has been shown to diminish trapping of erythrocytes in the microcirculation.[29] The effects of UW solution on protecting nonparenchymal cells in kidney, liver and pancreas may be important aspects of its abilities to provide extended preservation, rather than effects on the more easily preserved parenchymal cells.

Individual components of solutions for continuous perfusion and single-pass flush

COLLOIDS

A colloid of molecular size similar to or exceeding that of plasma protein, to provide oncotic vascular support, is essential for continuous machine perfusion preservation.

Omission of colloid leads to progressive trans-capillary passage of perfusate into the interstitium by ultrafiltration, causing explosive weight gain and oedema of the organ, increased resistance to perfusion, and early organ failure.

Machine perfusion has been applied most extensively to kidney preservation. Belzer and colleagues in the 1960s demonstrated the effectiveness of cyroprecipitated plasma thawed and passed through a membrane filter. Subsequently perfusates based on purified albumin (MW 68,000) were shown equally effective by Johnson and Toledo-Pereyra.[30,31]

Substitutes for these biological colloids sought to make perfusate preparation easier and to eliminate risk of transmitted viral infection. The most widely applied synthetic colloid, subsequently used by Belzer and Southard in developing UW solution, was hydroxyethyl starch (HES). A UW solution developed for continuous perfusion, containing HES, has been effective in extending preservation of canine kidneys by machine perfusion for seven days. Dextrans are synthetic macromolecules derived from bacterial digestion of starch, and occupy a spectrum of molecular size from 40,000 to 150,000 MW. Perfusates utilizing Dextran 70,000 or Dextran 40,000 have been successful in preservation of kidneys and other vascularized organs. Polyethylene glycol (PEG), an impermeable polymer also available in a range of molecular sizes, has been used by Collins to develop solutions suitable either for perfusion or for flushing. PEG also has been suggested to have additional effects in diminishing the intensity of the immune rejection response. Colloids, while essential for perfusion preservation, are neither necessary nor advantageous for simple hypothermic storage.

IMPERMEABLE SOLUTES

Solutes which are impermeable, or only slowly permeable to cellular membranes are a vital component of preserving solutions – whether used for cold-flushing or for continuous perfusion. Cooling below 15 °C paralyses the cell sodium pump, and cold cells imbibe water. Impermeable solutes are required at concentrations of up to 140 millimolar to counteract this effect. Large inert molecules, metabolically and nutritionally unavailable to cellular physiological processes, have been most effective. Gluconate (for perfusing and flushing solutions – MW 196) and lactobionate (for flushing solutions – MW 358) have been very successful. Other polysaccharides and disaccharides (raffinose – MW 594, sucrose – 134, maltose – 360, trehalose – 378) and chelates of citrate and magnesium have also been very effective. Metabolizable monosaccharides (glucose, fructose) have been less effective. Colloids alone are not effective.

Other agents, which diffuse into cells, but at a slower rate than do simple electrolytes, have also been used. Aminoacids (glutamine – MW 146, histidine – 155, tryptophane – 204, aspartate – 133, glycine – 75) and fatty acids (linoleic acid – 280) have also been used in successful solutions.

ELECTROLYTES

Originally based on 'extracellular' solutions high in sodium and chloride, the composition of solutions subsequently saw a swing of the pendulum to solutions high in potassium and phosphate ('intracellular'); and now back towards normal ECF concentrations of sodium and potassium as some of the potentially deleterious consequences of very high potassium levels were appreciated. Cold-induced paralysis of cell membrane pumps requires the chloride anion to be reduced and replaced by less permeable anions – gluconate, lactobionate, citrate, phosphate. Apart from these measures electrolyte content is of relatively little importance. Potassium in the original 'intracellular' concentrations of 125–135 millimolar is markedly vasoconstrictive and potentially harmful to the vascular endothelium. Solutions lower in potassium and higher in sodium (sodium 130 millimolar, potassium 5–30 millimolar) have been equally effective. Magnesium has been a useful additive in some solutions. Calcium tends to precipitate with phosphate and has been excluded from solutions or added in minute concentration. Successful solutions have been isosmolar or hyperosmolar.

The hydrogen ion concentration again has not been a critical factor, although extremes of acidity and alkalinity have diminished effectiveness. All successful solutions use one or more buffers – bicarbonate, phosphate, citrate and many others; but intracellular acidosis and its correction have not wholly correlated with effectiveness of solutions.

OXYGEN AND SUBSTRATE SUPPLY

During perfusion storage some nutrient substrate has been regarded as essential, together with continuous oxygenation. This is usually glucose (10 millimolar) often enriched with fructose, pyruvate, aminoacids, fatty acids, phosphates, adenosine and other high energy compounds or precursors. But even during optimal oxygenated hypothermic machine perfusion, aerobic metabolism may not be fully restored during storage.[32]

The role of nutrient substrates during static storage, and upon reperfusion, is also unclear. Substrates are not obligatory – solutions containing no utilizable metabolic fuel can give excellent preservation of most organs for 24 hours, and in the case of the kidneys for up to three days experimentally with PBS solution.

Static storage normally implies anaerobic conditions with progressive fuel depletion. Addition of oxygen during storage is possible by continuous bubbling of gas ('persufflation') delivered to the stored organ by its artery, vein or duct (ureter). Multiple fine needle punctures on the surface of the organ allow more even diffusion and escape of gas during storage. The organ needs tethering beneath the fluid, otherwise the increased gaseous buoyancy causes it to bob to the surface. Persufflation of the kidney during storage has shown some experimental advantages; but concerns that damage to the vascular lining may be induced, and the uncertainty of its effectiveness in larger human kidneys, have limited its use to the laboratory.[33]

Precursors of ATP added to flushing solutions (such as adenosine, adenine, inosine and phosphate), or ATP itself, have been shown helpful in some experiments. Whether these agents are acting by increasing availability of high energy compounds or in some other metabolic or haemodynamic manner, is uncertain. Nutritional depletion in experimental conditions can enhance tolerance to ischaemic damage. Glycogen-depleted livers from fasted rats show resistance to ischaemic injury and better survival after transplantation.[34]

Another means of enhancing oxygen supply is the use of synthetic perfluorochemical polyols to substitute for erythrocytes in facilitating oxygen transport. These agents have been used for continuous machine perfusion, and also for static storage. In simple hypothermic storage, they have been helpful in enhancing oxygen delivery to stored organs balanced at the interface between two solutions of different physical characteristics – one containing the synthetic fluorochemical. The method has been shown to enhance pancreatic storage in particular.[35,36]

OTHER COMPONENTS

Thus colloid is necessary for machine perfusion; but not for simple storage. Extended preservation times by perfusion require suitable nutritional precursors and oxygen supply. Otherwise prerequisites for machine perfusion solutions, and for solutions used for cold flushing prior to hypothermic storage, have been similar. The major requirements have been suitable impermeable solutes (anions or non-electrolytes) replacing permeable chloride, together with a suitable buffer system.

Small concentrations of added agents which quench or scavenge oxygen free radicals, anti-oxidants, calcium-channel blockers and other pharmacologically or metabolically active compounds can help as adjuvants in increasing ischaemic tolerance. In UW solutions allopurinol, glutathione and adenosine have been demonstrated to be helpful – insulin and steroids appear of no benefit. Other additives have included cytokines, prostaglandins, leukotrienes, lazaroids (aminosteroids) and their intermediaries, hormones, chlorpromazine and other benzodiazapines, nitric oxide, interleukins, tumour necrosis factor antagonists, and other growth factors.

Storage/reperfusion injury affects endothelial cells and activates hepatic macrophages (Kupffer cells). This injury leads to leucocyte and platelet adhesion and exacerbation of hypoxia/ischaemia. Kupffer cells are a main source of oxygen radicals and also release other proinflammatory mediators, such as tumour necrosis factor (TNF alpha), interleukins 1 and 6, eicosanoids, and nitric oxide. Agents suppressing TNF alpha formation include calcium channel blockers, pentoxifylline, adenosine, and prostaglandin E; whereas liposaccharides from bacteria are stimulatory to TNF alpha.

Organ-specific differences in the response to ischaemia are influenced by the relative sensitivity of parenchymal, endothelial and interstitial cells. Nonetheless, fortunately, a globally protective effect across all organs can be obtained during initial organ procurement by using a suitably modified universal flushing solution.

Importance of storage temperature

Organs surrounded by preserving solution, and stored and transported in a container of crushed ice, maintain a storage temperature of melting ice (0 °C). Storage temperatures above 0 °C aim to diminish cold-induced membrane paralysis while maintaining the protective effects of hypothermia. Temperatures above 15–20 °C have usually been considerably less effective than 0 °C. Temperatures for continuous perfusion have been maintained between 4 and 10 °C.

The optimal storage temperature of nonparenchymal cells may differ from that required by parenchymal cells, perhaps over quite a narrow range. In rat liver preservation 48-hour storage was possible by simple hypothermic storage at 4 °C but not at 0 °C.[37] Differences in effectiveness between various solutions (UW, sucrose, citrate) were less marked during storage at 4 °C than at 0 °C.

These results suggest a greater sensitivity of endothelial cells to cold damage than occurs with parenchymal cells; and temperatures around 4 °C may be optimal for hypothermic storage. Storage at 4 °C could be simply obtained by domestic refrigerator storage of organs surrounded by cold liquid rather than by ice (Figure 11.3).

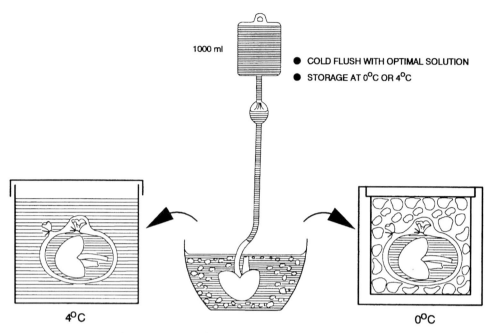

Figure 11.3 Methods of hypothermic organ storage.

Preventing reperfusion damage

The vascular endothelial cells are the first to face reperfusion damage. Technical complications of machine preservation can damage both major blood vessels feeding the organ, and even more significantly, its microvasculature. The liver vasculature with its complex sinusoidal fenestrated membrane guarded by endothelial cells and by macrophages (Kupffer cells), and its double blood supply, is particularly liable to reperfusion injury. Vascular complications have also plagued pancreatic transplants. Many additives found of value in organ preservation (benzodiazapines, calcium channel blockers, angiotensin-converting-enzyme inhibitors, prostaglandins, cytokines and their antagonists) have important rheological and vasodilating effects which may partly explain their benefit.

EFFECTS OF PREVASCULARIZATION RINSE: CAROLINA RINSE SOLUTION

A prevascularization flush-out or rinse at the conclusion of the storage period, just prior to reimplantation, has two main aims: firstly, to remove potentially toxic components of the preservation solution and accumulated waste products (for example, high levels of potassium and hydrogen ions) or to prevent a large intracardiac bolus of cold fluid on clamp release. Secondly, rinsing aims to smooth the transition between hypothermic storage and isothermic revascularization. Anastomoses can be tested prior to revascularization by intravascular wash-out before final clamp release. This has been most studied in relation to the liver. One method allows blood flow to the liver via portal vein, venting the initial effluent from the infrahepatic vena cava prior to completing this anastomosis, and

Table 11.3 Prevascularization rinse: Carolina Rinse (CR)*

Potassium (mmol/l)	5
Sodium (mmol/l)	130
Chloride (mmol/l)	112
Lactate (mmol/l)	28
Phosphate (mmol/l)	1
Fructose, glucose (mmol/l)	10
Adenosine (mmol/l)	1
Allopurinol (mmol/l)	1
Desferrioxamine (mmol/l)	1
Glutathione (mmol/l)	3
Osmolality (mmol/kg)	300
pH (at room temperature)	6.8

*Other additives include: calcium, magnesium, nicardipine, insulin, glycine, hydroxyethl starch.

finally releasing the clamp on the suprahepatic vena cava. Alternatively the organ can be reflushed via the portal vein or hepatic artery with cold or warm electrolyte solution (for example Ringer's-lactate) prior to clamp release.[38]

Carolina rinse (CR) solution, developed by Lemasters and colleagues at the University of North Carolina, maintained the basic electrolyte composition of Ringer's-lactate, but had added components; and gave better results than a simple electrolyte rinse in animals and man (Table 11.3).[39,40] The CR solution combines ECF electrolyte composition with a mildly acidic pH, oncotic support (albumin or HES), antioxidants (allopurinol, glutathione, desferrioxamine), energy substrates (adenosine, glucose and insulin, fructose), vasodilators (calcium channel blockers), and glycine. As with UW solution, it was not clear which of the several components were effective; and whether the effects were due to the beneficial effects of the CR solution or the potentially harmful effects of a Ringer's-lactate rinse.[41]

Analysis of the various components and techniques indicates that the temperature of the rinse solution is preferably above the required cold storage temperatures of 0–4 °C. Rinses at ambient temperatures (20 °C) and at body temperatures (37 °C) have given better results. The preferred reaction may be slightly acidic (pH 6.8). Components which seem unnecessary are colloid (HES or albumin), nicardipine and magnesium, glucose, fructose and insulin. Beneficial agents include adenosine, allopurinol, desferrioxamine, glutathione and glycine, possibly acting by free radical scavenging and quenching.[40,42,43,44]

Similarities between these findings and those with UW solution are apparent. Whether a CR rinse has advantages over a pre-vascularization second rinse with a UW-derived low potassium SLS solution used for storage is not yet established; nor is the effect of CR on other vascularized allografts.

Applied strategies of organ preservation

OPTIMAL ORGAN PROCUREMENT

Lessons in the procurement of organs from living donors are applicable to the cadaver donor. Ischaemia at body temperatures (warm ischaemia – WI) should be minimized

214 *Organ and tissue preservation*

by effective management of the heart-beating cadaver donor. Best practice for multiorgan procurement is outlined in Table 11.4.

In situ cold perfusion with balanced electrolyte solution is followed by meticulous but expeditious *en bloc* multiorgan excision. Further 'bench surgery' prepares individual organs cooled externally by iced saline-slush and by vascular washout with the chosen cold preserving fluid. The temperature of the organ rapidly falls to the desired temperature (from 0 to 5 °C) for subsequent maintenance of storage. Procurement is completed by surrounding the organ with a protective layer of cold

Table 11.4 Management of brain-dead cadaver donor for multiorgan retrieval

Monitor	Circulation, ventilation, urine, blood gases, haematology. Parenteral nutrition (glucose, insulin) if ICU stay prolonged. Individual organs RFTs, LFTs, CO, GFR
Ventilation	Pressure/volume – cycled ventilator 40–60% FIO_2, $PaO_2 > 150$ mmHg
Circulation	CVP 5–15 cm water, BP > 90–100 mmHg, Ht > 30% Balanced electrolyte/colloid infusion
Urine	Between 60–100 ml/hour Use desmopressin (DDAVP) for severe diabetes insipidus
Avoid	Dehydration and shock Nephrotoxic antibiotics Catecholamines unless arrhythmias Electrolyte disturbances Na, K Lactic acidosis Hyperthermia

Give during organ procurement surgery	Steroids 0.5–1 G Methylprednisolone
	Dopaminergic renal support 1–5 μg/kg/min
	Mannitol 20 G
	Heparin 5000–10000 units
	Chlorpromazine 12.5 mg/hour
	Verapamil 5–10 mg/hour
	In situ cold intra-arterial flush (4 °C)
	    2–6 L Hartmann's with venous runoff
	    1–2 L organ preservation solution
	En bloc removal

Bench surgery in refrigerated organ preservation solution
Individual organs – reflush with optimal cold flush solution
Store surrounded by cold flushing solution 2–4 °C
Do not store surrounded by ice-slush solution, avoid direct ice contact

Organ storage and transport
Use optimal solutions (UW-derived) and storage conditions (temperature 4 °C)

Transplantation into recipient
Minimize additional WI damage during anastomoses
Reflush livers before clamp release (CR)
Recipient treatment to mitigate reperfusion injury

Table 11.5 Current status of organ preservation by simple cold storage after a preliminary flush

	Experimental	Clinical	Optimal flushing solution
Kidney	5 days (120 h)	24 (to 60) h	HES-free UW, SLS
Pancreas	4 days (96 h)	12 (to 36) h	UW, SLS and others
Liver	2 days (48 h)	12 (to 36) h	UW, SLS and others
Small intestine	2 days (48 h)	8 (to 12) h	UW and others
Heart	1 day (24 h)	6 (to 12) h	CU
Lung	1 day (24 h)	8 (to 12) h	UW and others
Machine perfusion preservation			
Kidney	7 days (168 h)	48 to 72 h	HES-UW, other additives

Other organs: incremental extensions of storage periods achieved by perfusion

preserving solution within a sterile plastic container. The organ can then be transported in an insulated container surrounded by crushed ice or chilled liquid.

The current status of organ preservation by simple hypothermic storage is shown in Table 11.5, giving the limits of storage which have been achieved experimentally and clinically using optimal flushing solutions and static cold storage.

Preservation of individual organs for clinical transplantation

Currently, vascularized grafts are most commonly and optimally preserved by static hypothermic storage, which for kidney, liver and pancreas provides effective storage and likelihood of immediate function after 24-hours storage (Figure 11.3). Extended storage times beyond this are more liable to poor early function. Perfusion preservation for kidneys is relevant to such extended storage; and also for resuscitation of ischaemically damaged organs from asystolic cardiac donors with significant WI damage.

KIDNEY

Prospective, randomized clinical trials using preservation times averaging around 24 hours have shown no significant differences in early function or in long-term survival between machine preservation and simple cold storage.[14,45] Extended cold storage times of up to two and three days, although still capable of giving satisfactory results, are accompanied by an increased frequency of ATN as storage times lengthen beyond 24–48 hours. Although in experienced units, delayed initial function is compatible with good long-term results, the advantages of immediate good renal function with a maintained urine output and rapid correction of azotaemia after transplantation are striking. Clinical management is markedly facilitated, with earlier hospital discharge and earlier recognition and treatment of rejection episodes. Kidneys with early ATN, even when this is followed by adequate function, do less well in terms of long-term survival than kidneys functioning immediately, raising the likelihood that ischaemically damaged kidneys may be more prone to immunological rejection.

Thus even with the kidney, where dialysis can maintain patient survival in the face of poor early function, good early function remains the paradigm of clinical organ preservation. Simple hypothermic storage can give early function in the majority of grafts. Storage times over 48 hours are rarely obligatory except when organs are transplanted over very long distances, for example between continents.

UW-derived solutions (HES-free UW) have given successful static storage for five days[46] in dogs (Figure 11.4). The gold standard for cadaver kidney preservation thus remains UW solution. Comparable results are achievable by a variety of other solutions when preservation times do not exceed 24 hours; so for shorter anticipated preservation periods and for living donor transplantation, UW, sucrose, Collins', citrate and Bretschneider's solutions can give equivalent results. As expected, prospective clinical trials comparing these solutions with each other do not demonstrate major differences.

LIVER

The introduction of UW solution revolutionized liver transplantation. Previous solutions had given successful preservation for only 8–16 hours in canine, porcine or rodent models. UW solution allowed reliable and near 100% successful preservation in dog and in rat livers for 24 hours. Extended preservation to 30–48 hours was also possible experimentally, but with less adequate initial function.

Clinical practice requires immediate function for liver grafts. Currently 80% or more grafts function adequately; but about 10% show initially poor graft function (IPF), and

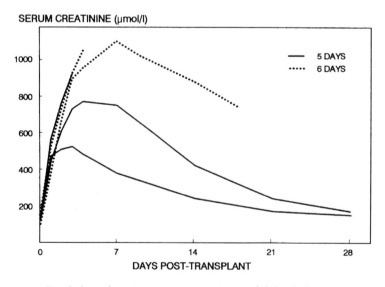

Figure 11.4 Simple hypothermic organ storage. Successful dog kidney preservation with HES-free UW solution (Monash).

about 5% have permanent non-function of the graft (PNF) necessitating life-saving retransplantation. Factors influencing and responsible for poor initial function include fatty change in the donor liver, older donor age, preliminary liver reduction surgery, and longer cold storage.[23] Although storage for 24 or 30 hours can give good early function, it is thus preferable to keep storage times to under 12 hours for optimal results.[47]

Benchmark clinical practice remains hypothermic storage using UW solution. Modifications without HES and with reversed Na/K ratio are equally effective. When using high-K UW solution for preservation, a pre-vascularization rinse with a further washout solution is desirable. Carolina Rinse solutions are preferable to a Ringer's-lactate rinse.[48] Important considerations, whatever the composition of the solution, are that a brief pre-vascularization rinse should be at room or body temperature; and that a basic ECF-like crystalloid rinse composition can be enhanced by addition of adenosine, allopurinol, desferrioxamine and glycine. Glucose, fructose or other nutrients seem unnecessary in either the cold preserving solutions or the pre-transplant rinse.

Other additions to UW solution have been pentoxyfilline, chlorpromazine and calcium-channel blockers, prostaglandins, eicosanoids and lazaroids. Future modifications are likely to incorporate these and other additives to storage and rinse solutions, together with glycine and reversed Na/K electrolyte content.

PANCREAS

UW solution was originally introduced to improve pancreas preservation. Preservation must preserve the essential insulin-producing beta cells. Exocrine cell function can provide a marker of viability in the form of amylase excretion. Another important aim of preservation is to minimize the occurrence of ischaemia-induced acute pancreatitis, which adds morbidity after transplantation. Preservation of the pancreas proved difficult using Collins', citrate and other solutions, and it was not until the advent of UW solution that reliable preservation for 24 hours was achieved experimentally and clinically. Experimental preservation has since been extended to 96 hours using UW with an added prostanoid inhibitor.[49] Other variations of UW can give satisfactory cold storage, including removal of HES or its replacement with dextran, and combinations of sodium lactobionate with histidine.[50,51,52]

The pancreas has also been studied in relation to the benefit of additional oxygen using a two layer storage technique with a perfluorochemical adjacent to UW solution. In small animals, the technique of cold storage, whether surrounded by the preserving liquid or wrapped in moistened gauze, has influenced efficacy of preservation; perhaps related to the filamentous and non-encapsulated nature of the pancreas resulting in a greater tendency for harmful oedema and weight gain to occur during storage immersed in liquid.[53,54]

Preservation needs to include the associated duodenal segment containing the duodenal papilla and pancreatic ducts. Fortunately, the duodenum has not proved to be more sensitive to preservation injury than the pancreas itself.

SMALL BOWEL

Small bowel grafts can be transplanted alone or combined as organ-cluster grafts with liver or pancreas. The main site of injury during small bowel preservation has been thought to be the endothelium and basement membrane of the highly vascularized mucosa. The small intestine has a high concentration of the enzyme xanthine oxidase, rendering it liable to reperfusion injury. Free radical damage has been implicated particularly after prolonged preservation and after WI injury. Experimental preservation is possible for up to 48 hours in the rat and dog; the advantages of any one preservation solution are less obvious than with other organs. Sucrose, Collins', UW, and Bretschneider's solutions, and solutions with added dextran and containing free radical scavengers have given more or less equivalent results.[55] Pre-transplant rinse has not been shown helpful in preventing reperfusion injury. Polyethylene glycol has been suggested as an additive both to enhance preservation and to modify rejection in experimental small bowel, heart and pancreas transplantation.[56]

HEART

Preservation of the vasculature together with myocytes and conductive tissue is vital in optimal cardiac preservation. Simple hypothermic storage has been used for clinical heart transplantation since the earliest days. As most of the energy consumed by the heart fuels contraction of the myofibrils, and as the ischaemic heart continues beating until all its fuel has been depleted, cardioplegic arrest has been an essential feature of heart procurement and preservation. The development of heart transplantation from open heart surgery influenced significantly early strategies. Cardioplegic solutions previously developed for cardiac surgery were applied to cardiac transplantation. The heart was excised after induction of hypothermic cardioplegic arrest by an *in situ* flush with one of several standard cardioplegic solutions. Stanford, St Thomas's, and Bretschneider's solutions provided effective preservation of myocardial function for up to 6 or 8 hours.

Flushing solutions used to preserve other organs are also cardioplegic, but generally have contained much higher concentrations of potassium (over 100 millimolar rather than 20 millimolar); and do not contain any calcium. Initial reluctance to use these solutions in heart preservation stemmed from studies which indicated that such levels of potassium were deleterious.[57] A 'calcium paradox' effect in cardiac preservation was also important: cardiac muscle when incubated in calcium-free medium undergoes severe and irreversible damage when reperfused with calcium-containing medium due to a massive influx of calcium. This influx is enhanced by WI damage. Calcium paradox has been seen in experimental cardiopulmonary bypass, but is possibly less influential during hypothermia.[58]

Following the development of UW solution and its introduction as a solution appropriate for multiorgan transplantation, studies of UW and UW-derived solutions have been applied to open heart surgery and to heart transplantation. Experimental models have ranged from heterotopic transplantation into the abdomen or neck in small animals, isolated perfused working heart models, metabolic tissue analysis and histology, nuclear magnetic resonance spectroscopy, and allograft function in large animals.[59,60,61] UW solution has given superior donor heart preservation over that obtained by Stanford or St Thomas's solution in randomized clinical trials.[62,63,64] Repeated cold flushing with UW or other solutions was detrimental to hypothermic storage. Generation of oxygen free radicals has been implicated in ischaemic heart disease and in reperfusion injury after transplantation. Several agents have been used to pretreat the donor and as additions to the preserving solutions. Addition of ATP or its precursors to preserving solutions, although sometimes improving cardiac function, has not necessarily been due to improved high energy phosphate levels. Many institutions preoxygenate the solution. Other additives thought to be of importance include adenosine, pyruvate and aspartate, prostaglandins to improve perfusion, pentoxifylline to diminish neutrophil-induced vascular injury and neutrophil adhesion, and lazaroids to prevent lipid peroxidation. Nitric oxide (NO) is a biological messenger with diverse effects. Failure of the NO pathway during preservation and transplantation results in formation of oxygen free radicals during reperfusion, which quench available NO. Augmentation of NO by cAMP-dependent mechanisms can enhance vascular function after ischaemia and reperfusion.[65]

A novel preservation solution (Columbia University Solution – CU – Table 11.6), developed primarily for cardiac preservation, incorporates many of the salient features of UW solution (potassium 120, magnesium 5, adenosine 5 millimolar) and a colloid (dextran 50 G/litre). Added components include glucose, verapamil, cysteine and heparin; together with agents enhancing vascular homeostatic mechanisms by repleting extracellular/intracellular messengers, and nitroglycerin to enhance nitric oxide-related mechanisms. This solution contains a cAMP analogue and has extended cardiac preservation beyond that observed with UW alone in both rat and baboon cardiac transplant models. Primate hearts have been stored for 24 hours in CU solution and have functioned normally after orthotopic cardiac transplantation.[66]

LUNG AND HEART–LUNG

Simple cooling has been used for short-term preservation for heart–lung transplants. Lungs were originally flushed with cold EuroCollins' solution via the pulmonary artery immediately after induction of cardioplegia. Flushing should maintain pressure below that normally found in the pulmonary artery. Hyperinflation of the lung prior to storage was shown to be beneficial.[67] Treatment of the donor with prostaglandin E1 followed by a hypothermic flush gave adequate 6-hour preservation.[68]

Table 11.6 Flushing solutions for cardiac preservation Columbia University solution (CU)

Potassium (mmol/l)	120
Magnesium (mmol/l)	5
Gluconate (mmol/l)	95
Phosphate (mmol/l)	25
Glucose (mmol/l)	67
Adenose (mmol/l)	5
N-acetyl cysteine (mmol/l)	0.5
Butylated hydroxyanisole (μmol/l)	50
Butylated hydroxytoluene (μmol/l)	50
Verapamil (μmol/l)	10
Nitroglycerin (mg/ml)	0.1
Dibutyryl cAMP (mmol/l)	2
Heparin (u/l)	10
Dextran (G/l)	50
Osmolality (mmol/kg)	325
pH	7.6

Single lung transplantation is now increasingly common and lung preservation using a EuroCollins' flush was only successful for 5–6 hours. Progressive deterioration with increased pulmonary vascular resistance and decreased compliance occurs by 12 hours, with interstitial haemorrhage and oedema by 24 hours. The disaccharide trehalose has been substituted for glucose in EuroCollins' solution with advantage.[69] Increasing experience with lung and heart-lung transplantation has demonstrated that the lung may be more robust in relation to its tolerance of extended storage than previously thought. UW and other solutions have given successful preservation for 24 hours.[70] Additions to the preserving solution which have been found useful have included prostaglandin E1, pentoxifylline, verapamil, and glutathione. Equivalent results to UW solution have been found with a dextran-based low potassium solution.[71]

COMPOSITE TISSUES, LIMBS

Reconstructive plastic surgery uses vascularized autografts of composite tissues to fill large defects. Such composite grafts involve skin, subcutaneous fat, muscle, bone, nerves and blood vessels. Severed limbs and digits can be replaced as autografts. These complex operative procedures can take many hours. Storage of the grafts relies predominantly on simple external cooling by refrigerated saline and wrapping the grafts in cold saline-soaked packs during implantation. Supplementary vascular flushing has not been widely used to augment hypothermic storage, for fear of damage to the small vessels requiring microvascular suture. Flushing does not give any significant improvement over simple hypothermic storage in cold preserving liquid. Simple hypothermic storage usually adequately covers the periods (usually less than 12 hours) required in clinical practice. Tolerance of the various tissues to ischaemia varies: muscle and nerve are more sensitive than skin and bone. It is prudent to remember that restoration of the circulation to reimplanted limbs or other composite grafts of large bulk may release a large bolus of potassium into the circulation and can induce fatal hyperkalaemic cardiac arrest. After extended storage the contained blood or preservation solution should be rinsed out with warm plasma or balanced electrolyte solution prior to release of the clamps.

Preservation of other tissues: non-viable scaffolding or viable graft?

Apart from the major vascularized allografts, many other transplanted tissues are used in clinical practice.

BLOOD VESSELS

Autografted blood vessels such as long saphenous vein and internal mammary artery have been extensively used in vascular and cardiac surgery. They provide vascular conduits with intact and maintained endothelial linings. Vascular allografts also provide valuable conduits for extending vascular access in multiorgan grafts. These are preferably stored in chilled preservation solution at 4 °C prior to use, and can withstand storage for over 24 hours with minimal deterioration. They act as viable grafts in these circumstances. Vessels can be removed from cadaver donors, treated with glutaraldehyde, and stored indefinitely. They then become essentially collagenous tubes without apparent antigenic activity. This allografted non-living tissue does not cause a rejection reaction but will slowly degenerate, and a thrombosis or aneurysm can develop. It is an inferior option to the use of fresh autogenous vessels.

BONE MARROW

Bone marrow transplantation is now the preferred treatment for aplastic anaemias and for many leukaemias. Invariably, living donors have been used. Marrow allografts thus usually require minimal preservation, other than that provided by simple hypothermia as for blood collection. Extension of bone marrow transplantation in the future using cadaver donors could involve cryoprotective techniques similar to those used for freezing of blood. There is also an increasing use of autologous marrow grafts in cancer, the patient's marrow being removed and preserved before intensive chemotherapy and/or irradiation is given. The preserved marrow is then returned to the patient.

PANCREATIC ISLETS

Preservation and transplantation of pancreatic islet cells have been extensively studied experimentally. Intraportal transplantation of islet extracts is now achieving some clinical success. Cellular preparations, after collagenase digestion and purification, have been successfully preserved by tissue culture for 24 hours at 37 °C, by simple ice cooling for 24 hours, or for more prolonged periods by freezing with addition of cryoprotectives.

SKIN

Harvested split skin grafts used as autografts can be preserved for approximately two weeks by simple wrapping and rolling in packs moistened with saline or tissue culture fluid. These are then stored in the refrigerator at 4 °C. Preserved allografts can give temporary cover of denuded sites, as can xenografts of pig skin. More prolonged preservation can be provided by freezing, or by culturing epidermal cells which can subsequently be reimplanted as a monolayer.

BONE

Autografts of fresh cancellous bone taken from iliac crest or other areas remain the most effective source of bone grafts. Autografts provide viable osteoblasts and stimulate osteogenesis. Bone is a complex and active metabolic tissue, containing osteocytes bearing histocompatibility antigens nourished within a hydroxyapatite matrix. Stored allografts of bone can be cryopreserved for 12 months or more to provide a sterile source of bone matrix without viable cells. Preserved allografts are less active in stimulating osteogenesis than are fresh autografts. Transplantation of large segments of cortical bone can provide a rigid non-viable bony matrix which is gradually replaced by creeping substitution of new bone. Transfer of viable cortical bone requires a vascularized graft which restores the bone's blood supply at its new site. Vascularized autografts of fibula, iliac crest or other sites can replace bony defects resulting from congenital anomalies, disease, or injury.

CARTILAGE

The metabolic needs of cartilage are chiefly met by diffusion of nutrients from synovial fluids; re-establishment of a blood supply is less critical than for bone. Chondrocytes also possess histocompatibility antigens and evoke an allograft rejection response. Intact cartilage survives better than isolated chondrocytes, showing the importance of the cartilage matrix. Experimentally, cartilage can be stored for 28 days in tissue culture medium at 4 °C. The morphology of the cells and concentration of nutrients and of collagen show no significant change, and the ability of cells to incorporate ^{35}S-sulphate into glycosaminoglycans does not diminish.

Cryopreservation at subzero temperatures with glycerol or dimethylsulphoxide produces a less favourable outcome, with loss of chondrocytes and conversion of hyaline cartilage to fibrocartilage. Isolated chondrocytes can be well preserved by freezing, suggesting poor penetration of cryopreservates into solid cartilage. Onlays or plugs of cadaveric articular cartilage fragments have been used as allografts in the management of degenerative arthritis. The methods remain experimental and do not provide viable chondrocytes.

The future

Preservation of organs and tissues for transplantation has entered an exciting phase with the development of global and organ-specific storage solutions, and the advent of additional rinse solutions for reflushing prior to revascularization of the graft. The basic requirements of preservation remain cooling and the use of colligative agents and buffers to mitigate cold-induced enzymatic paralysis. A multitude of adjuvant pharmacologically and metabolically active agents have given incremental extension of tolerated storage times of all organs and tissues. Even with heart and heart-lung transplants, experimental transplantation for 24 hours or more is now a reality; and effective preservation is restoring clinical order to these transplants, as it has done with kidney, liver and pancreatic grafts. Resuscitation of damaged organs, and an adequate means of testing viability for each organ prior to implantation, are important future requirements for preservation research.

References

1. Carrel A, Lindbergh CA. *In the Culture of Organs*, Hamish Hamilton, London, 1938, page 221.
2. Burg MB, Orloff MJ. Active cation transport by kidney tubules at 0 °C. *Am J Physiol* 1964, **207**: 983.
3. Levy MN. Oxygen consumption and blood flow in the hypothermic, perfused kidney. *Am J Physiol* 1959, **197**: 111.

4. Leaf A. Maintenance of concentration gradients and regulation of cell volume. *Ann NY Acad Sci* 1959, **72**: 396.
5. Collins GM, Bravo-Shugarman M, Terasaki PI. Kidney preservation for transportation. Initial perfusion and 30 hours' ice storage. *Lancet* 1969, **2**: 1219–22.
6. Sacks SA, Petritsch PH, Kaufman JJ. Canine kidney preservation using a new perfusate. *Lancet* 1973, **1**: 1024–8.
7. Collins GM, Hartley LC, Clunie GJ. Kidney preservation for transportation. Experimental analysis of optimal perfusate composition. *Br J Surg* 1972, **59**: 187–9.
8. Collins GM, Green RD, Halasz NA. Importance of anion content and osmolarity in flush solutions for 48 to 72 hr hypothermic kidney storage. *Cryobiology* 1979, **16**: 217–20.
9. Hardie I, Balderson G, Hamlyn L, McKay D, Clunie G. Extended ice storage of canine kidneys using hyperosmolar Collins' solution. *Transplantation* 1977, **23**: 282–3.
10. Andrews PM, Bates SB. Improving EuroCollins' flushing solution's ability to protect kidneys from normothermic ischemia. *Miner Electrolyte Metab* 1985, **11**: 309–13.
11. Bretan PN, Baldwin N, Martinez A et al. Improved renal transplant preservation using a modified intracellular flush solution (PB-2). Characterization of mechanisms by renal clearance, high performance liquid chromatography, phosphorus-31 magnetic resonance spectroscopy, and electron microscopy studies. *Urol Res* 1991, **19**: 73–80.
12. Grino JM, Castelao AM, Sebate I et al. Low-dose cyclosporine, ALG and steroids in first cadaveric renal transplants. *Transplant Proc* 1988, **20**: 18–20.
13. Ross H, Marshall VC, Escott ML. 72-hour canine kidney preservation without continuous perfusion. *Transplantation* 1976, **21**: 498–501.
14. Marshall VC, Ross HL Scott DF et al. Preservation of cadaver renal allografts: comparison of ice storage and machine perfusion. *Med J Aust* 1977, **2**: 353–6.
15. Jablonski P, Howden BO, Marshall VC, Scott DF. Evaluation of citrate flushing solution using the isolated perfused kidney. *Transplantation* 1980, **30**: 239–43.
16. Marshall VC, Howden BO, Jablonski P, Tavanlis G, Tange J. Sucrose-containing solutions for kidney preservation. *Cryobiology* 1985a, **22**: 622.
17. Lam FT, Mavor AID, Potts DJ, Giles GR. Improved 72-hour renal preservation with phosphate buffered sucrose. *Transplantation* 1989a, **47**: 767–71.
18. Lam FT, Ubhi CS, Mavor AID, Lodge JPA, Giles GR. Clinical evaluation of PBS140 solution for cadaveric renal preservation. *Transplantation* 1989b, **48**: 1067–8.
19. Lodge JPA, Perry SL, Skinner C, Potts DJ, Giles GR. Improved porcine renal preservation with a simple extracellular solution-PBS 140. Comparison with hyperosmolar citrate and University of Wisconsin solution. *Transplantation* 1991, **51**: 574–9.
20. Erhard J, Lange R, Scherer R et al. Comparison of histidine-tryptophane-ketoglutarate (HTK) solution versus University of Wisconsin (UW) solution for organ preservation in human liver transplantation. A prospective, randomized study. *Transplant Int* 1994, **7**: 177–181.
21. Groenewoud AF, Thorogood J. A preliminary report of the HTK randomized multicenter study comparing kidney graft preservation with HTK and EuroCollins solutions. *Transplant Proc* 1992, **5**: 429.
22. Wattiaux R, Wattiaux-De Conninck S. Trapping of mannitol in rat-liver mitochondria and lysosomes. *Int Rev Exp Pathol* 1984, **26**: 85–89.
23. Belzer FO. Evaluation of preservation of the intra-abdominal organs. *Transplant Proc* 1993, **25**: 2527–2530.
24. Belzer FO, D'Alessandro AM, Hoffman RM et al. The use of UW solution in clinical transplantation. A 4-year experience. *Ann Surg* 1992, **215**: 579–583.
25. Stein DG, Drinkwater DC, Laks H. Cardiac preservation in patients undergoing transplantation. A clinical trial comparing UW solution and Stanford solution. *J Thorac Cardiovasc Surg* 1991, **102**: 657–65.
26. D'Alessandro AM, Reed A, Hoffman RM et al. Results of combined hepatic, pancreaticoduodenal, and renal procurements. *Transplant Proc* 1991b, **23**: 2309–11.
27. Ploeg RJ, van Bockel JH, Langendijk PT et al. Effect of preservation solution on results of cadaveric kidney transplantation. The European Multicentre Study Group. *Lancet* 1992, **340**: 129–137.

28. Hefty T, Fraser S, Nelson K, Bennett W. Comparison of UW and Euro-Collins solutions in paired cadaveric kidneys. *Transplantation* 1992, **53**: 491–2.
29. Jacobsson J, Tufveson G, Odlind B, Wahlberg J. The effect of type of preservation solution and hemodilution of the recipient on postischemic erythrocyte trapping in kidney grafts. *Transplantation* 1989, **47**: 876–9.
30. Johnson RWG, Anderson M, Flear CT, Murray S, Swinney J, Taylor RMR. Evaluation of a new perfusion solution for kidney preservation. *Transplantation* 1972, **13**: 270–5.
31. Toledo-Pereyra LH. Kidney perfusion. In *Basic Concepts of Organ Procurement, Perfusion and Preservation for Transplantation*, Toledo-Pereyra, LH (Ed.) Academic Press, New York.
32. Pegg DE, Wusterman MC, Foreman J. Metabolism of normal and ischemically injured rabbit kidneys during perfusion for 48 hours at 10 °C. *Transplantation* 1981, **32**: 437–43.
33. Rolles K, Foreman J, Pegg DE. Preservation of ischemically injured canine kidneys by retrograde oxygen presufflation. *Transplantation* 1984, **38**: 102–6.
34. Sumimoto R, Southard JH, Belzer FO. Livers from fasted rats acquire resistance to warm and cold ischemia injury. *Transplantation* 1993, **55**: 728–732.
35. Kuroda Y, Tanioka Y, Morita A et al. Protective effect of preservation of canine pancreas by the two-layer (University of Wisconsin solution/perfluorochemical) method against rewarming ischemic injury during implantation. *Transplantation* 1994, **57**: 658–661.
36. Urushihara T, Sumimoto K, Ikeda M, Hong HQ, Fukuda Y, Dohi K. A comparative study of two-layer cold storage with perfluorochemical alone and University of Wisconsin solution for rat pancreas preservation. *Transplantation* 1994, **57**: 1684–1686.
37. Marshall VC, Howden BO, Jablonski P. Effect of storage temperature in rat liver transplantation: 4 °C is optimal and gives successful 48 h preservation. *Transplant Proc* 1994, **26**: 3657–8.
38. Emre S, Schwartz ME, Mor E et al. Obviation of prereperfusion rinsing and decrease in preservation/reperfusion injury in liver transplantation by portal blood flushing. *Transplantation* 1994, **57**: 799–803.
39. Gao W, Takei Y, Marzi I et al. Carolina rinse solution: a new strategy to increase survival time after orthotopic liver transplantation in the rat. *Transplantation* 1991, **52**: 417–24.
40. Gao W, Hijioka T, Lindert K, Caldwell-Kenkel J, Lemasters J, Thurman R. Evidence that adenosine is a key component in Carolina rinse responsible for reducing graft failure after orthotopic liver transplantation in the rat. *Transplanation* 1991, **52**: 992–998.
41. Egawa H, Esquivel CO, Wicomb WN, Kennedy RG, Collins GM. Significance of terminal rinse for rat liver preservation. *Transplantation* 1993, **56**: 1344–1347.
42. Bachmann S, Caldwell-Kenkel JC, Currin RT et al. Protection by pentoxyifylline against graft failure from storage injury after orthotopic rat liver transplantation with arterialization. *Transplant Internat* 1992, **5**: 345–350.
43. Post S, Palma P, Rentsch M, Gonzalez AP, Otto G, Menger MD. Importance of rinse solution vs preservation solution in prevention of microcirulatory damage after liver transplantation in the rat. *Transplant Proc* 1993, **25**: 1607.
44. Gonzalez AP, Post S, Palma P, Rentsch M, Menger MD. Essential components of Carolina rinse for attenuation of reperfusion injury in rat liver transplantation. *Transplant Proc* 1993, **25**: 2538–2539.
45. Heil JE, Canafax DM, Sutherland DER, Simmons RL, Dunniy M, Najarian J. A controlled comparison of kidney preservation by two methods: matching perfusion and cold storage. *Transplant Proc* 1987, **19**: 2046.
46. Marshall VC, Howden BO, Thomas AC et al. Extended preservation of dog kidneys with modified UW solution. *Transplant Proc* 1991, **23**: 2366–7.
47. Adam R, Bismuth H, Diamond T et al. Effect of extended cold ischaemia with UW solution on graft function after liver transplantation. *Lancet* 1992, **340**: 1373–1376.
48. Sanchez-Urdazpal L, Gores GJ, Lemasters JJ et al. Carolina rinse solution decreases liver injury during clinical liver transplantation. *Transplant Proc* 1993, **25**: 1574–1575.
49. Kin S, Stephanian E, Gores P et al. Successful 96-hr cold storage preservation of canine pancreas with UW solution containing the thromboxane A2 synthesis inhibitor OKY046. *J Surg Res* 1992, **52**: 577–582.

50. Sumimoto R, Dohi K, Urushihara T et al. An examination of the effects of solutions containing histidine and lactobionate for heart, pancreas and liver preservation in the rat. *Transplantation* 1992, **53**: 1206–1210.
51. Urushihara T, Sumimoto R, Sumimoto K et al. A comparison of some simplified lactobionate preservation solutions with standard UW solution and EuroCollins solution for pancreas preservation. *Transplantation* 1992, **53**: 750–754.
52. Morel P, Moss A, Schlumpf R et al. 72-hour preservation of the canine pancreas: successful replacement of hydroxyethylstarch by dextran-40 in UW solution. *Transplant Proc* 1992, **24**: 791–794.
53. Urushihara T, Sumimoto K, Ikeda M, Hong HQ, Fukuda Y, Dohi K. A comparative study of two-layer cold storage with perfluorochemical alone and University of Wisconsin solution for rat pancreas preservation. *Transplantation* 1994, **57**: 1684–1686.
54. Howden BO, Jablonski P, Marshall VC. A novel approach to pancreas preservation: Does the gaseous milieu matter? *Transplant Proc* 1992, **24**: 795–796.
55. Muller AR, Nalesnik M, Platz KP, Langrehr JM, Hoffman RA, Schraut WH. Evaluation of preservation conditions and various solutions for small bowel preservation. *Transplantation* 1994, **57**: 649–655.
56. Itasaka H, Burns W, Wicomb WN, Egawa H, Collins G, Esquivel CO. Modification of rejection by polyethylene glycol in small bowel transplantation. *Transplantation* 1994, **57**: 645–648.
57. Haljula A, Mattila S, Mattila I et al. Coronary endothelial damage after crystalloid cardioplegia. *J Thorac Cardiovasc Sur* 1984, **25**: 147–52.
58. Alto, LE, Dhalla NS. Hypothermia appears to protect against calcium parodox role of changes in microsomal calcium uptake in the effects of reperfusion of calcium-deprived hearts. *Circulation Res* 1981, **48**: 17–24.
59. Stringham JC, Paulsen KL, Southard JH, Mentzer RM Jr, Belzer FO. Prolonging myocardial preservation with a modified University of Wisconsin solution containing 2, 3-butanedione monoxime and calcium. *J Thorac Cardiovasc Surg* 1994, **107**: 764–775.
60. Lasley RD, Mentzer RM Jr. The role of adenosine in extended myocardial preservation with the University of Wisconsin solution. *J Thorac Cardiovasc Surg* 1994, **107**: 1356–1363.
61. Mertes PM, Burtin P, Caxteaux JP et al. Changes in hemodynamic performance and oxygen consumption during brain death in the pig. *Transplant Proc* 1994, **26**: 229–230.
62. Demertzis S, Wippermann J, Schaper J et al. University of Wisconsin versus St Thomas's Hospital solution for human donor heart preservation. *Ann Thorac Surg* 1993, **55**: 1131–1137.
63. Stein DG, Drinkwater DC Jr, Laks H et al. Cardiac preservation in patients undergoing transplantation. A clinical trial comparing University of Wisconsin solution and Stanford solution. *J Thorac Cardiovasc Surg* 1991, **102**: 657–665.
64. Jeevandandam V, Barr ML, Auteri JS et al. University of Wisconsin solution versus crystalloid cardioplegia for human donor heart preservation. A randomized blinded prospective clinical trial. *J Thorac Cardiovasc Surg* 1992, **103**: 194–198.
65. Pinsky DJ, Oz MC, Koga S et al. Cardiac preservation is enhanced in a heterotopic rat transplant model by supplementing the nitric oxide pathway. *J Clin Invest* 1994, **3**: 2291–2297.
66. Oz MC, Pinski DJ, Koga S et al. Novel preservation solution permits 24-hour preservation in rat and baboon cardiac transplant models. *Circulation* 1993, **88**: II291–297.
67. Puskas JD, Hirai T, Christie N, Mayer E, Slutsky AS, Patterson GA. Reliable thirty-hour lung preservation by donor lung hyperinflation. *J Thorac Cardiovasc Surg* 1992, **104**: 1075–1083.
68. Kirk AJ, Colquhoun IW, Dark JH. Lung preservation: a review of current practice and future directions. *Ann Thorac Surg* 1993, **56**: 990–1000.
69. Hirata T, Yokomise H, Fukuse T et al. Effects of trehalose in preservation of canine lung for transplants. *J Thorac Cardiovasc Surg* 1993, **41**: 59–63.
70. Kawahara K, Itoyanagi N, Takahashi T, Akamine S, Kobayashi M, Tomota M. Transplantation of canine lung allografts preserved in UW solution for 24 hours. *Transplantation* 1993, **55**: 15–18.
71. Steen S, Kimblad PO, Sjoberg T, Lindberg L, Ingemansson R, Massa G. Safe lung preservation for twenty-four hours with Perfadex. *Ann Thorac Surg* 1994, **57**: 450–457.

CHAPTER 12

Organization of donation and organ allocation

JOHAN DE MEESTER, BERNADETTE J.J.M. HAASE-KROMWIJK, GUIDO G. PERSIJN, BERNARD COHEN

Introduction 226
General organizational aspects 227
Organ exchange organizations: reasons for existence 227
Organ exchange organizations: additional functions 228
Organizational aspects of an organ exchange organization 229
Allocation of donor organs within an organ exchange organization 232
Summary 235
Organ allocation within the Eurotransplant organization 235
References 238

Introduction

Organ donation is possible from either living or cadaveric donors. The organization of each is quite different, mainly because living donation is restricted to the small circle of recipient, donor and physician. This is in contrast to cadaveric organ donation that has rapidly evolved to a much broader environment with intense international co-operation. This chapter will only deal with cadaveric organ donation and transplantation. Legislative issues, organ donation initiatives and donor selection criteria related to living donor transplantation are described in Chapters 7 and 10 respectively. In the first part of the chapter, we concentrate on the co-ordination infrastructure between the local donor hospitals and transplant programmes, cadaveric organ donation, allocation and transplantation: the so-called organ exchange organization (OEO). In the second part, we discuss the basic principles of organ allocation and some allocation models.

General organizational aspects

The need to organize cadaveric donation and organ allocation on a large scale is directly proportional to the number and/or the organ type of transplant programme located in a particular geographical area. National public health structures and services should be able to establish and maintain cadaveric organ transplant programmes. In addition, hospitals should be equipped with intensive care facilities and co-operate as donating hospitals with local transplant programmes. The inability to realize such organization explains the rarity of cadaveric organ transplant programmes in many countries in Latin and South America, Africa, South-east Asia and South-west Asia. In these countries, if transplantation is available, more interest is shown in living related and unrelated (kidney) transplantation, which can be organized more easily. The legal acceptance of brain death, as the total and irreversible loss of brain function, besides the more classical concept of 'absence of cardiac activity and cessation of respiration', is crucial to the development of transplant programmes. This concerns, in particular, the non-renal organs as successful transplantation of heart, lung, liver and pancreas are simply not possible with organs obtained after circulatory arrest. This factor hampered the establishment of non-renal transplant programmes in Denmark until 1990, the year in which the diagnosis of brain death was accepted by parliament.[1] At the beginning of 1995, no official approval of brain death has yet been given by the Japanese Congress. Therefore, besides living donor kidney transplantation, some Japanese centres have started living related liver transplant programmes.

Organ exchange organizations: reasons for existence

An OEO is an alliance of donor hospitals ('input') and transplant programmes ('output'), with their accompanying tissue typing laboratories. All of them follow the same set of general policies, standard operating procedures and organ allocation rules. Currently, such OEOs are found in particular in Europe,[2-7] North -America[8,9] and Australia.[10] The idea of supra-local co-operation dates from the early days of organ transplantation (kidney)[11] and was based on two conclusions, which were also relevant to the emergence of the non-renal transplant programmes.

REDUCTION IN THE WASTE OF DONOR ORGANS, BY MAXIMUM PLACEMENT OF ALL DONATED ORGANS

It was realized that the maximum usage of donated organs depends on the size of the area and/or the pool of transplant candidates to be served. Characteristics of donors show large variations, accentuated by the current relaxation of several donor criteria (age, size). The definition of appropriate donor organ quality is not constant among transplant programmes. The larger the OEO, the larger and more diverse the waiting lists of transplant candidates will be. The more transplant programmes, the higher the chance of ultimately using the organs suitable for transplantation within the service area, and thus, the lower the wastage of useful donor organs. Although the reasons for recipient pooling could justify a plea for wide international, or even worldwide, agreements, limited service areas of OEOs are demanded because of the restricted ischaemic tolerance of certain donor organs.

OPTIMAL USAGE AND TRANSPARENT ALLOCATION OF DONOR ORGANS

It was also believed that organ allocation within the area of an OEO should primarily aim at the transplantation of two categories of patients. The first category concerns patients with the best transplant outcome, achievable by good 'matching' between donor and recipient. The second category is patients in the most urgent need of a transplant. Subsequently, the benefit of co-operation was also seen in the approval of a uniform and justifiable organ allocation policy. Transparent execution of these policies, with the inevitable necessity of monitoring compliance, would eliminate doubts, voiced by government, public and patients, about the fairness of selection and the possible existence of special interests that take inappropriate advantage of donated organs for single transplant programmes or patients.[9] To achieve optimal usage and maintain allocation transparency, compliance from donor hospitals and transplant programmes is indeed obligatory. The transplant programmes should prospectively list all transplant candidates on the composite waiting list of the OEO, with the necessary data to match them with donors. The donor centres should report every cadaveric donor to the OEO, to match the donor against the waiting lists before an organ is offered and used for transplantation.

Organ exchange organizations: additional functions

Most OEOs have adopted one or more of the following activities.

SCIENTIFIC RESEARCH

Data are collected to assess the importance of donor and/or recipient related factors, which could affect recipient selection and transplant outcome. In addition the characteristics of used and unused donors and donor organs are monitored. Results are subsequently used to improve the existing organ allocation algorithms, to redefine the guidelines for donor organ suitability, and to formulate recommendations for donor management. In addition, individual transplant programmes are given the opportunity to evaluate their local policies. The transplant programmes should regularly provide the OEO[6,12] with all relevant donor, recipient, transplant and transplant follow-up data.

REFERENCE TISSUE TYPING LABORATORY

By establishing a central tissue typing laboratory, standardized procedures can be carried out at the tissue typing laboratories involved with organ donation and transplantation. This is achieved by organizing training courses and bench workshops, and by the distribution of standard reagents and material. A second goal, quality assurance and control, is accomplished by random post donation retyping at the central laboratory and/or by establishing regular specific proficiency testing schemes.[4] The aim is uniform and reliable donor and recipient HLA typing, recipient HLA antibody screening and cross-match results. In case of odd HLA-typing and/or cross-match results, the central laboratory can provide a second opinion service.

PUBLIC, PATIENT AND/OR PROFESSIONAL EDUCATIONAL PROGRAMMES

OEOs support a variety of informative and instructive projects to increase organ and tissue donation. Examples of these projects include the European Donor Hospital Education Programme (EDHEP), an initiative of the Eurotransplant International Foundation (ET) and Vital Connections, an initiative of the United Network for Organ Sharing (UNOS) in the United States of America. These educational programmes provide medical professionals with communication strategies to help them feel confident and effective in dealing with bereaved relatives and in making a request for organ and tissue donation. General donor awareness campaigns are conducted towards health care providers and the public. Transplant centres usually initiate patient educational programmes themselves. Some OEOs have established a telephone helpline, to answer any question about donation, organ procurement and transplantation (Spain, UNOS). This service is available for the public, health professionals and journalists.

PARTICIPATION IN TISSUE BANKING AND/OR BONE MARROW DONOR REGISTRIES

Several OEOs, for example, United Kingdom Transplant Support Service Authority (UKTSSA) and France also concentrate on cadaveric tissues, such as cornea, bone, skin and heart for valves.[2,3] The main tasks are tissue recovery, co-ordination of tissue processing for transplantation and tissue allocation. Sometimes, the tissue banking programmes evolve into separate organizations, maintaining a close working relationship with the founding organ exchange organization. An example is Bio Implant Services, founded by ET, which co-ordinates internationally several tissue banking programmes.[7] Some also maintain registries of unrelated living bone marrow donors e.g., Swiss Transplant, France.

Organizational aspects of an organ exchange organization

The structure of the modern OEO is a product of evolution rather than design.[10] Historical bonds, national culture and character, geographical and demographic characteristics, medical and immunological progress, changes in political and governmental control, managerial developments, telecommunication and computer networks have all led to a diverse spectrum of structures.[2,3,5,6,7,13,14]

INFRASTRUCTURE AND DEVELOPMENT

Most OEOs operate nationally, either directly since its foundation (France – 1969) or gradually over time (USA – 1986). For others, the co-operation was always international, ET – 1967 (Austria, Belgium, Germany, Luxembourg and The Netherlands), UKTSSA – 1972 (United Kingdom and Republic of Ireland), and Scandiatransplant – 1969 (Denmark, Finland, Iceland, Norway and Sweden). International collaboration for kidney and liver transplants exists between Australia and New Zealand.[10] In contrast, in Italy, several regional organizations continue to work independently. The service areas of the OEOs vary widely: UNOS, 9.4 per million square kilometres with a population of 253 million, ET 3.2 per million square kilometres with a population of 113 million,

Spain 1.5 per million square kilometres with a population of 38 million, and North Italy 0.4 per million square kilometres with a population of 1.8 million. The OEOs often started as private initiatives, founded on the recognition of the benefits widespread professional collaboration could attain in the field of transplant surgery. The voluntary co-operating transplant programmes were comparable to the members of a club who agree to abide by the rules set by their profession.[6] The OEO, as the official of the club, was the custodian and operator of those rules and had to strike a delicate balance in satisfying the demands of the individual transplant programmes while maintaining the integrity of the consensus framework representing all the members.

During the last decade, donation systems, transplant programmes and organ exchange organizations have received attention from governments, in particular, Ministries of Public Health. The reasons are the increasing burden of all types of organ transplantation on the organization of the healthcare system, the constitutional right of equal access to medical care, the shortfall of cadaveric donors, allegations of organ trading, medical tourism and professional non-compliance with the allocation rules. Some existing OEOs are attached to the Department of Health (UKTSSA, France), others are contracted by the government, remaining a private corporation (UNOS). In some countries the opportunity was taken by the government to create a formal national OEO, closely linked to the government (Spain, Saudi Arabia). Legislative measures have been introduced on brain death, organ procurement and transplantation (Austria, Belgium, Sweden, Denmark) and/or decrees were formulated concerning authorization of programmes to perform transplants (The Netherlands). ET, an international OEO, continues to operate on free will and consensus among the participants, achieving a true international collaboration.

POLICY DEVELOPMENT

OEOs have a hierarchical decision-making process.[15] Advisory committees, the cornerstones in this process, are usually related to the different transplantable organs, the organization of organ procurement and histocompatibility testing. An equal representation of the diverse regions and/or programmes in each committee is safeguarded. It is through these committees that recommendations for change of existing policies and the application of new regulations are prepared for approval by the OEO Board and/or government. In this policy development process, the transplant community is given the opportunity to act as a sounding board. In addition, all policies are regularly monitored and can be evaluated at the request of an individual transplant programme, the initiative of an advisory committee or on demand of the Board. With respect to policy making, all regulations are developed and/or sounded out by medical professionals actively involved in the field of transplantation. However, today, increasing appeals are made for participation in decision making by the public. UNOS has Board members who represent health organizations serving the interests of patients and of donor families and ethicists and economists.[12] Anticipating this trend of non-medical interest, several exchange organizations have standing committees like Patient Affairs and Ethics.

INTERNAL OPERATIONAL STRUCTURE

Central office

Frequently, many administrative and some operational tasks of an OEO are governed by a central office. The tasks vary widely. The central office usually operates the organ matching and allocation system. Sometimes, a central allocation office is established which participates, to a varying extent, in the organ allocation process within the territory of the OEO. In case of no suitable recipients being found, the central office offers the available organs to another OEO. A telecommunication network is operated to ease listing of transplant candidates, report of donors, registration of transplants and entry of transplant follow-up data. The central office collects and handles the requests for statistical analysis. The office organizes the Board and advisory committees' meetings and informs the transplant centres about new decisions and changes. The office also acts as the vocal point for the public and/or to the government.

Organ donation

The organization of donor organ referral and procurement is differently managed. In UNOS, organ procurement and transplantation are separated since the establishment of organ procurement organizations (OPO) which are frequently not hospital-based. Within an assigned geographical area, the OPOs promote donation in the hospitals. When notified of a potential donor they inform the central office of UNOS about the donor and assist in donor management and subsequent organ allocation. In contrast, in Europe, the transplant centres themselves act as the donor-reporting agency. The procurement area of a transplant centre consists either of a clearly-defined geographical area (e.g., Austria, The Netherlands, Spain, France) or of a conglomerate of affiliated donor hospitals (e.g., Belgium, Germany). The contact person between the donor/ transplant centre and the central office has changed from a surgeon or physician willing to co-ordinate the donor procedure, or who wished to be personally informed of an organ offer for one of his patients, to specially trained individuals. Chapter 16 expands on the introduction and many tasks of these, so-called, transplant co-ordinators.

Organ transplantation

Authorization of transplant programmes is a matter for the Ministry of Health rather than the OEO. Some OEOs have adopted corporate byelaws governing standards for the establishment or continuity of transplant programmes (e.g., UNOS, Saudi Arabia). Frequently, computer networks between transplant programmes enable the 24-hour-a-day listing of transplant candidates on the composite waiting list for any variety of organ transplant. Similar facilities are available for transplant registration and the entry of transplant follow-up data.

Organ matching/allocation and central allocation office

The differences in the achievement of organ matching and allocation between OEOs can be reduced to the presence or absence of a central organ allocation office. Some OEOs have opted for an organ matching and allocation procedure, fully initiated and executed by the donor centre (e.g., Scandiatransplant, Australia). Thus, no central allocation office is involved. The participation of a central organ allocation office in organ allocation depends on the allocation protocols, agreed by the transplant community, for the different donor organs. In one system, the central allocation office is the only one to co-ordinate the matching and allocation ('top-down' approach);[4] in the other, it only assists with the organ placement when needed ('bottom-up' approach).[8] The former option is in use in ET, while the latter exist in UNOS and France. Finally, the central office often arranges shipment of donor organs from donor centre to transplantation centre and/or transport of donor surgical teams.

FINANCIAL ASPECTS

The financial structure of an OEO is dependent on the legal entity of the organization and on the characteristics of the national health care system(s). As for the legal entity, a distinction needs to be made between a private non-profit foundation (e.g., ET), a governmental institution as a part of the Ministry of Health (e.g., UKTSSA), and a private non-profit corporation, contracted by the government (e.g., UNOS). ET is financed by the health insurance authorities in the participating countries, by means of the payment of a fee for every patient registration on the waiting list. The contribution is related to the type of transplant procedure and to the extra activities that ET provides (e.g., salary of transplant co-ordinators, HLA typing of donor and recipient). The fee is annually adjusted to meet the Board-approved budget. In addition, ET functions as a clearing house, the donor centre is reimbursed by ET for the costs related to donor management, donor removal and donor organ transport. At a later stage, ET charges these expenses to the transplant recipients or centres. The Department of Health is responsible for the operational maintenance of UKTSSA. Adequate funding as a cash-limited budget is provided and comes out of public funds raised through general taxation.[6] The UNOS contract is a cost-sharing arrangement in which the government pays only part of the cost accrued through contract performance. UNOS pays the remaining costs from patient registration fees.[12]

Allocation of donor organs within an organ exchange organization

PRINCIPLES FOR AN ALLOCATION MODEL

It is generally accepted that organ allocation should be equitable. The two principles important for policy decisions concerning allocation of donor organs for transplantation are *medical utility* and *justice*.[14,16]

Usually, these allocation principles depend on the individual transplant candidate. Patient selection based upon medical utility points towards the predictably best outcome, i.e. the maximum number of patient years of graft function. Selection based upon justice is related to giving an organ to those who are worst off, either when an organ is to be allocated or from the 'over-a-lifetime' perspective. Strictly speaking, these two principles are conflicting. However, in any allocation model, the decision to use an allocated organ remains the privilege of the transplant surgeon and/or physician responsible for the care of the patient. The physician evaluates both the donor organ quality and the circumstances of the proposed recipient patient before making a final decision. In addition, the patient reserves the right to decline the offered possibility for a transplant.

FACTORS USED IN AN ALLOCATION MODEL

Factors, used in the different organ allocation models, can be divided into *medical* and *non-medical*. Examples of medical factors are ABO blood group, HLA-typing (and thus HLA-matching between donor and recipient), cross-match result, body size parameters, virological results, recipient and/or donor age, cold ischaemia time, medical urgency of a recipient and HLA-sensitization. Time on the waiting list, donor organ exchange balance, geographical areas, logistical issues and centre transplant activity are categorized as non-medical factors. These factors and their ultimate use are related to either one or both allocation principles and to the type of organ to be allocated. The use of HLA-typing, cross-match results, ABO blood group, medical transplantability and body size parameters are supported by the 'medical utility' principle, where as medical urgency, waiting time, HLA-sensitization, likelihoods of finding a suitable organ in the future and geographical aspects come under the 'justice' principle. Factors, such as HLA-typing and HLA-sensitization, are almost exclusively used in kidney allocation models. Body size parameters are only used as a factor in thoracic organ and liver allocation models. Each allocation factor might not be applied in the same manner in the different OEOs. These decisions often depend on the results of the scientific analyses and/or consensus within the transplant community. As an illustration, we refer to the use of the HLA-typing and HLA-matching in kidney allocation. ET prioritizes patients without an HLA-antigen mismatch between donor and recipient on the HLA-A,B,DR loci. After that, HLA-A, B, DR mismatch classes are ranked per increasing number of mismatches. The highest priority is given to an HLA-DR antigen mismatch, followed by an HLA-B antigen mismatch, which, in turn, has more weight than an HLA-A antigen mismatch. In contrast, in UKTSSA, priority is given to zero HLA-A,B,DR mismatch recipients and the recipients who have only a single HLA-A or HLA-B antigen mismatch (called 'beneficial matching'). UNOS recently changed its policy regarding HLA-typing. Formerly, the first choice for a kidney was recipients who were HLA-A, B, DR identical with the donor. This has now been modified to include patients with no HLA-A, B, DR mismatch. After that, both in UKTSSA and in UNOS, several HLA-A, B, DR mismatch classes are grouped together and, subsequently, ranked or weighted. The definition of (highly) sensitized recipients also varies.[2,3] In ET, UKTSSA, France and Scandiatransplant, high immunization of a recipient means incompatibility with 85% or more of a standard donor lymphocyte panel. This is expressed as percentage of panel reactive antibodies (%PRA). In UNOS, 80% or more positive reactions are sufficient for the status of 'Highly Sensitized', whereas, in Swiss Transplant, patients with 50% or more positive reactions qualify as highly immunized. Even more variation is seen in the definition of sensitization, ranging from 1%PRA (UKTSSA), over 5%PRA (ET, Swiss Transplant) and up to 20%PRA (UNOS). Regarding geographical areas, the principle 'justice' is often quoted to give priority, for locally procured organs, to recipients listed at the local transplant programme associated with the donor hospital. The local programme should be rewarded for its efforts in initiating organ donation in its own area. Sometimes, such a local priority is also defended using the 'medical utility' principle, as reduction of cold ischaemia times will improve the outcome of the transplant. Lastly, the costs associated with organ exchange and budgetary constraints, increasingly, oblige geographical features to be considered. Therefore, many allocation protocols include geographical aspects, donor organs are first allocated locally, then regionally, then nationally and finally internationally (UNOS, France, UKTSSA). Some allocation models incorporate the so-called organ export/import balance of a transplant programme. This is the difference between organs locally procured but not transplanted locally and organs received (and transplanted) from outside the region (ET). Following the 'justice' principle, neutral organ export/import balances should be achieved. An alternate method to meet the loss of potential local transplants is the organ payback system. For example, if organ exchange was mandatory (UNOS), transplant centre A would return an organ to centre B that offered the organ for a patient listed at centre A.

ALLOCATION

Organ allocation can be patient-oriented or centre-oriented. In the former category, the organ offer is made to a specific patient, while, in the latter, the offer is made to a transplant centre that subsequently selects the best suitable recipient. Often, the two types are combined in a particular organ allocation model. The ultimate example of a patient-oriented offer is the, widely applied, idea of 'mandatory exchange'. Whenever, for a particular donor, transplant candidates appear in this 'mandatory exchange' category, the offer and/or exchange of the donor organ is obligatory. The definition of 'mandatory exchange' differs according to the donor organ and between the different OEOs. In ET, regarding kidney transplantation, zero HLA-A, B, DR mismatched recipients, current and non-current sensitized recipients selected by the Acceptable Mismatch Programme and eligible recipients in the Highly Immunized Trial protocol (HIT, conducted by Professor Dr G. Opelz, Heidelberg, Germany) come into the 'mandatory exchange' category. Patients with failure of a thoracic organ within the first week following transplant qualify for the thoracic 'mandatory exchange' category. In ET and in many other organizations (France, Spain), the 'mandatory exchange' category for liver transplantation consists of patients with an intractable acute *de novo* liver failure and patients with an irreversible acute liver graft failure. An allocation model should also use a specific method to generate the allocation lists, either a point's system, a 'stepwise' hierarchical system or a strict rotational system. In a points system, either per patient or per centre, several factors are weighted simultaneously and the sum is used for ranking the patients or the centres, e.g., kidney allocation in UNOS (patients) and liver allocation in ET (centres). In the 'stepwise' hierarchical (most often patient-oriented) system, the factors are used sequentially. One factor after another is sorted according to a preset pattern to generate the final recipient allocation list, e.g., kidney and thoracic organ allocation in ET. The pure rotational system is nearly always centre-oriented. All centres are placed in a list, every time a centre accepts an offer it is put at the end of the list, e.g., liver allocation in UKTSSA.

SPECIAL FEATURES

It is generally accepted that sensitized patients have a high chance of a positive cross-match with a donor. To reduce inappropriate offers to these recipients, some OEOs demand the regular shipment of sera of (highly) sensitized patients on the kidney waiting list to all tissue typing laboratories within the OEO. A kidney is only offered to (highly) sensitized patients who have a negative preliminary cross-match at the donor centre. At the recipient centre, the cross-match must be repeated, before transplantation (ET). In Australia, however, sera shipment involves all kidney patients on the waiting list, despite the level of sensitization. To reduce the cold ischaemia time the cross-match, performed at the donor centre, is not repeated at the recipient centre. Efforts made for immunized kidney patients also concern the development of special programmes by the OEOs reference tissue typing laboratories.[17,18] In ET, the Acceptable Mismatch Programme allows the incorporation of HLA-A and HLA-B mismatched antigens, which are expected not to result in a positive cross-match, in the HLA-typing of the recipient. Thus, the chance of finding a suitable donor kidney without a positive cross-match will be greater. If selected by the programme, the recipient has the priority for the kidney offer.

Summary

The organization of cadaveric organ donation and allocation differs worldwide due to governmental participation, financial support, historical bonds and socioeconomic factors. Hospital infrastructure and acceptance of the brain death are essential to the establishment of (renal and) non-renal cadaveric organ transplant programmes. OEOs were established to avoid wastage of donor organs, optimize the use of donor organs, execute equitable and transparent organ allocation and create an (inter)national alliance of donor hospitals and transplant programmes, all applying the same set of organ allocation rules. Differences between the existing OEOs are concerned with infrastructure and development, financial support, internal operational structure and decision making about organ allocation protocols. Medical utility and justice are used in the development of an organ allocation model. Allocation lists are generated by using medical and non-medical factors in a specifically defined method accepted by the transplant community involved.

Organ allocation within the Eurotransplant organization

As an example, the allocation for kidney, liver and thoracic organs within the Eurotransplant Foundation (status January 1996) will be described.

BASICS OF THE ALLOCATION PROCEDURE

The procedure can be divided into three levels (see Table 12.1). First, the category *mandatory exchange*, whose impact on the allocation varies according to the type of organ. The impact for kidney and liver transplantation is about 20%, whereas for thoracic transplantation it is negligible (<3%). The offer is always patient-specific and is made by the central allocation office of Eurotransplant. Secondly, the transplant programmes associated with the donor hospital, decide on the possibility of a *local transplantation*. The programme selects the best suitable recipient, according to a minimum set of rules. For all organs, this considers ABO blood group, and, for kidney, a minimum HLA-antigen sharing criterion. The third level consists of the allocation of donor organs, which could not be used on the second level. These donor organs are offered to the *ET recipient pool* and are placed by the 24-hour duty office following specific allocation rules.

KIDNEY ALLOCATION

First level

Donor kidney exchange is mandatory for patients with zero HLA-mismatches, patients selected by the Acceptable Mismatch Programme and patients selected by the HIT procedure (see page 234). Except the two latter programmes, the blood group type O kidneys are allocated only to blood group type O and/or B recipients. There is no priority for ABO identical versus ABO compatible blood groups in matching the donor.

Table 12.1 Eurotransplant Foundation: the basics of the allocation procedures (January 1996)

	Kidney	Heart	Lung	Liver
LEVEL 1		Mandatory Exchange Eurotransplant (patient specific)		
	HIT-AM '000' HLA-mismatch 20%	HU <3%	HU <3%	HU 20%
LEVEL 2		'Local' Transplant Centre (within constraints of selection rules)		
	50%	60%	20%	35%
LEVEL 3		Eurotransplant Pool		
	30% (patient-specific)	40% (patient-specific)	80% (patient-specific)	45% (centre-oriented)

The percentages, per organ, represent the average proportion of transplants, performed at each of the three levels of allocation. The levels have a hierarchical sequence: Level 1: mandatory exchange. Level 2: local transplant centre. Level 3: Eurotransplant pool

The zero HLA-mismatched recipients are sorted first according to the current allo-sensitization and then according to the time on the waiting list.

Second level

The programme selects the best suitable recipient(s) on the waiting list, respecting the ABO blood group and the minimum HLA-antigen sharing rules. The HLA-antigen sharing for a kidney transplant should be minimum either 2 HLA-DR antigen sharing or 1 HLA-B antigen + 1 HLA-DR antigen sharing between the donor and the recipient.

Third level

If donor kidneys are offered to the Eurotransplant pool, the recipients are sequentially sorted, first according to the medical urgency, secondly to the HLA-mismatch class, thirdly to the sensitization and finally to the waiting time.

Concerning medical urgency, priority is given to High Urgency patients, who have severe physical and/or psychological problems. The HLA-A, B, DR mismatch classes are sorted as follows: 100, 200, 010, 110, 210, 001, 101, 201, 020, 120, 220, 011, 111 and 211. Regarding sensitization, priority is given first to the highly immunized patients (85–100% PRA), then to immunized patients (6–84% PRA) and then to the non-immunized patients (0–5% PRA). No offer is made to sensitized patients with a positive cross-match at the donor centre, to patients below the minimal degree of HLA-sharing, and to the patients belonging to a transplant programme with a net kidney import of five or more.

LIVER ALLOCATION

First level

Donor liver exchange is mandatory for patients, accepted on the High Urgency waiting list. Donor size requirements, donor age ranges and ABO blood group matching should be indicated. The allocation is made after preliminary stratification according to these donor characteristics.

Second level

The programme selects the best suitable recipient. The ABO blood group is identical between donor and recipient.

Third level

The liver offer procedure is centre-oriented. Transplant centres that have transplant capacity and have ABO identical recipients on their waiting list for whom the donor weight is appropriate are sorted according to the position they have on the centre rank list at the time of the offer. The position on the rank list is influenced by the liver transplant activity, by the liver donor activity and by penalties. A penalty is given in case of a three-in-a-row decline of a donor liver that was subsequently transplanted.

THORACIC ORGAN ALLOCATION

First level

Exchange is mandatory for patients, accepted on the High Urgency waiting list. This status is assigned to patients with failure of a thoracic organ within the first week of transplantation. Donor size requirements, donor age ranges and ABO blood group matching, and, where applicable, the type of donor lung needed, must be indicated for all recipients.

Second level

The programme decides which thoracic organs will be transplanted locally, the type of thoracic transplant and selects the best suitable recipient(s). The ABO blood group rule for lung and heart–lung transplantation is ABO blood group compatibility, whereas, for a heart-only transplant, blood group type O hearts are allocated only to blood group type O and/or B recipients.

Third level

If thoracic donor organs are available to the ET pool, the allocation lists are made after a preliminary stratification according to the donor size, donor age and donor blood group. The thoracic organ offer procedure is always patient-specific. The recipients are sequentially sorted first according to the type of transplant, then according to the medical urgency. This is followed by ABO blood group and finally according to the time on the waiting list. As for the type of transplant, heart–lung transplant candidates have priority over heart-only or lung-only transplant candidates. Regarding the medical urgency, priority is given to the 'Special Urgency' recipients over the elective recipients. The inclusion criteria for a 'Special Urgency' request is permanent hospitalization and/or stay on intensive care unit, the presence of a life-threatening situation despite adequate therapy and inotropic support and/or mechanical ventilation. An annual fixed number of 'Special Urgency' requests is granted to the transplant programmes, based on the transplant activity during the previous calendar year. There is priority for ABO identical versus ABO compatible donor/recipients combinations. The waiting time in thoracic transplantation is always corrected for the periods of non-transplantability. In heart-only allocation, an additional sorting factor, the donor country, is placed between medical urgency and ABO blood group. This nationalized allocation is not applied in heart–lung and lung-only allocation procedure. With respect to lung allocation, no distinction is made between double lung and single lung transplant candidates.

References

1. World Health Organization, Legislative responses to organ transplantation. Martinus Nijhoff Publishers, Dordrecht, 1994.
2. Council of Europe, Matesanz R, Hors J, Persijn G et al. (Eds), *Transplant*, July 92, **04**.
3. Council of Europe, Matesanz R, Hors J, Persijn G et al. (Eds), *Transplant*, September 93, **05**.
4. Persijn GG, Cohen B. Eurotransplant part I, organizational aspects. In *Clinical Transplants 1986*, Terasaki P. (Ed.), UCLA Tissue Typing Laboratory, Los Angeles, 1987, pages 35–46.
5. Balderson R. UKTSSA – the support service for transplant units nationwide. *British Transplantation Society Newsletter* 1993, **1**: 7.
6. Pudlo P. European transplant programs – what can the U.S. learn? *J Transplant Coordination* 1993, **3**: 138–140.
7. Cohen B, Haase-Kromwijk B. Eurotransplant 1990 – progress report. *J Transplant Coordination* 1991, **1**: 5–10.
8. Ferree DM. Cadaveric organ sharing: the organ centre. In *UNOS: Organ procurement, preservation and distribution in transplantation*, Philips MG, (Ed.), The William Byrd Press, Richmond, 1991, pages 129–144.
9. Barnes BA. Experience of the New England organ bank. *Clinical Transplants 1986*, Terasaki P. (Ed.), UCLA Tissue Typing Laboratory, Los Angeles, 1987, pages 53–60.
10. Armstrong G. Organ donation in Australia – an overview. *ETCO Newsletter* 1994, **2**: 3–8.
11. Van Rood JJ. A proposal for international cooperation in organ transplantation: Eurotransplant. *Histocompatibility Testing* 1967, 451–452.
12. UNOS. New OPTN, Scientific Registry contracts foretell increased responsibilities. *UNOS Update* 1993, **9**: 2–12.
13. Saudi centre for organ transplantation. Regulations of organ transplantation in the Kingdom of Saudi Arabia. *Saudi J Kidney Dis Transplant* 1994, **5**: 37–98.
14. Milford EL. The end-stage renal disease transplant program: an experiment in participatory democracy and national health care. *Seminars in Dialysis* 1994, **1**: 69–74.
15. *Eurotransplant Newsletter* 1994, **10**: 2–6, 11–21.
16. 1991 UNOS Ethics Committee. General principles for allocating human organs and tissues. *Transplant Proc* 1992, **24**: 2227–2235.
17. Claas FHJ, De Waal LP, Beelen J et al. Transplantation of highly sensitized patients on the basis of acceptable HLA-A and HLA-B mismatches. *Clin Transpl* 1989, **20**: 185–190.
18. Special schemes for transplanting highly sensitized patients. In *Renal transplantation: sense and sensitization*. Gore SM, Bradley BA (Eds), Kluwer Academic Publishers, Dordrecht, 1988, pages 268–281.

CHAPTER 13

Bone marrow donation

KERRY ATKINSON

Background 239
Results of family member donor bone marrow transplantation 241
Results of HLA-matched unrelated donor bone marrow
    transplantation 242
Tissue typing techniques in selecting bone marrow donors 243
The search process 245
Donation from young minor donor to other family members 250
Differences in recruitment between related and unrelated bone marrow
    donors 250
Process of bone marrow donation – processing, labelling and
    transportation 253
Donor confidentiality 262
Second donation of marrow for the same patient 263
Recovery, counselling and long- and short-term physical outcome 263
Role of erythropoietin and colony stimulating factors 264
Potential for peripheral blood stem cell harvests 264
Potential for cord-blood harvests 266
References 266

Background

The body's blood forming tissue, marrow, is located inside the large bones of the body. All the formed elements of the circulation are generated from pluripotent haemopoietic stem cells and subsequently released into the circulation. A number of diseases including acute and chronic leukemia, aplastic anaemia and certain genetic and immunodeficiency diseases result in life threatening malfunction of the marrow. In order to eradicate the disease the patient receives supralethal doses of chemotherapy or chemoradiation. This treatment is aimed at eliminating the disease, but in consequence also destroys the patient's marrow. In order to survive, the patient must receive a marrow transplant from a healthy donor.

For transplantation between 500 and 1500 ml of marrow is removed from the donor's posterior iliac crests using percutaneous needle aspiration. This is done while the donor is under anaesthesia in an operating room and usually requires an overnight stay in the hospital. Once the marrow has been collected it is then infused intravenously to the recipient in a manner similar to a routine blood transfusion.

Each year many thousands are diagnosed with illnesses for which marrow transplantation is the treatment of choice. A successful marrow transplant, however, requires an HLA-compatible donor. Until the late 1970s the only patients who could take advantage of marrow transplantation were those with an HLA-compatible family member donor. Based on inheritance of HLA type and current family size this meant approximately only 30% of eligible candidates were able to receive such a transplant.

An additional small number of patients (1–5%) have a family member with whom they are one HLA antigen mismatched. While there is an increase in severity of graft-versus-host disease, patients who receive a one antigen mismatched family member transplant have a survival that is not significantly different from that of patients who receive HLA-identical sibling transplants. Survival for those who receive a two or three antigen mismatched transplant is significantly worse.

In 1979 the first marrow transplant for a patient with a haematological malignancy from an unrelated but HLA-matched donor took place and preliminary data on HLA-matched unrelated marrow transplants shows that while more risky than HLA-matched family member transplants, they are capable of curing otherwise incurable diseases.[1-4]

These early results provided the rationale for developing unrelated volunteer donor registries. The first national registry was the Anthony Nolan Research Centre in London. The biggest current registry is the National Marrow Donor Program in the United States; there are more than 2.5 million donors recruited world wide at the present time with over 30 countries contributing (see Chapter 24). Since HLA type is directly linked to ethnic origin it has been important for registries composed of others than Caucasian to be developed. There is now a Japanese registry and ethnic Chinese registries in Taiwan, Singapore and Hong Kong. A major effort is being made to recruit black Americans, Hispanics and Orientals in the United States. The first reported case of a phenotypically identical unrelated marrow transplant for haematological malignancy was reported in 1980;[5] the first for a patient with severe aplastic anaemia was reported in 1973.[6]

The diseases which can be considered for treatment by unrelated donor marrow transplantation are the same as those considered for family member donor transplantation, although many physicians would have reservations about proceeding to an unrelated donor transplantation as early as they would proceed to HLA-identical sibling transplantation. The clinical indications for unrelated and family member donor transplantation are shown in Table 13.1.

Currently the chance of finding a matched unrelated donor for a given individual is approximately 20%. This varies greatly from one individual to another, however, and depends on how common or rare that individual's HLA haplotypes are. For those with common haplotypes a donor will always been found. For those with very rare haplotypes there is virtually no chance of finding a donor at the present time. This may change as the global donor pool enlarges, and as the matching requirements for donor-recipient pairs becomes less stringent with the development of new treatment modalities to counteract the several clinical problems currently inherent in unrelated donor transplantation.

Table 13.1 Indications for HLA-identical family member or unrelated donor bone marrow transplantation

Haematologic malignancies	Hodgkin's disease
Myelodysplasia (preleukemia)	Non-Hodgkin's lymphoma
Acute lymphoblastic leukemia	Myeloma
Acute nonlymphoblastic leukemia	Chronic primary myelofibrosis
Chronic myeloid leukemia	Acute myelosclerosis
Chronic lymphatic leukemia	
Hairy cell leukemia	
Nonmalignant haematologic disorders	*Immunodeficiencies*
Fanconi's anaemia	Wiskott–Aldrich syndrome
Severe aplastic anaemia	Severe combined immunodeficiency (SCID)
Congenital aregenerative anaemias	Chediak–Higashi syndrome
Chronic neutropaenia disease	Adenosine deaminase deficiency
(Kostmann's syndrome)	Leucocyte adhesion molecule deficiency
Thalassaemia and other lethal haemoglobinopathies	
Diamond Blackfan syndrome	
Paroxysmal nocturnal haemoglobinuria	
Metabolic disorders	*Other diseases*
Mucopolysaccharidosis	Osteopetrosis
Wolman's disease	Neuroblastoma
Sanfillipo's syndrome	
Harber's syndrome	
Hunter's syndrome	
Hurler's syndrome	
Fabry's disease	
Gaucher's disease	
Metachromatic leucodystrophy	

Results of family member donor bone marrow transplantation

In the early 1970s most HLA-identical family member transplants were carried out in patients at the end stage of their diseases. It was during this time that most of the regimens employing total body irradiation or chemotherapy were developed. In 1977 the Seattle group reported results of HLA-identical sibling transplants in 100 patients with end-stage leukemia.[7] This now classical study illustrated the potentially curative effect of marrow transplantation in acute leukemia with 13 very long-term disease-free survivors. The enthusiasm over this remarkable result was tempered by the actuarial relapse rate of 70% and the high incidence of non-leukemic deaths due to treatment-related complications. Thomas and his group reasoned that transplants performed in early remission should fare better since patients would have a smaller tumour burden and would be in better clinical condition than end-stage patients who, besides having fulminant leukemia, were often infected. The report in 1979 by the Seattle group of transplantation of patients with acute non-lymphoblastic leukemia in first remission

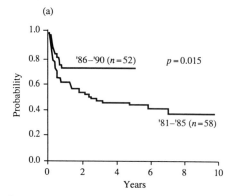
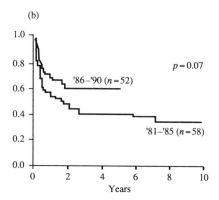

Figure 13.1(a) Survival of good risk patients transplanted during the period 1981–1985 or 1986–1990.

Figure 13.1(b) Leukemia-free survival of good risk patients transplanted during the periods 1981–1985 or 1986–1990.

confirmed their suppositions.[8] Indeed, 10 of the first 19 patients transplanted were alive, free of leukemia for greater than five years in the subsequent follow-up report.[9]

With better control of acute graft-versus-host disease by combination immune suppression utilizing cyclosporin in combination with methotrexate[10] and the prevention of cytomegalovirus pneumonia by the use of prophylactic ganciclovir,[11] a sequential improvement in the long-term results of HLA-identical sibling bone marrow transplantation for acute leukemia has been seen[12] (Figure 13.1a,b).

Similar results have been obtained with HLA-identical sibling transplantation in patients with chronic myeloid leukemia. For those transplanted in first chronic phase, and in particular within one year of diagnosis, long-term disease-free survival of between 60% and 80% can be achieved.[13,14] Similarly, excellent results can now be obtained with HLA-identical sibling transplantation for patients with severe aplastic anaemia.[15]

Results of HLA-matched unrelated donor bone marrow transplantation

In 1990 McGlave and colleagues reported 102 patients with chronic myeloid leukemia given unrelated donor transplants in Minneapolis, Milwaukee, London and Seattle between April 1985 and February 1989. Forty-four (44) received a 6/6 HLA-match whereas 58 received a transplant from a donor with a 1 HLA locus disparity or showing bi-directional mixed lymphocyte culture (MLC) reactivity. Forty-six (46) of the 102 were alive at the time of the report with a Kaplan–Meier actuarial survival rate of 29% at 2.5 years. Those receiving a fully matched transplant had a 39% ± 26% (95% confidence intervals), while those receiving mismatched transplants had a survival rate of 27% ± 15%.

In 1991, and subsequently in 1993, the first clinical results of unrelated donor transplants facilitated by the National Marrow Donor Programme, USA, were published.[16] This report detailed transplants for 459 patients with malignant ($n = 387$) or non-malignant disorders ($n = 72$). The transplants were performed in 29 centres and the transplant protocols varied considerably. The median age of the patients was 25.5

(range 0.3–54.5) years, while the median age for the donors was 37.9 (range 19–56) years. Some 304 patients received marrow from phenotypically HLA-A, -B, -DR identical donors, while 60 (13%) received marrow that differed by one HLA Class I cross-reactive antigen, and a further 20% received marrow that differed by either a major Class I or Class II HLA antigen. Ninety-three percent (93%) of patients engrafted. Acute graft-versus-host disease of moderate to severe degree occurred in 67% and chronic (late) graft-versus-host disease (GVHD) in 51%. The actuarial probability of disease-free survival at 1.5 years was $36 \pm 4\%$ for patients transplanted for acute leukemia in first or second complete remission or chronic myeloid leukemia (CML) in first chronic phase. This was significantly better than the disease-free survival of patients with advanced leukemia (beyond first remission acute leukemia or beyond first chronic phase of CML), which was $20\% \pm 3\%$. Factors affecting disease-free survival included patient age, degree of HLA disparity, recipient CMV serological status and time between diagnosis and transplant.

This large series thus indicated that unrelated donor transplantation is considerably more difficult than HLA-identical sibling transplantation; the problems are the same as those seen after HLA-identical sibling transplantation but are more common and more severe. It is clear from these reports that unrelated donor transplants are difficult, time consuming, expensive and fraught with physical and emotional difficulties for the patients, their families and the hospital staff looking after them. They should therefore be undertaken only by units experienced in family member allogeneic marrow transplantation at the present time, and probably should be restricted to large centres.

Tissue typing techniques in selecting bone marrow donors

There are four main methods of tissue typing techniques in use for selecting bone marrow donors at the present time; serological, cellular, biochemical and molecular (Table 13.2). Almost all marrow donor registries utilize serological testing for typing of HLA-A, -B and -DR antigens.

Mixed lymphocytes cultures are still commonly performed in selecting family member donor-recipient pairs, although several groups have recently reported a lack of correlation between MLC reactivity and subsequent outcome after *unrelated* donor transplantation. The cytotoxic T-lymphocyte precursor frequency has predicted the severity of graft-versus-host disease after unrelated donor transplantation in one centre[17] but not another.[18] Helper T-lymphocyte precursor frequency analysis is currently also being explored in this regard. The extent of HLA polymorphism has been further revealed by techniques such as isoelectric focusing gel electrophoresis; for example, the HLA-B27 antigen which cannot be subdivided serologically, is now definable at the amino acid sequence level as having at least six micro variants. This technique separates protein molecules on the basis of their isoelectric point (pI) which is determined by amino acid composition. The nomenclature for these micro variants is derived from the standard nomenclature for the original antigen. Thus micro variants of the HLA-B27 antigen are designated B*2701 to B*2706. The * indicates that the allele has been uniquely defined at either the amino acid or nucleotide level. The numbering system for the micro variants has established '01' as the most frequent micro variant, and the numbering of the remaining micro variants usually follows their relative positioning in a IEF gel from most basic to most acidic.

Such biochemical typing methods for defining HLA antigens are sensitive enough to detect serologically undetected subtypes (or micro variants), but they have the

Table 13.2 Tissue typing methods for selecting bone marrow donors

Technique	Comment
Serological	For routine HLA-A, -B, -DR typing
Cellular	
Mixed lymphocyte culture (MLC) or reaction (MLR)	Gives overview of HLA-D region compatibility
Primed lymphocyte test (PLT)	For HLA-DP typing
Homozygous typing cell testing (HTC)	For HLA-Dw phenotyping: a panel of T cell defined, HLA-D region-associated antigens (HLA-Dw) are recognized. Primarily HLA-DR specific
Cytotoxic T cell precursor frequency	Appears to predict severity of acute analysis (CTLp) graft-versus-host disease after unrelated donor transplantation
Helper T cell precursor frequency analysis (HTLp)	Appears to predict severity of graft-versus-host disease after allogeneic transplant
Biochemical	
Isoelectric focusing (IEF)	Can distinguish variants of a given allele differing by a single amino acid
Molecular	
Restriction fragment length polymorphism (RFLP)	Restriction enzymes digest genomic DNA into fragments by gel electrophoresis, blotted on to membranes, and hybridized with radiolabelled cDNA probes, to reveal intron region polymorphisms for HLA-DR, -DQ, and -DP genes. Most DR specificities can be distinguished by RFLP analysis, although some, especially of the DR4 family, cannot. Inferior to allele-specific oligonucleotide typing
Sequence-specific oligonucleotide (SSO) or allele-specific oligonucleotide (ASO) typing using polymerase chain reaction (PCR)	DNA is amplified by PCR using oligonucleotide primers specific for nucleotide sequences within known hypervariable regions within the first domain of class II alpha and beta chains. At least 11 alleles of DR4 have been identified by SSO-PCR typing. Currently, it is the molecular method of choice

limitation that two quite different protein molecules may have the same pI, since the pI is determined by the sum of the charged amino acid residues.

There is little doubt that the future of typing for both family member and unrelated marrow transplantation lies in molecular typing. A number of methods are in practice, of which the most basic is restriction fragment length polymorphism. The most acceptable at the present time is allele-specific oligonucleotide typing (ASO) using polymerase chain reaction technology. Molecular typing has largely replaced cellular typing for class II antigens and will soon do so for class I as well. The ultimate gold standard will be sequencing of MHC antigens. This will have to await the development of practical automated sequencing technology.

RFLP analysis can readily distinguish DR 1 to DR 18 haplotypes and most Dw subtypes. Sequence specific or allele-specific olygonucleotide (SSO and ASO respectively) typing systems have been developed for HLA-DR, -DQ and -DP. Such DNA typing represents a precise approach to the identification of allelic subtypes.

PCR-SSO typing currently represents the most accurate method for identifying novel allelic subtypes of Class II genes. Correlation between different Class II typing techniques is shown in Table 13.3. The application of SSO or ASO typing to Class I typing must await the completion of DNA sequencing of all Class I alleles. It appears likely that the complete polymorphism of HLA will be known at the DNA level within the next decade.

The search process

Once it is determined that a patient could be appropriately treated by allogeneic bone marrow transplantation, the patient and his/her siblings should be HLA-A and -B typed. Should a 4/4 HLA-A and -B match be found, the next step is to perform HLA-DR typing and in most centres a mixed lymphocyte culture. If a 6/6 HLA-A, -B, -DR identical or 5/6 HLA-A, -B, -DR identical family member donor is found, such a person will make the most appropriate marrow donor. Should such a donor not be available, a search should then be made of the extended family and for an unrelated donor simultaneously. The advantages of the extended family approach is that the parents, grandparents, children and in some cases aunts, uncles or cousins, can share one of the patient's haplotypes and there is a possibility that the other haplotype has been introduced into the family via spouses. Hence, there is a possibility that cousins or children may be HLA-identical or mismatched at only one HLA antigen with the patient. The probability of this occurrence is dependent on the frequency of the non-shared haplotype in the general population. In cases where the patient has two haplotypes of low frequency in the general population, the greatly reduced probability of finding an HLA matched donor in an extended family search gives this approach little hope of success. The frequency of the commoner haplotypes in a predominantly Caucasian population is shown in Table 13.4.

The strategy for an extended family search is to explore the side of the family from which the less common haplotype in the patient has been inherited, in the hope that the commoner haplotype has been introduced by marriage. It seems reasonable to conduct extended family studies on the basis of the 16 most prominent haplotypes (frequency >50/10,000) (Table 13.4).

An example of a family tree in which a successful search was performed is shown in Figure 13.2.

PERFORMING AN UNRELATED DONOR SEARCH

The easiest way to perform a preliminary search for a patient needing a matched unrelated donor is to utilize the listing of all donors worldwide which is published every four months by the EuropDonor Foundation at the instigation of the Immunology Working Party of the European Bone Marrow Transplant Group. This publication

Table 13.3 Correlation of serological and DNA typing technology for HLA-A, HLA-B, HLA-C, HLA-DR and HLA-DQ

Serology	DNA A*	Serology	DNA B*	Serology	DNA C*	Serology	DNA
A 1	A* 0101, 0102	B 7	B* 0702–0705	Cw 1	C* 0101–0103	DR 1	DRB1* 0101–0104
2	0201–0217	8	0801, 0802	2	0201, 0202	15 (2)	1501–1505
3	0301, 0302	13	1301, 1303	3	0302–0304	16 (2)	1601–1605
11	1101, 1102	64 (14)	1401	4	0401–0403	17 (3)	0301
23 (9)†	2301	65 (14)	1402	5	0501	18 (3)	0302–0303
24 (9)	2402–2406	62 (15)	1501, 1504–1508, 1515	6	0602	3	0304–0308
25 (10)	2501		1520, 1524	7	0701–0705	4	0401–0423
26 (10)	2601–2604	75 (15)	1502	8	0801–0803	11 (5)	1101–1127
29 (19)	2901, 2902	72 (70)	1503	—	1202–1203	12 (5)	1201–1204
30 (19)	3001–3005	70	1509	—	1301	13 (6)	1301–1324
31 (19)	3101	71 (70)	1510, 1518	—	1402–1403	14 (6)	1401–1425
32 (19)	3201	15	1511, 1521–1521, 1525	—	1502–1505	7	0701
33 (19)	3301–3303	76 (15)	1512, 1514, 1519	—	1601–1603	8	0801–0815
34 (10)	3401, 3402	77 (15)	1513	—	1701, 1702	9	0901
36	3601	63 (15)	1516, 1517	—	1801	10	1001
43	4301	18	1801, 1802				
66 (10)	6601, 6602	27	2701–2709			DRw 52	DRB3* 0101, 0201–0206, 0301
68 (28)	6801, 6802	35	3501–3510			53	DRB4* 0101–0103
69 (28)	6901	37	3701			51	DRB5* 0101–0104, 0201–0203
74 (19)	7401	38 (16)	3801, 3802				
—	8001	39 (16)	3901–3907			DQ 5 (1)	DQB1* 0501–0504

60 (40)	4001	6 (1)	0601–0609
61 (40)	4002, 4006	2	0201, 0202
40	4003–4005, 4007	7 (3)	0301, 0304
41	4101, 4102	8 (3)	0302
42	4201	9 (3)	0303
44 (12)	4402–4406	3	0305
45 (12)	4501	4	0401, 0402
46	4601		
47	4701		
48	4801, 4802		
49 (21)	4901		
50 (21)	5001		
51 (5)	5101–5105		
52 (5)	5201		
53	5301		
54 (22)	5401		
55 (22)	5501, 5502		
56 (22)	5601, 5602		
57 (17)	5701–5703		
58 (17)	5801, 5802		
59	5901		
67	6701		
73	7301		
78	7801		
–	8101		

† The number in parentheses is the broad serological antigen group, e.g. A23 and A24 are both splits of A9. Compiled in October 1996; due to new alleles constantly being identified this list will become rapidly out of date.

Table 13.4 List of haplotypes to be investigated in extended family studies

A	B	DR	Haplotype frequency per 100,000*
1	8	3	477
3	7	2 (15)	260
3	35	1	133
2	7	2 (15)	127
29	44	7	111
1	57	7	101
2	62	4	86
2	18	11	84
30	13	7	83
11	35	1	79
2	51	11	78
2	8	3	74
24	7	2	69
25	18	2	69
2	44	2	62
2	44	11	61

*Ninth International Histocompatibility Workshop data

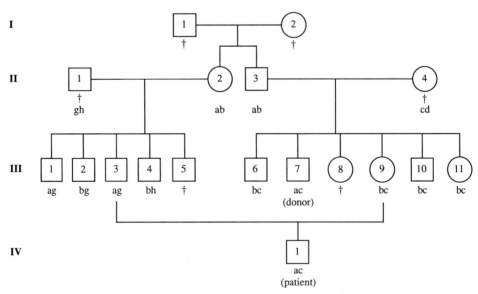

Figure 13.2 Family tree showing successful identification of an uncle as an HLA-compatible donor through extended family searching.

entitled *Bone Marrow Donors World Wide* (BMDW) is available on computer disk and on the Internet.

Every four months a circular is sent out to the collaborating national registries with a request to send an update of their respective donors. These are sent in on a floppy disk and are sorted on HLA phenotype number. The first phenotype is thus HLA-A1, B5 and the second HLA-A1, -B5, -B7 and so on. The files of the individual registries are combined in a single file. The phenotypes are again sorted on the number of the 'broad' HLA-A and -B alleles, directly followed by the splits of the respective broad alleles.

The computer program performs several functions. In the first place, a validity check on all tissue typing data received is performed. Secondly, the data are stored on the basis of HLA alleles and then the data from all centres are merged according to HLA phenotypes. Finally, the file is condensed (4 megabytes for 800,000 donors) for easy use of a matching program on a personal computer.

Such a search process is cost-effective and relatively speedy. It gives the clinician an immediate insight as to whether a given patient stands a chance of finding a suitable donor. If the HLA-A, -B, and -DR phenotype of the patient does not occur in the latest edition of BMDW, the chance that it will turn up in one of the national registries not linked to BMDW is less than 1 in 200,000. If the HLA-A, -B phenotype does not occur, the chance is less than 1 in 850,000.

DEGREE OF MATCHING BETWEEN RECIPIENT AND DONOR

For HLA-identical family member donor–recipient pairs, there should be genotypic identity at 6/6 or at worst 5/6 of the HLA-A, -B, -DR antigens. Additionally, MLC reactivity should show a relative response of <5%.

The situation is considerably different for matching between a recipient and an unrelated donor. Donor and recipients who are phenotypically identical may be either genotypically identical or show a micro mismatch for either Class I or Class II antigens (Table 13.5). They may display a minor mismatch of a cross-reacting group (CREG) (for example B7 in the recipient, B27 in the donor, defined serologically). It should be

Table 13.5 Definition of grade of tissue matching*

Class of HLA antigen	Match grade	Method	Example
Class I	Phenotypically identical	Serology	B27 vs B27
	(i) Genotypic identity	SSOP	B*2701 vs B*2701
	(ii) Micro mismatch	IEF/SSOP	B*2701 vs B*2702
	Minor mismatch	Serology	B7 vs B27 (CREG)
	Major mismatch	Serology	B7 vs B8 (non-CREG)
Class II	Phenotypically identical	HLA-D/MLC	Dw4 vs Dw4
	(i) Genotypic	SSOP	DRB1*0401 vs DRB1*0401
	(ii) Micro mismatch	SSOP	DRB1*0401 vs DRB1*0408
	Minor mismatch	SSOP/HLA-D	Dw4 vs Dw10
	Major mismatch	Serology	DR1 vs DR4

*Reproduced from Mickelson and Hansen. *Marrow Transplantation Reviews*, 1991, **1**, 8–13.

stressed, however, that cross-reacting groups simply reflect the presence of a shared epitope. They do not indicate the closeness of the overall amino acid sequence. They may show a major mismatch if they differ by at least one HLA antigen defined serologically (for example, recipient B7, donor B8: these representing noncross-reacting groups). The impact of phenotypic identity, minor mismatching and major mismatching on outcome of unrelated donor bone marrow transplantation is currently being defined clinically.

The ideal unrelated bone marrow donor would be young, CMV seronegative, with no history of pregnancies or blood transfusions and have a low frequency ($< 1 : 100,00$) of cytotoxic T-cells reactive against recipient cells. Acute graft-versus-host disease has been shown to be predisposed to by older donors, a female to male donor–recipient relationship and increasing parity in female donors. Cytomegalovirus disease has shown to be predisposed to by CMV seropositivity in the recipient as well as in the donor. Hence, as indicated above, the optimal donor would be young, CMV seronegative and not have been allosensitized by either preceding pregnancy or blood transfusion. In addition he/she should be a 6/6 HLA-A, -B and -DR match with the recipient.

If an HLA-identical donor is not available, individual bone marrow transplant teams may wish to utilize an unrelated donor mismatched for a single Class I locus or within the Class II region. It is not recommended that volunteer unrelated donors be used where there is more than one major antigenic disparity.

Donation from young minor donor to other family members

Successful HLA-identical family member bone marrow transplants have been performed using infants as donors for siblings or parents several decades older than the donor. Indeed, several sets of parents have specifically conceived additional children with the hope that the newborn child would be HLA-identical with a previous child from the same couple in need of a marrow transplant.

Informed consent is a process of the utmost importance in marrow transplantation as in other forms of medicine today. The adequacy of informed consent for young minors represents a potentially difficult problem. It is important for both the parents and the medical staff involved to feel satisfied that the donation by such a minor is not performed under duress, a situation legislated for in a number of countries.

Differences in recruitment between related and unrelated bone marrow donors

While selection of a family member donor is normally a straightforward, predominantly tissue typing process, the finding of an appropriately tissue typed unrelated donor is only the beginning of the assessment and selection process for a matched unrelated donor. Matched unrelated donors should be 18 to 55 years, and counselling is very important. Donors should be counselled on three separate occasions:

(a) When recruited to the registry.
(b) When selected for further tissue typing tests.
(c) When definitely selected as a donor.

Thus the final stages of recruitment begins by participation in an informational session. The purpose of the informational session is to provide donors with the knowledge to give informed consent to donate marrow. The donor is told exactly what will happen to them, what risks they face, why the patient needs a transplant, what are the chances the patient will be cured by the transplant, what other options that patient has and finally what are the patient's chances of survival without the marrow transplant.

The information session is conducted by the donor centre medical director and the donor centre co-ordinator. The donor is encouraged to bring their spouse, family and/or friends or anyone else who should have input into their final decision.

After contemplation the donor contacts the donor co-ordinator of the donor centre and gives his/her decision as to whether or not they want to continue in the final recruitment process.

Counselling should aim to cover the following topics:

- Emphasis on anonymity for donor and patient.
- Requirement for further blood samples before donation.
- Requirement for virological testing, especially HIV and HBs Ag.
- Risks of anaesthesia and harvest procedure.
- Loss of time from normal activities.
- Location of harvest procedure, i.e., near the donor's home.
- Requirement for collection of autologous blood unit.
- Possibility of need for allogeneic blood and associated risks.
- Donor's right to withdraw and consequences for the patient if this right is exercised after the transplant protocol has started.
- Possibility for second donation for the same patient.
- Details of compensation for loss of income and details of insurance cover.

After counselling and confirmation of the tissue typing, the donor should be examined by a physician not involved with the transplant procedure who should assess the donor's risk for undergoing general anaesthesia. The donor must be able to tolerate general anaesthesia or spinal anaesthesia and the removal of 1–1.5 litres of marrow over the period of only a few hours. All donors must be a Class I risk of the American Society of Anesthesiology for Anesthetic Risk (Table 13.6).

The physician performing the physical examination should not be employed by the donor centre or the marrow transplant team. Besides the physical examination, a chest X-ray, electrocardiogram, serum electrolytes and chemistry profile, complete blood count with differential white cell count and platelet count are performed. The results of the donor medical history, physical examination and tests are forwarded to the medical director of the donor centre.

In addition to the assessment by a third party physician (haematologist) the donor should be assessed by an anaesthetist, again at a different hospital to that at which the transplant will be performed. Donors with a Class II status of the American Society of Anesthesiology for Anesthetic Risk may be accepted at the discretion of the donor centre director, but any risk must be clearly understood by the prospective donor and communicated to the transplant centre before the donor centre releases donor details to the transplant centre.

PROTECTION OF RECIPIENTS

At recruitment prospective donors with the following conditions must be deferred:

- HIV and HTLV 1 seropositivity are absolute contraindications to marrow donation.
- Individuals exposed to human growth hormone or human pituitary gonadotrophin prior to 1985 cannot be accepted because of the risk of transmitting Creutzfeldt–Jakob disease.

Table 13.6 Criteria for American Society for Anesthesiology (ASA)

Class I Anesthetic Risk

'The donor has no organic, physiological, biochemical or psychiatric disturbance.'

Minimum Criteria for acceptance at Medical Director's discretion are those of ASA Class II Anesthetic Risk

'Mild to moderate systemic disturbance caused by pathophysiological processes such as non-limiting organic heart disease; mild diabetes; essential hypertension; anaemia; chronic bronchitis; morbid obesity; smoking' as described below.

Suggested criteria for exclusion of patients from the Bone Marrow Donor Register due to increased risks of anaesthesia.

1. Personal or family history of severe reactions to general anaesthesia. Include known susceptibility to malignant hyperpyrexia, porphyria etc.
2. Cardiovascular disease:
   - Any history of angina, myocardial infarction, atrial fibrillation or ventricular arrhythmia
   - Hypertension that is not well controlled
   - Congenital and valvular heart disease
   - Any episodes or congestive heart failure
3. Pulmonary disease:
   - Asthma leading to hospitalization in the past year
   - Chronic airways disease with FVC or FEV <70% of predicted value, secondary polycythaemia
4. Haematologic function:
   - Chronic anaemia
   - Polycythaemia
   - Coagulation disorders, including history of DVT in past year
5. Psychiatric disorders:
   - Treatment with anti-depressant drugs or major tranquillizers in the past year
6. Endocrine and renal function:
   - Diabetes – all insulin-dependent diabetes, those requiring hospital for control or with history of hypoglycaemia episodes in the past 6 months
   - Thyroid disease – unless stable on thyroid replacement
   - Steroid treatment in the past year
   - Renal impairment with elevated serum creatinine
7. Gastrointestinal function:
   - Hiatus hernia
   - Peptic ulcer, unless endoscopic evidence of healing or asymptomatic for 12 months
   - History of hepatitis or jaundice in the past year. Hep B Ag or HIV positive, abnormal liver function tests
   - Obesity, >40% over ideal weight

- Those with any form of malignant disease except for adequately localized skin cancers (basal cell carcinoma or squamous cell carcinoma, or carcinoma *in situ* of the cervix).
- Myeloproliferative diseases including polycythaemia vera, thrombocytopenia and myelofibrosis.
- Other evidence of serious bone marrow dysfunction (e.g. refractory anaemia, chronic neutropenia).

- Thalassemia major.
- Severe coagulopathy.

In addition, if the donor has had an infectious disease as listed below he/she should be referred for an expert opinion before being accepted on to the panel because of the risk of transmitting the disease:

- Bilharzia, brucellosis, filariasis, kala azar, Q fever.

CONSENT FORMS

Consent forms are required as follows:

1. At recruitment. Informed donor consent must be obtained for entry on to the registry indicating a willingness to be a bone marrow donor at the time of enrolment. An example of that currently utilized by the Australian Bone Marrow Donor Registry is shown in Table 13.7.
2. At the time of HLA-DR typing (if this has not been performed at the time of initial HLA-A and -B typing) and at MLC testing. Consent forms utilized for this purpose by the National Marrow Donor Program USA are shown in Tables 13.8a and 13.8b.
3. At time of third party physician examination. An example is shown in Table 13.9.
4. At the time of admission to hospital for bone marrow donation. An example is shown in Table 13.10.

INFECTIOUS DISEASE MARKERS

These apply to both HLA-matched family member and HLA-matched unrelated donors.
Infectious disease markers should be assessed at three stages in relation to the actual donation.

1. At the time of donor recruitment – as for blood donation, e.g. hepatitis B, hepatitis C, HIV, CMV, HTLV I, VDRL.
2. At the time of compatibility testing and before the final donor consent form is signed, the donor should be assessed for:
   (a) HIV II if she/he has been at risk as a result of travelling in Africa.
   (b) For Chagas disease if she/he has travelled in South or Central America.
3. Within 7 days of marrow harvest: testing for HIV, HBV, HCV, HTLV I and VDRL should be performed again.

The form utilized by the Australian Bone Marrow Registry at the time of the examination by the third party haematologist is shown in Table 13.11.

Process of bone marrow donation – processing, labelling and transportation

PROCESS OF UNRELATED DONOR HARVEST PROCEDURE

Whilst related family member donor harvests are almost always carried out at the transplant centre performing the transplant, this is quite different for unrelated donor

Table 13.7 Australian bone marrow donor registry

Consent Form for Bone Marrow Donor Registration

I,..(*please print*) hereby give consent for a blood sample to be taken for testing my tissue type and for me to be contacted for further tests should I be a possible donor for a patient needing a bone marrow transplant. I have read the information leaflet and understand that I shall be entered onto the Australian Bone Marrow Donor Registry.

I also give permission for my blood to be tested for evidence of a number of conditions including hepatitis and the AIDS virus and understand that I will be told the results if it is positive.

Address: ....

....

.... Postcode........................

Signed: .... Date: ....

Witness: .... Date: ....

Phone No: Home _____ Work: _____

It is of assistance when searching for a suitable donor to know if the donor and recipient are of similar ethnic background. Unless the request offends you, please tick the ethnic group in which you would class yourself.

_____ Caucasian   _____ Black

_____ Native American   _____ Oriental

_____ Hispanic   _____ Other (please state) ....

_____ Decline to answer

Table 13.8a Consent form – DR stage

The National Marrow Donor Program would like you to give a sample of blood in order to determine if your HLA type matches a patient who needs a bone marrow transplant. Your HLA-A and -B type have been found to be an adequate match to a patient. A sample of your blood is now needed for additional matching.

The white blood cells (lymphocytes) in your blood sample will be separated and HLA-DR typing performed. If your HLA-A, -B and -DR type adequately match the patient, you could be requested to give another blood sample for further matching. If your HLA type is not a good match, no further blood will be drawn regarding this search. However, your HLA-A, -B and -DR type will be searched to try to match other patients.

The patient has not yet begun the marrow transplant process, so a decision by you not to continue participation in the program will not place the patient in any immediate danger.

I consent to having _____ ml or _____ oz or _____ teaspoons of blood drawn for HLA-DR typing.

The blood sample is obtained by inserting a needle into a vein in your arm. This procedure is unlikely to cause complications, but there is a small chance of fainting, bleeding or developing an infection or haematoma (bruise) at the site of venipuncture.

My signature indicates that I agree to have my blood drawn for HLA-DR typing.

Signature: ………………………………………….. Date: ……………………………..

Witness: ……………………………………………. Donor ID: ………………………..

Donor Reference Number: ………………………………………………………………..

Of which ethnic group do you consider yourself a member? (Since certain HLA types are more common in each race, information on race will help in matching donors with patients).

_____ Caucasian   _____ Black

_____ Native American   _____ Oriental

_____ Hispanic   _____ Other (please state) ……………………..

_____ Decline to answer

Table 13.8b Consent form – MLC stage

The National Marrow Donor Program would like you to give another sample of blood in order to determine if your HLA type matches a patient who needs a bone marrow transplant. Your HLA-A, -B and -DR type have been found to be an adequate match to a patient. A sample of your blood is now needed for additional matching.

A portion of your blood will also be used for mixed lymphocyte culture (MLC) or mixed lymphocyte reaction (MLR) testing. In this test, donor's white cells and patient's white cells are mixed to test for compatibility. A portion of the blood will also be sent to researchers to perform further immunlogic tests. Your name, however, will not be revealed to these researchers.

If the mixed lymphocyte culture test shows that your cells are compatible with the patient's, it is very likely that the National Marrow Donor Program will ask you to donate marrow for this patient. If your cells are reactive with the patient's, you will probably not be asked to donate marrow.

The patient has not yet begun the marrow transplant process, so a decision by you not to continue participation in the program will not place the patient in any immediate danger.

Your blood will also be tested for evidence of infectious agents which could be transmitted to a marrow recipient.

These tests include cytomegalovirus (CMV), hepatitis B surface antigen, hepatitis B core antibody and serum ALT, and syphilis in addition to the AIDS testing. If any medically significant test, including the antibody for the AIDS virus, is positive, you will be notified confidentially of these results. If any of these tests are positive, you may not be used as a marrow donor. If local law requires it, the test results will be reported to your State health department.

I consent to having _____ ml or _____ oz or _____ teaspoons of blood drawn for the MLC tests and the abovementioned infectious disease tests and research studies.

The blood sample is obtained by inserting a needle into a vein in your arm. This procedure is unlikely to cause complications, but there is a small chance of fainting, bleeding or developing an infection or haematoma (bruise) at the site of venipuncture.

My signature indicates that I agree to have my blood drawn for MLC test and above infectious disease markers to determine if I might be a marrow donor for a specific patient.

Signature: .... Date: ....

Witness: .... Donor ID: ....

Donor Reference Number: ....

Table 13.9 Australian bone marrow donor registry intent to donate marrow

My signature indicates that I am interested in being a marrow donor and am willing to discuss this further with the staff of the Transplant Centre and hereby authorize the _____ Donor Centre to release my name to the transplant team so that I can discuss this further with them.

I understand that my blood cells have been found to be a satisfactory match with those of a patient who is a candidate for marrow transplantation. I have been asked to donate marrow.

I have been given the opportunity to review the information materials and the brochure on bone marrow donation.

Also I have received the information on marrow donation and life insurance coverage.

I have had the opportunity to ask questions of _____ (third party haematologist) and these have been answered to my satisfaction.

I understand that marrow transplantation with unrelated donors is an investigational medical therapy and that any information about me or any of my test results obtained by the Australian Bone Marrow Donor Registry may be used for research to improve the marrow transplantation and donation process.

The patient has not begun the marrow transplant process, so a decision not to donate will not place the patient in any immediate danger.

I understand that my signature here indicates an intent to donate, but is not my final consent to donate.

Donor Signature: .... Date: ....

Witness Signature: ....

Witness address: ....

Table 13.10 Consent form for donation of bone marrow

I understand the need to donate my bone marrow, specifically that this will provide to the recipient a potential chance of cure of an otherwise fatal disease. Donation of marrow will require a general anaesthetic, during which aspiration of the marrow is performed from the posterior (rear) part of the pelvic bone and sometimes the anterior (front) part of the pelvic bone. I understand that the cells removed represent only approximately 1% of the marrow and will be rapidly replaced by my body. I understand further that some bleeding will occur from the skin and the bone and that to compensate for this one unit of my own blood will be infused during the procedure. This unit will have been removed and stored on my behalf one week prior to marrow donation. Occasionally it may be necessary to give you blood from the Sydney Red Cross Blood Bank. I understand also that I may be required to take iron tablets for a period of three months after the donation to restore my red cell level (haemoglobin) to normal. The marrow donor and other relatives or friends may also be required during the 3–4 weeks after the transplant to donate platelets, approximately 2–3 times per week to prevent bleeding in the recipient. This explanation of the procedure is a summary only and has been fully explained by my doctor prior to my reading it. Side effects of marrow donation include discomfort at the operative site for hours to days after the procedure. Immobilization (by being in hospital) can predispose to deep vein thrombosis.

A general anaesthetic is required for the procedure. Any general anaesthetic carries serious risks including death or brain damage; however, these occur extremely rarely. Sputum retention, urinary retention, heart rhythm irregularities and lung infection can occur and are readily treated. Lesser complications include sore throat, headache, nausea and vomiting and pain at the site of intravenous cannulation occur more frequently but do not last long.

I ………………………………………… of ………………………………………
volunteer to undergo bone marrow donation.

Signature ……………………………………… Date: ……………………………

Witnessed by …………………………………………………………………………
Please print name

Signature of witness ………………………………… Date: ……………………………

Table 13.11 Australian bone marrow donor registry donor examination form

To be completed by third party Haematologist and sent to Donor Centre

Recipient name: _____
Donor ID: _____
Donor sex: _____
Date: _____

Proposed date of transplant: _____

1. Infectious diseases markers   Results sighted: [ ] Yes   [ ] No

                                        Date of Test _____ Lab _____

   (a) HBsAg   [ ] pos [ ] neg   _____
   (b) HIVI Ab   [ ] pos [ ] neg   _____
   (c) HIVII Ab   [ ] pos [ ] neg   _____
   (d) HCV Ab   [ ] pos [ ] neg   _____
   (e) HTLVI   [ ] pos [ ] neg   _____
   (f) CMV Ab (Elisa)   [ ] pos [ ] neg   _____
   (g) STS (Syphilis)   [ ] pos [ ] neg   _____
   (h) Other (specify) [ ] pos [ ] neg   _____

Creutzfeldt–Jacob Disease

1. Are you a woman who was treated with injections for infertility between 1968 and 1985?   [ ] Yes   [ ] No

2. Are you a man who received hormonal replacement treatment between 1968 and 1985?   [ ] Yes   [ ] No

3. As a child between 1968 and 1985 were you treated with injections for short stature?   [ ] Yes   [ ] No

The Donor Declaration Form has been completed to the satisfaction of the Donor Centre   [ ] Yes   [ ] No

Signature _____ Name _____
(Donor Centre Director or third party Haematologist)

Donor Centre to send copy to   (i) Transplant Centre on release of Donor details;
                                    (ii) Executive Officer, Australian Bone Marrow Donor Registry

marrow harvesting. The marrow collection should be performed at a Transplant Centre in a hospital experienced in marrow harvest, ideally near the donor's normal residence. A senior member of the harvest team should communicate with the recipient transplant centre if the donor is unusually small or if there are other factors that might influence the success of the marrow collection. General anaesthesia is recommended but spinal or epidural anaesthetic is acceptable if the donor and medical team so agree. The duration of the general anaesthetic should not exceed two hours. Marrow should be aspirated from the posterior and anterior iliac crests. The sternum should be avoided if possible, but may be used if a donor has agreed in advance.

CELL COUNTS

The harvest centre must provide cell counts for each bag of marrow and the total number of cells collected – the target to collect should be no less than 2×10^8/kg recipient weight of nucleated cells.

VOLUME

The harvest team should aim to aspirate a total volume of 1,000 to 1,200 ml. This figure may be increased to 1,500 ml at the specific request of the harvest centre. It should on no account be less than 500 ml (unless the recipient is a child).

ANTI-COAGULANT

The marrow should be anti-coagulated with ACD unless heparin is requested by the transplant centre. If the harvest centre routinely uses heparin they may use heparin for a volunteer donor harvest provided the marrow transit time is not expected to exceed 12 hours.

AUTOLOGOUS BLOOD TRANSFUSION

The harvest centre should aim to collect one or more units of autologous blood for transfusion to the donor during or after the harvest procedure. Every effort should be made to avoid the use of allogeneic blood.

MARROW COLLECTION

The volume of marrow aspirated at each puncture site should be kept to a minimum (e.g. 5 ml) to reduce the dilution of marrow with peripheral blood. Once aspirated the marrow must be mixed with tissue culture media containing the anti-coagulant without preservatives. The aspirated marrow must be passed through stainless steel screens of 0.3 mm openings initially, then 0.2 mm openings. Screens will remove clots or bone fragments and fatty particles, avoiding potential embolic problems to the recipient.

LABELLING

The marrow should be labelled clearly with the donor's unique reference number (but not the name) and the unique reference number (but not the name) of the intended recipient, together with the name and address of the recipient transplant centre. Further marrow manipulation will ordinarily be performed at the transplant centre.

TRANSPORTATION OF MARROW

Courier

The designated courier should be a nurse, laboratory technician, doctor or other person of comparable training or comparable level of responsibility. He/she should not be related to the donor or the patient. He/she must have no other obligations until after the marrow is delivered.

Travel arrangements

The courier must keep the marrow in hand or in sight at all times. He/she must communicate with the transplant centre if any change occurs in travel arrangements and must be prepared to improvise new travel arrangements if necessary (airline agents are always very co-operative with the arrangements for the marrow delivery).

Travel documents

The courier should carry documents confirming the nature of the material, its destination and the fact that it is HIV negative.

Irradiation

The marrow should not be subjected to any radiation in any airport security system.

Temperature

The marrow should be carried at room temperature (unless otherwise requested by the transplant centre) in an especially designed rigid container duly labelled. Transportation at 4 °C may be optimal for long distances. The marrow cells must on no account be cooled below 4 °C.

Donor marrow has now been transported from the northern hemisphere to the southern hemisphere with transit times of up to 40 hours. When carried by a courier in the cabin of the aircraft in a container that keeps the marrow at 4 °C, viability and cell number have not been adversely affected and engraftment has routinely occurred. Acid citrate dextrose is the preferred anticoagulant preferred, used in a ratio of one part to five parts of marrow, particularly for long flights. This is due to the short half-life of heparin, requiring replenishment in flight, a procedure difficult to achieve with sterility. Dry ice should not be used in transportation of the marrow.

DONOR INSURANCE

All medical centres involved in unrelated donor bone marrow transplantation (donor centre, tissue typing centre, marrow harvest centre, transplant centre) must provide general liability insurance.

Table 13.12 Recommendations for long distance transportation of marrow

1. Marrow should be collected in a minimum of two sealed plastic bags.
2. ACD should be the anticoagulant (to be used in a ratio of 1 part to 5 parts marrow) (although the marrow may be harvested using heparin).
3. Marrow bags should be labelled with:
   - Donor number
   - ABO and Rh type
   - Harvest centre – time and date of collection
   - Transplant centre – time and date of infusion (planned)
   - Patient name
   - Hospital
4. Marrow bags should be wrapped in surgical towel or water absorbent paper towel.
5. Freezer packs should be wrapped in bubble plastic. These are then placed in a rigid container with insulating properties. They line the side and base of the container. Freezer pack to be changed regularly to maintain desired temperature.
6. Marrow bags should be vertically stacked or so arranged to be equally chilled.
7. Marrow bags should be chilled to between 4 °C and 6 °C and monitored by a standardized alcohol thermometer. Temperature to be checked several times during flight. Thermometers are best placed in between wrapped marrow bags (ensure broken thermometers will in no way pierce marrow bags).
8. Container should be labelled:
   - 'Human bone marrow for transplantation. DO NOT X-RAY'
9. Peripheral blood samples to be collected:
   - EDTA 5 ml
   - Clot 10 ml
   - Heparinized 30 ml

   Should be labelled with:
   - Donor number
   - Harvest centre and date of collection

Tubes should be placed in a plastic specimen bag wrapped with bubble plastic.

Tubes should not be irradiated.

It is suggested that a pair of examination gloves be included in the transport container for the security guard to examine marrow bags if required.

Guidelines for transplantation of marrow are shown in Table 13.12 and for couriers in Table 13.13. Further details are published by the Executive Committee of the World Marrow Donor Association.[19]

Donor confidentiality

Maintenance of unrelated donor confidentiality is of paramount importance. In most cases such confidentiality should be maintained for life. Some centres accept the principle that the donor and recipient may meet six months or more after a successful transplant, provided both individuals have independently expressed a desire to do so.

Table 13.13 Guidelines for marrow couriers

When marrow is collected at a hospital which is not the requesting transplant centre, every effort must be made to ensure that the marrow arrives at the transplant site within 36 hours of collection.

While the transport of marrow seems straightforward, there are problems which may arise.

Courier responsibilities are to ensure that:

- Marrow is transported in a rigid container, 4 °C (unless requested otherwise by the transplant centre). Never use dry ice in the container.
- Marrow is NEVER subjected to the airport X-ray screening devises. Any security check must be done by hand under direct supervision of the courier.
- Alcohol is not consumed by the courier while transporting the marrow.
- No sedative drugs are allowed.
- Immediate, alternate flights are obtained in case of delays.
- All changes in original plans are communicated immediately to the donor and transplant centres involved.
- Marrow is delivered directly to the designated person and no one else.

When deciding who the courier will be, the following criteria must be assessed:

1. The courier must know and understand the significance of the marrow he/she carries and have the maturity to take the task seriously.
2. The courier must be an experienced traveller.
3. The courier must have a major international credit card.

Second donation of marrow for the same patient

A donor who has donated marrow in the past may on occasion be requested to repeat the donation for the same patient as a consequence of graft failure or relapse of the underlying condition.

The donor should be warned in advance at the original donation that there is a small possibility that he/she may be asked to donate again for the same patient. He/she may be asked soon after the first donation if he/she would be prepared to donate again and he/she should then be informed as to the timing of a possible second transplant. The donor should be asked to notify the donor centre of any extended travel or change of address.

If the need for further marrow cells arises the transplant centre should contact the national registry headquarters. It may be appropriate that a request for a second donation is considered by a small committee established especially for this purpose.

Recovery, counselling and long- and short-term physical outcome

All donors should be warned that they are likely to feel bruised and sore on recovery and that this may last for up to six weeks, although most commonly it has largely resolved by two weeks post donation. It is particularly important for those employed in physical work jobs.

In a recent review the acute complications suffered by the first 493 persons who donated marrow through the National Marrow Donor Program USA were described.[20] Acute complications related to the collection procedure occurred in 5.9% of donors, but a serious complication (apnoea) during anaesthesia occurred in only one donor. When donors were questioned approximately two days following discharge from hospitalization, most donors described symptoms related to the collection. Some 74.8% experienced tiredness, 67.8% experienced pain at the marrow collection site and 51.6% experienced low back pain. The mean recovery time was 15.8 days; however, 10% of donors felt that it took them longer than 30 days to recover fully. The duration of the marrow collection procedure and the duration of anaesthesia correlated positively with donor pain or fatigue following the collection. This demonstrated that marrow donation is well tolerated with few complications. To decrease further the incidence of donor discomfort and recovery time following donation, the duration of the collection procedure and probably the duration of anaesthesia and the volume of marrow collected should be kept to a minimum.

In a companion article, the psychosocial affects of unrelated marrow donation were described.[21] Donors were generally quite positive about the donation one year after donation. Some 87% felt that it was 'very worthwhile' and 91% said they would be willing to donate again in the future. Donors who had experienced low back pain or difficulty in walking as a result of the donation were more likely to register the experience as more stressful and painful than expected, but were no more likely to register it as less positive emotionally than donors who did not experience these side effects.

DONOR FOLLOW-UP

Donors should be contacted by the Donor Centre 72 hours after the marrow donation and then weekly until well. Contact should again be made at three months with a visit to the Donor Centre at some stage during that period to check on their health. Donors, in some registries, are retired from the registry after donation for 12 months. A letter in this case should be sent at 12 months to check on their general well being and to invite them to rejoin the registry again if they wished.

Role of erythropoietin and colony stimulating factors

Considerable interest is currently being generated by the use of recombinant erythropoietin to increase the haematocrit of the marrow donor prior to marrow donation with a view to avoiding the need for autologous blood collection storage and reinfusion to the recipient during the marrow harvest.

Likewise at the present time the use of myeloid colony stimulating factors such as G-CSF are being evaluated in an attempt to increase the cellularity and progenitor/stem cell content of the marrow inoculum in order to reduce the amount of marrow required to be harvested from the donor.

Potential for peripheral blood stem cell harvests

Under certain circumstances there may be advantages in using peripheral blood stem cells instead of bone marrow for allogeneic transplantation. Currently the indications could include inability or unwillingness for a donor to donate marrow or the treatment of failure to engraft of previously infused donor marrow. This, however, is a new and rapidly developing area and the indications for allogeneic peripheral blood stem cell transplantation may change with increased experience.

STEM CELL MOBILIZATION

Amongst the agents currently available for peripheral blood stem cell mobilization, granulocyte colony-stimulating factor (G-CSF) appears to be the safest, having only mild dose-related side effects consisting of bone pain and myalgia. G-CSF has been administered to a considerable number of healthy subjects including granulocyte donors, peripheral blood stem cell donors and volunteers, without long-term toxicity having so far been reported. G-CSF should be given subcutaneously. Although the optimal dose for mobilization is not yet known, successful mobilizations have been reported using daily doses between $5-10$ μg/kg/day given until completion of the stem cell harvest.

STEM CELL HARVESTING

Leukapheresis to harvest blood stem cells should be carried out in an appropriate setting following institutional blood bank centre guidelines governing the use of blood cell separators. Leukapheresis should be performed with an automated continuous flow blood cell separator using peripheral venous access with the objective of processing 9–12 litres of whole blood on each collection. In donors receiving $5-10$ μg/kg/day of G-CSF, CD34 progenitor cells peak in the blood by day 5 of G-CSF therapy and it is recommended that the first leukapheresis be performed at this time.

It should be remembered that unfractionated G-CSF mobilized peripheral blood stem cells will contain large numbers of $CD3^+$ T cells. Since these have the theoretical potential to cause considerable acute graft-versus-host disease, consideration should be given to T cell depletion of the peripheral blood stem cells.

A SECOND DONATION COMPRISING PERIPHERAL BLOOD LEUCOCYTES

Leucocytes (or, more specifically, mononuclear cells) from the peripheral blood of the original transplant donor have demonstrated activity in reversing relapse in patients previously allografted for leukemia, particularly those with chronic myeloid leukemia. For this reason a transplant donor on occasion may be asked to undergo leukapheresis for mononuclear cell collection and donation. This request may be made some months or years after the original donation.

The request for leucocyte donation should be handled in a manner analogous to the request for a second marrow donation (see above). If the donor agrees to donate leucocytes, they should be collected on a continuous flow blood cell separator; between two and five sessions should be performed within 5 to 9 days. The buffy coat should contain $2-4 \times 10^8$/kg nucleated cells. They should be labelled and transported as per bone marrow cells.

Potential for cord-blood harvests

Cord blood contains multilineage and multipotential haemopoietic cells which can be supported in culture systems. The cell content peaks between 32 and 41 weeks of gestational age. Cord blood contains 5–10 times more haemopoietic stem cells per ml than the peripheral blood of an adult and is comparable with adult bone marrow. At birth there are some differences between cord and adult blood in lymphoid cell content, the most significant being an increase in $CD4^+$ subsets. Furthermore, $CD3^+$, $CD4^-$, $CD8^-$ (double negative cells), which are implicated in the development of tolerance, are present. These differences may be relevant immunobiologically and may be associated with a lower frequency of GVHD with the use of cord blood as a source of haemopoietic stem cells for transplantation. (It is well recognized that increasing donor age is a risk factor for the development of acute graft-versus-host disease – the younger the donor, the lower the risk).

On average 100 ml of blood can be collected from the human umbilical cord taken while the placenta is *in situ* and from the placenta after expulsion. The procedure for cord blood collection is standard: the cord is doubly clamped at the fetal end within 15 seconds of birth: the umbilical vein is needled and blood collected into tissue culture medium containing heparin. The cells can then be either frozen unprocessed in 10% DMSO at a cooling rate of 1 °C per minute and stored in liquid nitrogen at −180 °C, or can be transported or stored at 4 °C for three days without significant loss of viability. Contamination with maternal cells has not been detected. The presence of a red cell lysate has not been a clinical problem in one cord blood transplant where ABO incompatibility existed.

The potential role for cord-blood transplants is in the situation where the patient has no other matched family member donor, but a poor prognosis with a limited life span, an increased risk of GVHD with other possible transplant options (such as matched unrelated donor transplantation), and the mother is pregnant. At the present time it is not clear what constitutes the minimum number of cord-blood stem cells required for engraftment.

A small number of both related and unrelated cord-blood transplants have been reported. Currently 44 related human umbilical cord-blood transplants have been reported in paediatric recipients.[22] Early experience in unrelated cord-blood transplants has included paediatric patients up to 70 kg in weight and high levels of HLA mismatch (3–6 antigen match). Not only has the level of engraftment been satisfactory but the degree of graft-versus-host disease has been surprisingly low. If a reduced incidence of GVHD with cord-blood transplantation can be substantiated, the use of matched unrelated or even partially mismatched related and unrelated cord blood may be associated with acceptable results. Thus a cord-blood bank, using a tissue resource that is currently discarded, may become an important resource for the future.[23]

References

1. Ash RC, Casper JT, Christopher RC et al. Successful allogeneic transplantation of T cell depleted bone marrow from closely HLA-matched unrelated donors. *N Eng J Med* 1990, **322**: 485–494.
2. Beatty PG, Hansen JA, Longton GM et al. Marrow transplantation from HLA-matched unrelated donors for treatment of hematological malignancies. *Transplantation* 1991, **51**: 443–447.

3. McGlave PB, Beatty P, Ash R et al. Therapy for chronic myelogenous leukemia with unrelated donor marrow transplantation: results in 102 cases. *Blood* 1990, **75**: 1728–1732.
4. Atkinson K, Downs K, Dodds AJ et al. Unrelated volunteer bone marrow transplantation: initial experience at St.Vincent's Hospital, Sydney. *Aust NZ J Med* 1993, **23**: 450–457.
5. Hansen JA, Clift RA, Thomas ED et al. Transplantation of marrow from an unrelated donor to a patient with acute leukemia. *N Eng J Med* 1980, **303**: 565–567.
6. Speck B, Swann FE, van Rood JJ et al. Allogeneic bone marrow transplantation in a patient with aplastic anaemia using a phenotypically HLA-identical unrelated donor. *Transplantation* 1973, **16**: 24.
7. Thomas ED, Buckner CD, Banaji M et al. One hundred patients with acute leukemia treated by chemotherapy, total body irradiation and allogeneic marrow transplantation. *Blood* 1977, **49**: 511–533.
8. Thomas ED, Buckner CD, Clift RA et al. Marrow transplantation for acute nonlymphoblastic leukemia in first remission. *N Engl J Med* 1979, **301**: 597–599.
9. Thomas ED. Marrow transplantation for acute nonlymphoblastic leukemia in first remission. *N Engl J Med* 1983, **309**: 1539.
10. Storb R, Deeg HJ, Whitehead J et al. Methotrexate and cyclosporine compared with cyclosporine alone for phrophylaxis of acute graft-versus-host disease after marrow transplantation for leukemia. *N Engl J Med* 1986, **314**: 729–735.
11. Atkinson K, Downs K, Golenia M et al. Prophylactic use of ganciclovir in allogeneic bone marrow transplantation: absence of clinical cytomegalovirus infection. *Br J Haematol* 1991, **79**: 57–62.
12. Atkinson K, Biggs J, Concannon A et al. Changing results of HLA-identical sibling bone marrow transplantation in patients with haematological malignancy during the period 1981–1990. *Aust NZ J Med* 1993, **23**: 181–186.
13. Thomas ED, Clift RA, Fefer A et al. Marrow transplantation for the treatment of chronic myelogenous leukemia. *Annals of Intern Med* 1986, **104**: 155–163.
14. Goldman JM, Apperley JF, Jones L et al. Bone marrow transplantation for patients with chronic myeloid leukemia. *N Engl J Med* 1986, **314**: 202–207.
15. Storb R, Longton G, Anasetti C et al. Changing trends in marrow transplantation for aplastic anaemia. *Bone Marrow Transplant* 1992, **10**: 45–52.
16. Kernan NA, Bartsch G, Ash RC et al. Retrospective analysis of 462 unrelated marrow transplants facilitated by the National Marrow Donor Program for treatment of acquired and congenital disorders of the lymphohematopoietic system and congenital metabolic disorders. *N Engl J Med* 1993, **328**: 593–602.
17. Kaminski E, Hows J, Man S et al. Prediction of graft-versus-host disease by frequency analysis of cytotoxic T cells after unrelated donor bone marrow transplantation. *Transplantation* 1989, **48**: 608–613.
18. Fussell ST, Donellan M, Cooley MA, Farrell C. Cytotoxic T lymphocyte precursor frequency does not correlate with either the incidence or severity of graft-versus-host disease after matched unrelated donor bone marrow transplantation. *Transplantation* 1994, **57**: 673–676.
19. Executive Committee, World Marrow Donor Association. Bone marrow transplants using volunteer donors – recommendations and requirements for a standardised practice throughout the world. *Bone Marrow Transplant* 1992, **10**: 287–291.
20. Stroncek DF, Holland PV, Bartsch G et al. Experiences of the first 493 unrelated marrow donors in the National Marrow Donor Program. *Blood* 1993, **81**: 1940–1946.
21. Butterworth VA, Simmons RG, Bartsch G et al. Psycho-social effects of unrelated bone marrow donation: experiences of the National Marrow Donor Program. *Blood* 1993, **81**: 1947–1959.
22. Wagner JE, Kernan NA, Steinbach M, Broxmeyer He, Gluckman E. Allogeneic sibling umbilical-cord-blood transplantation in children with malignant and non-malignant disease. *Lancet* 1995, **346**: 214–219.
23. Jenney MEM. Umbilical cord-blood transplantation: is there a future? *Lancet* 1995, **346**: 921–922.

CHAPTER 14

Tissue banking

STEPHEN CORDNER, LYNETTE IRELAND

Introduction 268
History 270
Obtaining tissue: policy and law 272
Technical regulation of tissue banking 275
Tissue banking organization and scope: an international view 282
Donor selection and technical aspects of tissue banking 287
Appendix 292
References 302

Introduction

Tissue banking has paralleled developments in organ transplantation but has generally been overshadowed by them. This is because of organ transplantation's close association with changing notions of death, the mystique of solid organ transplantation and the highly developed and complex medical science which surrounds it. One effect of this has been that the separate and distinct issues associated with the removal, processing, storage and use of tissues removed from asystolic (non-beating heart) donors have received less attention.

Much effort has been expended trying to define precisely the difference between organs and tissues. This is actually a futile exercise because it is only one of a number of distinctions that can be made that are important for policy and regulatory purposes.

The important distinctions which encompass the critical issues in human tissue or organ supply and usage are shown in Table 14.1. It is difficult to imagine a comprehensive legal scheme or regime which will cover the range of issues (and there are probably others) implied by some of these distinctions, let alone cope with the dynamic nature of the issues around the supply and demand for human tissue.

To clarify the scope of this chapter, human tissue is regarded as encompassing all parts of the human body including its products. (Organs are therefore a subset of human tissues.) To follow the scheme outlined in Table 14.1, the provider with which this chapter is concerned is the asystolic cadaver. The only use to which the tissue will be put, for the purposes of this chapter, is transplantation into another human being. The

Table 14.1 Critical issues in human tissue and organ supply

1. The provider; whether:
   - a living human being
   - a beating heart (systolic) brain dead cadaver
   - a non-beating heart (asystolic) cadaver
   - a foetus or embryo

2. The nature of the tissue or organ; whether:
   - regenerative or non-regenerative
   - single or paired
   - if single, whether vital
   - gametes
   - excretions, by-products or wastes (e.g. urine, faeces, tears)
   - sui generis (e.g. hair, teeth, fingernails)
   - altered or developed by human agency

3. The use to which the tissue or organ will be put; whether:
   - diagnosis
   - transplantation (with or without banking)
   - research – epidemiological non-identifying use
     – results of possible consequence to provider
   - education
   - commercial development and exploitation (e.g. cell lines)
   - other uses such as public display, cosmetics, information gathering (e.g. DNA from hair roots)

tissues that meet these criteria in practice are:

- Cardiovascular tissue (heart valves, blood vessels).
- Ocular tissue (corneas, sclera).
- Skeletal and related tissue (bone, ligaments, tendons, cartilage, fascia).
- Skin.

(Although some tissue banks in the United States collect and process dura mater[1] following reports of the transmission of Creutzfeldt–Jakob disease by dura allografts[2] the Donor Tissue Bank of Victoria does not collect, process or distribute dura mater. The authors are therefore not discussing the banking of dura mater allografts.)

Tissue banks are involved in the retrieval, processing, storage and distribution of these tissues. To be more precise, the American Association of Tissue Banks[3] defines a tissue bank as follows:

> A tissue bank is deemed to exist when any of the following criteria are satisfied: (a) viable or non-viable human tissues are collected, processed, stored, and distributed to clinicians who are not involved in the collection process; (b) tissues are collected, processed, and stored in one institution and made available to clinicians in other institutions ...

What is distinctive about this definition is the requirement that the tissue bank be organizationally separate from the users of the tissue.

A note of caution is appropriate in this introductory section. Enthusiasm for tissue transplantation must be tempered by respect for the unknown. Scott[4] for example, is enthusiastic about the success of the Australian National Pituitary Programme. By 1985, this programme had treated 664 children with short stature and supplied pituitary gonadotrophin for the treatment of 1447 adults with infertility. Recently a number of these patients have developed Creutzfeldt–Jakob Disease, and more will do so as time passes.[5] Even more so, the disastrous infiltration of HIV into the blood supply everywhere serves as a warning, if one was needed, about the humility with which tissue banking should proceed.

History

TISSUE TRANSPLANTATION

> Diseases desperate grown
> By desperate appliance are relieved,
> Or not at all.
> *Hamlet* IV: 3

It is notions such as this, which survive today in the concept of 'heroic' surgery, that characterize the history of tissue transplantation. Notwithstanding that the techniques and other advances enabling routinely successful tissue transplantation have only developed in the last century, its history stretches back into antiquity.

Skin

Woodruff claims that skin flaps were used by Indian surgeons, mainly for rhinoplasty, over 2000 years ago.[6] The first successful skin allograft was reported by Girdner in 1881.[7] The patient had suffered lightning burns to the left shoulder and arm. The donor had died from self-inflicted incised wounds to the neck and the skin had been removed 'several' hours post mortem.

It was Brown who established the routine use of skin allografts for burns and other situations in the early 1950s.[8]

Corneas

A good summary of the history of corneal grafting is given by both Rycroft[9] and Payne.[10] Rycroft credits the Frenchman Pellier de Quengsy with first having the idea, in 1789, of relieving blindness caused by a scar on the cornea by the replacement of the fibrous tissue. He actually attempted to replace the scarred cornea with a glass disc, embedded in a circlet of silver, and then stitched into the cornea. He also predicted the outcome: glaucoma, extrusion and leakage of the intraocular contents. This operation, however, is in the field of prosthetics, not tissue transplantation.

The first corneal graft involving a human was performed by Reisinger in 1817, when he tried to graft a fresh animal cornea onto a human eye. Sellerbeck (cited by Payne[10]) was the first to undertake a lamellar keratoplasty in a human being using human donor tissue in 1878. In 1905 Zirm reported the first successful corneal allograft in a human. The patient had sustained chemical burns to both corneas. The operation on his left eye was successful but that on the right failed. The donor of one cornea was a young boy whose eye had to be removed because of an intraocular foreign body.

The modern era of keratoplasty began in the 1950s with the advent of improved surgical technique, improved eye banking methods and the use of anti-inflammatory drugs to control graft rejection.

Bone

Lyons and Petrucelli refer to the legend of Cosmos and Damien, two 'well-known' surgeons around 500 AD who tried transplanting an entire lower limb.[11] Meekeren in 1682 filled a defect in a Russian soldier's head with a piece of canine skull.[12] Because the implant was from an animal the Russian was excommunicated. Wanting to restore his standing with the church, two years later he had the xenograft, which had now well taken, removed. Shortly thereafter, he was given a proper church burial!

Macewen is credited with performing the first human bone allograft in the modern surgical era in 1878.[13]

Aortic valves

The first aortic valve replacement allograft in a human was performed by Donald Ross at Guy's Hospital on 24 July 1962.[14] The operation became possible following the development of cardiopulmonary bypass machines. Brian Barratt-Boyes at the Green Lane Hospital in Auckland, New Zealand, independently developed the same concept of using aortic valve allografts and began in August 1962.[15]

TISSUE BANKING

Tissue banking is basically a post-Second World War phenomenon and followed the development of blood banking. An exception to this was the organized banking of human milk for therapeutic use in the 1920s at Winnipeg Children's Hospital.[16] Carrel wrote in 1912 about the banking of human tissue, but it was not until the advent of reliable refrigeration in the 1930s that thoughts of such banking could start to take clear shape.[17] The first blood bank was established at Cook County Hospital in 1937.

Skin

Webster was the first to report in 1944 the clinical use of skin grafts refrigerated for significant periods, in his case up to 21 days.[18] A chance observation by Polge discovered the cryopreservative properties of glycerol, making possible the freezing of tissue in a viable states.[19] This was not applied in the case of human skin until Cochrane reported the first clinical use of skin allografts which had been frozen and then thawed.[20] This led to the first skin banks at the Burns Unit of Massachusetts General Hospital and the Bethesda Naval Hospital.[21]

Corneas

In the first decade of this century, Magitot (cited by Payne[10]) in Paris began the first research on the preservation of donor corneal tissue. In 1911 he performed a successful lamellar keratoplasty on a 14-year-old boy with a dense corneal scar following a lye burn. He had preserved the donor eye, enucleated from a living patient with glaucoma, for eight days in a solution of haemolysed blood serum taken from another person. The graft had remained clear for a year when he reported the case. Magitot thus laid the foundations for eye banking, and he also emphasized the importance of homologous donor tissue to the success of keratoplasty. The first eye bank was probably that at the New York Hospital, which was regarded as experimental and established in 1944. In 1945 Paton started the Eye Bank for Sight Restoration Inc., with philanthropic support.

Bone

The first centralized bone bank was the United States Naval Tissue Bank established in Bethesda, Maryland in 1949. The work of Curtiss and Herndon rekindled interest in bone allografting in the mid-1950s, showing a reduction in donor bone rejection if it is preserved at very low temperatures.[22]

Aortic valves

Cryopreservation of allograft valves began in 1968 but was soon abandoned as it was thought that this process killed fibrocytes and fibroblasts and cell viability was regarded as an important factor in graft longevity. Refinements in cryopreservation occurred and O'Brien in Brisbane, Australia is the leader of a team which is regarded as having established the procedures leading to the first viable homograft cardiac valve bank in 1975.[23]

Obtaining tissue: policy and law

That transplantation is a desirable activity is no longer seriously debated. Furthermore, given the availability of a sufficient number of organs and tissues there seem to be no obvious moral reasons why the practice of transplantation should not increase. This increase, however, has been slower than it might have been because of a shortage of organs and to a lesser extent, tissue. Therefore it is issues of supply which have preoccupied the minds of policy and law makers worldwide more than any other aspect of transplantation.

Latterly, there has been particular concern also to prevent commerce in human tissues. Related to both these issues (supply and commerce) is the extent to which the law recognizes any property rights in human tissues.

HISTORY

The scarcity of human tissue to meet medical demand has precedence. Scott recounts how the town council of Edinburgh in 1504 granted a charter to the Guild of Surgeons and Barbers allowing them to claim the body of one executed criminal a year for dissection.[4] Similarly in England in 1540, Henry VIII granted to the United Company of Barbers and Surgeons the corpses of four executed felons 'yerely for anatomies'. Elizabeth I gave an equivalent grant to the College of Physicians in 1564, leading occasionally to unseemly squabbles between the physicians and barber surgeons over possession of the body. The activities of resurrectionists (or grave robbers) were well known on both sides of the Atlantic in the eighteenth and early nineteenth centuries to meet the need of medical schools for bodies for anatomy classes. It was the Massachusetts legislature that responded first, passing its Anatomy statute in 1831. The tale of Burke and Hare is infamous: they became multiple murderers in order to supply corpses for a fee. The British Parliament reacted to their outrages by the Anatomy Act 1832, and the Births and Deaths Registration Act 1836. The former regulated anatomy schools and provided a means by which citizens could agree to leave their bodies for anatomical dissection if they so wished. Similar acts were passed shortly afterwards in others of the American states and in the Australian colonies.

The anatomy acts worked well for over 100 years but they neither contemplated nor authorized the retention of tissues for transplantation, so that when corneal grafting became routinely available, legislative reform was required.

Europe

The more modern history of the legal regulation of how tissue for transplantation could be obtained seems to have begun with the French. In a decree of 20 October 1947, doctors in a limited number of listed hospitals were authorized to perform autopsies and to remove tissues from dead patients 'in the interests of science or therapeutics ... even in the absence of the family's authorization'. This decree was mentioned in Volume 1 of the WHO's International Digest of Health Legislation (1948) indicating that organization's interest, from its inception, in transplantation issues. In 1949, the French stepped back from their initial 'presumed consent' approach when Public Law No. 49.890 was enacted on 7 July. This authorized 'the practice of corneal grafting with the aid of voluntary eye donors'. In 1952, the English Corneal Grafting Act was passed, followed quickly by similar laws in Canada and Australia (e.g. Corneal and Tissue Grafting Act 1955 (NSW). These acts follow the same consent mechanisms for corneal donation as the anatomy acts used for whole body donation to anatomy schools. That is, consent in life to the removal was required or the person in lawful possession of the body had no reason to believe that the deceased's surviving spouse or relatives objected. In England, the Corneal Grafting Act 1952 was replaced by the Human Tissue Act 1961 which is still in force.

In 1975, the Council of Europe convened a committee which produced a report entitled *Ad Hoc Committee of Exchange Views and Information on Legislation in Member States Concerning the Removal, Grafting and Transplantation of Human Organs and Tissues*. As a result the Council produce a model legal code. This was adopted by Resolution 78(29) of the Council's Committee of Ministers on 11 May 1978 and the matters dealt with included:

1. The prohibition of commerce in body materials.
2. (Article 10) Presumption of consent provided there was no recorded objection by the deceased (so called 'opting out'). No inquiry of the relatives was needed.

The Council was persuaded, in coming to a conclusion favouring 'opting out', by:

(a) The invaluable importance of substances for transplantation.
(b) The shortage of substances available.
(c) The interests of sick people.

Shortly afterwards, the French enacted Law No. 76.1181 'concerning the removal or organs' which was regarded as instituting an opting out regime, except where the deceased was a minor or mentally handicapped. A decree in 1978 (Circulaire du 3 Avril 1978 J.O. 5 Avril 1978, p. 1532) set out the mechanism for objections to be recorded. In fact the decree explained that the family is the 'privileged depository' of the deceased's wishes and is in the best position to communicate them to the medical profession. This seems to represent the legislative method of having it both ways. As the Council of Europe points out in a recent document 'The Law of Organ Transplantation in Europe' (Council of Europe, Undated).

> Experience shows that this move (i.e. to 'opting out') is not beyond criticism ... some of the procedures involved, particularly in France, have led in practice to the family being systematically asked whether the deceased was opposed to the idea of organ removal.

This refers to instances where families were outraged that organs and tissues were used without their consent, whatever the law said. The consequent bad publicity effectively overturned the law.

The Council of Europe confirmed its position on presumed consent at the Third Conference of European Health Ministers in Paris on 16/17 November 1987.[24] It also noted in a preparatory document: 'The practice in most countries shows that the relatives are consulted and though in most cases its opinion is legally not overriding, none would go against the expressed refusal of the family.'[25]

USA

In the United States, the Uniform Anatomical Gift Act (UAGA) was approved by the Commissioners of Uniform State Laws in July 1968. Unlike any prior model law, the UAGA enjoyed extraordinary success, being approved in all 50 states and the District of Columbia in just 3 years.[26] Sadler[27] believes the main reason for this remarkable acceptance was its foundation on:

1. Principles of voluntarism.
2. Express consent (by donor in life or relatives after death).
3. Protection of individual rights.

The UAGA also streamlined consent procedures, eliminated interstate inconsistencies and, while recognizing the unique needs of transplantation also recognized the conflict of interest between those caring for the donor and those caring for the recipient.

In the 1980s, the gap between the demand for organs and tissue and their availability widened. This led to the following two initiatives.

Required request/routine enquiries

This policy presumes that the problem lies not with engendering altruism but with helping people to act on their good intentions and overcoming the reluctance of health professionals to approach families.[28] It imposes a duty on hospitals and physicians to actually asks relatives of potential donors for permission to use tissues. Twenty-six states enacted required request legislation in the late 1980s, and 18 states enacted routine enquiry legislation. In general, exceptions were made where:

1. The wishes of the deceased were already known.
2. Medical staff could not locate the family in time.
3. It seemed that the enquiry or request would add to the relatives' mental distress.

Over time the schemes do not seem to have had a major impact on organ procurement.[29,30]

However, increases in tissue procurement have been noted. A 135% increase in donor eyes was reported in Oregon[31] while the New York Skin Bank reported a 180% increase in skin donation.[32]

An administrative requirement for donor identification protocols
(Section 1138 Social Security Act)

As a condition of participation in Medicare and Medicaid, hospitals were required to establish written protocols to identify potential donors. Again, there is no published account assessing the impact of this requirement on organ or tissue donation rates.

PRESUMED CONSENT LAWS AND TISSUE AVAILABILITY

Presumed consent laws could increase enormously tissue availability from asystolic donors. This can be said with more confidence than in relation to systolic donors because of the vastly bigger donor pool and the relatively unexacting technical aspects of tissue removal. For example, in England in any one year there are approximately 600,000 deaths. All these deaths are potential donors of some form of tissue. However, the donor pool for solid organs is comparatively tiny. In the UK in 1989 there were only 1054 patients where brain death was confirmed and there were no general medical contraindications to organ donation.[33] Twenty-five percent of all deaths in England are reported to Coroners and most undergo post-mortem examination. These deceased are therefore in facilities where tissue removal for transplantation is at least theoretically possible. The volume of tissues potentially available from such cases, if a presumed consent regime prevailed, would be massive. However, the amount actually collected would be less than theoretically available because relatives would still need to be consulted about the deceased's medical and social history and therefore would effectively exercise a veto over tissue removal. This requirement alone is arguably sufficient to overcome proponents of presumed consent laws.

In the United States, several states have presumed consent legislation for the removal of corneas. In its guidelines for preventing HIV transmission in transplantation the US Centers for Disease Control[34] actually exempt such retrievals from the requirement to interview the next of kin of cadaveric donors. Informing such an exemption is the empirical observation that, in instances where corneas have been grafted from patients subsequently shown to have HIV infection[35] the recipients of corneas have not (yet) been infected. However, given the generally low infectivity of HIV, this could be mere chance and not a special feature of allograft corneas. It will be interesting to see if the Centers for Disease Control (CDC) continue with this exemption in future editions of the Guidelines. The Food and Drugs Administration (1993) in its Interim Rule entitled 'Human Tissue intended for Transplantation' endorses this exception by only requiring inspection of 'all available medical, coroner and autopsy records'.[36] The inference is, that even if such records are not available or are deficient, the corneas may still be removed and grafted.

Technical regulation of tissue banking

USA

The Food and Drugs Administration (FDA) has regulated blood and blood products for decades under the Federal Food, Drug and Cosmetics Act and the Public Health Service Act. The National Organ Transplant Act of 1984 as amended, provides for federal oversight of the organ transplant system. The Health Resources and Services Administration and the Health Care Financing Administration within the Department of Health and Human Services (DHHS) currently administer programmes related to organ transplantation. In June 1991, DHHS published proposed rules governing performance standards for organ procurement organizations (56 FR 28513, 21 June 1991).

Self-regulation replaced by FDA regulation

The American Association of Tissue Banks (AATB) was established in 1976 and in 1982 published its Standards for Tissue Banking. These standards are the basis of a system of accreditation of Tissue Banks by the AATB. Although it has been estimated that 90% of all human tissue used was distributed by AATB accredited Tissue Banks,[37] it was also claimed[38] that only about 50 of 400 known tissues banks (including sperm banks) were accredited with the AATB with another 50 awaiting accreditation. These two statements may not be conflicting but they are a particular example of the general problem that the precise scope of tissue banking is not known, as far as the authors are aware, anywhere in the world.

Under the Medical Device Amendments of 1976 (Pub L 94-295) the FDA had previously regulated dura mater allografts, heart valve allografts, and skin and bone materials that are processed in ways other than to only reduce infectivity or preserve tissue integrity. However, in December 1993 the FDA issued an Interim Rule consisting of minimum standards covering the field of tissue banking.[36] These standards are regarded as less stringent than the standards of the AATB. The Interim Rule (Table 14.2) was inspired by the discovery that significant quantities of allograft human bone were being imported from, amongst others, eastern European tissue banks. It seemed that much of this bone had been procured without proper consents, without adequate investigation of the donor's medical and social history and without adequate testing of the donor. The standards applied to the processing of the tissue were not known. To date, there is no disease transmission known to be related to the transplantation of this bone.

One effect of the Interim Rule is that it continues a drift in the USA towards developing different regimes for different groups of tissues, within the limited range of tissues being considered in this chapter. For example, the Interim Rule does not include tissues which had previously been regulated as devices such as aortic valves.

Furthermore the Interim Rule institutes a different and lower standard for donor screening (that is the discovery of the patient's medical and social history) in relation to corneas. The basis for this is the apparently lower risk of HIV transmission from corneal grafting compared with bone and skin. This different standard seems, at least in part, to have originated in presumed consent legislation for corneal removal in some states. A requirement to speak to relatives would have effectively nullified such legislation. The importance, however, of the medical and social history is of course a major argument against presumed consent in cadaveric organ and tissue donation.

The Interim Rule exempts from direct regulation establishments which only store tissue for use within the same facility. It is unclear if this is a matter of policy or a matter of jurisdiction of the FDA. If the former, there seems little basis for the exemption, as such a 'cottage industry' is likely to be a vulnerable part of tissue banking.

UNITED KINGDOM

The UK Tissue Banking Review

The absence of formal regulation in the UK has prompted the Departments of Health of England, Wales, Scotland and Northern Ireland to jointly commission the UK

Table 14.2 Summary of the FDA's Interim Rule: Human Tissue intended for transplantation

1. Objective – to provide adequate and clear regulatory authority to prevent the use in transplantation of banked human tissue, whether domestic or imported, which may not have been properly tested or otherwise has a potential for transmitting AIDS or hepatitis.

2. Scope – Persons engaged in the recovery, processing, storage or distribution of banked human tissue. Banked human tissue intended for transplantation includes bone, ligaments, tendons, fascia, cartilage, corneas and skin. The Interim Rule does not apply to:

    (a) Vascularized organs (livers, hearts etc.).
    (b) Tissues already subject to regulation by FDA as a drug, biologic or medical device.
    (c) Semen, other reproductive tissue, bone marrow and human milk.

    FDA will not directly regulate establishments which only store tissue for use within the same facility.

3. Testing requirements. A blood specimen from each tissue donor must be tested by a laboratory certified under the Clinical Laboratories Improvement Act 1988 (Pub. L. 100-578) using FDA approved tests. The blood must be negative for:

    (a) Antibody to HIV I and II.
    (b) Hepatitis B surface antigen.
    (c) Antibody to Hepatitis C virus.

4. Donor screening requirements. A medical history must be obtained for each tissue donor to assure freedom from risk factors or clinical evidence of Hepatitis B or C or HIV infection. A limited exception is provided for corneas, consistent with current practice and the laws of some states where the relevant medical history 'shall include all available medical, coroner and autopsy records' (S1270.5(e)).

5. Written procedures must be followed for the testing and screening described above.

6. Records, written or electronic, concerning donor testing and screening must be kept for 10 years. Tissue must either be held or identified as in quarantine or accompanied by records of testing and screening results.

7. Inspection. FDA has authority to inspect regulated establishments. (In the first 3 months of 1994 the FDA conducted 16 inspections. Deficiencies were discovered in 5 of these inspections.)

8. Recall and destruction of banked human tissue. FDA may order the retention and recall of tissue not meeting the standards of the interim rule. After appropriate administrative procedures, FDA may destroy, or order the destruction of, violative tissue. (The 5 inspections by the FDA uncovering deficiencies mentioned above resulted in the retention, recall and destruction of 9,000 pieces of tissue, representing foregone revenue of $US 5,000,000.)[39]

Transplant Support Service Authority to conduct the 'The UK Tissue Banking Review', (UK Tissue Banking Review).[40] The review regards its functions as:

1. To investigate whether tissue banking should be regulated to ensure common standards. If so, whether this should be by legislation or guidelines.
2. To investigate the most effective way of providing tissue banking services nationally.

The Review is also concerned to protect as far as it can the altruistic basis of tissue donation and the non-profit basis of the enterprise. The Tissue Banking Review is not concerned with the banking of blood, foetal tissue, bone marrow or reproductive tissue for therapeutic purpose or the banking or provision of tissue for research. In addition to the Tissue Banking Review, the Department of Health has established an Advisory Committee on the Microbiological Safety of Blood and Tissues.

EUROPE

The major concerns of Europe have been to:
1. Promote the availability of tissues by supporting a presumed consent approach to donation.
2. Prohibit trading (Council of Europe, 1978;[24] Third Conference of European Health Ministers, 1987).[25]

As far as more technical matters are concerned the European Health Committee (CBSP) of the Council of Europe met in Strasbourg from 30 November to 2 December 1993. At that meeting it approved a 'Draft Recommendation on Human Tissue Banking' (Council of Europe, 1993).[41]

In the draft, reference is made to Resolution (78) 29 on the harmonization of legislation of member states relating to the removal, grafting and transplantation of human substances. This resolution was adopted at the Third Conference of European Health Ministers in Paris 16–17 November 1987. The draft also notes the fact that human tissue is donated by the public for altruistic reasons and condemns the purchase and sale of tissues. The following draft recommendations about tissue banking were made to member states.

1. For the purposes of the recommendations the following aspects of tissue banking are considered: organization, processing, preservation, internal quality control, storage and distribution; nothing is being said about the collection of tissues.
2. Tissue banking functions should be carried out on a non-profit basis within institutions which are officially licensed by national health administrations or recognized by competent authorities.
3. 'That by way of derogation from paragraph 2, and in the case of unexpected and imminent public health need, the activities described in paragraph 1, with the exception of distribution of tissues may be carried out by an officially recognized profit making body.' (The authors envisage that this relates to skin in a situation of a major disaster involving many burns victims. However, it is difficult to imagine that there is a profit making body, waiting in the wings, that would be able to offer anything of significance in an emergency to the system that is already retrieving skin).
4. That tissue banks comply with local laws with respect to testing for transmissible disease.
5. That recognized and safe measures are used in the storage of tissues.
6. 'That records of all tissues retrieved and issued should be kept by the tissue banking organizations in such a way that their source and their destination are clearly identifiable, providing always that access to such records will be restricted to the extent necessary to protect confidentiality of information and individual privacy.'
7. 'That distribution should take place in such a way as to permit optimal use of the tissues on an equitable basis in accordance with national law and practice and objective selection criteria.'

8. That close mutual co-operation should be pursued by all officially recognized exchange and tissue banking organizations and that follow up data on donor/recipient combinations should be shared between relevant institutions within the framework of national guidelines and legislation providing always that the privacy of the person concerned is fully respected.

For the purposes of the recommendations, human tissue includes all constituent parts of the human body, including surgical residues but excluding organs. However, organs are not defined. Blood and blood products are excluded as well as reproductive tissue such as sperm, eggs and embryos. Hair, nails, placentas and body waste products (not defined) are also excluded. The distribution of tissue is defined as the supply of tissues for therapeutic purposes.

It seems that the policy goals behind these recommendations are:

1. To minimize commercialization and ban profit making by institutions banking human tissue (Recommendation 2), except in cases of unexpected health need (Recommendation 3).
2. The prevention of transmissible disease (Recommendation 4).
3. Tissue safety; that is, the structural properties of the tissue are fit for their purpose. This policy goal in only partly addressed in Recommendation 5, as there is no mention of tissue processing for example.
4. That record keeping must be such as to enable both tracing of transplanted tissue back to its source and research into tissue transplantation outcomes. At the same time principles of confidentiality and privacy must be respected (Recommendations 6 and 8).
5. That tissues be distributed equitably based on objective selection criteria.

As far as formal technical regulation is concerned it seems that only Belgium has legislated.[42] Legislation was enacted on 13 June 1986 'concerning the retrieval and transplantation of organs and tissues' which made presumed consent the basis of cadaveric organ and tissue provision. Article 3 also states that 'every retrieval ... must be performed by a medical doctor in a hospital ...'. Article 17 provides that infringement of Article 3 is punishable by imprisonment of 3–6 months and/or a fine of 500–5000 francs.

More specifically, on 15 April 1988, regulations under the 1986 Act were issued 'regarding tissue banks and the retrieval, storage, preparation, importation, transportation, distribution and deliverance of tissues'. The stated concerns behind these regulations were:

1. The need to guarantee the quality of the retrieved, stored and delivered tissues in hospitals.
2. The need to be able to determine the requirements for allografts and their availability to patients.
3. The need to evaluate all the costs related to retrieval, preparation, storage and distribution of allografts so that tissue banking takes place without profit.
4. The need to specially approve tissue banks and systematically follow their operations.

To receive approval from the Minister, the tissue bank, taking into account the tissues with which it is involved, must:

1. Have the necessary premises, staff, materials and hardware to be able to guarantee the quality of the tissues.

2. Have a medical director with requisite knowledge and experience responsible for its management and organization.
3. Maintain accounts which make it possible to determine the cost of its activities.

Also, tissues must come from donors who have been subject to clinical, biological, microbiological and immunological examinations that clearly show that the donor does not suffer from a non-infectious or infectious disease that would represent a danger to the recipient. The medical director must be personally satisfied that tissues distributed from the bank have no characteristics which could be harmful to a recipient.

Furthermore allografts can only be distributed to medical practitioners with the required knowledge and experience to use them, or to other approved tissue banks. It seems, however, that this process is informal and based simply on personal knowledge of the user surgeon by the bank. These doctors, in return, must provide in writing information about the recipient including the results of the procedure and any undesirable side effects.

The minister determines the 'price' at which the allograft may be delivered in such a way that profit is excluded. This and other accounting procedures, are subject to oversight by officials of the 'Service for Bookkeeping of Hospitals' of the Ministry of Public Health and Environment.

AUSTRALIA (VICTORIA)

The technical regulatory regime of tissue banking in the Australian state of Victoria includes the state's Health Act (1958), as amended and the Pathology Services Accreditation Act 1984. Shortly, the Therapeutic Goods Administration, a Commonwealth (or national) agency with the authority of the Therapeutic Goods Act (1989) will be auditing tissue banks according to criteria now being developed. The National Health and Medical Research Council[43] Guidelines for the Donation of Cadaveric Organs and Tissue completes the picture.

The Health Act (1958), as amended, in Victoria, provides a statutory defence in Victoria to any legal action brought by a person claiming to have been infected with HIV by tissue taken from the body of a dead person. This immunity applies to the person dealing with the tissue who is required to have performed approved tests for the presence of HIV and to have made reasonable enquiries about the behaviour of the dead person to find out if s/he was at high risk of being infected with HIV. This immunity relates only to HIV transmission and nothing else.

The Pathology Services Accreditation Act 1984 covers services 'in which human tissue, human fluids or human body products are subjected to analysis for the purposes of prevention, diagnosis or treatment of disease in human beings …'. Although probably not intended, this clearly covers the activities of a tissue bank. Under Section 40, the Governor in Council has wide powers to make regulations. The Schedules attached to the Pathology Services Accreditation (General) Regulations 1990 make detailed reference to staff, facilities, equipment and instrumentation, quality control, reporting and standards for specimens.

The Therapeutic Goods Act (1989) is a Commonwealth Act and represents Australia's approach to the regulation of therapeutic goods.[44]

The object of the Act is to provide 'for the establishment and maintenance of a national system of controls relating to the quality, safety and efficacy of therapeutic goods that are used in Australia whether those goods are produced in Australia or elsewhere' (Section 4). Therapeutic goods are classified as either drugs or devices. The definition of a device is:

> ... an appliance, material or other article ... which does not achieve its principal intended action by pharmacological, chemical, immunological or metabolic means ...

This clearly includes all forms of cadaveric tissue transplantation. Fresh viable human tissue (but not blood), human organs, parts of human organs or human bone marrow intended for direct donor to host transplantation have been excluded from the operation of the Act. The Act is administered by the Therapeutic Goods Administration (TGA).

The regulatory regime for which the TGA is responsible has 4 components:

1. Premarket evaluation.
2. The listing of goods on the Australian Register of Therapeutic Goods (ARTG).
3. The conduct of audits and then licensing the manufacturer.
4. Post market evaluation (problem reports, testing of market samples).

By May 1994, the TGA had received 15 applications from various tissue banks for licensing. The TGA have advised that it will be distributing for comment a draft of a 'tissue banking code' which will be the standards which tissue banks will be expected to meet. The code will be developed taking into account current accepted practices in Australia and overseas. In a recent TGA circular,[45] the following documents were listed as informing the development of the code.

1. The Food and Drug Administration Interim Rule.[36]
2. Policy and procedure manual checklist of the American Association of Tissue Banks.
3. European Association of Tissue Banks: General standards for Tissue Banking.
4. National Health and Medical Research Council Guidelines for the Donation of Cadaveric Organs and Tissues.
5. Australian Standards 3901 and 3902 (Quality Systems for design/development, production, installation and servicing, and Quality Systems for production and installation).
6. European Standard 46001.

Also considered important is compliance with Good Manufacturing Principles which incorporate 9001. The manufacturing principles outline the requirements for each of:

1. Place and method of manufacture.
2. Staff.
3. Quality assurance.
4. Documentation.
5. Complaints.
6. Sterile therapeutic goods.
7. Compliance with application for licensing.
8. Expiry dates.
9. Sub-contracting.

The last requirement is important because banks sub-contract various aspects of their operations (e.g. infectious disease testing, irradiation of bone). The sub-contracted laboratories will also be audited.

The power to audit is exercised by an authorized person within the meaning of Section 3 (1) of the Therapeutic Goods Act 1989. A condition of licensing is that the person or institution licensed will allow an authorized person to:

- Enter, at any reasonable time, premises at which the person deals with the goods; and while on those premises, to inspect those premises and therapeutic goods and to take samples, to inspect processes relating to the manufacture.
- If requested to do so, produce to the authorized person such documents relating to the goods as the person requires and allow the person to copy the documents (Section 28 (5), 40 (4)).

The same can be required of an applicant for a licence (Section 37 (2)). Sections 47–51 set out substantial powers of entry, search and seizure if an authorized person has reasonable grounds for suspecting 'that there is in or on the premises a particular thing that may afford evidence of the commission of an offence against the Act'.

The following aspects of tissue banking are not specifically regulated in Victoria:

1. The content of the interview conducted by Donor Transplant Co-ordinators with relatives with a view to assessing donor suitability.
2. Other than in the general terms of the NH&MRC (1994) Draft Guidelines, the exclusionary criteria setting out the circumstances in which tissue should not be taken from a potential donor.
3. The reimbursements claimed by tissue banks from health insurers and hospitals for tissue provided.
4. Whether or not a tissue bank is 'profit making'.

Tissue banking organization and scope: an international view

USA

The organization of tissue banking in the United States is complex and exists on a different scale to anywhere else in the world. It seems that while there are about 400 tissue (including semen) banks in the USA, it is estimated that 90% of banked tissue is distributed from the 50 largest banks. Some idea of the scale of tissue banking in the USA can be gained from the 1992 American Association of Tissue Banks survey of 46 tissue banks which that organization inspected or accredited[1] (see Tables 14.3 and 14.4).

Over 300,000 bone allografts were distributed in the USA in one year (Table 14.3). A large number of these would have been in the form of powder or gel and used in dental procedures. Nonetheless, assuming the population of the USA to be 250,000,000, more than 1 person in 1000 underwent a medical or dental procedure in 1992 involving the transplantation of human bone. There are 10 banks distributing more than 5,000 allografts annually, one of which is distributing more than 80,000. Just on this basis alone, tissue banking in the USA is a billion dollar industry (Table 14.4).

As far as heart valves are concerned, the numbers are less dramatic (see Table 14.5) though, of course, 1598 donors could produce up to twice as many valves (aortic and pulmonary).

The 46 tissue banks visited by the AATB in 1992 retrieved tissue from a total of 7,104 cadaveric tissue donors of which three (0.04%) tested positive for anti-HIV antibodies, 0.48% were hepatitis B surface antigen positive, 0.62% were hepatitis C antibody positive and 0.17% had positive syphilis serology. The three HIV positive

Table 14.3 Bone allografts distributed in the USA in 1992

Not sterilized	144 699
Irradiated	111 373
Ethylene oxide sterilized	46 923
Total	302 995

Table 14.4 Number of bone allografts distributed per tissue bank in the USA in 1992

Allografts distributed	Number of banks
0–500	3
501–1000	4
1001–1500	7
1501–2000	5
2001–3000	6
3001–4000	4
4001–5000	1
5001–10 000	4
10 001–20 000	1
20 001–50 000	4
More than 80 000	1

Table 14.5 Heart valve donors in the USA in 1992 (total: 1598 donors)

Number of donors	Numbers of banks
0–25	17
26–50	5
51–75	5
76–100	2
101–125	1
126–150	1
426–450	1
Total	31

cases were all from one bank which accepted injecting drug abusers on the basis that its tissue processing procedures inactivated the virus. One conclusion supported by this is the efficacy of donor screening routines designed, by assessment of the potential donor's medical and social history, to screen out donors at high risk for HIV.

There is a separate organization in the USA covering corneas: the Eye Banking Association of America. Over 28,000 corneal grafts were carried out in the USA in 1986.[32]

Tissue banks have varying relationships with Organ Procurement Agencies. These agencies were originally funded to improve organ availability from potential systolic cadaveric donors. However, they are now being required to give more attention to the needs of tissue banks as well. It is not possible to say what percentage of tissue supplied originates from asystolic versus systolic donors. A major issue in the organization of tissue banking in the United States is its commercialization. There are both non-profit and profit making tissue banks.

UNITED KINGDOM

There are an unknown number of tissue banks in the UK distributing an unknown amount of tissue. As at June 1992 there were over 60 members of the British Association of Tissue Banks but as this is individual membership, it does not reflect the number of banks in the UK. There are some units which have been operating continuously for a long time. The most prominent of these is the Yorkshire Regional Tissue Bank based at the Pinderfield Hospital in Wakefield. It is principally involved with bone banking.

From 1965–1992, the Yorkshire Regional Tissue Bank has issued 'in the region of 13,500 packets of bone implants ... without any reports of primary infection or excessive graft failures being attributable to the bone implants'. The Director of the Yorkshire Regional Tissue Bank, Dr Kearney, was responsible for the establishment of the British Association of Tissue Banks.

In 1993, a National Corneal Service was established in the UK. The service was the result of a union of the two strongest eye banks in the country in Bristol and Manchester. It is funded centrally from the Department of Health and provides ocular tissue, in so far as it is able, to any NHS Unit requiring it. All eye banks have been invited to join the national service and, hence, be centrally funded, but many have not taken up the invitation. The reasons for this are not clear.

Because of separate funding for health services through the Scottish, Welsh and Northern Ireland Offices, there is scope for variable service developments in health in the different countries of the UK. This is the case with tissue banking in Scotland. In November 1989, at the request of local orthopaedic surgeons, the Regional Transfusion Centres in Edinburgh and Aberdeen undertook the reorganization and running of bone banks in Lothian and Grampian regions respectively. Previously such bone banking in Scotland had existed in hospital orthopaedic units. Glasgow Regional Transfusion Centre (RTC) commenced bone banking in 1991 and it was intended that the Dundee RTC commence bone banking in 1992. In 1991 The South East Scotland Blood Transfusion Service issued a total of 119 allografts. Table 14.6 lists organizations in the United Kingdom known to be involved with tissue banking.

The document, *Cadaveric Organs for Transplantation: A code of practice including the diagnosis of brain death* (1983) lists all the units in the UK performing corneal grafts. A recent review of tissue banking in the United Kingdom has been conducted under the auspices of the UKTSSA (see page 276).

EUROPE

A key agency with responsibilities in tissue banking in Europe is Bio-Implant Services (BIS), which is based in The Netherlands. It is a non-profit organization founded in

Table 14.6 Organizations involved with tissue banking in the UK (BATB News 1, 1992)

BTS Birmingham	Bone, heart valves
Harefield Hospital, Bucks	Heart valves
Clwyd and Oswestry Tissue Bank	Bone and other tissues
BTS Southampton	Bone, valves
University College Hospital, London	Bone, valves, skin, corneas
Yorkshire Regional Tissue Bank	Bone, valves, skin, corneas
Killingbeck Hospital, Leeds	Heart valves
Heart Valve Bank, Oxford	Heart valves
BTS Oxford	Bone, heart valves
BTS Carluke	Bone
London Hospital Medical College	Cartilage
Royal Brompton, London	Heart valves
Leicester Bone Bank	Bone
Addenbrookes Hospital, Cambridge	Bone

(BTS: Blood Transfusion Service)

1989 on the initiative of Eurotransplant. The aim of BIS is to facilitate the preservation, distribution and exchange of human tissue for transplant purposes.[46]

BIS acts as a central co-ordinating and administrative body for an increasing number of tissue banks. Membership of the network signifies acceptance of common standards (but there is no system of accreditation) and enables the pooling of resources for strategies to increase donation rates. Furthermore, BIS acts as the middleman between the donor hospital and the tissue bank and the tissue bank and the user hospital. A user hospital accesses a Europe-wide tissue bank network by contacting BIS. BIS invoices the hospital, and after deducting its own charges, passes on the residue, which is at the level of cost recovery, to the tissue bank. BIS also supports research and development projects in tissue banks.

There are clearly numerous tissue banks which are not members of BIS, especially in France, Spain, Portugal and Italy. 'It is also true at the moment in much of Europe, that any surgeon wishing to set up a tissue bank can do so without having to satisfy any minimum professional criteria'.[47] The attractiveness of the BIS model is that it operationally ties the tissue banks together. The communication and administration involved in this will promote adherence to uniform standards and facilitate formal regulation. Although there is no formal accrediting process, BIS claims that adherence to uniform standards is promoted because:

1. Joining the network signifies a commitment by the bank to follow BIS standards.
2. If BIS becomes aware of a failure to comply then the bank can be dropped from the network. Donor hospitals will cease to be referred to the bank and user hospitals in the BIS network will not be referred to the bank.
3. Later formal regulation will be facilitated because of the already functioning lines of communication. With a strong central co-ordinating body there is a better chance that the regulating agency will understand what it is that is being regulated.

An additional benefit of the European model is that BIS deals centrally with the finances. Individual banks are not making individual reimbursement arrangements with individual hospitals or patients; this task is performed centrally, with uniformity. Given the clear policy position in Europe against commercialization, and hence profit making in relation to human tissues, this approach minimizes the opportunity for the creative approaches to non-profit making that exist elsewhere.

AUSTRALIA

Tissue banking in Australia is dislocated; the vast distances involved for a relatively small population complicates national approaches to many activities. However, despite this, there is an Australian Corneal Graft Registry and the Australian Orthopaedic Association has developed its standards for Bone Banking. Professor O'Brien at the Prince Charles Hospital in Queensland has probably the longest continuous experience worldwide in aortic valve banking. The development of the Donor Tissue Bank of Victoria (DTBV) in the late 1980s represented a new phase, being the first organization in Australia that procured, processed, stored and distributed more than one type of tissue. Furthermore, this was done more from a donor base (as part of an organization responsible for forensic autopsies) as opposed to being a development arising out of a user (i.e. a hospital) surgical unit.

The Australian Corneal Graft Registry collects reports of corneal grafts performed in Australia. As at July 1994, it had a total of 6031 grafts registered.

Table 14.7 lists all tissue banks in Australia with interim licensing from the Therapeutic Goods Administration. There is an unknown number (possibly 20–30) of

Table 14.7 Organizations with interim licensing from the therapeutic goods administration to provide human tissue for transplantation in Australia (August 1994)

Cardiac valves and related tissue
 Donor Tissue Bank of Victoria
 Prince Charles Hospital, Chermside, Queensland
 Royal Perth Hospital, Western Australia
 Cardiothoracic Surgery, St Vincent's Hospital, Darlinghurst, NSW

Ophthalmic tissue
 Donor Tissue Bank of Victoria
 Lions NSW Eye, Bank
 Lions Eye Bank – Melbourne
 Queensland Eye Bank
 The Lions Eye Bank of Western Australia

Skeletal and related tissue
 Donor Tissue Bank of Victoria
 Perth Bone and Tissue Bank (Western Australia)
 Queensland Bone Bank
 St Helen's Private Hospital, Hobart, Tasmania
 The Avenue Private Hospital (Victoria)

other small tissue banks, probably bone banks in orthopaedic units, scattered throughout the country.

There is no national body representing tissue banks. The Eye Banks have formed a loose network and, as mentioned above, have a common focus at the Australian Corneal Graft Registry.

Donor selection and technical aspects of tissue banking

The procedures described below are based on those in place at the Donor Tissue Bank of Victoria. Application of these procedures elsewhere should take into account the requirements of relevant legislation, regulations and local user surgeons. The approaching of relatives is a sensitive (and stressful) undertaking and should be governed by clear ethical guidelines.

Documents and publications from the following have informed the development of the procedures in place at the Donor Tissue Bank of Victoria:

- Transplant Resources and Services Center, Dallas, Texas.
- Cardiothoracic Surgery Unit, Prince Charles Hospital, Brisbane, Queensland.
- Cardiothoracic Surgical Unit, Green Lane Hospital, Auckland, New Zealand.
- South Eastern Organ Procurement Foundation, USA.
- Lions Eye Bank of Western Australia.
- Lions Eye Bank of South Australia.
- American Association of Tissue Banks.[3,49]
- American Federation of Clinical Tissue Banks.[50]
- US Centers for Disease Control and Prevention, Atlanta, Georgia.[34]

INITIAL ASSESSMENT OF POTENTIAL TO DONATE

Potential asystolic tissue donors are initially selected on the information received about the circumstances of the death. Information may be obtained from:

- Police.
- Coroner's or medical examiner staff.
- Hospital staff, if person died in hospital.
- Medical records.
- Funeral and mortuary staff.

This information includes:

- Age.
- Estimated time of death.
- Time of admission to mortuary.
- Possible cause of death.
- Likelihood of risk factors.
- Existence of a donor card.

A pathologist may need to be consulted regarding the condition of the body and for an opinion as to which tissue can be retrieved in the light of the whole circumstances of the case.

INFORMATION REGARDING NEXT OF KIN

Details about the family need to be obtained in order to approach the senior available next-of-kin. Information may be obtained from:

- Police.
- Coroner's or medical examiner staff.
- Medical records or hospital staff.

This information includes:

- Name, address and telephone number.
- Current location of the next-of-kin and/or where they may be during the next few hours.
- Whether and who has been informed of the death to prevent the possibility of approaching an uninformed relative.
- Cultural and social details and family relationships.
- Whether there has been any prior approach for organ and/or tissue donation by any other organization.

APPROACHING RELATIVES TO DISCUSS TISSUE DONATION CONSENT

Phone contact

This contact should be made by a suitably trained and experienced person with a knowledge and understanding of the issues in transplantation and the needs in general of recently bereaved families. Whoever makes the contact should regard him or herself as looking after the interests of the deceased and his or her family. Any potential conflicts of interest should be declared prior to obtaining consent.

Nearly all initial approaches to relatives take place on the telephone for the following reasons:

1. Relatives' rights and the need for privacy at a tragic time, and when making such an important decision.
2. Time factors involved in visiting relatives could complicate tissue retrieval.
3. Staff safety.

When speaking to the next-of-kin:

1. The reason for telephoning needs to be explained. The reason is to allow the family the opportunity to consider tissue donation from the deceased.
2. If the relative consents or shows interest in donation, the ensuing discussion should include:
    (a) An appropriate explanation of heart valve, bone, skin and/or corneal transplantation including the nature of the donor evaluation process (e.g. medical and behavioural history, blood tests, autopsy results).
    (b) Specific information about the deceased's lifestyle and any risk factors, in order to ascertain whether the option of donation should be eliminated immediately (see Table 14.8).
    (c) Reference to confidentiality and anonymity generally but particularly as between the donor and the recipient.
    (d) The feelings of other family members about tissue donation from the deceased.

Table 14.8 Contraindications to tissue donation from aystolic cadaveric donors

(a) Any communicable disease.*
(b) Systemic infection or sepsis.*
(c) Malignancy. (Exceptions exist for primary central nervous system tumours and localized basal cell and squamous cell skin cancers).*
(d) History of use of human pituitary extracts (for dwarfism or fertility treatment).
(e) Past or present nonmedical intravenous intramuscular or subcutaneous injection of drugs.
(f) Clinical or laboratory evidence of HIV, Hepatitis B or C infection.
(g) Male homosexual activity.
(h) Haemophilia or related clotting disorders treated with human-derived clotting factor concentrates.
(i) Any engagement in sex for money or drugs.
(j) Sexual involvement with a person defined in (e), (g), (h) and (i), or with a person known or suspected to have HIV infection.
(k) Exposure in previous 12 months to known or suspected HIV infected blood through percutaneous inoculation or through contact with an open wound, non-intact skin, or mucous membrane.
(l) Inmate of correctional systems.
(m) Unexplained death

NB If the donor has received either:

- 4 units of blood or blood products within 48 hours of death
  or
- 3 litres of fluid within 1 hour of death.

then a pretransfusion blood sample needs to be tested for infective agents.

These contraindications include the 'Behaviour/History Exclusionary Criteria' recommended by the US Centers for Disease Control and Prevention[34]

*The existence of these contraindications implies that a full autopsy will need to be performed, the results of which determine whether or not these contraindications exist.

(e) The appropriateness of a visit from the donor transplant co-ordinator. If such a visit is not wanted or cannot occur, a means of signing a consent form should be agreed (e.g. by fax or mail or by signing it when attending to formally identify the deceased if this is required). In all cases it is desirable to have written consent in addition to verbal consent although legally the latter may be sufficient. The signed consent form, however, is usually good evidence that a consent was given. Exact understanding of local legislation is clearly important.

Visiting the next-of-kin

1. Take consent forms, business cards, relevant literature about tissue donation and grief support groups to give to the family.
2. Meet relevant family members. Depending on who is present, it may be necessary gently to guide family members involved in making the decision away from a large group.

3. Because each family is different, their needs must be assessed individually. As they have allowed the donor co-ordinator into their home, the family may already have some idea about tissue donation and how the deceased felt about it. Conversation will ensue about the deceased, the death, the autopsy, the organization of funerals or other arrangements. Discussion should include:
   (a) Specific questions about medical and social risk factors.
   (b) The value of donation.
   (c) How other family members feel about donation.
   (d) The opportunity to raise questions about any aspect of donation or, if relevant, coronial or medical examiner procedures. Anonymity of donors/recipients should be mentioned.
4. Specify and confirm which tissue is to be taken and for what purpose.
5. Have the consent form signed. The consent form should clearly state and specify which tissues are being donated. In relation to the donation of skeletal tissue, each bone should be specified. Have it signed by senior available next-of-kin. Other family members may like to sign if they have contributed to the decision made. The donor transplant co-ordinator may sign as a witness or have it witnessed by another person present.
6. Offer to send the family a copy of the consent form.
7. Tell the family that a letter of appreciation will be sent in a few days. Offer ongoing support as necessary.
8. Answer any questions in regard to donation and/or other matters that may be raised, e.g. funeral arrangements, autopsies etc. Refer relatives to the correct resource if necessary.

CRITERIA RELATING TO THE SUITABILITY OF PARTICULAR TISSUES

Corneas

1. Age limits: lower limit is 6 months – there is no upper age limit.
2. Time limits: retrieval of corneas should occur no more than 12 hours after death, preferably sooner.
3. All the general contraindications apply for corneal donations as well as the existence in the donor of:
   - Alzheimer's disease.
   - Amyotrophic lateral sclerosis.
   - CNS disease of unknown aetiology.
   - Congenital rubella.
   - Creutzfeldt–Jakob disease.
   - Crohn's disease.
   - Jaundice.
   - Multiple sclerosis.
   - Parkinson's disease.
   - Rabies.
   - Subacute encephalitis.
4. Diabetes mellitus is not necessarily a contraindication to donation but a special assessment should be made to ascertain the extent of any corneal deterioration due to the disease.

5. The following conditions are not contraindications but should be discussed with the local ophthalmic surgeon if such conditions are detected:
   - Evidence of prior eye disease, surgery, recent conjunctivitis, glaucoma, uveitis and diabetic eye disease.

Heart valves

1. Age limits: from birth to 55 years.
2. Time limits: retrieval of valves should be completed within 24 hours of death if the body is cooled within 6 hours of death. If the body is kept at 20 °C, retrieval should occur within 12 hours of death. (This requirement seems to have a more historical than biological basis. Arguably the criteria should relate more to the microbiological assessment of the valve, and if considered important, leaflet viability as measured by fibroblast culture results).
3. Factors that may contraindicate valve donation include chest trauma and chest surgery, history of valve disease (e.g. floppy mitral valve syndrome, rheumatic valve disease).

Skeletal tissues

1. Age limit: women: 18–50 years; men: 18–65 years.
2. Time limit: if refrigerated within 6 hours, bones may be taken up to 24 hours from time of death. If the body is kept at 20 °C retrieval should occur within 12 hours of death.
3. In addition to the general contraindications, the following are also contraindications to bone donation:
   - Evidence of slow virus infection (e.g. presenile dementia) based on history.
   - Immune complex disease.
   - Chronic renal failure.
   - Osteoporosis.
   - Irradiation therapy.
   - Steroid treatment.

Skin

1. Age limit: 55 years.
2. Time limit: if refrigerated, skin may be taken up to 24 hours after death, however, it is preferable if the allograft is taken as soon as possible after death.
3. In addition to the general contraindications, the following are also contraindications to skin donation:
   - Any skin malignancy including localized basal cell carcinoma and squamous cell carcinoma.

SEROLOGICAL TESTING REGIME FOR POTENTIAL DONORS

Blood from the potential donor is tested as follows:

- Antibodies to HIV-1, HIV-2.
- Hepatitis B surface antigen.
- Antibodies to hepatitis C virus.

- Antibodies to HTLV-1.
- Syphilis serology.

In addition, HIV-1 DNA testing by PCR is performed and blood grouping is undertaken.

APPENDIX

Removing and processing tissue for transplantation

CARDIAC TISSUE RETRIEVAL

The following procedures are almost completely based on those developed by O'Brien in Brisbane, Australia.[48]

1. Shave the chest with a razor if necessary.
2. Dry area with a towel.
3. Prepare area with Betadine.
4. Scientist/technician scrubs up:
   - Put on mask and hat.
   - Wash hands with Hibiclens and dry with a towel from the sterile pack.
   - Don sterile gown.
   - Put on sterile gloves.
5. Dry prepared chest area with towel from the sterile pack (from centre outwards then downwards).
6. Drape area with sterile towels/drapes.
7. Cover area with sterile opsite.
8. Using a sterile scalpel make a midline incision extending from the sternal notch to just below the xiphisternum, taking care not to puncture the peritoneum.
9. Cut through the sternoclavicular joint and extend down the lateral border of the costal cartilage. The sternum can then be removed exposing the heart and lungs.
10. Examine the pericardial sac and heart for bruising, ruptures, effusions or any abnormalities. If any abnormality is present a pathologist should be contacted for advice.
11. Making a thin transverse cut open the pericardial sac near the diaphragm. Take a swab of the internal pericardial surface and epicardium. Completely expose the heart by making a midline incision through the pericardial sac to its upper limit.
12. Check heart again for any abnormalities.
13. Initially the heart will be removed *in toto*, and in a sterile manner. This removal will include the ascending aorta to just beyond the origin of left subclavian (last major thoracic branch) and the whole of the pulmonary artery and its left and right branches to their first divisions if possible. By cutting through the superior and inferior vena cavae the excision of the heart is completed.

Removal of the valve block

The excision of the valve block takes place in a sterile manner on a sterile towel:

1. Place the heart on its posterior aspect on the towel with the aorta facing the scientist/technician. The right side of the heart should now be on the scientist/technician's left.
2. Using a finger, locate the transverse sinus and through this the junction of the left atrium with the pulmonary vein.
3. Cut through this junction taking care not to cut the pulmonary vein.
4. Raise the left atrial appendage and make a small cut through the left atrium.
5. Extend the cut to the right keeping 1 cm above the mitral valve annulus. The cut should be extended to just above the antero-lateral commissure of the mitral valve.
6. Similarly, cut to the left keeping 1 cm above the mitral annulus. The cut should be extended to above the postero-medial commissure. This will leave about half of the left atrium still attached.
7. The right atrium is similarly cut about half way around its circumference.
8. A cut is now made down through the atrial septum and then the ventricular septum. The cut must be made through the posterior leaflet behind the postero-medial commissure of the mitral valve.
9. A cut is made down the opposite side of the left atrium through the posterior leaflet of the mitral valve and just behind the antero-lateral commissure. The cut is continued down 1–2 cm into the left ventricular wall.
10. The left atrium is grasped and pulled gently down towards the apex of the heart exposing the anterior leaflet of the mitral valve and its chordae.
11. Using the point of chordae attachment to the anterior leaflet of the mitral valve cut down through the left ventricle and the ventricular septum.
12. After cutting through the entire left ventricular free wall and septum gently pull up the heart block away from the ventricles.
13. By looking into the right ventricular outflow tract the pulmonary valve can be visualized.
14. Cut the right ventricle approximately 2 cm below the valve all the way around.
15. The block should now be free. If not it will most likely be still attached via part of the atrium. Gently cut to remove fully.
16. Place the block in a sterile bag, then in a second sterile ensuring that both are tightly sealed.
17. Weigh the block with the remaining heart (minus weight of containers). If there is to be any delay in processing place heart block in Medium 199 with antibiotics and keep at 4 °C.

Trimming of the heart valve

The heart valve is to be trimmed as soon as possible after collection.

Equipment

The following sterile materials are required for trimming of one valve pair.

- 1 tray, containing
- 2 pairs of dissecting scissors
- 2 pairs of forceps

- 1 metal ruler
- 1 coarctation clamp
- 1 packet ties
- 2 pairs of measuring calipers (1 small and 1 large)
- 1 250 ml blood bag
- 1 extension tube
- 1 feeding tube
- 1 3-way tap
- 4 100 ml bottles M199 with antibiotics (50 μg/ml streptomycin, 30 μg/ml penicillin)
- 1 100 ml bottle M199 without antibiotics (rinsing solution)
- 2 disposable 50 ml syringes
- 2 disposable 10 ml syringes
- 2 mixing cannulae
- 3 specimen bottles
- 3 mixing bowls
- 2 incubating jars

1. Keeping the valve moist using a 50 ml syringe filled with M199 solution, begin to trim the valve. Dissect the adventitia away from the aorta using small untoothed forceps and small curved scissors.
2. The coronary arteries are identified then trimmed to no less than 5 mm from the aorta (and preferably 10 mm) for paediatric valves and 10 mm for adult valves.
3. All fat must be removed.
4. Dissect between the aorta and the pulmonary artery. Then identify the septal myocardium and cut in a downward direction, ensuring there is ample myocardial muscle attached to each valve.
5. Trim the myocardial muscle to 1.5 cm below the aortic valve, with a thickness of 4 mm. Leave the anterior mitral valve attached, but trim off the tricuspid valve.
6. The diameter of the aortic valve is measured using a pair of measuring calipers. To measure the valve using calipers, a blood bag is fitted with an extension tube attached (via a three-way tap) to a feeding tube. The bag is 1 metre above the valve for adult valves and 0.5 metre for paediatric valves. 100 ml M199 with antibiotics is injected into the blood bag using the three-way tap. The feeding tube is inserted through one of the coronary arteries, and the end of the tube is clamped at the top of the aorta.
7. Both coronary arteries are then ligated, and the valve is turned upside down and flooded with M199. The internal diameter of the valve is measured using the calipers, and valve competence may be checked. The external diameter of the aorta just below its first branch and the conduit length are also measured and recorded on an observations sheet.
8. For the pulmonary valve the bag is 0.5 metre above the valve. The valve is flooded with medium and the internal diameter of the valve, the length of the conduit, and the distance to the main pulmonary arteries is measured. The external diameter at the midpoint across the bifurcation is also measured and recorded.
9. The size and number of any fenestrations present on either valve are to be noted on the observation sheets.
10. Following dissection, each valve is rinsed in 100 ml of sterile M199 (without antibiotics), to remove any traces of blood.
11. The procurement and rinsing solutions, and a specimen of tissue after trimming, are sent to microbiology for culture.

Incubation with antibiotics

One leaflet of the tricuspid valve, is cut into three pieces. The aortic and pulmonary valves are each placed in 100 ml of M199 with antibiotics. A specimen of the tissue trimmings (approximately the same thickness as the valve), and one piece of tricuspid valve leaflet, are also placed into each of these bottles, and the valves are incubated at 37 °C for 6 hours. The third piece of tricuspid valve leaflet tissue is placed in a bottle of Kreb's medium for viability studies.

Freezing the valves

After the 6 hours incubation with antibiotics, the aortic and pulmonary valves are prepared for freezing, under sterile conditions. The following sterile materials are required for preparation of one valve pair for freezing.

- 180 ml M199 without antibiotics (at 4 °C)
- 20 ml Dimethylsulphoxide (DMSO)
- 1 disposable 20 ml syringe
- 1 disposable 50 ml syringe
- 2 mixing cannulae
- 2 pair dressing forceps
- 2 pair of dressing scissors
- 1 packet containing the appropriate numbered metal identification tags and ties (4 tags per packet, two for the aortic valve and two for the pulmonary valve)
- 1 packet containing two thin nylon tubing bags, two wide nylon tubing bags, 2 mylar foil pouches
- 3 specimen bottles

The scientist/technician responsible for the preparation of the valves for freezing scrubs up, gowns and gloves. With the help of an assistant, 20 ml DMSO is added to 180 ml of cold M199 without antibiotics, and mixed thoroughly. A small sample of this is taken for sterility testing, and half of the remaining solution (approximately 100 ml) is dispensed into thin nylon tubing bag. The aortic valve, along with a portion of the tricuspid valve leaflet, is placed in this bag. The temperature of this bag is kept at approximately 4 °C, because DMSO is toxic to tissue cells at temperatures above 10 °C.

The top of the bag containing the valve is enclosed in a piece of nylon foil, and the bag is sealed through the foil. The foil is then removed and discarded. The sealed bag is then placed into the wider nylon tubing bag and sealed. This is placed into a mylar foil pouch with the appropriate identification tag and sealed as above. The bag package is handed to the assistant, who places it in a Coolite container packed with ice.

The above procedure is repeated for the pulmonary valve.

The tissue trimmings which had been incubated with the aortic and pulmonary valves are transferred to separate dry, sterile specimen bottles.

In a non-sterile manner, duplicate identification tags are attached to the outer bag of both valve packages using the metal ties. If there is any delay in cryopreservation, the bags are kept on ice to keep the temperature below 10 °C.

Both valve packages are placed in a controlled rate liquid nitrogen freezer, which will freeze the valve at a rate of 1 °C per minute. This freezer should monitor both the sample temperature and the chamber temperature. To obtain the sample temperature a control package should be made up with the probe inserted into the control valve. This can be assumed to mimic the temperature of the valve for transplantation.

A chart recording must be obtained to ascertain whether the sample temperature has decreased at an average of 1 °C per minute.

After freezing, the valve packages are transferred to a liquid nitrogen tank for long-term storage in the vapour phase, i.e. the liquid nitrogen should be 2–3 cm below the valves. The valves are placed in the inventory system and their location in the storage tank and other details are recorded in an inventory book.

The antibiotic solutions from which the valves were removed, the tissue trimmings, and a sample of the freezing solution are sent to microbiology for sterility testing.

Testing

- The solutions and tissue trimmings are cultured for aerobes and anaerobes.
- The pericardial swab is cultured for aerobes, anaerobes and fungi.
- Written results of all tests are kept in the donor file.
- No valve is released from the cryopreservation bank before all bacteriological cultures have been completed and the results assessed by a senior scientist and a pathologist.

CORNEAL RETRIEVAL

Enucleation

1. Swab the area surrounding the eye in Betadine. Pull the lids back and use sterile saline to moisten the eye.
2. Scientist/technician scrubs up:
   - Put on mask and hat.
   - Wash hands with Hibiclens and dry with towel from the sterile pack.
   - Don sterile gown.
   - Put on sterile gloves.
3. The area surrounding the eye is draped in a surgical drape. Use lid retractors to hold back the lids.
4. Remove the surrounding conjunctiva using fine toothed forceps and iris scissors to allow for easy access to the muscles.
5. Using a muscle hook locate each muscle, then cut the muscle with fine scissors leaving the lateral rectus for last. When cutting the lateral rectus, cut the muscle as far away from the sclera as possible to enable forceps to grip the muscle. With these forceps pull the eye slightly out to allow the curved blunt ocular scissors to slide behind it and cut the optic nerve.
6. Trim excess fat and excess muscle off the eye and place in a sterile glass jar containing a swab moistened with sterile saline. Place two drops of Neosporin on the eye. The eye may be stored for up to 24 hours at 4 °C.

Removal of cornea

Removal of cornea must take place in a sterile environment (preferably a Class II laminar flow cabinet) with particular attention on disinfection of the eye.

1. Place eye in Betadine for 2 minutes then rinse in sterile saline.
2. Take a swab from the eye and send for culture.
3. Wrap eye in a sterile gauze swab for easy handling. Make a small incision approximately 2 mm below the cornea with a scalpel blade.

4. Make small snips around the cornea using corneal scissors with the points upwards. It is important not to pierce the vitreous cavity.
5. Using iris forceps carefully remove the cornea by lifting it clear from the eye. Avoid bending or creasing the cornea which may render it useless.
6. Place the cornea in a sterile bottle containing corneal storage medium.
   Note: The cornea storage medium *must* be at room temperature.
7. Depending on the medium used the cornea may be stored for up to 10 days at 4 °C.

Removal of cornea with eye in situ

1. Swab the area surrounding the eye in Betadine. Pull the lids back and use sterile saline to moisten the eye.
2. Technician scrubs up:
   - Put on mask and hat.
   - Wash hands with Hibiclens and dry with towel from the sterile pack.
   - Don sterile gown.
   - Put on sterile gloves.
3. The area surrounding the eye is draped in a surgical drape. Lid retractors are used to hold back the lids.
4. Carefully remove the surrounding conjunctiva using fine-toothed forceps and iris scissors.
5. Make a small incision approximately 2 mm below the cornea with a scalpel blade. Make small snips around the cornea using corneal scissors with the points upwards. It is important not to pierce the vitreous cavity.
6. Using iris forceps carefully remove the cornea by lifting it clear from the eye. Avoid bending or creasing the cornea which may render it useless.
7. Place the cornea in a sterile bottle containing corneal storage medium.
   Note: The cornea storage medium *must* be at room temperature.
8. Depending on the medium used, the cornea may be stored for up to 10 days at 4 °C.

Samples for microbiology

The following samples are required for further testing:

- Eye swab (if both corneas are taken – left eye swab and right eye swab).
- If the eye has been stored in the fridge before removal of the cornea, the storage media must be submitted for testing.

Testing

The eye swab is cultured microbiologically for aerobes, microaerophiles, and anaerobes. If the eye is stored in the fridge before the removal of the cornea, the storage medium should be cultured for aerobes, anaerobes and fungi.

Evaluation of corneas

Evaluation of corneas should include:

1. Wide beam slit lamp evaluation of the entire epithelium for surface abrasions or lacerations, vascularity and other defects.

2. Narrow beam slit lamp evaluation for a critical examination of the epithelium, stroma, Descemet membrane, Bowman's membrane and the endothelium.
3. Specular microscopy to provide a total cell count, assessment of the regularity of cell distribution and size, and assessment of any cellular abnormalities such as dystrophy which should then be noted.

SKELETAL AND RELATED TISSUE RETRIEVAL

Removal of bones

1. Using a soapy lather, shave procurement area.
2. Rinse with water.
3. Scrub area for 5 minutes using Betadine scrub solution (diluted 1:2 with sterile saline). Rinse, repeat scrub.
4. Scientist/technician scrubs up:
   - Put on mask and hat.
   - Wash hands with Hibiclens and dry with towel from the sterile pack.
   - Don sterile gown.
   - Put on sterile gloves.
5. Drape with sterile U-bar drape.
6. Apply 2% alcoholic iodine.
7. Make initial incision and extend far enough (proximally and distally) to provide adequate exposure. This initial incision is not to be made through abrasions or sites of previous IV lines. It should penetrate the epidermis and dermis but not the subcutaneous fat.
8. Discard blade, retract incision, spray incision site with alcohol.
9. Using a second sterile blade, proceed with procurement.
10. Immediately on exposure of bone, take anaerobic swab of exposed bone.
11. Using sharp dissection expose bone and strip off soft tissue. As bone is further exposed, take care to preserve major ligaments, tendons and capsular structures that may be needed to facilitate future implant in the recipient.
12. After bone is sufficiently exposed, cut bone to desired specifications using a sterile saw or osteotome. In the case of long bone procurement a midshaft osteotomy should normally provide adequate recovery length for most allografts. Immediately after cut is complete, take a deep swab of the bone marrow cavity for culture.
13. Gently lift bone from incision while carefully cutting tendons and ligaments that are still intact. Take care to avoid contamination by blood and tissue fluids and damage to articular cartilage and condyle surfaces, for example by instruments.
14. After procurement is complete take a swab of entire shaft and condyle surfaces. If X-ray unavailable, take specimen of lumbar spine or iliac crest for histology.
15. Place bone in sterile polyethylene bag and irrigate with antibiotic solution containing gentamicin. Expel excess air from bag, tie bag tightly. Maintaining sterile conditions, place bag with bone into a second sterile bag, expel air, close tightly. Label outer bag with donor number, type of bone, date of procurement and initials of technician. Transport on wet ice for processing and storage.
16. Ensure procurement worksheets are properly filled out.
17. Refrigerate the bone at 4 °C until processing and final packaging is to be carried out.
18. Reconstruct donor and close incision aesthetically.

Samples

Swabs should be taken during all steps of the retrieval.

- Exposure swab: a swab of entire surface immediately the bone is exposed.
- Marrow swab: a swab of the marrow immediately the bone is cut transversely.
- Distal swab: a swab of the entire surface of the distal portion.
- Proximal swab: a swab of the entire surface of the proximal portion.
- These swabs are required for all bones procured.

Processing long bones (fresh frozen irradiated allograft)

1. Activate lamina flow cabinet 30 minutes prior to the commencement of processing.
2. Establish sterile field using drape sheet in the laminar flow cabinet.
3. Scientist/technician scrubs up and dons sterile gown and sterile gloves.
4. Bones are removed from bags and passed to sterile field.
5. Take swab of all surfaces.
6. Using osteotomes, scissors, blades *carefully* remove remaining muscle, connective tissue.
7. Take care not to damage, score, stain cartilage surfaces.
8. When cleaning is complete take swab of all surfaces, then spray bone with antibiotic solution. Place small sterile polyethylene bag over cut shaft then insert bone into sterile polyethylene bag. Roll bag upon itself making sure to expel air. Heat seal bag. Place bag in a second bag and roll bag as before. Place bundle in sterile spun bonded olefin (Tyvec) bag. Seal bag using heat sealer. Label with identification number.

    The bones are radiographed, either immediately after packaging or after freezing.
10. Refrigerate graft for 1 hour, then place in quarantine freezer at −80 °C until all culture, serology and autopsy reports are final and the review procedure is complete.
12. If skeletal or related tissue is to be irradiated:
    (a) Arrange details of sterilization with gamma irradiation facility.*
    (b) Pack bones with plenty of dry ice.
    (c) Transport packs to facility.
    (d) Packs should be collected as soon as possible after treatment.
    (e) Bones are removed from packs and placed in a quarantine freezer.

Samples

Initial and final processing swabs must be taken from each piece of bone that is processed. Hence when bone is removed from antibiotic solution and placed on sterile field an initial swab is taken of the entire surface before any trimming or cleaning is performed. Just prior to final packaging a final swab is taken of the entire surface of each bone.

Testing

All swabs are cultured for aerobes and anaerobes; written results of all tests are kept in the donor file.

---

*It should be noted that validation studies of the dose received by the tissue would need to be performed in collaboration with the irradiation facility. This should be done before any tissue is released for use.

Other processing

As well as fresh, frozen irradiated allograft, bone may be prepared in other ways. For example, the bone may be milled to provide cancellous chips; bone may be cut into dowels, blocks or wedges. The tissue can be treated to provide demineralized bone; or the final processed bone may be lyophilized to produce freeze dried bone that can be stored at room temperature. Each of these procedures is specialized and require research, development and validation studies before they can be routinely provided to surgeons.

Milling

The bone is to be milled within 24 hours of processing; the bone should be stored at −75 °C during this time.

1. Turn on laminar flow cabinet.
2. Create a sterile field in the cabinet.
3. Scientist/technician scrubs up, dons sterile gown and gloves.
4. Open all packs, including bagged bone.
5. Set up sterile bone mill.
6. Place bone in clamp.
7. Take a swab of external surface for microbiological culture.
8. Using a file remove cartilage.
9. Cut bone to appropriate size pieces using a bone saw.
10. Mill bone, collecting it into a kidney dish.
11. Check milled bone for uniformity of size.
12. Place milled bone into sterile jars.
13. If bone is to be supplied freeze-dried proceed to lyophilization procedure. If to be supplied as fresh, frozen, irradiated allograft double bag the sterile jars.
14. A small sample of milled bone should be sent for microbiological culture.

Milled bone is usually produced from cancellous bone. The remaining cortical bone shaft may be cut into struts and packaged separately.

Demineralization

1. Place cleaned bone into a solution of 0.5 M HCl. Cover and place on rotator for a minimum of 24 hours at 4 °C.
2. Wash with copious amounts of deionized water.
3. Restore neutral pH by adding phosphate buffer (pH 7.4).
4. Rinse with deionized water.
5. Wash with 100% ethanol for 15 minutes, three times.
6. Remove ethanol and allow residual ethanol to evaporate.
7. Dry bone in dehydrator.
8. Package bone for lyophilization.
9. Sterilize by gamma irradiation.

Freeze drying

This method is based on that used at LifeNet Tissue Services, Virginia Beach, Virginia.

1. If tissue is to be freeze dried bottles containing grafts are placed under a laminar flow cabinet. A swab of each tissue is taken for microbiological culture.

2. Stopper is placed lightly on top of bottle and bottles are placed on tray.
3. Load freeze drier.
4. Cycle for small bottles of cancellous bone or corticocancellous bone:
    (a) Incubate overnight at −50 °C (shelf temperature −40 °C)   Day 1
    (b) Close baffle   Day 2
    (c) Set shelf temperature to +30 °C, turn on heat   Day 3
    (d) Observe   Day 4
    (e) Stopper each shelf   Day 5

SPLIT SKIN THICKNESS SKIN RETRIEVAL

Skin removal

1. Remove hair using a scalpel or razor.
2. Scrub area with Hibiclens for 5 minutes.
3. Remove Hibiclens by wiping with a sterile cloth.
4. Using a soft scrubbing brush scrub area with Betadine for 5 minutes.
5. Rinse area with sterile saline.
6. Apply a light layer of Betadine.
7. Record any unusual skin rash, abrasions, bruises, nevi or abnormal skin alteration.
8. Technician scrubs up and dons sterile gown, gloves.
9. Drape with sterile linen.
10. Remove Betadine with 70% isopropanol.
11. Remove skin using a sterile dermatome set to between 0.03 and 0.05 cm (0.012 and 0.018 inches). This will give medium thickness split thickness graft. The skin is removed in strips of 5 to 7 cm widths of varying lengths, but preferably approximately 30 cm. The skin is to be examined for actual thickness and appropriate adjustments made. Settings will vary with individual cases and differences in operator pressure.
12. Place procured skin in a bowl containing sterile Medium 199. Skin from each procurement area is to be kept separate. An aliquot of skin is to be cultured at the time of retrieval, together with the holding solution.
13. Discard original solution and place skin in sterile container containing Medium 199 with antibiotics (penicillin 30 μg/ml and streptomycin 50 μg/ml). Skin may be held in this solution at 4 °C until cryoprotectant is added for freezing. This should occur 1.5 hours prior to freezing and as soon as feasible after procurement NOT more than 72 hours after procurement.

Processing

1. Decant Medium 199 with antibiotics, and replace with sterile Medium 199 with antibiotics and 15% glycerol. Reseal bottle and allow between 30 minutes and 2 hours at 4 °C for glycerol to penetrate the cell matrix. Processing MUST be carried out within 2 hours of adding glycerol since prolonged exposure to glycerol prior to freezing may be harmful to cell viability.
2. Back skin onto sterile gauze or nylon netting dermal side up.
3. Trim skin to allow most even and uniform strips possible. Record grading of skin.
4. Strips are folded on themselves and placed in sterile nylon tubing which is heat sealed. This is then placed in a mylar foil bag and heat sealed. Express all air from

bags prior to sealing. Package two pieces of non-transplant quality skin for a QC sample and label as such.
5. Place bags in controlled rate freezer and freeze at 1 °C per minute. Store in vapour phase of liquid nitrogen.
6. An allquot of each solution must be submitted for microbiological culture.

Consider options that make the most efficient use of allograft and maximize the amount that can be graded as A. The allograft skin may be stored at −80 °C; or frozen in a controlled rate freezer and stored in the vapour phase of liquid nitrogen.

References

1. Eastlund T et al. *The prevalence of infectious disease markers in cadaveric tissue donors: 1992 AATB Survey results and a review.* AATB Annual Meeting, San Francisco, August 1994.
2. US Centers for Disease Control. Update: C.J. Disease in a second patient who received a cadaveric dura mater graft. *Mortality Morbidity Weekly.* Rep **38**: 37–43.
3. American Association of Tissue Banks. *Standards for Tissue Banking*, 1982, 1991.
4. Scott R. *The Body as Property.* Viking Press, 1981.
5. Allars M. *Report of the inquiry into the use of pituitary derived hormones in Australia and Creutzfeld–Jakob Disease.* AGPS, Canberra, 1994.
6. Woodruff MFA. *The Transplantation of Tissues and Organs*, CC Thomas, Springfield, USA, 1960.
7. Girdner JH. Skin-grafting with grafts taken from the dead subject. *Med Record* 1881, **20**: 119–120 (Cited by Spence RJ. The banking and clinical use of human skin allograft in trauma patients: History. *MMJ* **35**: 1: 50–52.)
8. Brown JB et al. Post mortem homografts as biological dressings for extensive burns and denuded areas. *Ann Surg* 1953, **138**: 618.
9. Rycroft PV. A recently established procedure: Corneal transplantation. In *Ethics in Medical Progress: with special reference to transplantation.* Wolstenholme GEW, O'Connor M (Eds) 1966, pages 43–53.
10. Payne JW. New Directions in Eye Banking. *Tr Am Opth Soc* 1980, Vol. LXXCVIII: 983.
11. Lyons AS, Petrucelli R. *Medicine: An illustrated history.* HN Abrams, New York, 1978.
12. Meekeren J Van (1688) Heeien geneeskonstige aanmerkingen, Amsterdam, Commelijn. (Cited by Decker ML. Bone and Soft tissue procurement. *Orthopaedic Nursing* 1989, **8**: 31–34. *See also*: De Boer H. The history of bone grafts. *Clinical Orthopaedics and Related Research* 1988, **226**: 292–298.)
13. Macewen W. Observations concerning transplantation of bone. *Proc R Soc London* 1891, **32**: 232.
14. Ross DN. Homograft replacement of the aortic vale. *Lancet* 1962, **2**: 487.
15. Barratt-Boyes BG. Homograft aortic valve replacement in aortic incompetence and stenosis. *Thorax* 1964, **19**: 131.
16. Law Reform Commission of Canada. Procurement and transfer of human tissues and organs. Working Paper 66, 1992.
17. Carrel A. The preservation of tissues and its application in surgery. *JAMA* 1912, **59**: 523. (Cited by Spence RJ. The banking and clinical use of human skin allograft in trauma patients: History. *MMJ* **35**: 1: 50–52.)
18. Webster JP. Refrigerated Skin Grafts. *Ann Surg* 1944, **120**: 421.
19. Polge C et al. Revival of spermatozoa after vitrification and dehydration at low temperature. *Nature* 1949, **164**: 666.
20. Cochrane T. The low temperature storage of skin: A preliminary report. *Br J Plast Surg* 1968, **21**: 118.
21. Trier WC, Sell KW. US Navy Skin Bank. *Plast Reconstr Surg* 1968, **41**: 543.

22. Curtiss PH, Herndon CH. Immunological factors in homogenous bone transplantation. *Annals of the New York Academy of Sciences* 1955, **59**: 434.
23. Watts LK et al. Establishment of a viable homograft cardiac valve bank: A rapid method of determining homograft viability. *Annals of Thoracic Surgery* **21** (3): 230–236.
24. Council of Europe Resolution (78) 29. Harmonization of legislations of member states relating to the removal, grafting and transplantation of human substances. Strasbourg 1978.
25. Council of Europe, Conference of European Health Ministers, Organ Transplantation: Current legislation in Council of Europe Members States and Finland. Strasbourg 1987, page 26.
26. Sadler AM Jr et al. Transplantation and the law: Progress towards uniformity. *NEJM* 1970, **282**: 717–723.
27. Sadler BL. Presumed Consent to Organ Transplantation: A Different Perspective. *Transplantation Proceedings* 1992, **24**: 5: 2173.
28. Hastings Center: Ethical Legal and Policy Issues pertaining to solid organ procurement 1985, page 15.
29. Mozes MF et al. Impediments to successful organ procurement in the required request era: an urban centre experience. *Transplantation Proceedings* 1991, **23**: 5: 2545.
30. Ross SE et al. Impact of a required request law on vital organ procurement. *J Trauma* 1990, **30**; 7: 820–824.
31. Burris EE et al. Impact of routine enquiry legislation in Oregon on eye donation. *The Cornea* 1987, **6**: 3: 226.
32. New York State Task Force on Life and the Law. Transplantation in New York State: the procurement and distribution of organs and tissues 1988, page 155.
33. Kings Fund Institute. New B et al. A question of give and take: Improving the supply of donor organs for transplantation. *Research Report* 1994, page 18.
34. US Centers for Disease Control and Prevention. *Guidelines for preventing transmission of HIV through transplantation of human tissue and organs*, MMWR **43**: RR–8. May 20, 1994.
35. Simmonds RJ et al. Transmission of HIV type 1 from a seronegative organ and tissue donor. *NEJM* 1992, **326**: 726–732.
36. Food and Drug Administration. Human Tissue intended for transplantation, *Federal Register* 1993, **58** (236): 65514–65521.
37. Thompson D. Office of Legislative Affairs, FDA, *FDA Position in Tissue Legislation* AATB Annual Meeting, San Francisco, August 1994.
38. Stroever BW. *Regulatory and Legislative Issues affecting Tissue Banking*. AATB Annual Meeting, San Francisco, August 1994.
39. Blumenschein F. *FDA Inspections of Tissue Bank Establishments under the Interim Rule*. AATB Annual Meeting, San Francisco, August 1994.
40. UK Tissue Banking Review, PO Box 96, Patchway, Bristol, BS12 6RT. Executive Officer: Mr Paul Pudlo. Personal communication.
41. Council of Europe, European Health committee. 'Draft Recommendation on Human Tissue Banking' Appendix III, Report CVSP (93) 24 (33rd meeting).
42. Lechat A. *Regulation of Tissue Banking in Belgium*. AATB Annual Meeting, San Francisco, August 1994.
43. NH&MRC, Health Care Committee. *Draft Guidelines for Donation of Cadaveric Organs and Tissues for Transplantation*, May 1994.
44. Woodruff C, Senior GMP. Auditor, Compliance Branch, Therapeutic Goods Administration, PO Box 9848, Melbourne, Victoria, 3001. Personal communication, 1994.
45. TGA *Tissue Banks Circular 2*, May 1994.
46. *BIS Annual Report* (1992). de By TMMH et al. (Ed.). Central Office: University Hospital, Leiden, The Netherlands (Tel: 31 71 182888; Fax: 31 71 149480).
47. *BIS Handbook*. BIS Central Office: University Hospital, Leiden, The Netherlands, 1994.
48. O'Brien MF et al. Protocol for human aortic and pulmonary valve collection. Cardiothoracic Surgery Unit, Prince Charles Hospital, Brisbane, Australia, 1990.
49. American Association of Tissue Banks. *Technical Manual for Tissue Banking*, 1992.
50. American Federation of Clinical Tissue Banks, Richmond, Virginia. *Operational Standards*, 1989.

CHAPTER 15

The donor family experience: sudden loss, brain death, organ donation, grief and recovery

SUE C. HOLTKAMP

Introduction 304
The donor family's experience 305
Meeting potential donor families' immediate needs 307
Factors influencing donor family grief 311
Aftercare: reducing the risks associated with donor family grief 314
Summary 320
References 320

Introduction

Transplanting human organs to save and enhance lives has captured public attention more than any other recent medical phenomenon. It has also become 'one of the most sociologically intricate and highly charged events in modern medicine'.[1] One reason for these intricacies is that 'Organ transplantation involves not only putting organs in human beings to help them live: it also necessitates taking organs from the bodies of others... .[2] This procedure is possible because of the category of patient who has been declared 'brain dead' due to the complete and irreversible cessation of brain function, including brainstem. This non-traditional way to die creates another new category of people who are closely related; they are family members, or significant others of the uniquely deceased, brain-dead patient. These survivors far outnumber donors for it is estimated that there are eight to ten individuals affected by each death.[3] During the past five years in the United States alone that would mean that more than 200,000 people have been directly affected by the deaths and the donation of organ donors.[4] Approximately, the same numbers hold true for the Eurotransplant community.[5] Tissue donor families add hundreds of thousands to these numbers. The miracle of transplantation begins when deeply traumatized families reach beyond their pain and consent to donation. Even in countries with presumed consent, the gift of life is framed by the grief of the attending family. Traditionally these families have faded into anonymity once the decision to donate was made, but that is no longer true. We recognize that transplantation is about reverence for life and to be truly meaningful, this reverence must extend to the lives of family members whose loved one made transplantation possible. Throughout the world, transplant communities are responding with increased awareness by offering the donor family ongoing care as they process their grief.

This chapter explores what these families have taught us about their experience with sudden loss, brain death, donation, grief and recovery. Donor family needs and the transplant community's response to those needs are also discussed.

The donor family's experience

Donor families come from every walk of life, rich and poor, educated and illiterate, religious and irreverent, the young and not so young. However, they share a common experience: the violation of their assumptive world.

> Each of us maintains an internal assumptive world, ... we walk through doors with confidence because we have learned at the deepest level of our mental processes that doors set off one region of solid footing from another region of solid footing. ... It is unlikely that we shall meet a door that looks like any other door but leads into an elevator shaft or an empty space where a now-demolished floor once existed.[6]

Organ donor families encounter empty space where there was once solid footing. Within this empty space these families wrestle with the sudden, untimely deaths of a loved one who, (a) experienced complete cessation of brain function, and (b) became a donor. Often, tissue donor families must also contend with the sudden loss of their loved one. Nearly 70% of these families are parents who experience the loss of a child.[7] Of course, every donor is someone's child; and since the loss of any child, regardless of age, leaves devastating sequelae in its wake,[3] it seems appropriate to consider this factor as part of the donor family experience.

DIVERSITY WITHIN UNIVERSALITY

Response to loss is determined by myriad factors.

> Death and grief, though they are universal, natural and predictable experiences that occur within a social milieu, are deeply embedded within each person's reality. ...The myths, mysteries, and mores that characterize both the dominant and non-dominant groups directly affect attitudes, beliefs, practices, and cross-cultural relationships.[8]

It is beyond the scope of this chapter to address all the variations of grieving that exist even within dominant cultures, and it should be noted that there is no one theory or psychology that applies to everyone. With various cultures, even within the same families, there will be differences which must be respected. It is the individual mourner who can best teach others about their experiences and their needs. With that understanding, the following general explorations are offered.

SUDDEN LOSS

Donors are often young, previously healthy individuals who have experienced a sudden, unexpected and often violent trauma. Family members can be overwhelmed when faced with the shock of sudden death and may be consumed with trying to master strong feelings of helplessness and the rush of affect (e.g. shock, fear, anxiety, terror and vulnerability).[3] That shock is described by one donor mother:

> When a death is sudden, the events surrounding it are cushioned for the victims. The mind and body unite to shelter the psyche from devastating knowledge. Like the yolk of an egg inside the shell, the albumin and the calcimined shell cushion the shock waves. ... I think this shocked period lasts far longer than we assume.[9]

In most cultures sudden loss is characterized by a difficulty in believing in the full reality of the event.[6] This denial, which is essential, allows the individual to absorb the reality of their loss without being totally overwhelmed. To the extent that reality penetrates the individual's denial there may be strong expressive responses such as crying, screaming or sobbing. Initial grief may also involve intense disorganization. There may be somatic responses such as chest or stomach pain, tight throat, or sighing.

The organ donor family experience with sudden loss is even more unique in that, although the death is unexpected and sudden, there is often some period of forewarning. This period of time between the trauma and the declaration of brain death may be hours or days, but it does have implications for the family and staff, for it permits certain kinds of anticipatory processes. Without this period of brief forewarning there would be no organ donation for there would be no declaration of brain death. 'For major organs ... must come from brain-dead patients who are maintained on ventilators.'[2]

BRAIN DEATH

In the beginning it may be difficult for the family to grasp the reality that the death has occurred – regardless of how this information is relayed to them. Yet, the manner in which the family is informed of their loved one's death plays a critical role in how that news is accepted and may have lasting effects on how a family adjusts to their loss.[10] The 'unreality' of the situation is often compounded when families visit their loved ones whose organs are being supported by a ventilator. These families report that acceptance of brain death is difficult when there is absence of external injuries and when there are signs of viability and normal body functions (the body is warm, colour is good and there is urine output).[11]

'I thought I understood the term brain death. But when they told me that my son was brain dead, I discovered that you don't really know what you believe until it becomes a matter of life and death.' – A donor father. 'She looked so peaceful, she was warm and pink. It looked as though she was breathing.' – A donor mother.

During this period when families are facing the greatest stress in their lives, when they are in shock, confused, disoriented and vulnerable, they must make critical decisions. For the family of potential organ donors, one critical decision is whether or not to donate. Every effort must be made to reduce further stress on these families by providing clear communication regarding the status of their loved one.

ORGAN DONATION

Donor families have indicated that donation offered them comfort in the face of a senseless death.[12-15]

'Organ donation was the only positive thing to come out of this tragic event. I am grateful that my wife was able to be a donor.' – A donor husband.

Although nurses involved in the request process perceive it to be emotionally draining,[16] it is not perceived by families as either burdensome or more stressful.[12,13,17] Indeed, Pelletier's study[15] even found that organ donation served as a coping strategy for some families. In Bartucci's 1987 study,[12] families offered the following reasons for donating:

- It felt good to help someone else.
- It offered some comfort knowing the death was not in vain.
- Donation gave some meaning to the untimely death of the loved one.

Another study[13] reported that 41% of the families felt donation made something positive come from the death. While 40% felt it would help someone (the recipient) live a better life, 19% of the participants in this study donated because their loved one had discussed their desire to donate.

Within this complex, paradoxical reality of life-giving processes and devastating grief, we have a chance to meet the important needs of donor family that, when met, can greatly enhance their ability to cope effectively with their loss. Meeting these needs must begin the moment the family enters the hospital.

Meeting potential donor families' immediate needs

Learning of the life-threatening accident or illness of a loved one can be devastating. Family members often feel helpless and vulnerable. Willis and Skelley[16] have outlined the potential donor families' immediate needs as shown in Table 15.1. Another study[15] indicates that families perceive information as vitally important. These families desired information about their loved one's condition, the diagnosis of brain death, about the recipient, and the time of complete organ removal. Other needs included emotional support, frequent visits to their loved one and opportunity to donate.

SUPPORT

Emotional support is essential and encompasses all other needs. Support is considered the most important variable in grief recovery[3,6,18] and that support should be in place

Table 15.1 The needs of the families of potential donors

1. Accurate information regarding patient's prognosis.
2. Information about determination of 'brain death'.
3. Empathy and sensitivity from both hospital and organ procurement organization.
4. Time for reconciliation to loss of loved one.
5. Time to say 'good-bye' to loved one.
6. Support for spiritual needs – hospital clergy or personal minister.
7. Assistance with arrangements such as choice of funeral homes, transportation of body or living will.
8. Reassurance and support regardless of donation decision.
9. Follow-up information about results of donation.
10. Access to additional bereavement support services.

when the family reaches the hospital.[10,19] Every effort made on behalf of the donor family weaves a safety net of emotional support, which continues when the family returns to their home.

While families listed a number of outside supports such as friends, family members, ministers, chaplains and other families,[15] intervention by hospital staff is primary to the family members' recovery[20] and should include practical as well as emotional considerations.

Concern for the family's health and welfare, contacting other family members, offering a cup of coffee or fruit juice, helps to remind the individual that others are present and attending. In spite of the shock experienced by the family there is the paradoxical element of hypersensitivity which allows traumatized family members to remember in detail how they were treated.[17,19] Families especially remember opportunities to be with their loved one.

VISITING

One of the most supportive gestures that hospital staff can offer is an opportunity for families to visit their loved one as much as possible.[10,16,19,21,22] Families will treasure these last moments.

One donor mother explains:

> As his Godmother was visiting, one of the nurses came in and told her she had to leave because she, the nurse, had a task to do with Byron. I had talked with the organ donation representative, the doctor, and my pastor and all the paperwork had been completed for organ donation. That one incident with the nurse almost made me change my mind. She seemed more task-oriented than sensitive to my family's need to be with Byron.[23]

Preparation for visual changes are also important. While Jones[24] found that family expectations are usually worse than what actually existed, each family must be prepared in advance for bodily changes or for the presence of medical equipment.[19] In one trauma unit a mother was startled to find her teenage daughter nude.

> I know it was necessary to remove her clothes to work on her but she would have been so embarrassed to be left there with only towels covering her. Medical personnel seemed to think it made no difference because she was dead, but it did.[25]

Willis and Skelley[16] urge that families be allowed to spend time with the patient outside of normal visiting hours. 'Restricting access of the family to the donor may serve to engender mistrust and bitterness toward the hospital and staff.[16] Becoming fixed on such unpleasant peripheral events or interactions may complicate the mourning process.

INFORMATION

Providing information regarding the patient's condition and prognosis on a continuous basis in a sensitive and consistent manner is another way hospital staff can be supportive of the family.[26] Information represents power and offers family members a sense of participation and control over a world that has ceased to make sense.

> Families can be much less reactionary and unco-operative when they are kept informed, in terminology understandable to them, of their 'significant other's' progress. Even when a patient's recovery seems very uncertain ... taking a few moments to talk with the family

and listen to their perceptions of what is happening can have tremendous impact on the family's acceptance of the patient's status, (including) the inevitability of death. Although it is simplistic, it is most essential that families and patients know that they are truly cared about as people.[27]

Without exception families confirm that it was vital that they receive information about their loved one's condition and their prognosis.[15,19] This information must be presented in understandable terms and may need to be repeated depending on the family member's emotional status. Informants should periodically question family members to determine what they know and how much they are comprehending.[6]

The family must be assured that everything possible is being done for their loved one.[28] Yet, they must also be advised as soon as possible of the possibility of a terminal outcome.[6,10] This information must be relayed to families in ways that will not overwhelm their already overextended capacity to understand. One simple yet effective technique involves preparing the family for the terminal outcome by communicating gradually the status of their loved one.[6] Relaying such information to the family should be done in a private and informal setting[29] for even the surroundings influence how bad news is received. 'It takes time to break bad news'[6] and nowhere is the time factor more critical than in organ donation.

Although information is a helpful coping strategy it may also seem to contribute to distress.[30] This is to be expected. Distress in this situation is not abnormal or even undesirable. Strong emotional response is to be expected as information penetrates the individual's denial and they begin to accept the reality of their loved one's condition. This acceptance is a necessary task of grieving[31] and their emotional distress is a normal response to an abnormal situation.

Information, and timing of that information in relation to the declaration of brain death is absolutely vital, not only to the success of the donation process but also for facilitating the donor family's grief.

UNDERSTANDING BRAIN DEATH

One donor mother who stated, 'When I saw Megan on the machine, I knew she was dead and only her shell remained' also confirmed the need for factual information regarding her daughter's status: 'For us, a clear explanation of brain death when we were talking to the medical personnel was helpful.'[9]

Donor families need a clear, understandable confirmation of brain death that conveys the unequivocal message that their loved one has died. The characteristic inability to accept the full reality of unexpected death is magnified when families are confronted with brain death; for the body, maintained by a ventilator, appears to be alive. Cognitive dissonance may occur when family members have to accept 'brain death' as congruent with personal knowledge of, belief about, and experience with, traditional death (i.e. absence of cardiac and respiratory function).[11]

People 'absorb information most effectively when it fits their existing views. Any information that requires radical reorientation and is, in addition, highly unpleasant, is apt to be distorted, suppressed, or exaggerated.'[6] Therefore, it is absolutely essential that donor families be given clear, understandable, consistent information regarding the true status of their loved one, including an understandable explanation of brain death.[11] One way to affirm the reality of the event for the family is to give an exact time of death.[2,32] It is, in fact, psychologically imperative for the family to be given a precise time of death.[33] Equally important to information about brain death is the timing of the request for consent to donation. The patient's status must be understood before any question of donation is proffered to the family.[12,26]

Much, if not all that has been written about timing the request for donation has been from the perspective of increasing donations. Understandably, that is a major concern for the transplant community because of the shortage of suitable donors. However, separating the declaration of brain death from the request is also in the best interest of donor families. It is both cruel and obscene to inform someone of a loved one's death and approach them for donation within the same conversation. Simply being informed of brain death is not enough; families must have time to assimilate this information, to absorb and understand the implications of what they are being told.[22,29,34,35]

Numerous problems in this area can be avoided when the family is prepared from the beginning; when they are provided honest, gradual information – which includes reassurance that everything possible has been done for their loved one – and by separating the declaration of 'brain death' from the donation option. During this time of processing the information family members must have time to view the loved one.[10,19]

Every opportunity to say 'goodbye' will help facilitate the family members' grief. If the patient is a child the parents may wish to hold it (if possible); a lock of hair or finger prints may prove very meaningful to a grieving mother or father.

Meeting donor family needs immediately after consent to donate

For too long donor families have been the forgotten factor in the miraculous equation of transplantation.[36] One family described what happened after they signed the consent form.

> From the moment we signed the paper, we felt everything change. It was if they's gotten what they wanted and we were merely in the way. ... there were no words of consolation, no acknowledgment of the horrible thing [we] were going through, and no words of thanks for the gracious gift. There was no offer of a bowl of soup or cup of coffee.[36]

It is essential that someone be with the family to provide whatever support might be needed. Some families need only a caring attitude on the part of the hospital and procurement staff while others may need practical as well as emotional support. Designated personnel should be available to assist families who choose to remain at the hospital until the donor surgery is complete.

The option for the family to see, be with, talk to, and touch their loved one before and after the donor surgery is critically important.[28] When a family accepts this opportunity to be with their loved one following donor surgery, they must be prepared for any changes in the donor's physical appearance. Families can generally accept changes of this nature when they are expecting it.

While families appreciate the option of remaining at the hospital, it may be equally important to grant them permission to leave. Care should be taken to reframe this option as 'leaving the hospital experience' rather than leaving the loved one.

During all this time families continue processing all that has happened.

> Even though bereavement is an established reality, the bereaved still need time and the chance to talk through the implications of their loss and to react emotionally. They have to prepare themselves for an event that has already taken place.[6]

This appears particularly true for families who experience sudden loss.

Attention must now turn to whether or not families have adequate emotional support on their return home and options to provide that care should be in place and offered to the family.[16] In order to provide responsible aftercare when the family leaves the hospital, a clear understanding is needed of what the donor family experiences when the shock and denial fade and they are left with the finality of their loss.

Factors influencing donor family grief

Many factors influence the individual grieving process. Some of these factors include family dynamics, prior coping skills, previous experience with loss, personality and temperament, the age and gender of both the mourner and the deceased.[3,18] The unique characteristics of both dominant and non-dominant groups within each culture must also be considered as influencing factors.

The same factors, sudden loss, combined with parental loss of a child, which influenced the family's experience prior to donation, may place family members at additional risk for complicated mourning.[3] Confrontation with brain death and consenting to organ donation, while not known risk factors of complicated mourning, are unique to the donor family experience and an understanding of how these issues impact on the grief process is warranted.

PARENTAL LOSS OF A CHILD

'The death of a child is perceived as a death out of season, a monstrosity, an outrage against the natural order of things ...'[37] This loss is one of life's most difficult experiences. The suffering generated by the death of a child is intense and long lasting.[38,39] It is the only grief that may increase rather than abate with time.[40] In one study Rando[38] reports an upsurge in parental grief during the third year, while another study[39] reports unabated pain through the first thirty months of bereavement following the death of a child. The leading cause of death for those children and all other donors is some form of sudden, unexpected trauma.[4]

SUDDEN LOSS

Those who suffer sudden, unexpected loss may experience a number of potentially complicating issues which are listed in Table 15.2.[41] The mourner's response to sudden loss must be understood within the context of the mourner's experience and background. Some cultures and some religions place a high value on stoicism, while others regard external displays of emotions as a vital part of their grief.[8] Some individuals are extraordinarily accepting while others meet their fate with considerable emotional expression. Each will respond in keeping with their own frame of reference.

Usually it is extremely difficult for individuals to accept a world where tragedy can occur so arbitrarily. This difficulty in accepting the death may lead members of the donor family to question the concept of brain death or the decision to donate.

In most cultures death following a long and productive life is considered more timely than when a person dies in their youth or in the prime of their life. Most donors die

Table 15.2 Factors complicating bereavement after sudden loss

1. The survivor must often deal with a pervasive sense of unreality about the loss.
2. There is a need to blame someone for the death.
3. There is often an exacerbation of guilt feelings.
4. Involvement with legal and medical systems is common. This involvement often keeps the mourners from dealing with their grief on a firsthand basis.
5. There may be a sense of helplessness linked with an incredible rage.
6. Sudden death may trigger a 'fight or flight' response in a person.
7. When death is unexpected there is more likely a chance that the survivor will have to deal with unfinished business.
8. There is often an increased need to understand; not only the need to ascribe cause but to blame.

untimely deaths which creates an added risk for complicated mourning for donor families.[3,31]

Another possible complicating factor for the donor family involves the mode of death. In any tragedy some individuals cope better than others. However, the 'nature and circumstance of a death can be so powerful as to bring complications to even the healthiest mourner.'[3]

The nature and circumstance of the deaths that lead to organ donation are often that powerful. During the period from 1988–1993, UNOS[4] reported that, in the United States, almost twice as many donors died from some form of violence compared with the number who died from what might be perceived as natural, albeit, untimely forms of death. These unnatural deaths leave the donor families feeling violated, victimized, and vulnerable.

BRAIN DEATH AND ORGAN DONATION

While organ donation is a remarkable means of bringing something of value from an otherwise tragic event[11,35] and the act of donation offers some solace to the grieving family, it also involves a paradoxical phenomenon. Two donor mothers poignantly express that reality:

> It gives me great joy to know that the recipients are doing fine and my son's death gave life to others. The death was one of the most heartbreaking, painful experiences I could ever have ... and I still feel an empty hurt.[23]

> There has been no protection from the agonizing pain of our loss, yet we found enormous comfort from the knowledge that Jamie's last gift brought life and joy to another family. [A donor mother.]

These mothers touched on a difficult concept: that two, apparently contradictory realities can co-exist. That one can have 'great joy' and an 'empty hurt', 'agonizing pain' and 'comfort'. To understand fully the donor family experience we must accept this paradox. We must not romanticize the donation process nor be cavalier about its benefits. While most families are grateful that at least something positive comes from their loss, that 'something positive' also means that their child, mother, spouse or other loved one is dead. Organ donation does not eliminate the pain of grief.[25,42] Nor should it, for in most cultures, grief pain is considered a necessary part of the healing process.[8,31]

While donation can have a part in that healing process, it is only a part. Indeed, for some families, donation is only a peripheral aspect of their experience.[22] For others the donation of a loved one's organs becomes an integral issue in the mourning process. For instance, many families report how important it was for them to receive a 'Thank you' from the recipient or his/her family. 'I was bitter', writes one family member, 'that I didn't receive a personal thank you. I spent a day writing thank yous for stupid flowers. You'd think the gift of life would be appreciated more than flowers.'[43] Another mother wrote, 'I had doubts that I had done the right thing until I started receiving thank yous from recipients.'[44]

Donation sometimes provides a sense of immortality for the loved one.[12] One donor mother summed up her feelings by referring to ongoing knowledge about the recipient as 'that tiny immortality for Meg, in some small way, to live out her allotment of time'. This, she added had 'a comforting quality.'[9]

The 'comforting quality' associated with donation is reinforced when donor families receive information about recipients of their loved ones' organs and tissue.

Ongoing information regarding the recipient's well-being appears to be of great importance to the donor family[13] and should be relayed to the family as soon as possible.[44] Updates on the recipient's condition should be available to donor families on request. This thirst for affirmation of meaning within the context of donation may be more prevalent within cultures that emphasize altruism. While information concerning the recipient's well-being can be a source of comfort, it can also be fertile ground for disappointment and added pain should the recipient die.

Confusion regarding brain death

While there is no clear indication that donation complicates mourning, studies indicate that some families donate without fully understanding the concept of 'brain death'.[11,13] This confusion has the potential to complicate the normal grief process if family members become fixated on perceived guilt which is often an attempt to 'undo' the loss.[31]

One mother who praised the explanation given at the hospital later experienced doubt:

> I have to say that I have wondered since whether or not keeping her on the machines and hoping that something amazing would happen might have taken place, and that we might have made the wrong choice. People are fallible. I don't spend a lot of time wondering about it and I don't have nightmares about it, and I think that, by and large, I am able to process the probable fact that nothing could ever have been done. However, those darkest days bubble it up.[9]

A written explanation of brain death and ongoing contact with the hospital or transplant community may lessen the possibility of serious complications in this area.

Prolonged attachment to the recipient

While there are no known studies regarding the risk of prolonged attachment to the recipient, we do know some families continue their interest in the recipient for a year or longer and anecdotal information indicates some potential for risk.[44,45] A concept developed by Vamik Volkan, a psychiatrist at the University of Virginia Medical Center, who has written widely on the subject of pathological grief may be worth exploring on behalf of the donor family.[46] Dr Volkan describes 'linking objects' as inanimate objects which the mourner has invested with symbolism that establishes a link between the mourner and the deceased. Volkan distinguishes these types of objects from mementos or keepsakes by the intensity with which the mourner is attached to them. Linking objects are invested with much more meaning and 'cause a great deal more anxiety when they are lost'.[31]

Since this phenomenon 'can hinder satisfactory completion of the grieving process'[31] it is important to consider whether recipients might, in some cases, qualify as linking objects. Though recipients are not inanimate, they are, in a unique and important sense, a link to the deceased. It is conceivable that their loss could be met with considerable anxiety by donor families who invested them with great symbolism. Prolonged attachment to the recipient could also lead to chronic or unresolved grief. The 'gift of life' may have the potential to 'fetter as well as free'.[45]

Premature restitution

Donation may be thought of as a form of restitution – the act of bringing something of value from tragedy. Yet this particular restitution comes before most families have had time fully to accept their loved one's death. While restitution can be made at any time it is generally thought of as belonging to the recovery phase of grief.[47] Because of the timing of the donation, the anonymity of the recipient, the possibility that the recipient might die, and because restitution is, in reality, an ongoing process, it may be particularly important for donor families to find additional forms of restitution.

Situations involving serious complications such as these must be referred for clinical intervention; they cannot be dealt with in the supportive setting. However, these potential risks, along with other complicating factors, may be reduced to some degree by providing supportive aftercare for the donor family.

Aftercare: reducing the risks associated with donor family grief

There are a number of reasons for transplant communities to develop aftercare support programmes for their families. One compelling reason is the growing sensitivity to the devastating sequelae that follow a traumatic loss. As previously discussed, these sequelae have special significance for donor families.

Sudden, unanticipated and untimely loss can injure functioning so severely that uncomplicated recovery may no longer be expected.[6] Bereaved parents have been described as having the 'appearance of individuals who suffered a physical blow [which] left them with no strength or will to fight, hence totally vulnerable'.[48] Lack of support during such traumatic experiences may leave lasting marks on the individual's adjustment and functioning.[49] 'One strand appears in the fabric of all the cultural groups in relation to death: the importance of community or group support'.[8] This support is especially needed because many families have fewer natural supports to draw on as they attempt to cope with one of life's most devastating events. The lack of interaction with the extended family, the mobility of families with the resulting lack of community, and the secularization of religion are some reasons for the unavailability of natural support systems to modern families facing trauma.[3]

This universal need for support following loss[8] and the lack of such support in the modern world[3] offers the transplant community an opportunity to provide ongoing care for the donor family that makes an influential statement to the entire community. For just as the donor's gift 'honors important human values'[12] so does offering the donor's family 'compassionate assistance in coping with the patient's death and the subsequent bereavement process'.[26] The most compelling reason of all for offering ongoing care to donor families may be found in the simple words of a donor mother: 'I donated because it was the right thing to do'. Addressing the needs of donor families is also the right thing to do.

GOALS AND LIMITATIONS

Providers of aftercare services must be ever mindful of the limitations inherent in such programmes. Aftercare programmes are generally support programmes. They do not attempt to offer therapy but instead, provide information, education, a forum for discussion and ongoing support.

These programmes affirm the gift of the donor, remind donor families that they have not been forgotten, and offer the transplant community an opportunity to 'act as though you care'.[6]

APPROACHES USED TO MEET DONOR FAMILY NEEDS

Bereavement support programmes will vary even within cultures. Some may be quite comprehensive while others will consist of a simple gesture, such as a referral service. The type of programme chosen by a sponsoring agency will be determined, in part, by the geographical area, population density, ethnic make-up, and resource availability of the area served by that agency. Most programmes will begin by designating or hiring someone to develop and implement the programme.

The bereavement specialist

When an organization makes a commitment to provide bereavement care, a person or persons must be designated to provide this service. There are important requisites for the bereavement specialist, whose attitude and commitment will influence the effectiveness of the entire programme. Whether this individual is recruited from within the sponsoring organization or hired to provide bereavement care on a contracted basis, this individual must meet a number of criteria which have been listed in Table 15.3.

Written material

Because newly bereaved individuals often have difficulty concentrating and remembering, they usually appreciate written material, which allows them to read and reread helpful information. This literature also offers support without requiring a response on the part of the reader. Appropriately designed and written material also introduces the bereavement specialist to the family. Most importantly, written material offers the most effective way to reach isolated families who may be unable to secure other services.

Bereavement literature should provide accurate information about the normality of the grief process and offer realistic encouragement and reassurance for the grieving

Table 15.3 The bereavement specialist – criteria for success

1. He or she must have come to terms with death and at least have started to come to grips with his/her own mortality.[50] Grief is not only difficult to experience, it is difficult to witness.[51] The bereavement specialist must be comfortable with this reality.
2. He or she must have effective listening skills and the ability to respond appropriately.[50]
3. He or she must have a commitment to the programme. Bereavement programmes require a considerable amount of time and hard work.
4. He or she must exercise sound judgment and awareness of limitations. What can and can't be done for families; when to take time away from grief and grieving and how to avoid burn-out.[50]
5. He or she must be knowledgeable about the grief experience. In order to reassure the mourner and normalize their experience, the helper must understand the intensity of grief pain. Understanding the attitudes, beliefs, and practices of both the dominant and non-dominant groups within a culture[8] also establishes rapport between the bereavement specialist and the donor family member.

    While this familiarity with the grief process is necessary for rendering care, it is also essential for developing policies and procedures, and for developing and timing any written component of the programme.

---

individual. It should be written in simple, easy-to-understand language that is appropriate for the people for whom it is intended. The visual presentation of bereavement literature makes a powerful statement before the first sentence is read. The appearance of the material should reflect that it was chosen and prepared with discernment and care. The timing for sending this material to the family will depend on what types of literature are being sent and on the cultural content of the organization sending it.

Who receives support?

Ethically and morally, the programme should be offered to families even when the patient is medically unsuitable for donation or when the family declines organ donation. In addition to the ethics and the helpfulness for the families this practice serves as a positive public relations gesture within the community.

Telephone support

Many programmes offer telephone support. The initial call is usually made two to three months after the death. In addition to offering supportive concern, this call allows the specialist to assess the effectiveness of the written material, and whether there are any problems that require intervention from the specialist.

Memorial programmes

Memorial programmes are valuable ways of remembering the donor and the exquisite gift that saved or enhanced another's life. These programmes serve as rituals which offer donor families a unique and structured way to process their grief. Memorial programmes may include special tributes, music, and expressions of appreciation. Such programmes may be combined with the dedication of a monument, a plaque or a park in honour of the donors. Memorial programmes should be offered regularly to include new donor families and should focus on the donor and the donor's contribution to the well-being of others. To be truly effective the ritual must be meaningful for the families and their inclusion in planning the memorial is essential.

Memorials reinforce the message to the donor family that their loved one has not been forgotten, and that their death has meaning on a societal as well as personal level.

Private counselling

In 1986 Tennessee Donor Services (TDS) in the United States began offering donor families individual grief counselling. While this programme has been well-received and effective, it is one of the most expensive forms of intervention.[44] However, aftercare programmes should offer referrals to necessary outside resources, including private counsellors. Ideally, the names of several counsellors should be offered in order to provide choices for the family.

Support groups

'Support groups are an invaluable source of information and emotional reassurance for individuals who have experienced loss'.[52] The bereavement support group meets with the common purpose of finding ways in which members can help themselves and each other. When there are not enough donor family members to constitute an ongoing group, the group may be opened to all in the community who have suffered a sudden, untimely loss. In other instances, it is in the best interest of all concerned to refer donor families to established support groups in the community. By making referrals to already existing groups, donor programmes have another opportunity to network with the community while assuring adequate care for their own families. Each programme must assess the needs of their families, the resources of their communities, and their own resources, and then attempt to meet those needs in the most effective and efficient manner.

EXISTING DONOR FAMILY SUPPORT PROGRAMMES

Approximately 85% of the organ procurement organizations in the United States offer some form of aftercare programme[53] and similar programmes are being developed in many other countries. Some offer long-term (one to two years) care with a wide range of services, while others limit their programmes to sending a donor family booklet or other pieces of bereavement literature.

The Organ and Tissue Donor Services at the Medical College of Georgia is probably the earliest organization of its kind to offer bereavement care for donor families.[54] The programme is called, 'People Supporting People' and is offered to families of organ, tissue and whole body donors. This programme covers a wide geographical area and uses written material, appropriately timed telephone support, and referral services to follow their families. A Family Support Co-ordinator manages the programme with help from volunteers.[54]

Another of the earliest aftercare programmes for donor families began in the United States at Hartford Hospital, Hartford, Connecticut.[55] This programme covers all families in the area who have experienced sudden loss. Families are followed by trained volunteers who have experienced a similar loss. These volunteers call the family two or three months post-death and with the permission of the family, make additional calls every one or two months as well as sending notes or cards on special occasions. A 24-hour 'hot line' is also available to these families and families are furnished information about resources such as local support groups and professional counsellors.[55]

Two well-received and valuable pieces of bereavement literature used by this group were developed by the National Kidney Foundation in the United States: *For Those Who Give and Grieve* written by and for donor families, and the National Donor Family Council's *For Those Who Give and Grieve* newsletter. The National Donor Family Council is made up of donor families and experts in death education, organ procurement and support services. The Council 'seeks to help families through a variety of activities including advocacy, support, research, and education.'[43] This Council, along with bereavement experts, developed the Donor Family Bill of Rights: a timely and valuable advocacy statement.

In 1986, Tennessee Donor Services (TDS), an organ procurement organization in the United States, contracted with a private bereavement programme, 'Something More', to develop a multifaceted programme that offers written material, community workshops, professional education, and individual/famly counselling for donor and referral families.[14]

Certified grief councellors who have experienced the loss of a significant loved one, follow the donor families. Because of the devastating sequelae that often follow sudden loss, the potential risks for complicated mourning and the availability of qualified counsellors, TDS also chose to offer individualized counselling for donor families. A booklet entitled, *Something More for the Donor Family*, and several other pieces of bereavement material are used. Bereavement material is sent throughout the first twelve months and each family is telephoned personally. A freephone number for the grief counsellor is offered to each family.[14]

The TDS programme also provides workshops for the community, a support group, workshops for healthcare professionals and an annual 'Something More for the Holidays' memorial programme each December. Like most other bereavement programmes a card is sent to each family near the anniversary of the donor's death.[14]

Improving donor family support by educating those healthcare professionals involved in the request and donor process is a part of the TDS programme and the Australian Donor Awareness Program for Transplantation (ADAPT) as well. This is an excellent method for providing bereavement care when the area served by an organization is widespread. Not only can such a method improve the initial care of the donor family, but these programmes have the added benefit of addressing the needs of healthcare professionals.

The ADAPT programme aims to 'increase knowledge and awareness of organ donation and grief counseling amongst health professionals dealing with brain-dead patients and their families so as to facilitate a positive outcome for all involved'.[56] The course covers assessment of normal and abnormal grief, reactions to acute grief and needs of bereaved people (see Table 15.4).[56] This type of programme is particularly important because the grief process begins in the hospital and the family's experience there influences the outcome of their grief.

Another programme in New South Wales, Australia[57] includes bereavement literature and a 24-hour telephone number which families are encouraged to use to reach the transplant co-ordinator. The bereavement co-ordinator sends each family an initial letter aimed at normalizing the grief experience and encouraging contact with the bereavement co-ordinator. Cards are sent to donor families on the anniversary of the death of their loved-one and a well-established, yearly Ecumenical Thanksgiving Service has also been enormously well-received.[57]

Table 15.4 The Australian Donor Awareness Programme for Transplantation

The ADAPT programme aims to assist health professionals working in critical care areas:
1. To promote healthy adaptation to grief and loss for adults and children.
2. To develop personal and general methods of helping relatives of brain-dead patients.
3. To develop their skills and knowledge of the processes involved in organ donation and transplantation.[56]

While each of the programmes discussed here is slightly different, each has been designed to provide care for donor families. As other programmes have developed throughout the transplant communities of the world, some have duplicated the older, more established programmes outlined here. But these newer programmes are also bringing innovative ideas to the care for their families. This aspect of aftercare is critically important. While organizations must build on a solid foundation of knowledge about the donor family and the grieving process, there must also be great creative latitude for programmes to develop their own style. Those working directly with families breathe life and vitality into their programmes which make them authentic; this quality is especially important when working with bereaved individuals.

Just as donation does not 'take away' grief, neither do aftercare programmes. These programmes do serve the vital function of reducing isolation, providing accurate information, and offering the donor family members ongoing support, but they are as effective as the participants perceive them to be.

EFFICACY STUDIES OF DONOR FAMILY SUPPORT PROGRAMMES

Assessing the effectiveness of bereavement programmes is difficult, in part because grief never moves in a linear fashion. Dr Vachon, a nurse counsellor with extensive experience in support and research of the bereaved, frames the reality of efficacy studies of bereavement programmes with candour and wisdom.[58]

> There is no definitive way in which to measure the exact helpfulness of a specific support attempt. Therefore, social support may be construed as a property of the individual, since intervention must be seen as helpful by the recipient in order to be supportive. While structural and functional characteristics of a social network will influence the potential availability of support, it is the individuals' appraisal of network transactions that determine whether help has been provided in the face of threat.[58]

One of the most common research methodologies, the descriptive method, is often used to learn from 'the personal experience of a given individual as reported by that individual'[59] and has the ability to determine the 'effects of treatment'.[60] Such a study was used to evaluate the Tennessee Donor Service Aftercare Program.[14] This and other studies of various donor family programmes have indicated that aftercare programmes have been perceived as helpful by an overwhelming number of participants.

While respecting the knowledge that is beginning to accumulate, we must not limit our evaluations to traditional research methodology. For wherever donor families have an opportunity to expressed their complex, often ambivalent, sometimes contradictory feelings, they weave a rich tapestry of valuable information as they express both joy and pain.

A closing observation regarding outcome studies of bereavement programmes may be particularly fitting:

> Life is ongoing and so are outcomes. The program ('Widow to Widow Project') was successful to the extent that it helped people through times of distress; showed them they could carry on with some joy, pleasure and excitement; and supported them as they met the problems of daily living with a somewhat different point of view.[61]

Bereavement aftercare programmes for donor families can also help donor families at times of great distress, encourage them to believe that they can carry on, and support them in such a way that leaves no doubt that their sorrow is shared, their gift appreciated.

Summary

While clinical technology associated with transplantation has made remarkable strides, much remains to be learned about the psychosocial impact of donation on families of donors. Organ and tissue transplantation forces the biomedical world to truly confront the whole person: the psychological, the social, and the physical being. It is in this arena of life and death that we become startlingly aware of human values, beliefs and concerns that often resist being reduced to mere analytic study.

Organ and tissue transplantation is about life, yet it is also about loss. There would be no transplantation without that loss and the inevitable grief that follows. The generosity of donor families must not be forgotten, and their loss must be acknowledged. Responding to their loss:

> ... demands an openness of mind to what is observed and what is unseen, what is known and what is implied, to the logical and to the irrational, to the concrete and to the symbolic, to the practical and to the mystical aspects of life. It calls forth in the helper the poetic as well as the scientist, the personal as well as the professional, and it bridges the gap between helper and the one who needs help. In recognizing the universality of loss and grief, we come to appreciate the commonality of all peoples. Loss removes barriers of race, color, creed, class, religion, sex; it is a leveler.[47]

By offering the 'gift of our presence' through supportive care for donor families, we share in the remarkably complex and creative process of loss, transplantation, and recovery.

ACKNOWLEDGMENTS

With grateful appreciation to Helen Barrett, donor family member and editor, Lyn Simpson, typist and editor for their able assistance, and to staff members from organ procurement organizations and tissue banks who contributed their knowledge and encouragement to this project.

References

1. Swazey J. Transplants ... the gift of life. *Wellesley* (Alumni Magazine) 1986, **70**: 15.
2. Youngner S. Organ Donation and Procurement, In *Psychiatric Aspects of Organ Transplantation*. Craven J, Rodin GM (Eds), Oxford University Press, Oxford, 1992, pages 121–130.

3. Rando TA. *Treatment of Complicated Mourning*. Research Press, Champaign, Illinois, 1993.
4. *United Network of Organ Sharing Update*. United States, 1994, **10**: (2).
5. Cohen B, Persijn G. European transplant programs – what can we learn? *J Transplant Coordination* 1993, **3**: 128–133.
6. Parkes CM, Weiss RS. *Recovery from Bereavement*. Basic Books, New York, 1983.
7. Kirste G, Muthny FA, Wilms H. Psychological aspects of the approach to donor relatives. *Clinical Transplantation* 1988, **2**: 67–79.
8. Irish DP, Lundquist KF, Nelsen, VJ. Conclusions. In *Ethnic Variations in Dying, Death, and Grief: diversity in universality*. Irish DP, Lundquist KF, Nelsen VJ (Eds), Taylor and Francis, Washington 1993, pages 181–190.
9. Barrett H. Personal correspondence with S. Holtkamp PhD, 1994.
10. Johnson C. The nurse's role in organ donation from brain dead patients: management of the family. *Intensive and Critical Care Nursing* 1992, **8**: 140–148.
11. Pelletier M. The organ donor family members' perception of stressful situations during organ donation experience. *J Advanced Nursing* 1992, **17**: 90–97.
12. Bartucci MR. The meaning of organ donation to donor families. *Anna Journal* 1987, **14**: 369–371, 410.
13. Savaria DT, Rovell MA, Schweizer RT. Donor family surveys provide useful information for organ procurement. *Transplant Proceedings* 1990, **3**: 316–317.
14. Holtkamp S, Nuckolls ES. Completing the gift exchange: a study of bereavement service for donor families. *J Transplant Coordination* 1993, **3**: 80–84.
15. Pelletier M. The needs of the family members of organ and tissue donors. *Heart & Lung* 1993, **22**: 151–156.
16. Willis R, Skelley L. Serving the needs of donor families. The role of the critical care nurses. *Critical Care Nursing Clinics of North America* 1992, **4**: 63–77.
17. Signa Report. Marketing study commissioned by Tennessee Donor Services. Nashville, TN. Strategic Marketing Services, 1989.
18. Raphael B. *The Anatomy of Bereavement*. Basic Books, New York, 1983.
19. Back K. Sudden, unexpected paediatric death: caring for the parents. *Pediatric Nursing* 1991, **17**: 30–41.
20. Buschbacher BI, Delcampo RL. Parents response to sudden infant death syndrome. *J Pediatric Health Care* 1987, **1** (2): 85–90.
21. Coolican M. Caring for donor families: a unique perspective. *Making the critical difference*. The National Kidney Foundation, 1990, Video.
22. Coupe D. Donation dilemmas. *Nursing Times* 1990, **86**: 34–36.
23. Norton V. Life ... even after the tragedy. *The Forum Newsletter* 1994, **XX**: 1, 17.
24. Jones WH. Emergency room sudden death: what can be done for survivors? *Death Education* 1978, **2** (1): 231–245.
25. Holtkamp S. Private communication with donor families, 1994.
26. Partnership for Organ Donation, Inc. and The Annenberg Washington Program. *Solving the donor shortage by meeting family needs: a communications model*, 1990.
27. Wright J. Toward a common goal. *Heart Lung* 1978, **7**: 978–979.
28. Gyulay JN. Sudden death – no farewells. *Issues in Comprehensive Pediatric Nursing* 1989, **12**: 71, 102.
29. Perkins K. The shortage of cadaver donor organs for transplantation. Can psychology help? *American Psychologist* 1987, **42** (10): 921–930.
30. Pelletier M. Emotions experienced and coping strategies used by family members of organ donors. *Canadian Journal of Nursing Research* 1993, **25**: 63–73.
31. Worden JW. *Grief Counseling and Grief Therapy*. Second edition, Springer Publishing, New York, 1991.
32. Kinney S. Care of the donor family. *Confederation of Australian Critical Care Nurses Journal* 1990, **3** (3): 2.
33. Jasper J, Harris D, Jackson R, Lee B, Miller K. Organ donation terminology: are we communicating life or death? *Health Psychology* 1991, **1** (10): 34–41.

34. Lange S. Psychosocial, legal, ethical, and cultural aspects of organ donation and transplantation. *Critical Care Nursing Clinics of North America* 1992, **4**: 25–42.
35. Batten HL, Prottas JM. Kind strangers: the families of organ donors. *Health Affairs* 1987, Summer: 35–46.
36. Helmberger PS. *Transplants: unwrapping the second gift of life.* Chronimed Publishing, Minneapolis, 1992.
37. Viorst J. *Necessary losses.* Simon & Schuster, New York, 1986.
38. Rando TA. *Parental Loss of a Child.* Research Press, Champaign, Illinois, 1986.
39. Miles MS. Emotional symptoms and physical in bereaved parents. *Nursing Research* 1985, **34** (2): 76–81.
40. Pine VR, Bauer C. Parental grief: a synthesis of theory, research, and intervention. In *Parental Loss of a Child.* Rando TA (Ed.) Research Press, Champaign, Illinois, 1986, pages 59–96.
41. Worden JW. *Grief Counseling and Grief Therapy.* First edition, Springer Publishing, New York, 1982.
42. Couch NS. A divided person, part nurse, part parent. *Nursing* 1991, August: 48, 49.
43. Coolican M. After the shock: support for donor families. *The Forum Newsletter* 1994, **1**, 17.
44. Holtkamp S. *A study of the perceptions of donor families toward the bereavement aftercare program offered by Tennessee Donor Services.* Unpublished Doctorial Dissertation, 1991.
45. Fox R, Swazey C. *The Courage to Fail.* University of Chicago Press, Chicago, 1978.
46. Volkan DV. *Linking Objects and Linking Phenomena: a study of the forms, symptoms, metapsychological, and therapy of complicated mourning.* International Universities Press, New York, 1981.
47. Simos B. *A Time to Grieve.* Family Service Association of America, New York, 1979.
48. Sanders CM. A comparison of adult bereavement in the death of a spouse, child and parent. *Omega* 1979–80, **10** (4): 303–320.
49. Van der Kolk BA. *Psychological Trauma.* American Psychological Press, Washington DC, 1987.
50. Rando TA. *Grief, Dying and Death.* Research Press, Champaign, Illinois, 1984.
51. Bowlby J. *Attachment Loss: sadness and depression.* Basic Books, New York, 1980.
52. Garrett JE. Multiple losses in older adults. *J Gerontological Nursing* 1987, **13** (8): 8–12.
53. Coolican M, Savaria D. Professional correspondence to United States organ procurement organizations, 1992.
54. Durham D. Professional communication, Georgia Medical College, 1994.
55. Coolican M. Caring for donor families: a unique perspective. *Making the critical difference.* The National Kidney Foundation, 1990, Workbook.
56. Bullen P. Personal correspondence, 1994.
57. Barnwell A. Personal correspondence, 1994.
58. Vachon MLS. The role of social support in bereavement. *J Social Issues* 1988, **44** (3): 175–190.
59. Francis JB. *The proposal cookbook.* Action Research Associates, Naples, Florida, 1979.
60. Lazarsfeld S. The courage for imperfection. *J Individual Psychology* 1966, **22**: 163–165.
61. Silverman PR. *Widow to widow.* Springer, New York, 1986.

PART 3

Methods of increasing organ and tissue donation

CHAPTER 16

Transplant coordinators

BARBARA A. ELICK

Introduction 325
Definition of transplant coordinator 325
Procurement transplant coordinators 326
Clinical transplant coordinators 327
Growing professionalism of transplant coordinators 329
Conclusion 330
Appendices 331
References 342

Introduction

Organ and tissue transplantation is one of the most extraordinary and rapidly growing medical technologies of the twentieth century. It has given rise to novel healthcare structures and complex issues. In just the last three decades, transplantation has become increasingly available to the general population, due in large part to universal efforts to increase organ and tissue donation. Crucial to this expansion are the efforts of health care professionals known worldwide as transplant coordinators.

This chapter begins by defining the multifaceted role of transplant coordinators from both a historical and a geographical viewpoint. It then differentiates between two distinct but interrelated job emphases: *procurement* (largely concerned with donor awareness and organ and tissue recovery) versus *clinical* (oriented toward care of recipients). Finally, it discusses the growing professionalism of the field, as evidenced by the advent of organ sharing systems, a network of national organizations, a formal certification process, and a peer-reviewed journal.

Definition of transplant coordinator

'Transplant coordinator', broadly defined, has been the title given to persons who facilitate the transplant process – from donor identification and organ procurement to implantation and follow-up recipient care. When transplantation first became a reality in the early 1960s, the coordinator role often took on a meaning of its own, depending on the employee's transplant center, past medical experience, and educational background. Frequently the job description was all-inclusive, encompassing the dual responsibilities of procuring organs and caring for recipients.

In the United States during the 1960s, assistants in the transplanting process did not have the title of transplant coordinators. In the early 1970s, with the rising popularity of perfusion machines to maintain viability of cadaver kidneys, coordinators were sometimes labelled preservation or pump technicians. At the same time, those working with potential kidney recipients were called recipient, clinical, or nurse coordinators. But in a number of neophyte programs, they were one and the same person.

As transplant programs grew in size and graft survival rates increased, so did the demand for transplants. To keep up with this demand, the responsibilities of transplant coordinators expanded. Donor or procurement coordinators began to concentrate more on the elaborate donation and procurement process, including community and professional education. Likewise, those working with transplant candidates took on additional patient education tasks and helped assess ever-evolving immunosuppressive protocols (such as total lymphoid irradiation and thoracic duct drainage), various pharmacologic therapies (such as antilymphocyte globulin), and operative techniques (such as pre-transplant splenectomies).

In Europe the first known coordinator position was established in the Netherlands in 1979.[1] Soon thereafter, transplant coordinators were appointed in the United Kingdom, and in 1983, in Sweden.[1,2] During this same period, a number of newly created coordinator positions were filled in Australia, South America, and other countries with developing transplant programs.[3-7] In 1986, full-time transplant coordinators began working in Singapore, where a subsequent rise in kidney transplants was attributed, in part, to these appointments.[8] Responsibilities of these early transplant coordinators centered on organ donation. But as late as 1987, some of them were still functioning in dual roles. working with donors as well as recipients, while other programs had already delineated two areas of expertise handled by different people.[2,3,7]

But whatever extra adjectives have been used, in whatever part of the world, the word 'coordinators' has always implied the act of working together harmoniously. Ensuring the most efficient collaboration between the donor and recipient transplant teams was and is crucial to the success of transplantation.

Procurement transplant coordinators

In the 1970s, as transplants became more frequent, additional people were hired to assist with organ procurement as well as to prepare recipients for transplantation. Before then, because of the controversy surrounding the definition of brain death and the lack of legislation supporting it, kidneys were typically removed after cardiac arrest in cadaver donors. Kidneys then suffered from warm ischemic damage, compromising initial function. But with the creation of hypothermic pulsatile perfusion pumps and cold solutions that sustained kidney viability, and with the enactment of the Uniform Anatomical Gift Act of 1971, more organs could be utilized.[9,10,11] More staff members were needed to expedite this growing effort.

Often, these new appointments were the technicians working in the animal laboratories. Others were nurses or health paraprofessionals. They usually worked on a part-time basis, primarily when a potential organ donor had been identified. They had to be available around the clock to respond to the call that would identify potential donors. Their first task was to meet with the family of a potential donor to obtain permission to 'harvest' (as it was commonly called) the kidneys. If permission was granted, the coordinator might remove lymph nodes at the donor's bedside, deliver them to the immunology laboratory, schedule the organ removal, assist in the surgery, and ultimately flush and cool the kidneys. The coordinator would then either place the kidneys on a portable perfusion pump or package them for transport to another destination. When the pump was used, the coordinator would monitor perfusate flow and electrolyte balances – sometimes with little sleep for a number of days – until the kidneys were transplanted. The coordinator usually accompanied kidneys on perfusion machines to their destination, often meeting the accepting center's transplant coordinator. Coordinators frequently conversed over the phone at odd hours of the night and met in airports after being awake for long hours. Because of this networking, coordinators grew familiar with each other's programs and developed friendships. By the time a small group of them met in Los Angeles in 1974, many of them already had long-standing professional relationships.

During those early surgical organ retrievals, transplant coordinators became managers, not by choice but by necessity. They were called on to coordinate multiple teams in removing corneas, skin, bone, and other tissues. Coordinators schooled in the organization of *en bloc* nephrectomies soon found themselves developing protocols to optimize removal of extrarenal organs. It became commonplace to coordinate the efforts of numerous surgical teams. In Europe, coordinators faced the additional challenge of dealing with the many different languages spoken by various surgical teams. Developing leadership skills often afforded the opportunity for experienced coordinators to move into management positions and direct overall efforts to increase organ and tissue donation.

Procurement coordinators were employed, for the most part, by the transplant programs that used the organs. But early on, the issues surrounding donors and their families were recognized as separate from those involving recipient care. Independent organ procurement agencies have now become the norm in the USA.[12] Worldwide, however, many transplant programs continue to operate procurement protocols out of hospital-based transplant programs.[13]

Clinical transplant coordinators

The first clinical transplant coordinators in the USA tended to be nurses working with dialysis patients or in the operating room. In the 1960s, many of them began their careers in organ procurement, but soon acquired the additional responsibility of maintaining the waiting list of potential recipients. Often, when a kidney became available, the coordinator chose and called the most appropriate candidate – frequently in the middle of the night. In that early era, many candidates were transplanted according to blood typing alone since histocompatibility markers were uncommon. Only a small number of HLA antigens had been determined at that time, but soon thereafter, became a mainstay for patient selection.[14,15]

As transplantation flourished, clinical transplant coordinators were increasingly called on to manage the complicated care of patients, both in preparation for and after surgery. And because transplantation was a surgeon's domain, care of recipient patients was largely the responsibility of the watchful transplant coordinator who consulted a multitude of related specialists to provide ongoing post-transplant care. It was not uncommon for clinical coordinators to monitor and adjust immunosuppressive therapies, validate laboratory results, admit patients to hospitals for treatment of complications, and interpret clinical findings. Most of these coordinators were registered nurses with specialties in extended nursing roles. At that time, medical providers skilled in the art of caring for immunosuppressed patients were scarce, so clinical transplant coordinators were the reliable sleuths. Increasingly, their responsibilities began to center on recipients, rather than on donors.

At the same time, it became evident that separation of donor issues from recipient issues was ethically and politically justified. Centralized procurement agencies specializing in organ and tissue recovery were formed. They launched community education efforts to increase donation, develop hospital working relationships, expedite donor referrals, and educate health care professionals. These organ procurement organizations (OPOs), as they were formally named, soon became an integral part of the US transplant structure.

The increasing success of kidney transplantation launched successful efforts in transplantation of the pancreas, liver, and heart. As transplant numbers grew, clinical transplant coordinators were in demand to develop programs to accommodate these extrarenal patients.

With the increased application of extrarenal transplantation, multiple organ recoveries became commonplace, programs multiplied, and more patients were transplanted.[16] As a result, clinical transplant coordination became more specialized.

By the early 1980s, many coordinators, especially in large programs, specialized in one organ subfield or perhaps even in one age group within one organ subfield. The degree of specialization and the number of clinical transplant coordinators within a given program became directly proportional to the number of transplants performed and the number of patients being followed.[17] In smaller programs, however, one coordinator often worked with both kidney and pancreas recipients, usually because they were transplanted simultaneously into one recipient. This also was the case with some heart and lung recipients.

With increasing graft and patient survival rates, the field of transplantation opened up to include more living donors in addition to cadaver donors.[18,19] Nurses-turned-coordinators were needed to help evaluate potential living related and unrelated donors. They organized tissue typing and an array of other tests, and ultimately arranged for both the donor and the recipient surgery. They also provided long-term follow-up health care, usually in the outpatient clinic setting or by telephone.[17] In addition, coordinators maintained a working relationship with physicians in the community and with patients and staff in various dialysis units.

Although surgeons, urologists, and nephrologists were also involved in patient care, coordinators were often the educational catalysts. They were most actively involved in teaching recipients about self-care and making them aware of the signs and symptoms of rejection. They developed expertise in creating patient care protocols and educational materials for recipients facing lifelong immunosuppression. They often monitored recipients for the mainstay problems of transplantation – rejection, infections, and recurrence of disease – for the life of the graft.

Providing care for immunosuppressed patients after transplantation proved to be a bigger, more complex task than preparing them for the surgery while they were still on dialysis. As new therapies evolved, clinical transplant coordinators were valuable resources in the numerous research protocols designed to improve graft and patient survival rates. Their expertise was sought whenever new procedures were implemented.

Growing professionalism of transplant coordinators

Transplantation requires the close collaboration of a multitude of care providers. In the early years of transplantation, coordinators developed working relationships that knew no boundaries. It was their cooperative spirit that inspired other transplanters to come together for the good of the patient.

ORGAN SHARING

Transplant coordination accomplished international organ sharing before there was legislation to support it, and developed this network with a diverse force of players. In 1982, the professional transplant coordinators organization, the North American Transplant Coordinators Organization (NATCO), implemented the first computerized system to support sharing of extrarenal organs. Called 24-ALERT, it was designed to place extrarenal organs in a timely and equitable way. Established by transplant coordinators, it displayed a mutual cooperative effort to provide organs for dying patients. Two years later, the National Organ Transplant Act of 1984 made provisions for a national organ procurement and transplantation network and a computerized effort for organ sharing. Closely aligned with this US system are similar national systems throughout the world, all designed to facilitate transplantation.[6,7,13,18] All along, transplant coordinators have served on numerous policy making task forces, and have been instrumental in instigating legislative changes.

PROFESSIONAL ORGANIZATIONS

Given the dramatic history of transplant coordination, a collegial atmosphere has always existed between procurement and clinical transplant coordinators. Their job responsibilitles are critically intertwined. This interdependence has resulted in the formation of a number of professional organizations: the North American Transplant Coordinators Organization (NATCO), established in 1975; the European Transplant Coordinators Organization (ETCO), established in 1982;[7] the Australasian Transplant Coordinators Association (ATCA) incorporated in 1990;[3] the Canadian Association of Transplantation founded in 1980.

These organizations share the goals of enhancing transplant coordinator education, increasing the donor pool, and improving the quality of recipient care. Annual meetings are sometimes held in conjunction with the scientific meetings of national or international transplant physician societies. A major objective of these organizations is to provide continuing education for their members.[20] In 1981, NATCO offered a weeklong course of didactic and practical training for procurement coordinators. In 1984, an additional course was offered for clinical coordinators. Today, these comprehensive courses are held biannually. Similarly, in other parts of the world, training courses continue to prepare practitioners for the transplant coordinator profession.[7,21]

These growing organizations have become instrumental in moulding public opinion and affecting legislation regarding transplant concerns. Transplant coordinators were appointed to the first task force that led to the initial legislation on transplantation in the US. Coordinators worldwide have helped implement laws that support and monitor national transplant efforts.

CERTIFICATION OF PRACTITIONERS

In the US, the National Organ Transplant Act of 1984 (Public Law 98-507) stimulated transplant coordinators to assess and monitor their own practice and to develop competency guidelines. In 1985, NATCO set up a 'credentialing' task force and launched a job task analysis in preparation for a national certification program. This difficult task was magnified by the fact that procurement and clinical transplant coordinators come from diverse backgrounds. In the US, most coordinators are nurses or medical paraprofessionals, but some have minimal scientific education (unlike our counterparts in Europe, where many are nurses or physicians). Requiring and teaching a basic level of medical or scientific education proved to be a daunting task. The overwhelming problem is that transplant coordinators have had to provide formal education and training for themselves, instead of relying on other educational institutions.

In 1987, the American Board of Transplant Coordinators was established formally to oversee professional standards and academic credentials. This academic body developed an examination assessing basic skills in transplant coordination and the first examination was set in 1988. Today, over 700 transplant coordinators are certified in procurement transplant coordination, and over 600 in clinical transplant coordination.[22] Professional excellence of this magnitude guarantees high quality service for everyone involved in the transplant process. The expected outcome criteria for standards of practice of procurement transplant coordinators is included in Appendix A; for clinical transplant coordinators in Appendix B.

JOURNAL OF TRANSPLANT COORDINATION

Since the beginning, transplant coordinators have been an integral part of transplant teams worldwide, in the scientific arena as well as the procurement and clinical setting. Their names have appeared as authors and co-authors of historic articles in major journals attesting to their contributions to transplant advancement. In 1991, in the US, the inaugural issue of the *Journal of Transplant Coordination* was published, providing a mechanism for sharing knowledge and research. This journal has given further opportunities to deepen and preserve the interdisciplinary progress of the profession. Incorporating an international editorial board, it is published three times a year with subscribers numbering over 2500.

The journal's timely entry has made it essential reading for transplant coordinators worldwide. Recent articles have addressed diverse topics such as the viability of cryopreserved menisci, transplantation of the small intestine, public attitudes toward organ donation in Saudi Arabia, and minority donation rates in the US.

Conclusion

Transplant coordinators have played a major and humane role in the medical miracle of transplantation. They have mandated sweeping as well as subtle changes and constantly refined their unique, diverse specialty. Moving from the role of assistant to facilitator, they have always championed a team spirit and kept the needs of patients uppermost. Coordinators have made profound contributions to legislation, education, research, and patient care. Their voices have been heard.

APPENDICES

Appendix A

OUTCOME CRITERIA FOR PROCUREMENT COORDINATORS

I Public education

A. *Outcome*   The public should be knowledgeable about current information regarding organ donation, transplantation, and related issues.

B. *Assessment*
   1. Assist as needed with assessment of population demographics, knowledge base, and attitudes regarding organ donation and transplantation.
   2. Assist as needed with assessment of local media outlets, schools, libraries, public service organizations, legislative and regulatory contacts and key community leaders regarding willingness to promote public education efforts about organ donation and transplantation.
   3. Examine barriers and specials issues about organ donation that are of concern to the public.
   4. Examine effectiveness of existing education programs and resources for organ donation and related issues.

C. *Intervention*
   1. Establish and utilize media contacts as appropriate.
   2. Establish contacts/relationships with service organizations, libraries, schools, legislative/regulatory agencies and key community leaders.
   3. Develop, coordinate and/or conduct public education activities/programs for target groups.
   4. Communicate and/or coordinate availability of speakers (procurement/transplant professionals, recipients, donors and/or donor families) to service organizations, schools, libraries, legislative/regulatory agencies and key community leaders.
   5. Market education programs on organ donation and transplantation to key public target groups.

D. *Teaching*
   1. Develop and conduct education presentations appropriate to assessed needs of population for use by service organizations, schools, libraries, legislative/regulatory agencies, key community leaders and media.
   2. Maintain availability to media to assist with stories that can positively impact attitudes regarding organ donation and transplantation.
   3. Assure appropriate distribution of written education materials through service organizations, schools, libraries, hospitals, etc.

II Professional education

A. *Outcome*   Healthcare professionals should actively participate in a referral system for donation and transplantation.

B. *Assessment*
   1. Identify current and potential referral sources.
   2. Initiate and maintain communication with referral source personnel.
   3. Assess referral source knowledge base, attitudes, commitment, and education needs regarding organ/tissue donation process.
   4. Maintain personal visibility as a resource for information on organ donation process and transplantation with referral sources.
C. *Intervention*
   1. Develop a system of planned communication based upon needs and desires of referral sources.
   2. Develop a system of planned communication to inform key personnel of referral sources about policies/procedures relevant to organ donation process.
   3. Develop written and audio-visual teaching materials suitable to needs, attitudes, etc. of referral sources.
   4. Coordinate education activities/program with other organizations as appropriate (i.e. AOPO, NKF, AACN, Tissue Banks, etc.).
   5. Maintain awareness of personnel changes in referral sources.
D. *Teaching*
   1. Conduct education programs regarding aspects of the organ/tissue donation process with key personnel of referral and potential referral sources.
   2. Seek opportunities to collaborate with other organizations to provide education activities/programs on organ donation.

III Hospital development

A. *Outcome* A referral system for organ/tissue donation should be established with referral hospitals.
B. *Assessment*
   1. Identify and evaluate existing policy and practices regarding donation process of referral and potential referral hospitals.
   2. Assess donor potential.
   3. Assess donor referral patterns and actual donor activities.
   4. Compare donor potential to actual donor activity.
   5. Determine and characterize donor gap.
   6. Initiate and maintain communication with referral hospital personnel.
   7. Assess referral hospital knowledge base, attitudes, commitment, and education needs regarding organ/tissue donation process.
C. *Intervention*
   1. Develop a system to prioritize referral hospitals according to donor potential versus donor gap.
   2. Develop a system of planned communication to report results of donor referral patterns, actual donor activities, and needs and desires of referral hospitals.
   3. Develop a system of planned communication to inform key personnel of referral hospitals about policies/procedures relevant to organ donation process and to evaluate effectiveness of hospital development activities.
   4. Develop written and audio-visual teaching materials suitable to needs, attitudes, etc. of referral hospitals.
   5. Coordinate education activities/programs with other organizations as appropriate (i.e., AOPO, NKF, AACN, Tissue Banks, etc.).

6. Maintain awareness of personnel changes in referral hospitals.

D. *Teaching*
   1. Develop and conduct education presentations appropriate to key personnel about policies/procedures relevant to assessing donor potential, to evaluating donor referral patterns, to identifying actual donor activities.
   2. Conduct education programs and distribute education materials to key hospital personnel.

IV Suitability of potential vascular organ donors

A. *Donor family consent*
   1. *Outcome*   The donor family/next of kin should have the opportunity to make an informed decision regarding donation.
   2. *Assessment*
      (a) Assess family/next of kin dynamics, decision making process, and identify legal next of kin.
      (b) Assess family/next of kin knowledge regarding medical status of potential donor.
      (c) Assess family/next of kin understanding of donation process, barriers for understanding the donation process, and willingness to donate.
      (d) Evaluate family/next of kin capability of making an informed decision regarding donation.
   3. *Intervention*
      (a) Identify the family/next of kin understanding of the potential donor's current medical condition.
      (b) Answer questions asked by the family, next of kin, and healthcare giver.
      (c) Support the family/next of kin decision regarding the donation.
      (d) Provide information on support services for donor families.
   4. *Teaching*
      (a) Discuss with the family/next of kin the different donation options available.
      (b) Provide information regarding the donation and transplantation.
      (c) Explain the donation process.
      (d) Explain and review the consent form.

B. *Donor suitability*
   1. *Outcome*   Suitable donors should be identified through meeting the current criteria for each organ system.
   2. *Assessment*
      (a) Identify and examine national policy/regulations to ascertain donor criteria for each organ system.
      (b) Identify local and national donor medical criteria for each organ system.
      (c) Examine existing industry wide practices that will facilitate donor suitability determination.
      (d) Review current medical documentation and evaluation of potential donor to determine medical suitability for each organ system.
   3. *Intervention*
      (a) Implement procedures for collecting and recording necessary medical information to determine medical suitability for each organ system.
      (b) Evaluate health status of potential donor by obtaining current and past medical history which includes: physical assessment, hemodynamic status,

laboratory studies, serologic and infectious diagnostic studies, social history, and medical records review.
    (c) Organize and obtain appropriate consults and testing as necessary.
4. *Teaching*
    (a) Develop and conduct education presentations to appropriate key personnel about policies and procedures relevant to identification and determination of donor suitability and the criteria for the donation process.
    (b) Mentor donor facility personnel about the aspects of donor identification, evaluation and donation activities during the process of donation.

V Management of potential vascular organ donors

A. Medical–legal requirements
    1. *Outcome* All medical–legal requirements should be met prior to initiating organ recovery.
    2. *Assessment*
        (a) Identify and examine local, state, and federal legal and regulatory requirements regarding organ donation and the donation process.
        (b) Identify donor hospital policies regarding death, donation, and post-mortem care.
        (c) Identify the key legal decision makers for the donor hospitals and organ procurement organizations (OPOs).
    3. *Intervention*
        (a) Develop a planned system of communication to assure that the donation is in compliance with local, state, and federal legal and regulatory requirements and in compliance with donor hospital policies.
        (b) Facilitate communication between regulatory, legal, hospital, and OPO resources when issues or questions arise regarding donation.
        (c) Assure appropriate documentation of death.
        (d) Assure that informed consent has been obtained from the legal next of kind.
        (e) Assure that the medical examiner (ME)/coroner reporting requirements have been met and consent of ME/coroner obtained when appropriate.
        (f) Assure that there is no conflict of interest between the donor hospital, organ procurement organization (OPO), and transplant center staff.
        (g) Contact donor hospital and OPO legal counsel when appropriate.
    4. *Teaching*
        (a) Conduct education programs regarding the medical–legal requirements of organ/tissue donation with key personnel of donor hospitals, OPOs and ME/coroner offices.
        (b) Mentor donor facility personnel about the medical–legal aspects of donor identification, evaluation and donation activities during the process of donation.

B. Donor management
    1. *Outcome* Donated vascular organs should be considered viable for transplantation at the time of recovery.
    2. *Assessment*
        (a) Assess medical–legal status of potential donor prior to medical management interventions.
        (b) Evaluate current physiologic status of donor.

(c) Identify parameters for optimal function of each organ system.
(d) Identify mechanism for implementing management intervention.
3. *Intervention*
   (a) Verify that medical–legal requirements have been satisfied to enable donor management intervention.
   (b) Develop and implement donor management procedures to facilitate/ensure haemodynamic stability through maintenance of adequate hydration, oxygenation, and blood pressure.
   (c) Organize and obtain appropriate consults and testing as necessary.
4. *Teaching*
   (a) Review with donor facility staff appropriate donor management principles and techniques.
   (b) Review with donor facility staff the lines of communication and responsibilities needed to manage the donor for optimal function of each organ system.

VI Organ distribution

A. *Outcome* Organs will be allocated to recipients according to local, regional, and national sharing policies.

B. *Assessment*
   1. Review local, regional, and national sharing policies.
   2. Review local recipient waiting list for accuracy.

C. *Intervention*
   1. Determine local, regional, and national recipient needs for vascular organs.
   2. Match available organs with recipients according to priority need based on local, regional, and national policies.
   3. Provide complete documentation of donor information to prospective recipient center(s).
   4. Coordinate timely transportation of organ(s), tissue typing materials, and donor documentation to recipient center(s).
   5. Maintain accurate documentation regarding organ allocation efforts.

D. *Teaching*
   1. Conduct education programs on organ distribution policies and procedures with transplant hospital staff.
   2. Conduct education programs or organ distribution procedures with OPO staff.

VII Surgical recovery

A. *Outcome* Organs should be recovered to maximize utilization in conjunction with adherence to hospital policies.

B. *Assessment*
   1. Review hospital operating from procedures for cadaveric organ recovery.
   2. Review hospital's 'credentialing' procedures for recovery teams.
   3. Identify recovery teams.
   4. Evaluate special needs for recovery teams.
   5. Evaluate needs of the hospital to accomplish multiple vascular organ recovery.

C. *Intervention*
   1. Contact appropriate surgical recovery teams.
   2. Coordinate transportation of recovery teams as appropriate.
   3. Prepare the operating room and anesthesia staff for organ recovery.
   4. Prepare donor for transport to the operating room.
   5. Assure availability of proper equipment and supplies.
   6. Review donor medical records with the leader of each recovery team.
   7. Facilitate proper communication among recovery teams and the operating room staff.
   8. Assure proper intra-operative donor management.
   9. Assure proper documentation of surgical recovery.
   10. Assure appropriate postmortem care.
D. *Teaching*
   1. Provide instructions on donor transportation to the operating room.
   2. Develop and conduct education programs to operating room and anesthesia personnel on surgical recovery procedures and supplies for vascular organs.

VIII Organ preservation

A. *Outcome* Organs will be preserved according to current methods of preservation practices.
B. *Assessment*
   1. Review protocols for organ preservation.
   2. Assess special needs for surgical recovery teams.
C. *Intervention*
   1. Prepare preservation solutions and equipment per protocols.
   2. Implement preservation procedures per current standards.
   3. Inspect organs and document anatomy and organ condition in donor record.
   4. Assure adequate tissue typing materials.
   5. Assure proper packaging and labeling of organs and tissue typing materials.
D. *Teaching*
   1. Provide instructions to hospital operating room staff on preservation techniques and procedures.

IX Maintain records

A. *Outcome* Permanent confidential records of donation should be maintained.
B. *Assessment*
   1. Review policies and procedures for appropriate documentation of all donation activities including but not limited to: laboratory results per organ system, current and past donor medical and social history, hemodynamic status, results of serologic and infectious disease studies, record of surgical recovery, medical-legal requirements (determination of death and consent of legal next of kin), organ distribution documentation, and compliance with organization and OPTN policies.
   2. Review policies and procedures for appropriately accessing records.
C. *Intervention*
   1. Implement standardized donation packet and records.
   2. Establish donation packet quality assurance checklist.

D. *Teaching*
   1. Instruct OPO personnel on importance of good record keeping.
   2. Implement programs to educate OPO personnel on proper record keeping.

X Follow-up communication

A. *Outcome*   Appropriate hospital personnel, OPO, transplant center, and donor family will be notified of the outcome of donation in a timely manner.

B. *Assessment*
   1. Assess the desire of the donor family to receive follow-up information.
   2. Identify which family members will receive follow-up information.
   3. Identify which healthcare members involved with the donation process will receive follow-up information.
   4. Review procedures for required follow-up to transplant centers and OPTN.
   5. Review procedures for identifying donor hospital charges and processing of those charges so they are not sent to the donor family.

C. *Intervention*
   1. Implement established procedures for timely verbal and written communication to families.
   2. Implement established procedures for timely verbal and written follow-up to healthcare members.
   3. Implement established procedures for timely verbal and written follow-up to OPTN.
   4. Identify all appropriate hospital charges to be billed to the OPO.
   5. Implement established procedure for timely verbal and written communication to transplant centers on donor laboratory and culture results, donor autopsy reports, outcome of other organs transplanted from the same donor which may affect transplant outcome.

D. *Teaching*
   1. Instruct personnel on appropriate follow-up communication with health care members, OPOs, transplant centers, donor families, and the OPTN.
   2. Educate personnel on appropriate and timely communication with transplant centers.
   3. Conduct educational meetings to assist personnel in identification of appropriate billing charges from the donation.

Appendix B

OUTCOME CRITERIA FOR CLINICAL COORDINATORS

I Public education

A. *Outcome*   The public should be knowledgeable about current information regarding organ donation and transplantation.

B. *Assessment*
   1. Assist as needed with assessment of population demographics, knowledge base, and attitudes regarding organ donation and transplantation.

2. Assist as needed with assessment of local media outlets, schools, libraries, and public service organizations regarding willingness to promote public education efforts about organ donation and transplantation.
3. Examine barriers and special issues about organ donation and transplantation that are of concern to the public.
4. Examine effectiveness of existing education programs and resources.

C. *Intervention*
1. Establish media contacts as appropriate.
2. Establish contacts with service organizations, libraries, and schools as appropriate.
3. Communicate availability of speakers (transplant professionals, recipients, living donors, and/or donor families) to service organizations, schools, and media.
4. Coordinate educational activities with other organizations as appropriate.

D. *Teaching*
1. Develop and conduct educational presentations appropriate to assessed needs of population for use by service organizations, schools, and media.
2. Assure appropriate distribution of written educational materials through schools, libraries, etc.
3. Maintain availability to media to assist with stories that can positively impact attitudes regarding organ donation and transplantation.

II Professional education

A. *Outcome* Healthcare professionals should be able to actively participate in a referral system for donation and transplantation.

B. *Assessment*
1. Identify current and potential referral sources.
2. Initiate and maintain communication with referral source personnel.
3. Assess referral source knowledge base, attitudes, and commitment to transplantation in conjunction with educational needs.
4. Maintain personal visibility within institution of practice as a resource for information on donation and transplant.

C. *Intervention*
1. Develop a system of planned communication with referral sources based upon needs and desires of each referral source.
2. Develop written and audio-visual teaching materials suitable to needs, attitudes, etc. of referral sources.
3. Coordinate education activities and/or programs with other organizations as appropriate, e.g. AACN, ANNA, ALF, etc.
4. Maintain awareness of personnel changes in referral sources.

D. *Teaching*
1. Meet with referral sources or invite referral sources personnel to institution of practice for in-services, workshops, or seminars utilizing educational materials developed for professional education.
2. Seek opportunities to collaborate with other education activities or organizations both inside and outside institution practice.

III Suitability of potential living vascular organ donors

A. *Outcome* The potential living donor should be able to make an informed decision regarding organ donation.
   1. *Assessment*
      (a) Identify potential living donor's knowledge, attitude, and commitment toward organ donation.
      (b) Identify potential living donor's decision making process.
      (c) Determine willingness of potential living donor to donate, motives for donation, and long-term expectations.
   2. *Intervention*
      (a) Present potential risks, benefits, and contraindications of living organ donation.
      (b) Provide education to the living donor and transplant recipient regarding the surgical procedures and possible short- and long-term complications.
      (c) Evaluate living donor's psychosocial needs and provide information regarding available support systems.
      (d) Educate potential living donor regarding expected emotional reactions associated with donation.
      (e) Review living donor evaluation process from initiation to completion.
         (1) Provide rationale for donor testing and criteria used to determine suitability.
         (2) Explain procedure for final review mechanism by the transplant physician or committee.
      (f) Introduce transplant team, discuss roles, and available support services.
      (g) Review hospital routines and procedures.
   3. *Teaching*
      (a) Review ways in which donation may affect donor/recipient relationship.
      (b) Plan individual and group meetings.
      (c) Provide education to potential living donor utilizing written material, audiovisual aids, workshops and seminars as appropriate.
      (d) Provide access to network(s) allowing the potential living donor the opportunity to speak to other post-operative living donors.
      (e) Assess potential living donor's recall of the donor process.

B. *Outcome* The potential living donor should complete the evaluation process with suitability determined.
   1. *Assessment*
      (a) Identify medical, socioeconomic, and legal findings which are consistent with or contradict living donor suitability according to established criteria.
      (b) Establish and implement procedure for decision making procedure regarding living donor suitability.
   2. *Intervention*
      (a) Establish guidelines for medical criteria for each organ system.
      (b) Screen potential living donor based on established medical guidelines.
         (1) Evaluate physiological and emotional status.
         (2) Assess decision making process.
         (3) Schedule tests/procedures to evaluate organ function including histocompatibility as appropriate.
         (4) Review medical–social history and physical assessment.
         (5) Organize the evaluation results for final determination of suitability.

3. *Teaching*
   (a) Review evaluation results with potential living donor.
   (b) Provide access to transplant physician/committee as necessary to answer any questions potential living donor may have regarding donation or surgical procedure.

C. *Outcome* The living donor should be in optimal health at the time of donation.
   1. *Assessment*
      (a) Evaluate current physiological and emotional status of living donor.
      (b) Obtain appropriate consultation as needed.
      (c) Recommend clinical interventions to maintain optimal organ function in the potential living donor.
   2. *Intervention*
      (a) Verify living donor/recipient compatibility.
      (b) Interview living donor regarding current health status.
   3. *Teaching*
      (a) Provide pre-operative counseling for potential living donor.
      (b) Reaffirm potential living donor's knowledge, attitude, and commitment toward organ donation.

D. *Outcome* The living donor should be discharged from hospital in optimal health.
   1. *Assessment*
      (a) Review physiological and emotional status of living donor.
      (b) Determine knowledge base for home care.
   2. *Intervention*
      (a) Recommend clinical interventions to facilitate recovery from living donor surgery.
      (b) Provide post-discharge follow-up and support as necessary.
   3. *Teaching*
      (a) Instruct living donor on post-discharge care and rehabilitation.

IV Suitability of recipients

A. *Outcome* Recipients candidacy for transplantation should be identified through meeting the established criteria for each organ system.
   1. *Assessment*
      (a) Review of current medical documentation to ascertain patient's eligibility for transplant.
      (b) Establish and implement procedures for meeting the criteria.
      (c) Establish a procedure for the final review mechanism to meet the accepted criteria.
   2. *Intervention*
      (a) Evaluate overall health status; review medical, social, compliance history, and perform physical assessment.
      (b) Obtain and review laboratory results including histocompatibility testing when indicated.
      (c) Review results of diagnostic procedures.
      (d) Organize and obtain additional consults and testing as necessary.
   3. *Testing*
      (a) Identify knowledge base, attitude, education level and commitment to transplantation.

(b) Review previous education and instructions.
(c) Identify patients who are at risk for poor comprehension and compliance with medical regimen.
(d) Develop appropriate level educational programs and materials.
(e) Identify and review patient's responsibility.
(f) Evaluate effectiveness of educational material and programs.

B. *Outcome* Potential transplant recipient should be able to make an informed decision regarding transplantation.
   1. *Assessment*
      (a) Identify patient's decision making process.
      (b) Identify current knowledge of the transplant process.
      (c) Identify the recipient's long-term expectations.
   2. *Interventions*
      (a) Review the educational information with patient.
      (b) Review the risks, benefits, and complications of transplantation.
   3. *Teaching*
      (a) Reinforce the necessary educational information.
      (b) Emphasize the importance and necessity of the patient commitment to transplant process.

C. *Outcome* The recipient will be in optimal physiological and psychological condition at the time of transplantation.
   1. *Assessment*
      (a) Review the current physiological and emotional status of the recipient.
   2. *Intervention*
      (a) Consult referral source regarding recipient's current suitability for transplantation.
      (b) Verify donor–recipient compatibility.
      (c) Evaluate current physiological and psychological status of the recipient.
      (d) Obtain appropriate consults as necessary.
      (e) Make appropriate recommendations to assure optimal physiological and psychological status of recipient.
      (f) Provide emotional support and counseling for recipient/family.
   3. *Teaching*
      (a) Review, and explain pre-operative procedures and activities to include final crossmatch.
      (b) Review, explain, and describe peri-operative procedures, and activities.
      (c) Review, and explain post-operative procedures and activities to include intravenous therapy, catheters, wound drains, pain therapy, immunosuppressive therapy.
      (d) Review potential for ATN, rejection, and/or graft loss.

V Recipient post-operative care

A. *Outcome* Patient, family, or significant other will be given adequate discharge teaching prior to discharge of hospitalization stay.
   1. *Assessment*
      (a) Identify patient's, family, or significant other's knowledge base.
   2. *Interventions*
      (a) Provide information, by teaching and written instructions, necessary to safely monitor home care and to recognize and appropriately react to organ dysfunction and other problems.

3. *Teaching*
    (a) Provide education regarding signs and symptoms of infection and/or rejection, diet, and activity.
    (b) Provide education regarding immunosuppressive medications to include side effects, time schedule, dosage, and how to obtain refills.
    (c) Provide education regarding importance of clinic follow-ups and who to contact regarding problems.
B. *Outcome* Recipient will be monitored and assessed post-operatively, in order to be in optimal health at hospital discharge.
    1. *Assessment*
        (a) Monitor and assess recipient physiologically, psychologically, and spiritually.
    2. *Interventions*
        (a) Monitor and assess laboratory values and procedure outcomes.
        (b) Collaborate with physicians daily regarding patient's progress.
        (c) Provide for social and spiritual support as needed; provide access to appropriate counseling as needed.
    3. *Teaching*
        (a) Initiate post-operative education regarding self-care.

VI Recipient outpatient care

A. *Outcome* Recipient should maintain monitored follow-up care in order to achieve optimal physical and rehabilitation potential.
    1. *Assessment*
        (a) Review knowledge base from prior teaching, assess physiological and psychological statuses.
    2. *Interventions*
        (a) Plan for physical and laboratory assessments during outpatient clinic.
        (b) Assist with and/or provide further consultations as needed.
    3. *Teaching*
        (a) Reinforce prior teaching to recipient and family, if necessary, that which had already been implemented and provide new teaching as needed.

References

1. Wight C. Transplant Coordinators and Organ Procurement in Western Europe. *J Transplant Coordination* 1991, **1**: 39–41.
2. Bergstrom C, Gabel H. Organ Donation and Organ Retrieval Programs in Sweden, 1990. *J Transplant Coordination* 1991, **1**: 47–51.
3. Fitzgerald LM, Martyn BN. The Evolution of Transplant Coordinators in Australia. *Transplantation Proceedings* 1992, **24** (5): 2051.
4. Garcia VC, Hoelfmann N, Bittar AE, Goldani JC. Transplant Coordinators in Rio Grande do Sul, Brazil: Initial Analysis. *Transplantation Proceedings* 1991, **23** (5): 2519–2520.
5. Milanes CL, Bellorin-Font E, Weisinger J, Pernalete N, Urbina D, Paz-Martinez V. Need and Demand of Kidneys for Transplantation in Venezuela. *Investagicion Clinica* 1993, **34**: 15–27.
6. Matesanz R. Organ procurement in Spain: the importance of a transplant coordinating network. *Transplantation Proceedings* 1993, **25** (6): 3132–3135.

7. Roels L. Transplant coordination in Europe or the charms and challenges of a hodge-podge. *NATCO Newsletter*, March/April 1994, 8.
8. Soh P, Lim SM, Tan ED. Organ procurement in Singapore. *Ann Academy of Medicine, Singapore* 1991, **20** (4): 439–442.
9. Smith SL. Tissue and Organ Transplantation. In *Implications for Professional Nursing Practice*, Mosby, New York, 1990, page 105.
10. Belzer FO, Ashby BS, Dumphy JE. 24- and 72-hour preservation of canine kidneys. *Lancet* 1967, **2**: 536–539.
11. Collins GH, Bravo-Shugarman MB, Terasaki PI. Kidney preservation for transplantation: initial perfusion and 30-hour ice storage. *Lancet* 1969, **2**: 1219.
12. Schaeffer MJ, Alexander DC. System for organ procurement and transplantation. *American J Hosp Pharm* 1992, **49** (7): 1733–1740.
13. Colpart JJ, Noury D, Cochat P, Kormann P, Moskovtchenko JF. Organization of organ transplantation in France. *Pediatrie* 1991, **46** (4): 313–322.
14. Alexandre GPJ et al. *History of Transplantation: Thirty-Five Recollections*. UCLA Tissue Typing Laboratory Publishers, Los Angeles, 1991, page 341.
15. Gjertson DW, Terasaki PI, Takemoto S, Mickey MR. National allocation of cadaveric kidneys by HLA matching: projected effect on outcome and costs. *New Engl Med* 1991, **324**: 1032.
16. Davis FD. Coordination of cardiac transplantation: patient processing and donor organ procurement. *Circulation* 1987, **75**: 29–39.
17. Heyl AE, Staschak S, Folk P, Fioravanti V. The patient coordinator in a liver transplant program. *Gastroenterology Clinics of North America* 1988, **17** (1): 195–206.
18. Swerdlow JL, Cate FH. Lifesaving Connections–Communications, Coordination, and Transplantation. *Transplantation* December 1990, **50**: 992–996.
19. Rapaport FT. Living donor kidney transplantation. *Transplant Proceedings* 1987, **19**: 169.
20. Kory L, Bocchino CA. North American Transplant Coordinators Organization. *J Transplant Coordination* December 1993, **3** (3): 114–115.
21. Karbe T. Standardized education program in organ procurement for transplant coordinators. *Transplantation Proceedings* 1991, **23** (5): 2539–2540.
22. Principe A et al. Board of Governors, American Board of Transplant Coordinators. *J Transplant Coordination* April 1993, **3** (1): 44.

CHAPTER 17

Informed or presumed consent legislative models

PAUL MICHIELSEN

Definition: the basic difference and the additional stipulations 344
The geographical differences 346
Is presumed consent more efficient than informed consent? 349
Individual autonomy and transplantation laws 354
Acceptability of the law 355
Conclusion 358
References 358

Definition: the basic differences and the additional stipulations

According to the Guiding Principles of the World Health Organization (WHO)[1] and the directives of the Council of Europe,[2] organ removal for transplantation may not take place against the expressed or presumed will of the deceased person. Consent of the individual is in all transplantation laws prerequisite for removal of organs after death. This general statement can be implemented in two different legal ways: presumed or informed consent. Presumed consent is synonymous with contracting out or opting out, and informed consent with contracting in or opting in. The basic difference between these two forms of consent has been excellently described by Margaret Sommerville.[3] 'The fundamental difference is in the initial presumption which governs. In a contracting in system, the initial presumption is of a "no ... unless" nature; that is, *No*, organs may not be taken *unless* certain conditions are fulfilled. In a contracting out system the initial presumption is of the "yes ... but" kind; that is, *Yes*, organs may be taken, *but* there are exceptions'. In both presumed and informed consent laws the explicit decision of the deceased during his lifetime is the crucial point. Both laws differ only on what is presumed in the absence of a known wish of the deceased: under presumed consent organs may be removed post-mortem with some exceptions, under presumed refusal, improperly termed informed consent, organ removal is not allowed unless some conditions are fulfilled.

Laws on organ transplantation are not concerned only with the consent of the potential donor, additional provisions will deal with the ways in which this consent can be expressed, the rights of the family and the duties of the medical profession.

The ways in which the willingness or the opposition to donate can be expressed will determine to a large extent the efficacy of the law. Compliance with the wishes of the deceased, which is theoretically the basis of all transplant laws, is only possible if they can be known rapidly and easily at the moment of death. If not, the individual autonomy remains a polite fiction, and this goes for both types of law. Although the need for a registry has been stressed in several publications,[4,5] it is not provided for in most informed consent laws. Private initiatives to make up for this absence produced disappointing results. Since 1983 a computerized kidney donor registry has been set up in Manchester and extended to Wales in 1986.[6] In 1994 14% of the adult population voluntarily registered, and 14 people who had registered died and became organ donors. In none of these cases was the registry consulted. In the US, the task force on organ transplantation assessed in 1986 the results obtained by three current and past donor registries. They found no information demonstrating clearly that registries are effective in increasing the number of donors and concluded: 'because so much emphasis is placed on next-of-kin consent, querying the registry is an unnecessary step'.[7] The task force failed to consider that the primary role for a registry is not to increase the number of donations, but to know rapidly and reliably the wish of the deceased. If, anyway, the decision of the deceased is subordinated to the wishes of the family, a registry of these decisions is, of course, superfluous. If, however, more than lip-service is to be paid to the individual autonomy an adequate registry is essential. In the absence of a computerized registry, the chances are real that the expressed decision of the deceased will not be honoured. It is interesting to note that, even in the setting of presumed consent laws, the need for a registry is progressively recognized. A transplantation law based on the informed consent principle was voted recently in The Netherlands and will be effective in 1998. It contains a provision for the establishment of a central registry. In Denmark an active registry has been in existence since 1990 and in the United Kingdom an informal registry was started by UK Transplant in 1995.

In contrast with most informed consent legislations, the set-up of a registry is included in several presumed consent laws. This registry is the cornerstone of the transplantation law in Belgium, where since 1986 every citizen has the possibility to record confidentially his will to donate or to oppose donation of organs, with the guarantee that this decision will be honoured. This computerized registry is accessible by modem only to the transplant teams and must be consulted before any organ removal. Such a registered decision can be modified at any time and cannot be overruled by the family of the deceased. In Singapore a register of objections is to be maintained by the Director of Medical Services.[8] In France and in Spain, no central registry existed, but all hospitals licensed to prelevate organs post-mortem keep a registry where the individuals or their family can record their decisions. It is clear that, in the absence of a registry, the transplant team will have to turn to the family for information on the wishes of the deceased, which will fundamentally modify the way in which the presumed consent principle is implemented. In October 1994 Portugal implemented a non-donor registry. After having adopted an informed consent law in 1988, Sweden reverted to a presumed consent law with a central registry in July 1996. In France the principle of a central registry has been recently approved and it is scheduled for 1997. Austria adopted a central registry in 1996.

Other modalities for expressing the wish or the objection to donation are as numerous as they are inefficient: donor cards, oral communication, recording in hospital files etc. Donor cards when filled in are seldom carried and are rarely found when needed. Decisions mentioned on the driving licence lack the confidentiality essential for a free decision. The present chaotic situation is summarized in a recent report of the UNOS ethics committee:

> The policy status quo is a state-centred approach relying on the use of the back of driver's licences, applications for driver's licences, or the distribution of donor cards to be carried with or attached to the driver's licence. The approach is unco-ordinated across the states: not only is there no centralized collection of donation preferences but not even the same data points are collected... .[9]

The rights of the family are another essential part of the laws on organ procurement. In the UK Human Tissue Act 1961, a central role is played by the family, which may override a positive, but not a negative decision of the deceased. In American law the permission of the family is limited in theory to the situation in which the wish of the deceased is unknown.

It is in the presumed consent laws that the greatest diversity concerning the rights of the family is to be found. In the Austrian law of 1982, sometimes referred to as a pure presumed consent law, the family is not supposed to express its opinion on donation. The relatives of the deceased do not have the right to be consulted and their consent is not required before an organ may be removed.[10] In France there is no explicit obligation to inform the family, but the transplant team should collect any information on the will of the deceased. If, as is usually the case, this is not mentioned in the ad hoc local registry in the hospital, there is no alternative to asking the family about the wishes of the deceased. In theory the family cannot object, but organ removal is forbidden if the family feels that the deceased opposed donation.[11] In Belgium there is no legal obligation to inform the family, but the family may take the initiative to object organ removal. In Norway and Finland the next of kin must be informed 'whenever possible' and if 'circumstances permit' about the intention to remove organs.[12] The Israeli law requires that the doctors inform the family about the intention to remove organs, and the family has the right to object.[13] Spain has a presumed consent law, but when presumed, the consent must be ratified by a written consent of the family.[14]

These additional provisions can modulate the effects of the law to such extent that it is not justified when discussing the efficacy, to consider all presumed consent laws as one single entity; this is also the case for the informed consent laws, although to a lesser extent.

The geographical differences

In the late fifties, transplantation of cadaveric organs started as an experimental procedure in the United States and in Western Europe. When a specific legislation was enacted in the pioneer countries, most English speaking countries adopted laws based on the informed consent principle, while in continental Europe most countries adopted presumed consent laws. When in the eighties, organ transplantation spread worldwide, countries with legal systems based on the Anglo-American tradition followed suit and most adopted informed consent laws. With the exception of Colombia,[15] all Latin American countries also adopted informed consent laws.[16] In the Middle East and in

Asia informed consent became the rule, with the exception of Singapore that switched from informed to presumed consent in 1987.[17] In continental Europe, Russia and the former eastern countries adopted presumed consent. Among the countries with an active transplant programme, Germany and The Netherlands followed the informed consent principle without a specific law. In The Netherlands a formal informed consent legislation which will be effective in 1998 was recently voted. Two countries, Denmark and Sweden, switched recently from presumed to informed consent. Sweden, however, already returned to a presumed consent law in 1996.

These geographical differences are not due to chance, but a law is the response of society to a problem or a need experienced at a given moment. As such it reflects a general attitude of the population, influenced by cultural and religious traditions. The laws on transplantation are no exception to this general rule. 'If one is to fully appreciate the law governing organ transplantation it may not be sufficient to ask 'what is the law?' One must also ask 'why was it enacted?'.[3] The two main informed consent laws were enacted relatively early. In the UK the Human Tissue Act was passed in 1961 and in the USA the Uniform Anatomic Gift Act in 1968. 'At that time the concern was to establish the legality of organ donation, but it was not to ensure an adequate supply of organs or the greatest degree of access to organs compatible with public sentiment on the matter'.[3] Logically a restrictive law of the 'no ... unless' type was adopted.

With the exception of Denmark in 1967, most presumed consent laws were enacted later. It is likely that in continental Europe the need for a specific transplant law was not immediately felt due to a tradition of more permissive practices on autopsy. The often quoted Austrian and Belgian laws were enacted only in 1982 and 1986 at a time when advances in organ transplantation made multiorgan donation necessary and the shortage of transplantable organs became a major concern. This resulted logically in a permissive 'yes ... but' kind of law. The presumed consent law officialized and modulated the current practice.

Differences in religious and cultural background influence the ritual surrounding death and mourning, with consequences on the acceptability of organ donation. Different attitudes concerning the dead body are illustrated by the embalming procedure, which is almost unused in continental Europe, but is a routine in the US, Canada and Australia, while becoming more common in England.[18] Curiously this geographical distribution of embalment coincides with the distribution of informed consent laws for transplantation.

In Islamic countries the use of cadaveric donors met with considerable resistance. Islamic law prohibits mutilation of the cadaver,[19] but Muslim scholars, belonging to various schools of Islamic law elaborated exceptions to this general principle. Although in 1987 a unified Arab Draft Law on Human Organ Transplants was adopted by the Council of Arab Ministers of Health,[20] it is clear that 'both religious beliefs regarding bodily resurrection and cultural norms about the treatment of the dead with respect and consideration make the donation of bodily organs loathsome to Muslim sensibilities'.[19] As a consequence all predominantly Muslim countries presumed refusal in the absence of explicit consent and adopted informed consent laws, with strong emphasis on the permission of the next-of-kin. In Singapore, which has a 15% minority of Muslim population, a presumed consent law was enacted in 1987, applicable only to non-Muslim accident victims.[17] This is the first legislation of this kind adopted in the Asia-Pacific region.

In Japan organ retrieval is even more difficult as in the Shinto concept injuring the dead body is a serious crime, and injury to a dead body was sometimes performed as a punishment.[21] In addition, the traditional Japanese notion of person has a communal, not an individual basis, and the right to dispose of the dead belongs to the family, not the individual.[22] As a consequence organ retrieval based on the individual decision of the deceased is conflicting directly with the cultural and religious tradition. In contrast, in the Buddhist view of oneness of humankind and the universe, the same resistance against organ donation does not exist.[23,24]

In Europe and the Americas, religion does not seem to have influenced the choice between presumed and informed consent, as in the Christian tradition, both Catholic and Protestant, there is no objection to organ donation. In the medically developed countries of these regions, organ transplantation became a routine treatment for kidneys in the sixties; heart, liver and pancreas came later and lung transplantation became only recently a common procedure. In the early days of transplantation the rules and the practice governing the post-mortem examinations were the only legal basis for organ removal. Not only the laws on the practice of autopsy, but also the history of anatomical dissections and autopsies in continental Europe and in England are fundamentally different. While anatomical dissections were common in continental Europe since the fourteenth century, '... in the English speaking countries of England, Scotland, Ireland and North America anatomical dissection languished'.[18] The main reason was the association, not between dissection and advancement of medical science, but between dissection and punishment, which remained until the 1832 Anatomy Act. This goes back to sixteenth century Royal Enactments by James IV in Scotland, and by Henry VIII in England, granting the British company of Barbers and Surgeons the annual right to the bodies of four hanged felons.[25] These public dissections were recognized by law as a punishment, an aggravation to execution, a fate worse than death.[25]

> The shortage of cadavers for dissection led to the buying and selling of human bodies. The holding of the English common law, that a body after burial was the property of no one, encouraged the grisly practice of body snatching – rifling the grave – to which the overcrowded condition of London churchyards and the need (and the lack of legal provision) for cadavers for anatomical study contributed.[26]

In 1829 a man was executed (and subsequently dissected) for the murder of 15 people, sold to an Edinburgh school of anatomy, and in 1831 two men were hanged for murdering about sixty people for sale to the anatomists in London. In the nineteenth century England also imported bodies for anatomic studies, 'a brisk trade did develop in bodies from the parts of Europe where corpses were plentiful and inexpensive'.[18] This practice was ultimately ended in the nineteenth century by the vote in 1832 of the Anatomy Act. The person lawfully in possession of the body became the depositor of the right to authorize dissection, autopsy and organ removal. The intention of the law was to put an end to body snatching by providing for dissection organs from the poor. In practice this meant that workhouse keepers would thenceforth be free to sell for dissection the bodies of people who died in their institutions. Hostility of the poor to the measure led to workhouse unrest and riot, to political demonstration and disobedience. This complicity with the hangmen and the body snatchers led to a widespread hostility and distrust of the medical profession. According to the fascinating book of R. Richardson[25] on this subject, a change in attitude of the public towards organ bequest was only observed since the Second World War. In line with the 1832 Anatomy Act, in England and in the US consent of the person lawfully in possession of the body was necessary for an autopsy and this principle was also applied to organ removal in the early days of transplantation.

In continental Europe, formal courses in anatomy with human dissection started in 1340 in Montpellier and in 1435 in Vienna. Modern practice of pathology was started by Carl von Rokitansky (1804–1878), who established a pathology school in Vienna, which became a model for teaching hospitals.[27] He performed or supervised approximately 60,000 autopsies on the basis of the legislation elaborated by Von Swieten, the personal physician to empress Maria Theresia, and dating from the second part of the eighteenth century. In this law no family consent to autopsy is required,[28] and a tradition of obligatory autopsy in public hospitals was established. It is on the basis of this same legislation that Austria has been active in organ transplantation since 1965.[10] In a recent review Svendsen found that 'in major teaching hospitals in Austria the autopsy rate is still near 100%, while in private hospitals, it is about 60%'.[27] The Austrian regulations were the original model for many other European countries and could have influenced the practice in Belgium, which lived under Austrian rule during most of the eighteenth century. Although Belgium still has no autopsy legislation, there was a tradition that most persons dying in University hospitals are autopsied without any necessary consent nor information of the family. From 1963 to 1986 active transplantation programmes were conducted in all Belgian Universities on the basis of this tradition, and without a specific law. Transplantation laws of the presumed consent type were adopted in Austria in 1982, and in Belgium in 1986. The cultural and legal background concerning the handling of dead bodies, with a strong emphasis on the rights of the family of the deceased in Britain and also in the United States, is thus different from continental Europe and explains the often quoted saying of D. Ogden: 'Presumed consent is not quite the American way'.[29]

Is presumed consent more efficient than informed consent?

The effectiveness of transplant laws is usually evaluated by comparing the retrieval scores per million inhabitants obtained in countries with different legal systems. Interesting as they could be, such comparisons must be interpreted with circumspection. Meaningful comparisons can only be made between countries with a similar degree of medical development and cultural traditions. Even then, such comparisons must be taken with great caution for two reasons:
1. The number of potential donors can be extremely variable from country to country and depends on the number of fatal accidents, the population density and age structure and the number of intensive care beds.
2. A transplant law can only create a framework within which organ retrieval has to be organized. The organ retrieval rate is the final result of a chain of events in which the general attitude of the population, the motivation and organization of the transplant teams, the extent of the collaboration of the medical profession and of the hospital administrations, the reimbursement of the cost of procurement etc. will play a role. Any of these can become a limiting factor and the final result will be determined by the weakest link in this chain of events.

In a recent study of transplant laws and organ donation in Europe, Land and Cohen[30] concluded to the absence of obvious correlation between high post-mortem organ removal rates and the existence of presumed consent laws. The absence of correlation is, however, due to the inclusion in this study of countries with an abnormally low transplant activity as Poland, Hungary, Greece and Italy, all with less than six donors per million population (pmp). Other factors than the type of law were obviously limiting the organ supply in these countries. Land and Cohen also included in their study data, which did not concern countries but regions such as Madrid, Southern Bavaria and Catalania. As acknowledged by the authors, this can be misleading as concentration of brain-injured patients in some specialized clinics in these regions makes the calculation per million inhabitants unreliable.

In the USA, with an informed consent legislation, UNOS reported 17.9 cadaveric donors pmp for 1992 and 19.24 for 1993.[31] The number of organ donors obtained in the countries of Europe with a significant transplant activity (more than 10 donors per million population and per year) is given in Table 17.1. Bold type indicates countries with a presumed consent legislation. Countries in italic adopted informed consent, or follow informed consent rules in the absence of a specific transplant law. From the 14 countries listed in Table 17.1 the 7 best results were obtained in countries with a presumed consent law. This suggests that the type of transplant legislation could greatly influence organ retrieval.

The data available thus indicate that successful organ retrieval programmes are found more often in countries with presumed consent laws than with informed consent laws. Presumed consent does, however, not automatically lead to optimal results, it can only provide an environment more or less favourable for the development of the initiatives to increase the number of donations. In addition not all presumed consent laws are identical and some administrative provisions can make the law more of an obstacle than an incentive to donation, as exemplified by the Italian law.[32] In France, a presumed consent law was voted in in 1976. At that time, this did not change the retrieval rate in a significant way [33] and this has often been quoted as illustrating the ineffectiveness of the presumed consent approach. On closer examination of the data, a clear cut improvement is seen since 1982, after the establishment of regional co-ordinators. The inefficient organization of the organ retrieval was in fact limiting the effect of the law. In addition, at that time the French law did not provide for a central registry, but stressed the obligation to take 'all precautions to find out what the wishes of the deceased were'.[34]

Table 17.1 Number of organ donors per million population and per year in countries (**in bold type**) with and (*in italic type*) without a presumed consent law. (Data from ETCO Newsletter 11, 1993 and 12, 1994)

	1992	1993	Mean 1992–93	% Multiorgan
Austria	20.8	27.2	24	67.1
Spain	21.7	22.6	22.15	70.5
Luxemburg	20	22.5	21.25	77.7
Belgium	18.3	22.2	20.25	79
Finland	19.4	19.2	19.3	57.1
France	17	17.1	17.05	80
Portugal	18.5	14.7	16.6	29
Switzerland		16.6		69
UK and Ireland	16.3	15	15.65	67.5
Norway	14.2	16.7	15.45	52.8
Denmark	15.8	15	15.4	
Netherlands	15.1	14.9	15	52.8
Sweden	14.5	14.9	14.7	65.6
Germany	13.1	13.5	13.3	58.4

In practice this led to asking the family about their own opinion and their own feelings,[35] a situation not far remote from asking for the permission. A somewhat similar evolution is seen in Spain which adopted a presumed consent transplant law in 1977, with the obligation to obtain a written consent of the family.[14] During the ten following years the organ retrieval rate increased steadily to a plateau of 15 donors pmp. After implementation of a decentralized network of co-ordinators in 1989 the retrieval rate increased further to the present value of 22.6 donors. It is interesting to note that in Spain the refusal rate has been between 20 and 30%,[14] much higher than in France, where Benoit found 8.6% refusals and in Austria. The Spanish data have been interpreted as showing that transplant co-ordination, not the law was determinant.[36] The data only indicate that in 1989 the suboptimal organization of the organ retrieval had become the limiting factor.

Considering the change in organ retrieval rate coinciding with the introduction of a transplant law eliminates the role of some of the interfering factors. In this respect it is interesting to compare the evolution with time of the number of organ donors in the four countries collaborating in Eurotransplant.

Belgium adopted a presumed consent law in 1986. The Netherlands did not have a specific transplant law until 1986 but followed a strict informed consent policy, based on their legislation for post mortem examination. As mentioned earlier, Belgium has no specific law on autopsy but there is a tradition for University hospitals to perform autopsies routinely without formal consent of the family. Both countries are similar in size, population density and medical development. Figure 17.1 illustrates the evolution of organ donation in the two countries. From 1976 to 1986, the organ retrieval rate improved slowly and at the same pace. When Belgium adopted a presumed consent law in 1986, this was followed by an immediate increase in retrieval rate, while the trend was unchanged in The Netherlands. This evolution is paradoxical as the law did not modify, but only officialized the presumed consent practice in Belgium. A closer analysis of the data of Leuven, which has the largest procurement activity in Belgium and which saw its number of donors more than doubled since the introduction of the law, can explain this paradox. In 1978, the nephrology department of Leuven University had started a close collaboration with 19 nephrology units from the Dutch

Figure 17.1 Evolution of the number of organ donors in Belgium and The Netherlands. (Data from the Eurotransplant Annual Reports.)

speaking part of Belgium, with the aim to improve organ donation and to co-ordinate the post-transplant follow-up.[37] Although this collaboration was active since 1978, the contribution of these non-university hospitals to organ procurement remained limited until the vote of the law in 1986. At that time there was a five-fold increase in the number of organs coming from these collaborating hospitals.[38] It is likely that the law which was largely publicized by the Ministry of Health in an effort to increase the donation rate, provided the legal safety and positive psychological background for the collaboration of a large number of non-university hospitals. An excellent procurement organization is not sufficient in itself, as illustrated by the example of The Netherlands, which is probably the country with the most developed decentralized network of transplant co-ordinators. Even the implementation in 1991 of the educational programme EDHEP did not increase the organ retrieval rate. This does not demonstrate that transplant co-ordination and training is not important, but that these were not the limiting factor for organ retrieval in The Netherlands.

It has sometimes been argued that not the law in itself, but the attention given by the media during the discussions preceding the vote of the law, could be responsible for the improvement. If this is true one would expect a similar influence on all the transplant centres in the country. As could be expected the vote of the Belgian presumed consent law was preceded by a long and passionate debate. Also the medical profession was divided. In Antwerp the transplant centre opposed the presumed consent principle and campaigned vigorously against the law, while the centre of Leuven was strongly in favour. When the law was finally voted, a strict 'opting-in' policy was continued in Antwerp, together with a maximal effort for an efficient transplant co-ordination operation and supply of information to the public. In contrast, the centre of Leuven made full use of the possibilities offered by the law. While there was a massive increase in the number of donors procured in Leuven, the number of donors procured in Antwerp continued to follow the disappointing pattern observed in The Netherlands. This observation eliminates the publicity as responsible for the increase in retrieval rate. It illustrates also how the vote of a presumed consent law does not necessarily result in a uniform attitude among the medical profession. To evaluate the impact of a presumed consent law in a given country it is essential to first examine the way and the extent to which its essential principles are applied.

As mentioned earlier, systematic autopsy was the traditional rule in Austria for all people dying in the hospital, whereas in Germany consent of the family was required.[27] In contrast to what happened in Belgium, the vote of the presumed consent law in Austria was not followed by an immediate increase in the number of donors. In 1984 a decentralized model of donor procurement was, however, developed progressively in Vienna and Linz. This coincided with a large and sustained increase in total donor numbers, with a 100% increase in these two centres.[10,39] When in 1992 a decrease in the number of donors occurred, an opinion poll was conducted among the collaborating intensive care units.[40] This indicated relation problems due to the overwork in the transplant unit. Appropriate measures increased the donor rate to its previous value. Both the Belgian and the Austrian experience confirm the importance of an efficient and decentralized organ procurement with a continuous motivation of the collaborating centres. Such a decentralized approach became, however, only productive in Belgium when the law had provided an adequate environment. In Austria the law preceded the reorganization of organ procurement and when a decentralized donor procurement was introduced, it resulted in an immediate increase in donation rate (see Figure 17.2).

Since 1989 the donation rate stabilized in all four Eurotransplant countries; in the countries following informed consent, Germany and The Netherlands at a mean value

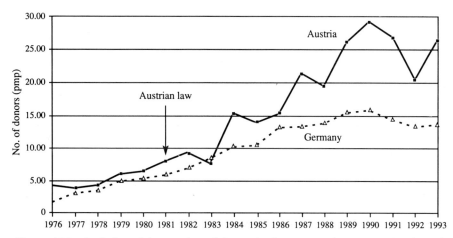

Figure 17.2 Evolution of the number of donors in Germany and Austria. (Data from the Eurotransplant Annual Reports.)

of 15 donors pmp, in Belgium at 20, and in Austria at 26. This evolution is interesting as it concerns four closely collaborating countries with similar medical development, and divergent evolution since the implementation of the law. The only interfering factor could be the number of traffic accidents, which is higher in Austria and Belgium, but this is unlikely to explain the observed differences.[41]

The evolution of organ donation in the Scandinavian countries is of interest as Sweden and Denmark changed to an informed consent law in 1988 and 1990 respectively. Until 1988 all four Scandinavian countries had adopted a presumed consent law with the additional obligation to inform the next-of-kin, thus allowing for any objection to be raised.[42] Only Finland and Norway however had accepted the definition of death based on destruction of the brain. In 1988 a bill was passed in Sweden, changing the definition of death from circulatory to brain death criteria. This new definition of death raised a controversy leading to the adoption in the same year of a modified transplant law based on informed consent.[43] In Denmark a debate concerning brain death coincided with a dramatic decrease in the number of donors in 1988.[44] In 1990 a law was passed accepting brain death and changing from a presumed to an informed consent law. In 1987 Denmark and Sweden had more donors than Finland and Norway, notwithstanding the absence of brain death criteria (see Figure 17.3). With all four countries now accepting brain death, Finland and Norway who had maintained their presumed consent legislation increased progressively their organ retrieval rate and in 1993 they were leading, ahead of Denmark and Sweden. Compared with the mean number of donors in 1987–1988, the mean of the number of donors in 1992–1993 decreased by 13.7% in both Denmark and Sweden, while it increased in the same period with 25% in Finland and Norway. This dramatic decrease after adoption of informed consent could have been influenced by the controversies concerning the definition of death. It must, however, be noted that these countries share the same cultural background and collaborate within a common organ exchange system. All have now accepted cerebral death, and the type of law is the only difference between them. An analogous phenomenon has been observed with the number of autopsies. Introduction in several countries of legislation requiring consent of the relatives has been associated

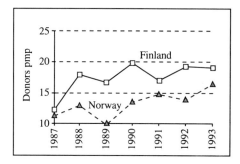 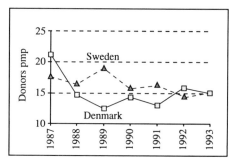

Figure 17.3 Number of donors per million population and per year in the Scandinavian countries. (Data from (44), (30) and the national procurement organizations.)

with a rapid decline in the number of autopsies.[27] In July 1996 Sweden reverted to a presumed consent law.

Taken together these data suggest that at least in Europe and among countries with active transplantation programmes, there is a correlation between the existence of a presumed consent law and a higher organ retrieval. The number of refusals is inversely correlated with the degree of responsibility given to the family in the decision to prelevation. To attribute this higher efficiency to a lesser number of objections alone would, however, be simplistic. The Spanish presumed consent law is efficient notwithstanding the obligation to obtain the consent of the family in writing. It is likely that the presumed consent law contributes to the motivation of the medical profession by increasing the legal safety and by creating a consensus that organ donation is the rule.

Belgium, Austria and most continental European countries have a presumed consent tradition for autopsies. Would a presumed consent law be equally efficient in countries with other autopsy traditions? A positive answer is suggested by the experience with cornea donation in the USA. From a worldwide survey, Lee et al.[12] concluded that evidence exists that movement away from requiring consent of surviving family members is associated with a higher yield of organs and tissues. In the USA eighteen states have adopted a presumed consent law for removal of corneas. As a result of these statutes a marked increase in the number of available corneas was observed, as contrasted with both other states and their historical experience: the increase was from 25 to more than 1000 in Georgia and from 500 to more than 3000 in Florida.[45]

Individual autonomy and transplantation laws

As opposed to informed consent which is assumed to better preserve the individual autonomy, presumed consent is often represented as a statist coercing law, where the state could use bodies without consent of the bereaved families. In fact, the decision of the individual is considered essential in both types of law, but in practice the individual autonomy is often jeopardized. In the USA, the Uniform Anatomical Gift Act states: 'An anatomical gift that is not revoked by the donor before death is irrevocable and does not require the consent or concurrence of any person after the donor's death'.[45] In practice, however, this guarantee is not honoured. Although signed donor cards constitute legally effective consent, physicians are reluctant to retrieve organs on the basis of these cards alone, and almost always require the consent of the next of kin.[6] According to a recent

UNOS publication[7] 'All organs for transplantation donated in the USA are obtained after explicit consent has been given by the family of the deceased person'. This means that in practice the will of the deceased to donate, even if expressed explicitly in writing, is subordinated to the decision of the next of kin. The reasons for this could be a reluctance of medical staff to offend the family's feelings. It would, however, need to be considered whether sometimes the motive for asking relatives for consent when there is a signed donor card reveals some ambivalence on the part of physicians concerning the taking of cadaver organs, whatever the source of this ambivalence.[85] Whatever the reasons, this practice common to most countries with informed consent laws is in contradiction with the principle of individual autonomy. The decision of the deceased is more likely to be honoured under presumed consent laws, especially if a registry is established by law as in Belgium.

In the absence of an expressed will of the deceased is it more ethical to presume consent or to presume refusal? Both can lead to non-respect of the wishes of the deceased, but this is less likely to occur if consent is presumed, if as most opinion polls indicate there is a significant majority in favour of donating organs after death.[5] Overall, when confronted with practice, the assumption that individual autonomy is better preserved by informed consent laws does not stand. In addition, in most countries with a presumed consent law, the significance of the abstention is largely publicized. In the Belgian law, in the absence of a registered will to donate or to oppose, organ removal can be decided, but the family keeps the right to object. In the same manner as it is an individual right not to write a testament, an individual can with full knowledge of the implications, decide not to register his decision. Mandated choice has been proposed in order to reduce the uncertainty about the wish of the deceased.[46] Mandated choice, however, conflicts, with the deeply rooted human attitude of denial of one's own death. Although most people are willing to write their testament in due time, most die without having done so. Coercing into an explicit choice does not seem to be the best way to preserve the individual autonomy.

Acceptability of the law

A recent excellent review by the King's Fund Foundation came to the conclusion that there was good evidence for the effectiveness of presumed consent legislation.

> Nevertheless, the medical profession, the transplant community and public opinion are split over the ethics of such a law, and so it would be inappropriate to recommend an immediate change in the law. If such a change provoked an acrimonious public debate it could damage the reputation of, and public confidence in, the transplant technology as a whole.[47]

Besides the patients waiting for a transplant, three other groups of persons, with at times conflicting interests, are concerned directly: the citizens, the family of the deceased and the medical profession. An efficient transplant law has to achieve an acceptable compromise between these conflicting interests. It is therefore important to examine the acceptability of the existing laws to the different categories concerned.

PUBLIC OPINION

In spite of a continuous negative publicity, representing presumed consent as the state deciding on the use of dead bodies, it is remarkable that in most public opinion polls a significant number considered presumed consent as an acceptable alternative. In a Canadian survey the majority thought this was a good system.[48] In two recent British polls, public opinion was evenly divided on the issue. When, however, in a third survey presumed consent was described in clear and simple terms, twice as many favoured as opposed it.[47] In a recent UNOS survey,[49] 39% of the responders said 'yes' and 52% 'no' to the question if doctors in the USA should be able to act on presumed consent. Here, also, there was evidence that the difficult notion of presumed consent had not been fully understood by some responders.[49] In Belgium the vote of the presumed consent law was preceded by bitter controversies in the media, as not only public opinion, but also the medical profession was divided on the issue. Here also there were grim predictions of catastrophic consequences on the public attitude against transplantation. When, however, the law was passed with a large majority in the House of Representatives and in the Senate, the controversies immediately subsided and the law is no longer debated. It is likely that the easily accessible and confidential registry greatly contributed to this acceptance by the public opinion, as those who did not agree with presumed consent for whatever reason could object. Only 2% of the population did so.

THE FAMILY OF THE DECEASED

A family confronted with the unexpected death of a relative is specially vulnerable. In addition, the family has at best to gain some increase in self-esteem, at worst an overwhelming sense of guilt for taking the decision to permit the mutilation of the body. Must the mourning process be interrupted by a request for donation? Should the family just be informed of the decision to proceed with organ removal? Should organ removal be hidden from the family? There is no easy answer to these questions. Under informed consent legislation at least 30% of the families will object, and this number is an underestimation as many doctors will refrain from asking if they expect an objection. Presumed consent laws leave the doctors some freedom to choose the attitude which seems most appropriate. Under the Austrian presumed consent law, which has no provisions for the rights of the family, 35% of the doctors in the Intensive Care Units inform the family[40] and in practice the refusal rate is 5%. In Belgium, as required by the law the central registry is consulted by modem before starting the donation procedure. In the absence of a will of the deceased, some transplant centres will always ask for permission of the family. In Antwerp this has led to a 23.7% refusal rate (M. De Broe, personal communication). From a recent informal survey conducted by the author among the centres collaborating with Leuven, it appeared that organ removal without information of the family was exceptional. It occurred when it was impossible to contact the family, or in circumstances suggesting a disinterest of the family. The smaller centres tended to ask systematically for the permission. The usual practice was, however, to inform the family to give them the possibility to object to donation, but without asking explicitly for the permission. All centres reported a refusal rate of less than 10%. Two factors could have contributed to this low refusal rate. The first is that merely informing the family, instead of asking for the permission, spares the family the responsibility for what can be perceived as a mutilation of the body. The second could be that in the Leuven Collaborative Group, the family is not approached by the transplant co-ordinator but by the doctor in charge of the patient or sometimes by the priest attached to the hospital. A close and trusting relationship between the asking person and the family of the patient could be more likely to lead to donation than contact with an unknown person, whatever his skills. Some presumed consent laws provide that the family should be informed and in Spain there is no choice, written consent of the family must be obtained. This led to a refusal rate of 26.7%.[50] Whatever the law, there are no examples of doctors removing organs against the explicit will of the family. In accordance with medical ethics, the right of the family to oppose donation is always respected. This certainly avoided emotional upheavals in the bereaved family, which would certainly have damaged severely the image of transplantation.

THE MEDICAL PROFESSION

Obviously the medical profession has no difficulty in accepting the principle of informed consent, although it limits the possibility to adapt their attitude to what is perceived to be the best interests of the patient's family. If they fear to confront the family with the request for donation, they will abstain, and even required request laws did not increase substantially the number of donors.[51] In addition legal security is guaranteed if all the legal requirements for authorization are fulfilled. Presumed consent on the other hand seems more difficult to accept. One of the reasons could be that it shifts the responsibility for the decision from the family to the doctor. This can create a fear of litigations. Within the medical profession the opinions on the acceptability of presumed consent vary with the position. Some 40% of doctors working in the transplantation field in Britain prefer presumed consent and 31% prefer informed consent. A large majority of transplant co-ordinators and doctors working in ICUs are, however, opposed.[52] In contrast presumed consent is usually well accepted in the countries where it was voted. In Austria 83.3% of the doctors working in intensive care units were in favour.[40] In Belgium it was well accepted, although a minority of doctors still oppose the concept that consent can be presumed. This acceptance is based essentially on two factors: the first is the legal safety. The existence of a computerized registry eliminates all discussions about the will of the deceased and the statement that the family must take the initiative to oppose donation, prevents all later controversies on how and to whom the information was given. The second factor is the flexibility permitted by the law, allowing all doctors to act according to their ethical principles and to what they estimate to be in the best interest of the bereaved family.

Two observations raise the question if there is something more than ethical considerations behind the opposition to presumed consent by some members of the medical profession. The first is the adoption in several states of the USA of presumed consent laws for cornea removal. The acceptance of these laws without any discussion of the need to inform the family or to obtain permission for donation stays in strong contrast with the discussions when removal of other organs is considered. As the only difference is in the time between death and organ removal, this could suggest a reluctance to take the responsibility for early organ removal in a brain dead patient without explicit consent of the family. This interpretation is supported by the recent events in Sweden and Denmark. In these two countries presumed consent was accepted as long as death was diagnosed on cardiac criteria. The change from cardiac to brain death criteria induced a change to informed consent. It could reveal on the side of the medical profession a reluctance to take the responsibility for removal of organs early after death, or an uneasiness with the diagnosis of brain death. By asking the authorization for organ removal the responsibility for the decision is shifted from the doctors to the family of the deceased and litigation becomes less likely.

Conclusion

The bulk of evidence indicates that presumed consent laws create an environment more likely to lead to higher rates of organ retrieval than informed consent. The real obstacles to the general adoption of such laws in the Western countries are not ethical, but lie in the attitude of the medical profession, which is determined in part by autopsy traditions, in part by the fear of litigation or other considerations. Any attempt to change the law should be preceded by a reflection on these issues. For all countries the law will have to be adapted to the different fundamental traditions of the population. In multicultural countries, different provisions for each community will have to be implemented, according to the example of Singapore. In Europe a trend away from presumed consent has been observed,[42] influenced by local difficulties in Germany and The Netherlands, and by the trend towards unification of the European health legislation. Those responsible for this trend should keep in mind that the price for this change is likely to be a decrease in the number of donors and additional suffering and unnecessarily death for those desperately in need of a transplant.

ACKNOWLEDGMENTS

The author is indebted to L. Roels, transplant co-ordinator, for providing data on organ retrieval rate. He thanks the doctors who agreed to discuss and to provide data on the procedure for organ donation in their unit, especially V. Bosteels and L. Janssens, E. Matthys, M. Segaert, S. Vanneste, Professor M. De Broe and Professor H. Van Aken.

References

1. WHO Guiding Principles on Human Transplantation. *International Digest of Health Legislation* 1991, **42**: 390–413.
2. 3rd Conference of European Health Ministers. *International Health Legislation* 1988, **39**: 274–278.
3. Somerville MA. 'Procurement' vs 'Donation' – Access to Tissues and Organs for Transplantation: Should 'Contracting Out' Legislation Be Adopted? *Transplantation Proceedings* 1985, **17**: 53–68.
4. Evers S, Farewell VT, Halloran PF. Public awareness of organ donation. *CMAJ* 1988, **138**: 237–239.
5. Cohen C. The case for presumed consent to transplant human organs after death. *Transplantation Proceedings* 1992, **24**:2168–2172.
6. Salaman JR, Griffin PJA, Ross W, Haines J. Lifeline Wales: Experience with a computerized kidney donor registry. *Br Med J* 1994, **308**: 30–31.
7. Task Force on Organ Transplantation. *Organ Transplantation. Issues and Recommendations.* US Department of Health and Human Services, 1986, page 51.
8. The Human Organ Transplant Act, Singapore 1987. *International Digest of Health Legislation* 1990, **41**, 257–261.
9. Dennis JM, Hanson P, Hodge EE, Krom RAF, Veatch RM. An Evaluation of the Ethics of Presumed Consent and a Proposal Based on Required Response. *UNOS Update* 1994, **10**: 16–21.
10. Mühlbacher F. Donor Recruitment in Austria. In de Charro Fth, Hessing DJ, Akveld JEM (Eds). *Systems of Donor Recruitment.* Kluwer, Deventer, 1992, pages 65–71.

11. Maroudy D, Maillard N, Mourey F, Eurin B. Approche juridique des prélèvements d'organes en France. *Soins Chirurgie* 1989, no. 104, 24–28.
12. Lee PP, Yang JC, McDonnell PJ, Maumenee E, Stark W. Worldwide legal requirements for obtaining corneas: 1990. *Cornea* 1992, **11**: 102–107.
13. Aiallam General discussion. In *Organ replacement therapy: Ethics, justice, commerce*. Land W, Dossetor JB (Eds), Springer Verlag, Berlin, Heidelberg, New York, 1991, pages 319–320.
14. Valderrabano F. Cadaver transplantation as an ethical and cost-effective alternative to living donor transplantation: The Spanish experience. *Transplantation Proceedings* 1992, **24**: 2103–2105.
15. Colombia. Law No. 73 of 20 December 1988. *International Digest of Health Legislation* 1990, **41**: 436–443.
16. Fuenzalida-Puelma HL. Organ Transplantation: The Latin American Legislative Response. *Bulletin of Pan Am Health Org* 1990, **24** (4): 425–445.
17. Soh P, Lim SML. Opting-out law: A Model for Asia – The Singapore Experience. *Transplantation Proceedings* 1992, **24**: 1337.
18. Iserson KV. Help for the living. In Iserson KV (Ed.) *Death to Dust*. Tuczon, AZ, Galen Press Ltd, 1994, page 85.
19. Sachedina AA. Islamic Views on Organ Transplantation. *Transplantation Proceedings* 1988, **20**: 1084–1088.
20. Daar AS. Organ Donation – World Experience; The Middle East. *Transplantation Proceedings* 1991, **23**: 2505–2507.
21. Namihira E. Shinto Concept Concerning the Dead Human Body. *Transplantation Proceedings* 1990, **22**: 940–941.
22. Nudeshima J. Obstacles to brain death and organ transplantation in Japan. *Lancet* 1991, **338**: 1063–1064.
23. Tsuji KT. The Buddhist View of the Body and Organ Transplantation. *Transplantation Proceedings* 1988, **20**: 1076–1078.
24. Sugunasiri SHJ. The Buddhist View Concerning the Dead Body. *Transplantation Proceedings* 1990, **22**: 947–949.
25. Richardson R. *Death, Dissection and the Destitute*. Penguin Books, London, 1988.
26. Corpse. *Encyclopaedia Britannica*, 1967, Vol. 6, page 542, William Benton Publishers, Chicago, London.
27. Svendsen E, Hill RB. Autopsy legislation and practice in various countries. *Arch Pathol Lab Med* 1987, **111**: 846–850.
28. Kokkedee W. Kidney Procurement Policies in the Eurotransplant Region 'Opting In' versus 'Opting Out'. *Soc Sci Med* 1992, **35**: 177–182.
29. Ogden D. Another View on Presumed Consent. *Hastings Center Report* 1983, **13**: 28. (Quoted in reference 9.)
30. Land W, Cohen B. Postmortem and Living Organ Donation in Europe: Transplant Laws and Activities. *Transplantation Proceedings* 1992, **24**: 2165–2167.
31. *UNOS Update*. 1994, **10**: 37.
32. Ponticelli C, Pellini F, D'Amico G. Renal transplantation – the dilemma in Italy. *Nephrol Dial Transplant* 1994, **9**: 746–748.
33. Colpart JJ, Vedrinne J, Revillard M, Moskovtchenko JF, Vialte P, Traeger J. Obstacles médico-légaux aux prélèvements d'organes. Influence des législations en France et en Europe. *Agressologie* 1986, **27**: 765–768.
34. Hors J. Presumed consent in practice in France. In de Charro FT, Hessing DJ, Akveld JEM (Eds). *Systems of Donor Recruitment*, Deventer: Kluwer, 1992, pages 121–128.
35. Benoit G, Spira A, Nicoulet I, Moukarzel M. Presumed consent law: Results of its application/Outcome from an epidemiologic survey. *Transplantation Proceedings* 1990, **22**: 320–322.
36. Matesanz R, Miranda B, Felipe C. Organ procurement and renal transplants in Spain: the impact of transplant coordination. *Nephrol Dial Transpl* 1994, **9**: 475–478.
37. Vanrenterghem Y, Waer M, Roels L, Lerut T, Gruwez J, Vandeputte M, Michielsen P.

Shortage of kidneys, a solvable problem? The Leuven experience. In *Clinical Transplants*, Terasaki P. (Ed.) UCLA Typing Laboratory, Los Angeles, California, 1988, pages 91–97.
38. Michielsen P. Organ shortage – What to do? *Transplantation Proceedings* 1992, **24**: 2391–2392.
39. Gnant MFX, Wamser P, Goetzinger P, Sautner T, Steininger R, Muehlbacher F. The impact of the presumed consent law and a decentralized organ procurement system on organ donation: Quadruplication in the number of organ donors. *Transplantation Proceedings* 1991, **23**: 2685–2686.
40. Wamser P, Goetzinger P, Barlan M, Gnant M, Hoelzenbein T, Watschinger B, Muehlbacher F. Reasons for 50% reduction in the number of organ donors within 2 years – opinion poll amongst all ICUS of a transplant centre. *Transpl Int* 1994, **7**: (S1) 668–671.
41. Michielsen P. Effect of transplantation laws on organ procurement. In: Touzaine JL (Ed.) *Organ Shortage: The Solutions*. Kluwer Academic Publishers, Dordrecht, Netherlands, pages 33–39.
42. Wolfslast G. Legal Aspects of Organ Transplantation: An overview of European Law. *J Heart and Lung Transplantation* 1992, **11**: S160–163.
43. Report by the Swedish Committee on Transplantation. Stockholm 1989. The Swedish Ministry of Health and Social Affairs.
44. Gabel H, Ahonen J, Södal G, Lamm L. The Procurement of Kidneys for Transplantation in Scandinavia. *Transplantation Proceedings* 1990, **22**: 330–332.
45. Uniform Anatomical Gift Act (1987). National Conference of Commissioners on Uniform State Laws. 676 North St Clair Street, suite 1700, Chicago, Illinois 60611 ((312) 915–0195).
46. Spital A. Mandated choice. The preferred solution to the organ shortage? *Arch Intern Med* 1992, **152**: 2421–2424.
47. New BN, Solomon M, Dingwall R, McHale J. *A question of give and take. Improving the supply of donor organs for transplantation*. King's Fund Institute, 14 Palace Court, London W2 4HT, page 82.
48. Corlett S. Public attitudes toward human organ donation. *Transplantation Proceedings* 1985, **17**: 103–110.
49. Kittur DS, Hogan MM, Thukral VK, McGaw LJ, Alexander JW. For the United Network for Organ Sharing ad Hoc Donations Committee. Incentives for Organ Donation? *Lancet* 1991, **338**: 1441–1443.
50. Matesanz R. Letter to the Editor. *Eurotransplant Newsletter* 1993, **111**, page 5.
51. Virnig BA, Caplan AL. Required request: What difference has it made? *Transplantation Proceedings* 1992, **24**: 2155–2158.
52. Taylor RMR. Opting in or out of organ donation. *Br Med J* 1992, **305**: 1380.

CHAPTER 18

The Spanish experience in organ donation

RAFAEL MATESANZ, BLANCA MIRANDA

Background   361
Development of the ONT network   362
Organ donor evolution   364
Transplantation figures   368
Possible gaps and future trends   368
Conclusion   371
References   372

Background

Spain is a European Community country with 38.5 million inhabitants divided into 17 autonomous regions. The National Health System comprises all facilities and public services devoted to health. Health counsellors in the 17 autonomous regions constitute the Interterritorial Council for the National Health System. The body is presided over by the Minister of Health and Consumer Affairs and is specifically in charge of the coordination of health policies. Today public health assistance is available for 99% of the population. The Spanish transplant story, however, started in 1965 with the first renal grafts performed in Barcelona and Madrid.

The transplant law approved by the Parliament in 1979 is technically quite similar to the laws in other western countries. Brain death is defined as 'the total and irreversible loss of brain function' and must be certified by three doctors (one of them neurosurgeon or neurologist) unrelated to the transplant teams. Signs of brain death must be explored clinically and documented by a silent EEG recorded for 30 minutes and these tests must be repeated twice with an interval of no less than six hours. The diagnosis is valid unless the patient is hypothermic or exposed to drugs with known brain depressive action. Organs can be retrieved only after informed consent of the family. The law also states that no compensation can be paid for donation, nor for grafted organs.

After the law was approved the transplant activity with cadaveric organs increased progressively during the eighties. After a maximum of 1182 kidney grafts in 1986, the number decreased by 20%. Then the annual number of renal transplants plateaued around 1000 during 1987–1989, leading to the exponential growth of the waiting list.

Development of the ONT network

In September 1989, the Organización Nacional de Trasplantes (ONT) was started as an organization attached to the Spanish Department of Health. From the very beginning it was presumed that the shortage of organs is the principal limitation to organ transplantation and that probably the problem was not a lack of suitable donors but, rather, a failure to turn potential into real donors.

Donation/transplantation is a complex process involving different steps that cannot be left to evolve in isolation (Figure 18.1). From the beginning the ONT emphasized the necessity that one person or group of few persons must be responsible for the organ and

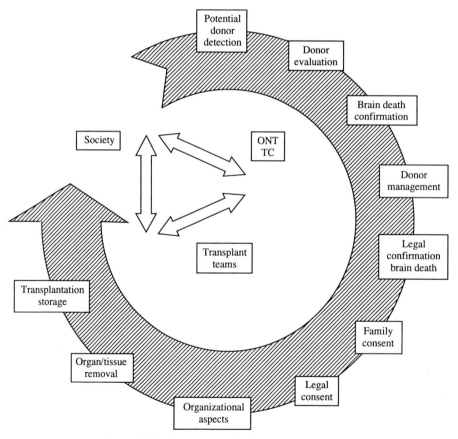

Figure 18.1 Donation and transplantation process.

tissue procurement and transplantation or storage for each potential donor in each hospital. The national transplant co-ordinating network was then conceived at four levels: national, regional and local or hospital, the development of which has been described previously.[1] The ONT has developed a formal but flexible management structure ensuring that the transplant co-ordinators who work at the 'grass roots' have a sense of involvement and accountability for performance. Most co-ordinators are qualified doctors and are mainly intensive care specialists or nephrologists, dedicated part-time to transplant co-ordinating tasks. In 1989 there were scarcely 25 transplant co-ordinating teams in Spain. At the moment 126 teams are active in our country, one in each hospital with the potential to have organ donors. Figure 18.2 shows the number, composition, origin and full- or part-time status of the transplant co-ordinators. The profile of the Spanish transplant co-ordinator is summarized in Table 18.1. As can be seen they report to the medical director of the hospital and not to the chief of the transplant unit, and they are increasingly involved in administrative tasks and relations with social groups other than those involved directly in hospitals but that could have potential influence on transplant activity and organ procurement. Media relations are carefully managed with specifically designed educational programmes to offer to

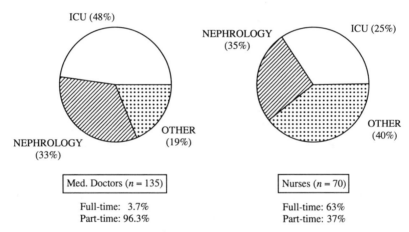

Figure 18.2 Co-ordinating teams – Spain 1994 ($n = 126$).

Table 18.1 The profile of the Spanish transplant co-ordinator

Qualification:	Physicians and nurses
Status:	Part-time
Report to:	Medical Director of the hospital
Origin:	Intensive Care Units
Continuity:	Temporary (3–4 years)
Location:	Within the hospital
Main aim:	Organ procurement
Increasingly involved in:	Educational programmes
	Management of resources
	Relations to media
	Administrative tasks

the transplant co-ordinators the best strategies to transmit messages to media professionals. Close contacts and successful collaborations are also carried out with patient associations, judges, coroners and other social group indirectly related with organ donation.

The main aim of all ONT professionals and transplant co-ordinators in Spain is organ procurement, others are secondary. In this setting the central office of the ONT acts as a service agency. It not only deals with the organ sharing and arranges transport of organs or transplant teams but also maintains the waiting lists and registries of transplant activity. It elaborates and updates the official statistical data and keeps interested groups informed about them. It maintains an open telephone line working 24 hours a day, seven days a week, to answer any doubt or question about organ procurement or transplantation. The direct link between the ONT and the Spanish Department of Health facilitates the flow of information to the health authorities about any problem or subject related to transplant activity. Last but not least, ONT is the support for the organ and tissue procurement and transplantation process in Spain, both at hospital and supra-hospital levels, guaranteeing the transparency of the entire procedure. The development of committees to discuss matters of transparency and possible conflicts with Health Authorities and transplant teams representatives has been very important. These committees can ensure compliance with all previously agreed regulations in matters such as organ distribution, criteria for inclusion in waiting lists etc.

Transplant co-ordination is a new discipline and is still developing. The great 'boom' in transplants did not occur until well into the eighties and the need for a system of co-ordination was not apparent until somewhat later when the growing complexity of the transplant process became obvious. The functions of the professionals dedicated to the retrieval and distribution of organs for transplant are thus not clearly defined. Like everything, the role of the transplant co-ordinator may be learned. The ONT has promoted the development of training programmes for health professionals, especially those involved in the organ retrieval, including all those themes involved in the process (Figure 18.1), from how to draw up a register of potential donors to the way to approach the grieving families, including the donor maintenance, the diagnosis of brain death etc. These training programmes are included in general intensive courses or in smaller single-subject seminars, i.e. those dedicated to 'general managing strategies', 'donation interview', 'relations with mass media' etc.

Organ donor evolution

From 1990, when the basic human infrastructure was developed in order to identify potential donors and obtain organs, the number of donors rose progressively (Figure 18.3). A 10% additional increase in the organ donation rate was observed during the first 6 months of 1995. Not only did the number of donors increase considerably but also the number of organs retrieved (Figure 18.4). The percentage of multiorgan retrieval soared from 30% in 1989 to 77% in 1994, enabling a two-fold rise of possible solid organ transplants and the decrease of the total solid organ waiting lists since 1991 (Figure 18.5). We have observed some variations in the characteristics of these organ donors. The average age has been increasing each year, making a difference of more than five years between 1992 and 1994 (34.5 ± 17 to 39.6 ± 18). More than 20% of our current donors are over 60 years. Although classically the most frequent source of potential donors was patients with cranial trauma, Spain has recently seen a reversal in the percentages of death caused by trauma and by cerebrovascular accidents (CVA) among organ donors. In 1992, 52% of deaths were caused by cranial trauma (43% road traffic trauma and 9% other traumas) and 39% by CVA. In 1994, those figures were 37% and 53% respectively, moreover only 27% of all donors died due to road traffic accidents. These changes are directly related to the 40% decrease in fatal road traffic accidents

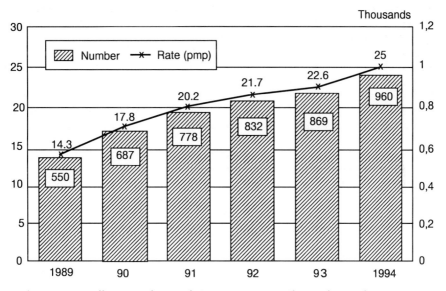

Figure 18.3 Effective cadaveric donors in Spain: number and annual rate (pmp).

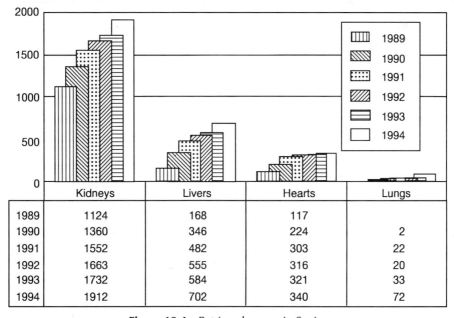

Figure 18.4 Retrieved organs in Spain.

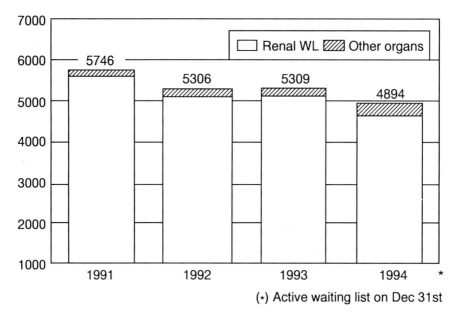

Figure 18.5 Solid organ waiting list: Spain 1991–1994.

Table 18.2 Average numbers of donors in spain stratified by the type of hospital for the years 1989–1996

Type of hospital	1989	1990	1991*	1992*	1993*	1994*
Type I	13.5 ± 7.8	16 ± 9.9	16.5 ± 8.2	18.5 ± 9.4	18.8 ± 9.5	20.5 ± 10.2
Type II**	3.8 ± 3.8	4.8 ± 3.8	6.3 ± 4.6	6.3 ± 4	6.7 ± 4.7	7.8 ± 6
Type III**	0.4 ± 0.8	0.6 ± 0.9	1.1 ± 1.9	1.1 ± 1.5	1.4 ± 1.6	1.5 ± 1.4

Type I Hospitals with neurosurgery and transplant units * $p < 0.05$ vs 1989
Type II Hospitals with a neurosurgery unit BUT without transplant unit ** $p < 0.05$ vs Type 1
Type III Hospitals with neither neurosurgery nor transplant unit

Table 18.3 Relationship of number of actual donors for the years 1989–1994 in Spain presented as an index of donor number per 100 hospital beds

Type of hospital	1989	1990	1991*	1992*	1993*	1994*
Type I	1.2 ± 0.7	1.5 ± 1	1.6 ± 0.8	1.7 ± 0.8	1.7 ± 0.8	1.9 ± 1
Type II	0.6 ± 0.6	0.9 ± 0.8	1.2 ± 1	1.2 ± 0.8	1.2 ± 0.8	1.4 ± 1
Type III	0.1 ± 0.2	0.15 ± 0.2	0.3 ± 0.4	0.3 ± 0.3	0.4 ± 0.4	0.4 ± 0.4

* $p < 0.05$ vs 1989

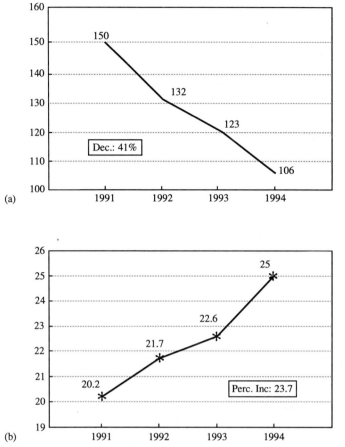

Figure 18.6 (a) Deaths from road traffic accidents in Spain (annual rate pmp) and (b) cadaveric organ donors in Spain (number pmp).

since the enactment of a new Road Security Law in June 1992. In spite of this fall no parallel decrease of donation rates was noticed. On the contrary, donor rates continue to rise (Figure 18.6).

There are 112 hospitals in which the organ procurement is allowed and from which at least one effective donor has been reported to the donor registry during the last few years. Among these hospitals 32 have transplant and neurosurgery units (Type I hospitals), 30 have neurosurgery but no transplant unit (Type II), and 50 have neither neurosurgery nor transplant units (Type III). Table 18.2 shows the annual average of donors and Table 18.3 the average index of donors/100 hospital beds in each type of hospital. Type I hospitals accounted for the largest number of donors but the percentage increase observed was higher in the other types of hospitals: 100% and 300% vs 50%. It is very important to note that this index is within the range of estimated theoretical capacity of donation in Type II and III hospitals (1.3 to 2.2 and 0.3 to 0.7 respectively) but Type I hospitals are below their theoretical capacity (3 to 4 donors/100 hospital beds).[2] In 1989, 24.7% of all donors were provided by hospitals without a transplant unit and this percentage increased to 31.6 in 1994 ($p < 0.05$) (136 vs 303 donors).

Transplantation figures

Table 18.4 shows the evolution of the solid organ transplantation overall activity during the last few years in Spain. As can be seen the contribution of living related organs to the general transplant activity is very low with only 1% of all kidney transplants compared to figures from USA (25%) or the Scandinavian countries (26%).[3] This transplantation activity allows a high turnover on the waiting lists: 77% of heart patients are grafted within 2 months and 74% of liver patients within 3 months. The renal waiting list has been decreasing progressively from 5593 patients at the end of 1991 to 4603 at the end of 1994. The mortality rate on the waiting list is also very low when compared with other published data:[4] 6% for liver patients and 9% for heart recipients during 1994.

Possible gaps and future trends

As we have said, donation/transplantation is a complex process involving different steps both starting and ending in the community (Figure 18.1). As Caplan testified in 1990

Table 18.4 Organ donation and transplantation activities in Spain* (Source: ONT 1996)

Years	1989	1990	1991	1992	1993	1994	1995
Cadaveric donors	569	681	778	832	869	960	1037
pmp	14	17.7	20.2	21.7	22.6	25	27.0
% Multiorgan donors	30%	51%	64%	69%	70.5%	77%	83%
Cadaveric kidney transplants	1021	1224	1355	1477	1473	1613	1765
pmp	26.5	31.8	35.2	38.4	38.4	42	46.0
Living related kidney	18	16	16	15	15	20	35
pmp	0.5	0.4	0.4	0.4	0.4	0.5	0.9
Liver transplant	170	313	412	468	495	614	698
pmp	4.4	8.1	10.7	12.1	13	16	18.1
Heart transplant*	97	164	232	254	287	292	278
pmp	2.5	4.6	6	6.6	7.5	7.6	7.2
Heart–lung transplant	5	4	2	1			
Single lung – double lung		2	3	10	16+4	17+19	17+28
pmp					0.5	0.9	1.1
Pancreas transplant	9	19	21	26	24	16	24
pmp	0.2	0.5	0.5	0.7	0.6	0.4	0.6

* Population 38.4 million
pmp = per million population

before the US Congress:

> ... What is truly distinctive about transplantation is not technology or cost, but ethics. Transplantation is the only area in all of health care which cannot exist without the participation of the public. It is the individual citizen who while alive, or in the case of vital organs, after death, who makes organs and tissues available for transplantation. If there were no gift of organs or tissues, transplantation would come to a grinding halt.

We should always keep this in mind: the donation/transplantation process starts and ends in society; the citizens are its motor and its main beneficiaries. Health professionals are obviously necessary since donors and recipients cannot keep in contact by themselves, but do not forget we are there only to help and be useful, we will never be the main protagonists. Our overall aim must be the optimization of the donation/transplantation process in order to alleviate the organ donor shortage.

Following the proposed schema (Figure 18.1) and after the conclusions of different studies,[6,7,8] three main areas of need or gaps can be identified in Spain: donor detection, donor management and family consent.

POTENTIAL DONOR DETECTION

This constitutes the starting point and is probably the most difficult step to make routine. The gap at this point has been estimated in ten lost potential donors pmp.[6]

Knowledge of the environmental characteristics in our area of work is mandatory: infrastructure of our hospitals, mortality of the area, incidence of road traffic accidents, CVA and cerebral tumours, health resources etc. We need an approach to the potential donor pool in our scope of influence with prospective registries of brain death cases complemented with retrospective medical record reviews to avoid errors such as over- or underestimation of brain death pool. Once a lack of donor detection is demonstrated in a particular hospital or area, it is mandatory to start a detection programme there.

This would be carried out by the transplant co-ordinator including:

- Educational programmes for health staff about the donation/transplantation process.
- Daily revisions of the list of patients admitted to the hospital.
- Follow up of patients admitted with Glasgow score <7.
- Daily visit to all critical care units that could have potential donors.

DONOR MANAGEMENT

During the time that it takes to evaluate the donor, to achieve all legal requirements and consents, and to organize the logistics of the retrieval, it is necessary to maintain the viability of the organs. This phase, depending on other requirements can be prolonged up to 24 hours or more. It is the direct responsibility of the doctor in charge at the intensive care unit, but it also depends on the surgical constraints and we cannot forget that time runs against us in the entire process. In the studies mentioned above[7,8] between 10 and 14% of donors presented with haemodynamic impairment or uncontrollable sepsis, contraindicating the donation. This problem could represent a loss of more than five donors pmp.[6] The promotion of research and educational programmes in this field should be mandatory to minimize such problems. New techniques or therapeutic strategies that could be helpful should be widely spread. Nevertheless there will always be a gap at this point that can probably only be minimized with the development of non-heart beating donor programmes.

APPROACHING THE FAMILY

Approaching the family represents a key point of the process and the most sensitive, since it occurs during the human drama that the death causes. In Spain, family refusal rates slightly decreased from 27.6% of all interviews in 1992 to 23.6% in 1994, with a wide range between regions (10–40%). This represents more than 300 potential donors that do not become real donors due to the lack of family permission (an estimated loss of 8 donors pmp). In most countries family consent is mandatory and family refusal rates are very high.[9,10,11] This has led to a great debate between authors sustaining the absolute necessity of strict presumed consent laws and authors supporting the consultation of relatives [12,13,14] (see Chapter 17). In a recent Spanish survey it was stated that most citizens (66%) are against a change in the current practice and only 6% believe that retrieval of organs should be performed without requesting the family's wishes,[15,16] noting that it could represent an abuse of authority or even an insult to the deceased person.

More than 90% of the general Spanish population has a favourable opinion about donor's families and transplant procedures.[16] There is thus a difference in attitude between this hypothetical situation evaluated in general surveys and the real situation when the relative has died and the donor option is presented. The main causes for family refusal to organ donation in Spain when the real option is presented are summarized in Table 18.5. The first problem arises when there is a lack of information about the wishes of the deceased person. People should thus be encouraged to speak about donation with the family and to transmit their opinion to their relatives, in order to facilitate the decision in this particular and difficult moment. This has also been stated in general surveys in Spain and the USA. In the absence of information only 54% of the Americans and 50% of the Spanish would donate the organs of their relatives, but when the problem has been discussed the percentages rose to 93% and 94% respectively.[16,17]

The family's answer does not depend only on their own attitude but also on the way in which the option is presented. The causes of denied consent do not vary very much between countries. Available data from the USA[11,17] emphasized the main reasons for failure of consent were:

- Doubts about brain death (40% of people do not believe that brain death equals death).
- Conflicts of interest (58% of people are sure that poor people do not have the same opportunities to get a transplant as the rich).
- The person requesting the organ donation is not comfortable presenting the option.
- Timing and place were not appropriate. The requesting interview must be decoupled from the communication of the death (this point has also been made in a French study).[18]

After our own experience, there are some further points that need special attention regarding the approach to the family.

- The approach should be made by specially trained staff. It is then mandatory to develop training programmes specifically designated to this purpose with the help of psychologists and communication experts.
- It is necessary to make a complete approach and offer of help, not simply an interview requesting organ donation.
- The first approach should be carefully prepared, gathering all available information about the family members and deciding both the optimum timing and place. Before starting with the interview it is mandatory to ensure that the family have already understood their relative is dead.
- We should never show that we are in a hurry.

Table 18.5 Reasons not to donate organs

	Study A	Study B
Centres	1	12
Timing	1989–93	1993–94
Interviews	205	618
Refusal rate	19.5%	16.6%
• Lack of/inaccurate information provided to family		
– Brain death	17.5%	6%
– Corpse integrity		5%
• Family opposed	20%	24%
• Lack of information about donor's wishes	15%	15%
• Social claims	10%	4%
• Previous negative wishes of the donor	25%	30%
• Religious causes	2.5%	4%
• Problems with hospital staff	5%	8%
• Other	5%	4%

Reproduced with the permission of C. Santiago and P. Gomez (unpublished data)

- All interviews need to be evaluated in a follow-up by the co-ordinating team, and all previous errors avoided. The most frequent errors are: to get angry, not to follow the rhythm of assimilation of the relatives, to interrupt the family etc.
- We should never forget we are there to help and be useful and never to disturb anyone. We should remember that most families believe that donation provides a positive outcome from death and helps with the grieving process, and that 100% of donor families would donate again and 30% of families that denied the donation would have changed their mind a year after.[19]

Conclusion

The Spanish organ procurement rates mainly result from the efforts to overcome obstacles such as untrained or undertrained staff, unidentified donors and reluctance to approach grieving families. This means professionalization of 'Organ Donation' and the need for an organization focusing on promotion and facilitation of all these actions. It is true that Spain has made a lot of progress in the organ donation field during the last few years, but more must be achieved. The co-ordinating network has been able to identify more and more donors in all types of hospitals, demonstrating that the problem is not a lack of potential donors but, rather, a failure to turn potential into real donors. The overall analysis of the organ donation process in the state, or the circumstantial evaluation of problems in each region or hospital, has provided us with the means to find appropriate solutions. Keeping records of all potential donors and brain deaths enable us to find when, how and where donors were lost and to define when and where efforts are needed. If we continue to succeed in this, Spanish people will continue to benefit from a secure and confident transplantation service.

References

1. Matesanz R, Miranda B, Felipe C. Organ Procurement in Spain: The impact of transplant coordination. *Clin Transpl* 1994, **8**: 281–286.
2. Darpon J, Texeira JB, Martinez L, Olaizola P, Lavari R, Elorrieta P, Aranzábal J. Potencial generador de donantes en diferentes tipos de hospitales (CAV). *Rev Esp de Trasplantes* 1995, **4**: 9–13.
3. Transplant Report of the Council of Europe. Vol 6, December 1994, pages 274.
4. Annual Report of the US Scientific Registry for Organ Transplantation and the Organ Procurement and Transplantation Network. UNOS, 1991.
5. Caplan AL. A market proposal for increasing the supply of cadaveric organs. Commentary on Cohen. *Clin Transpl* 1991, **5**: 471–474.
6. Aranzábal J, Texeira JB, Darpon J, Martinez L, Olaizola P, Lavari R, Elorrieta P, Arrieta J. Capacidad generadora de órganos en la CA del Pais Vasco: Control de Calidad. *Rev Esp de Trasplantes* 1995, **4**: 14–18.
7. Cabrer C. Aplicación del diagrama ASME al proceso de Obtención de Organos para Trasplante. Tesis Doctoral, Universidad de Barcelona, 1994.
8. Navarro A. Potential donors and brain death epidemiology in the region of Madrid. In *Organ Shortage: The Solutions*. Tourraine JL et al. (Eds) Kluwer Academic Publishers, The Netherlands, 1995, Chapter 20.
9. Miranda B. Aspects Juridiques et Législatifs des Tranplantation D'organes: Tranplantations D'Organes en Europe: Science et Conscience. European Postgraduate Training Institute of Health Care EURO COS. European Parliament, Strasbourg, France, 1994.
10. Nethen H, Jarrell B, Broznik B, Kochik R, Hamilton B, Stuart S, Ackroyd T, Nell M. Estimation and characterization of the potential organ donor pool in Pennsylvania. *Transplantation* 1991, **51**: 142–149.
11. Garrison R, Beutley F, Raque G, Polk H, Sledeck L, Evanisko M, Luces B. There is an answer to the shortage of organ donors. *Surgery Gyn & Obst* 1991, **173**: 391–396.
12. First MR. Transplantation in the Nineties. *Transplantation* 1992, **53**: 1–11.
13. Roels L, Varenterghem Y, Waer M, Gruwez J, Michelsen P. Effects of a Presumed Consent Law on Organ Retrieval in Belgium. *Transp Proc* 1991, **23**: 2546.
14. Matesanz R. El Consentimiento Informado en la Donación de Organos. In *Coordinación y Trasplantes. El Modelo Español*. Matesanz R, Miranda B (Eds), Ed Aula Médica, Madrid, Spain, 1995, pages 161–164.
15. Martin A et al. Donación de Organos y Trasplantes. Aspoectos psicosociales. *Nefrologia* 1991, **XI** (Sup 1): 62–68.
16. Martin A, Martinez JM, Lopez Martinez S. Encuesta Sociológica sobre la Donación de Organos. In *Coordinación y Trasplantes. El Modelo Español*, Matesanz R, Miranda B (Eds), Ed Aula Médica, Madrid, Spain, 1995, pages 143–160.
17. Gallup Survey. *The USA public's attitudes toward Organ Transplantation – Organ Donation*. Gallup Organization (G087073) Princeton, New Jersey, 1987.
18. Pottecher T, Jacob F, Pain L, Simon S, Pivirotto ML. Information des familles de donneurs d'organes. Facteurs d'acceptation ou de refus du don. Résultâts d'une enquête multicentrique. *Ann Fr de Réanim* 1993, **12**: 478–482.
19. Frutos MA, Blanca MJ, Rando B, Ruiz P, Rosel J. Actitudes de las familias de donantes y no-donantes de órganos. *Rev Esp de Trasp* 1994, **3**: 163–169.

CHAPTER 19

The European Donor Hospital Education Programme (EDHEP)

CELIA WIGHT, BERNARD COHEN

The origins of EDHEP   373
The development of EDHEP   374
Programme objectives and content   375
Adaptation and implementation of EDHEP   376
Evaluation of EDHEP   377
Conclusion   381
References   381

The origins of EDHEP

The Eurotransplant Foundation (ET) is an Organ Exchange Organization (OEO) based in Leiden, the Netherlands. ET provides a service for The Netherlands, Germany, Austria, Belgium and Luxembourg, which covers a population of over 110 million people. In 1990 the ET region experienced a fall in organ donation that resulted in a further increase in the gap between the supply and demand for donor organs.

Seeking a reason for these trends, researchers in ET discovered that many medical professionals within the five countries were still unaware of the needs of transplantation and in many hospitals protocols and procedures for organ donation were not available. Additional data showed that as many as 30% of potential donors were lost because of a refusal by the family.[1] This was often the result of difficulties in understanding the concept of brain death and misconceptions about the procedures of organ donation. The researchers felt that part of these misconceptions may have been the result of inadequate communication between the doctor and nurse and the relatives.

Following this train of thought a literature search was undertaken on the attitudes of doctors and nurses towards this subject. This review revealed that the difficulties staff say they have can be grouped under three headings:

1. The important role that the professional's own feelings towards death and organ donation can have in inhibiting the donation request.

2. Worries about intruding on a family's grief.
3. Lack of experience and training in approaching bereaved relatives.

These findings are supported by research from the United States.[2]

The development of EDHEP

Eurotransplant wanted to tackle the problem of the growing donor shortage and is the driving force behind the development of EDHEP. In 1991, with the generous support of the Dutch Kidney and Heart Foundations, ET developed a professional education programme for the Netherlands. Later, appreciating the value of a prototype that could eventually be adapted for use throughout Europe, Sandoz Pharma in Basel, Switzerland also agreed to support the development of what became EDHEP. The programme is part of a multifaceted process aimed at closing the gap between the supply and demand for organs and tissue for transplantation. It was created to meet the widely perceived need to help doctors and nurses feel effective in dealing with the bereaved and in requesting organ donation. EDHEP was produced by professionals from different specialties including medical doctors and clinical psychologists from the University of Maastricht, transplant co-ordinators (TCs) and communications specialists in close collaboration with The Rowland Company (Figure 19.1). The prototype was designed to be easily

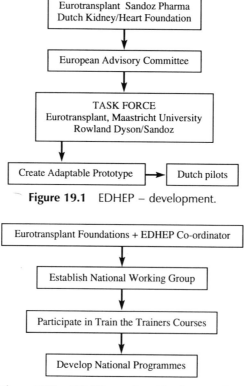

Figure 19.1   EDHEP – development.

Figure 19.2   EDHEP – national implementation.

translated into different languages, adjusted to meet national circumstances and cultures and to complement other existing educational programmes. With the help of the regional TCs a Dutch version of the programme was piloted in three regions in the Netherlands in 1991. The participants were relatively experienced doctors and nurses, although some had never been involved in requesting organ donation. Following minor adjustments EDHEP was introduced as a professional educational tool in the Netherlands and in 1992 was made available to other interested countries (Figure 19.2).

Programme objectives and content

The objectives of EDHEP are to:

1. Raise professional awareness of the problem of the donor shortage and to provide some possible solutions.
2. Develop an awareness of the needs of the bereaved between medical professionals and the needs of the professional in dealing with the bereaved.
3. Provide 'hands-on' communication skills training in dealing with grief reactions and requesting organ donation.
4. Provide guidance on setting up hospital protocols for dealing with bereaved relatives.

EDHEP is produced in two parts. EDHEP Part 1 is called 'Meeting the Donor Shortage'. The aim of Part 1 is to raise the awareness and understanding of organ and tissue donation and transplantation among all levels of medical, nursing and paramedical staff in all hospital departments. 'Meeting the Donor Shortage' is an informative slide presentation that covers the history and current state of the art in organ and tissue transplantation. It also outlines some reasons behind the growing donor shortage and puts forward some possible solutions to the many problems that surround this area of medicine.

Many countries already have TCs who regularly provide this type of information to hospital staff. The material is available both to those countries who wish to supplement their existing educational programmes and countries who are not yet providing this type of educational service. Slides and suggestions for overhead transparencies, which can also include statistics and information of local interest, are provided. The presentation takes about one hour and is intended to be given in hospitals, at an appropriate time, by TCs or other transplant personnel.

EDHEP Part 2, is called 'The Grief Response and Donation Request'. Part 2 takes the form of a highly interactive one-day skills awareness workshop, conducted outside the hospital setting, well away from bleeps and telephones. 'The Grief Response and Donation Request' workshop is organized and introduced by the local TCs but is moderated by communications skills experts. The workshop is targeted at professional staff working in any area of medicine that brings them in contact with bereaved relatives. It aims to help participants improve their communication skills and heighten their sensitivity to the needs of the bereaved. In addition, participants learn ways to break bad news and discuss organ donation with the bereaved. With the help of exercises and teaching video tapes the EDHEP workshop encourages participants to:

1. Examine their own feelings aroused by loss and separation.
2. React to different grief reactions with an appropriate professional response.

3. Listen to bereaved relatives describe their feelings of loss and donation.
4. Analyse problems of miscommunication.
5. Participate in skills training with simulated bereaved relatives.
6. Discuss ways of establishing hospital protocols for dealing with the bereaved.

Adaptation and implementation of EDHEP

To ease the national adaptation and implementation of EDHEP a co-ordinator was appointed in 1992, based at ET. The role of the co-ordinator is to identify countries with an interest in adopting EDHEP and to provide advice and assistance in five areas:

1. The formation and composition of national working groups. The national working groups (Table 19.1) are responsible for adapting the EDHEP educational material to their own language and needs. The group is also responsible for the national setting up and evaluation of the programme.
2. The acquisition of workshop moderators and simulated bereaved relatives. It is not necessary initially that the moderators are familiar with donation procedures, the ideal moderator being described in Table 19.2. However, this knowledge is important when running the EDHEP Part 2 workshops and can be provided under the supervision of the national working group. Simulated bereaved relatives are used to enhance the authenticity of the separate role play exercises when participants are asked to break bad news and request organ donation.
3. Organize 'Train The Trainers' (TTT) courses. The TTT courses are international training courses for national working group members and potential EDHEP workshop moderators. The courses are run, in English, by the EDHEP international co-ordinator and clinical psychologists from the University of Maastricht, The Netherlands. The main purpose of these courses is for participants to gain familiarity with the content and format of the programme. Training in moderating the different modules is provided for potential EDHEP Workshop moderators. National working group members are made aware of the variety of activities needed to adapt and set up the programme in their national environment. Finally, the TTT course aims to establish a close dialogue between EDHEP national working groups and the EDHEP international working group.
4. Provide guidance and assistance with the national adaptation of the EDHEP educational material.

Table 19.1 Composition of national EDHEP working groups

The ideal working group could comprise members of the following organizations:

- National transplant societies
- Intensive care/neurosurgical societies
- National medical/nursing associations
- Transplant co-ordinator's organizations
- Communications skills trainers
- Public relations specialists
- Religious advisors/community leaders

Table 19.2 Workshop moderators and simulated bereaved relations

The ideal EDHEP Workshop Moderator should have the following background experience:
- Knowledge of behavioural skills
- Experience in communication skills training
- Familiarity with the psychological aspects of bereavement
- Familiarity with the nature of the medical professional's work in a hospital environment
- Familiarity with donor procedures
- Experience with using a variety of educational material

Suitable EDHEP Moderators can be found among:
- Bereavement counsellors
- Clinical psychologists
- Crisis care councillors
- Educational psychologists
- Communication skills training experts

Simulated Bereaved Relatives
The University of Maastricht has a tradition of using local volunteers as simulated patients during the training of medical students. A simulated bereaved relative is a non-professional actor, trained in playing a prescribed role and giving feedback. Countries who do not have such a pool of volunteers available to them have chosen to use amateur actors as simulated bereaved relatives.

5. Ensure a degree of programme quality assurance. Countries are asked to sign a contract with ET. The purpose of the contract is to preserve the spirit and quality of EDHEP during its adaptation and implementation and to ensure that adequate training, help and guidance is provided by the international EDHEP co-ordinator.

The costs involved in the translation, adaptation and application of EDHEP have, for the main part, been borne by the national Sandoz affiliates. Although, some countries, have from the beginning, sought financial support from other sources. All countries hope that once EDHEP has become a regular part of professional education programmes the running costs of the EDHEP workshops will be funded from local budgets.

Evaluation of EDHEP

Since 1992, over 330 participants have attended sixteen international EDHEP TTT courses held, either at Eurotransplant's home base in Leiden, or another host country. By the end of 1995 EDHEP was running in over 33 countries in Western Europe (Figure 19.3), the Middle and Far East and Latin and South America. National Working Groups from these countries confirm that EDHEP adapts easily to all national legal, religious, cultural and educational needs. The programme generates a more favourable attitude among critical staff to organ donation, teaches more confident communication, stimulates national and international professional collaboration and increases donor referral. Participants in the EDHEP workshops report that as a result they feel more confident in their ability to talk with the bereaved and are more willing to do so.

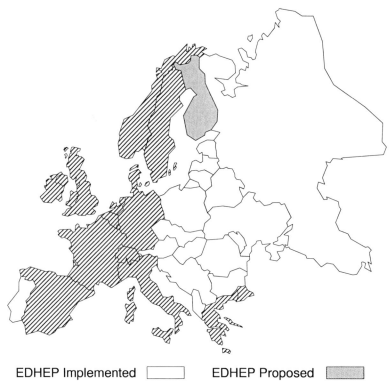

EDHEP Implemented ☐   EDHEP Proposed ▨

Figure 19.3 European Donor Hospital Education Programme 1995.

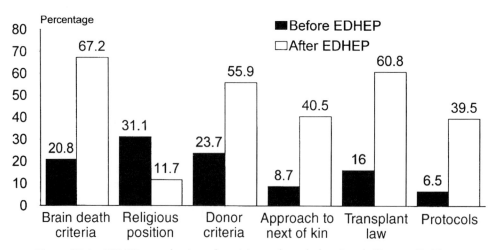

Figure 19.4 EDHEP – evaluation of participants knowledge: Israel. (Source: Dr Pierre Singer, Director ICU, Beilinson Medical Centre, Peta-Tikva, Israel.)

Do you think you learned how to communicate with grieving families?

Is this programme useful in your workplace?

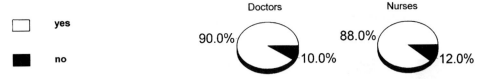

Figure 19.5 Pilot evaluation of EDHEP Kobe, Inuyama, Fukuoka: Japan. (Source: Uchida et al. Development and assessment of methods to educate medical professionals and increase organ transplantation in different prefectures. Japan.)

Countries operating in a 'presumed consent' system have also found the major elements of the EDHEP workshop very relevant. Communication skills are always needed when informing a bereaved family of the intention to remove organs from a deceased loved one. Specific evaluations of 163 participants in pilot programmes in Israel show an increase in knowledge on brain death criteria, the law, the supposed negative position of the Jewish religion in organ donation and the criteria for donation.

These data also showed a significant improvement in participants' ability and willingness to approach bereaved relatives for organ donation (Figure 19.4). Results from Japan show that 95% of doctors and 92% of nurses felt they had learned how to communicate better with the bereaved and 90% and 88% respectively felt that EDHEP was a useful educational tool (Figure 19.5). A one year controlled research project of matched professionals has been carried out in The Netherlands (Blok et al.). Both groups were tested two weeks before the experimental group participated in an EDHEP workshop. Both groups were retested at two weeks and six months following the experimental group's participation in the workshop. The results show an increase in knowledge of the experimental group on grief reactions, communication skills and donation procedures. In addition, participants show an increase in self-confidence in dealing with the bereaved. The experimental group demonstrated a reduction in the perceived difficulty in requesting donation and at six months follow-up had made more donation requests that the control group (Figure 19.6). All EDHEP participants rated the contribution of EDHEP to their current practice positively. EDHEP seems to have a positive effect on current practice in general not only relating to the care of the bereaved and donation procedures, but also to working together as a team (Figure 19.7). All these findings are statistically significant. A controlled evaluation by the Universities of Liverpool, Manchester and Maastricht is underway in the UK that will, in addition to the above, analyse relatives' satisfaction in the professional care they receive.

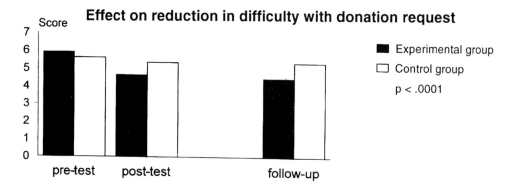
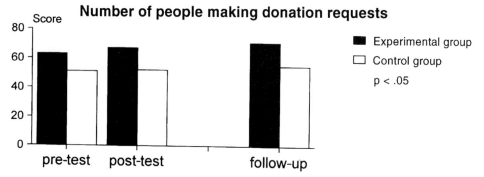

Figure 19.6 Effect of EDHEP on participants self-confidence: The Netherlands. (Source: Blok, GA et al. Effect of the European Donor Hospital Education Programme, University of Maastricht, The Netherlands.)

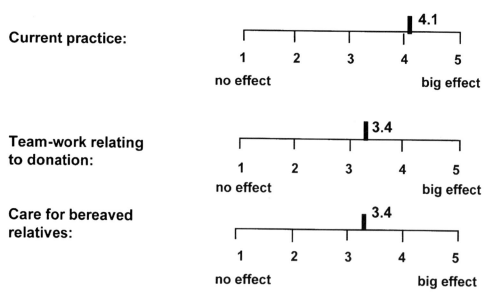

Figure 19.7 Effect of EDHEP on current practice: The Netherlands. (Source: Blok, GA et al. Effect of the European Donor Hospital Education Programme, University of Maastricht, The Netherlands.)

Conclusion

EDHEP is a bold initiative designed to tackle the lack of training or experience that exists within the medical profession in dealing with the bereaved and requesting organ donation. Its primary function is to establish and monitor a nationally appropriate education programme throughout Europe that will teach relevant medical staff to become confident, if not comfortable, in requesting organ donation from distressed relatives. The outstanding international interest in EDHEP displays almost universal concerns over the shortfall in donor organs and the lack of communication skills training in medical and nursing schools. There can be no doubt that the skills awareness workshop goes some way towards meeting the current needs of professionals working in critical care areas of hospitals. The long-term goal of EDHEP is that these staff should acquire a positive attitude towards the benefits of transplantation that will, perhaps, help mitigate the distress of the relatives and increase the rate of organ donation.

References

1. Gore S, Taylor RMR, Wallwork J. Availability of transplantable organs from brain stem dead donors in intensive care units. *Br Med J* 1991, **302**: 149–143.
2. Malecki MS, Hoffman MC. Getting to yes: How nurses attitudes affect their success in obtaining consent for organ and tissue donation. *Dialysis & Transplantation* 1987, 276–278.

CHAPTER 20

'Making the Critical Difference'™: Education, motivation, donation

GIGI POLITOSKI, JAN BOLLER, KATHLEEN CASEY

Introduction   382
Programme description   383
Conclusion   388
References   388

Introduction

Due to extraordinary medical advances, organ and tissue transplantation as a method of treatment has evolved from experiment to reality, creating an increasing need for organ and tissue donors. Currently, in the United States, more than 43,000 patients wait for life saving or life enhancing organ transplants. Eight to nine patients die each day waiting.[1] The healthcare professionals involved in transplantation, including procurement agencies, hospitals and other organizations, are constantly examining the possible causes for this severe shortage and prescribing actions to increase the number of families who participate in organ/tissue donation.

In addition to public education campaigns, supporting collateral material, public opinion polls and analyses of the potential donor pool, procurement agencies and hospitals continue to work collaboratively in developing comprehensive programmes that are targeted to those healthcare professionals most intimately involved in the donation process.

There is a public perception in America that donation is a positive act of goodwill, yet less than 30% of those who could donate become donors.[2] What happens during that critical time to influence the family's decision to participate in organ/tissue donation?

Doctors and nurses who care for dying patients and their families are the 'gatekeepers' in determining whether or not a family is provided with the opportunity to make a decision about donation. In particular, critical care nurses have been identified as pivotal participants in the entire donation process.[3] In the United States, 'required request' legislation came into effect in the late 1980s requiring hospitals to offer the option of donation to every family whose loved one met the criteria. In most states, families of potential donors have to agree to the donation by signing a consent form before tissue or organ donation can occur. Nurses suddenly became one of the key players in the donation process, where frequently there was little or no training that defined an appropriate role for them in the process.

This chapter describes an innovative programme that educates nurses about their roles and motivates them to be involved in the organ and tissue donation process.

Programme description

'Making the Critical Difference™: Education, Motivation, Donation' (MCD) is an eight-hour continuing education workshop that incorporates interactive videos, slide presentations, emotional exercises and group discussions to facilitate discussion about the underlying feelings and barriers that may hinder their participation in the organ donation process. Three positive roles are identified that nurses can incorporate into their daily routines. At the same time, it strengthens their skills in providing families with the option of donation.

One of the main objectives of the programme is to position making the option of donation available as part of the continuum of care nurses offer their patients and families. This extension of caring may often be the last positive thing they can do for the family.

HISTORY

To further the cause of organ and tissue donation among critical care nurses, the National Kidney Foundation (NKF) developed 'Making the Critical Difference'™ in collaboration with the American Association of Critical Care Nurses (AACN) and the North American Transplant Coordinators Organizations (NATCO). The programme's sensitive approach to learning by overcoming personal barriers took more than two years to develop.

The collaborative effort is continually reinforced and demonstrated as MCD workshops are presented in communities around the United States by procurement organizations, trained to use the programme, working with local Chapters of AACN and local Affiliates of the NKF. Over 200 transplant professionals have been trained and licensed by the NKF to provide the programme in the past seven years.

In 1989, before developing the programme, the NKF surveyed hundreds of nurses at AACN's premier educational meeting, the National Teaching Institute™ (NTI) to determine the effectiveness of existing programmes in motivating participation in the organ and tissue donation process and to determine what was lacking in this education.

Results of the survey indicated that while much of the training was effective in helping nurses with the technical aspects of identifying and caring for potential donors, few programmes dealt with nurses' needs, feelings and day-to-day anxieties affecting their willingness to become more actively involved in the donor process. Specific barriers which nurses repeatedly mentioned were:

- Unsupportive physicians.
- Difficulty of explaining or re-explaining brain death.
- The 'why me' attitude.
- Reluctance to intrude on a family's grief.
- Fear of rejection.
- Inability to form the words needed to approach a family about donation.[3]

Responses demonstrated that before nurses could successfully participate in the donation process, they needed to identify, reflect and acknowledge their feelings about death, not only their patients', but their own. While at the same time, nurses needed to identify positive roles they could comfortably play in the donation process.

WORKSHOP DESIGN

After consultation with a variety of experts, for example transplant professionals, critical care educators, physicians, bereavement counsellors, psychologists, critical care nurses, donor families and marketing professionals, a programme design was developed which provided the content detailed under the following headings.

Assessing personal beliefs and attitudes

Each workshop begins with an emotional and thought-provoking exercise that encourages quiet thought and reflection. By putting death in a personal perspective, it begins the process of breaking down barriers. To be effective with potential donor families and peers, nurses are asked to look at their own vulnerability and acknowledge the grief they sometimes experience working with these families.

Video scenarios (a case study)

Interactive video scenarios are then shown which address these barriers. Created along the lines of a mini soap opera, the video scenarios depict a family finding out their loved one is brain dead and the professionals involved in offering them the option of organ/tissue donation.

After each scene, participants are asked to reflect on personal feelings and share their experiences, if they wish, with their group. The intensity of the videos varies to stimulate and provoke participant involvement emotionally and intellectually and to encourage participants to talk about themselves and their own feelings about what is taking place in each scene. Through viewing these interactive videos and sharing experiences, nurses realize there is commonality in their experience. Through this awareness the learning process begins.

Initiator, facilitator, supporter

The roles of Initiator, Facilitator and Supporter are introduced and presented as flexible, interchangeable and equal in importance. Each nurse decides what role he or she will take in each situation, both with families and colleagues.

An exercise in loss

Nurses attending the programme also participate in an exercise simulating personal loss. This interactive exercise provides an opportunity to begin to experience the emotions that families experience with the sudden loss of a loved one. Through participation in this exercise, they can more freely understand the care and support families need.

A donor family perspective

To provide a deeper understanding of what the donation process means from the perspective of a donor family, a moving and memorable video is shown that dramatizes the real life experiences of a donor mother.

The donor mother is also a registered nurse and certified bereavement counsellor. After relating her personal experience with organ and tissue donation, she explores the nursing implications of supporting, comforting and communicating with grieving families throughout the donation process, from initial interaction at the hospital through bereavement aftercare to the personal anxieties many nurses feel as a result of losing a patient. To reinforce the video, a donor family member from the local community also participates in the presentation.

Understanding brain death and the donation process

After exploring personal feelings about death and donation, and identifying needs of donor families, participants learn how they can participate more fully in the donation process from a transplant professional. This presentation provides an overview of organ and tissue transplantation from inception to current practices and procedures. The presentation is localized to meet the needs of the nurses and procurement organizations in the community including a discussion about local laws and regulations, national issues and organizations that serve the industry, donor criteria, and legal and medical requirements for determining brain death.

The time spent discussing offering the family the option of donation, resembles the 'consent workshops' that procurement agencies routinely provide to educate nurses about their policies regarding organ/tissue donation procedures. The difference with MCD is that the focus is on helping nurses understand the process of offering the option of donation, and how this process relates to the feelings of the grieving families.

In the United States, the relationship between the hospital and the procurement organization differs from community to community. Due to geographic, cultural or other considerations, different healthcare professionals may be designated to speak with the family about donation. The transplant professional has an opportunity to explain his/her facilities' policies and practices and present an ideal model that meets that community's needs.

It is stressed throughout the presentation and the workshop, that the outcome of offering the option of donation is not as important as ensuring that the needs of the family are met effectively. Allowing the family to make this important decision, no matter what they say, may be beneficial to them.

Group discussion

Group discussions conclude the workshop, giving nurses an opportunity to address individualized concerns and questions. As a possible activity, role playing gives participants practice in what they have learned, and provides an opportunity to express aloud many things said in real life situations. It is through role playing that nurses identify unique ways to contribute to the donation process.

THE OUTCOME

To date, over 400 local programmes have reached an audience of over 10,000 critical care nurses. The programme consistently receives the highest participant satisfaction ratings among all AACN programmes, including the National Teaching Institute™, with 78% of the participants rating the programme quality as excellent and another 21% rating it as good. Some 99% of the participants indicate an increased willingness to participate in the donor process and 98% feel they are more capable to participate after attending a workshop.

MCD has built stronger partnerships at the local, regional and national levels, among transplant professionals and critical care nurses. In this case, the consumer was the catalyst.

OUTCOMES STUDY

Method

During the first two years of 'Making the Critical Difference'™, a longitudinal outcomes study was conducted to determine changes in participants' practice, values and beliefs after attending the programme. Developed specifically for this programme, the instrument was a scannable questionnaire consisting of 33 questions. The survey questions related to participant demographics, institutional information, hospital involvement in transplantation institution, the process of donation, and personal attitudes and skills relating to the donation process itself. The instrument was administered three times. The first time was the morning of the workshop, before the programme began. Participants were mailed the same questionnaire six months and twelve months after the programme. The outcomes study was completed in December of 1993.

Participants

Fifty-nine out of a potential 70 sites participated in the survey, representing workshops held between 1 January 1991 and 30 September 1992. Over 3,500 nurses took part in the initial survey, 650 nurses (18%) participated in the six-month survey, and 406 participants (11%) returned the third (twelve-month) survey. Although conclusions that can be made from the study are limited due the low response rate, some interesting observations can be made that warrant further exploration.

The typical survey respondent was between 37 and 43 years old, had worked in critical care between six and nine years, cared for adult patients, worked in a hospital greater than 300 beds, had a bachelor's degree, and was in a staff nurse position.

Results

No substantial change in organ transplantation was reported in the first six months by survey respondents. Nor was there a change in the mean number of potential donors or families who were approached. However, the reported affiliation with organ procurement organizations rose from 91% to 97%. Respondent participation in the organ donation process rose from 76% to 84% and the percentage of those who had participated in more than ten organ/tissue donation processes rose from 14% to 21%.

Another interesting change over the first six months was the change in the nurses' role in the donation process, and this is demonstrated in Table 20.1. These results may indicate that the participants may have become more actively involved in the process of offering families the option of donation.

In examining the process of approaching families, the critical care nurse was first at all stages of the study, the ICU physician was second and the family physician third. The transplant co-ordinator was a distant fourth, though they are most often thought of as the 'initiators'.

Probably the most striking changes reported by the survey respondents related to their own levels of knowledge and confidence and actions taken as potential donors themselves. In the question relating to personal reasons for lack of involvement in the donation process, lack of knowledge fell from 76% in the preliminary survey group to 48% in the six-month survey group and 38% at twelve months. Interestingly enough, 'not a priority' rose from 15% to 43%, most likely now that 'lack of knowledge' is no longer a personal barrier. In the question relating to words that describe your feelings if you were to approach a family about donation, 'confident' rose from 24% to 41% and 'unprepared' fell from 20% to 2%. The descriptors of 'apprehensive' and 'uncomfortable' remained unchanged. The percentage of respondents who had signed organ donor cards rose from 57% to 73% and the percentage of those who had discussed donation with their own family members rose from 76% to 93%.

These outcomes, though not conclusive, indicate that this programme may be influencing the confidence and active involvement of critical care nurses in the donation process with increased collaboration with transplant professionals. The desired outcome of increased donation is not reflected. Whether or not the workshop has increased the effectiveness of critical care nurses in making the donation process more positive for the grieving family may never be known.

CANADIAN AND EUROPEAN APPLICATIONS OF THE PROGRAMME

When the programme was first introduced by the NKF and the AACN in November 1990, transplant professionals from a number of countries, including Canada, New Zealand and

Table 20.1 Change in nurses' role in the donation process

Role in the donation process	Original	Six-month	Twelve-month
Identification of donor	55%	72%	75%
Initiation of request	39%	48%	53%
Accompanying requestor	32%	42%	44%
Supporting the family	54%	69%	72%
Observer of asking process	32%	38%	36%
No experience	28%	0.1%	0.1%

The Netherlands, were invited to explore foreign adaptations of the programme. The Eurotransplant Foundation has been extremely successful in using the programme model and its individual components to develop its European Donor Hospital Education Programme (EDHEP). It is also hoped that the original programme, 'Making the Critical Difference'™, will be tested in Canada.

FUTURE APPLICATIONS OF THE PROGRAMME

The wide acceptance of 'Making the Critical Difference'™ as an important resource in the education of nurses has stimulated much discussion about its usefulness and adaptability for the education of other healthcare professionals, including physicians, hospital chaplains and social workers. The programme developers are examining various components of the existing programme to adapt and test with other professional disciplines. A number of different modules for the original programme are being developed so that the programme can be easily adapted to a variety of donation practices. In addition, a programme that promotes a multidisciplinary approach to offering families the option of donation is also being investigated.

Conclusion

In 1994, the transplant community saw almost a 9% increase in the number of organ transplants. The number of cadaver organ donors increased from 4,521 in 1992 to 4,853 in 1993.[4]

While this is extremely gratifying, this increase is not sufficient in meeting the growing demand for more donor organs. The authors believe that the ongoing development and delivery of programmes such as the one described will positively impact on the professionals' level of awareness, knowledge and commitment to the process of organ and tissue donation. As relationships between hospitals and the transplant community continue to improve it is hoped that more families will be provided with the opportunity to make a decision about organ and tissue donation.

To learn more on how you can get involved in 'Making The Critical Difference'™ the programme, either as a participant, or a member of the faculty, please contact the National Kidney Foundation at 1-800-622-9010, or American Association of Critical Care Nurses at 1-800-899-2226.

References

1. United Network for Organ Sharing, Recipient Waiting List, March, 1995.
2. National Kidney Foundation, *Controversies in Organ Donation. A Summary Report*, June, 1993.
3. National Kidney Foundation, Survey of critical care nurses, May, 1988.
4. United Network for Organ Sharing, *UNOS Update*, November, 1995, pages 32–33.

CHAPTER 21

The Partnership for Organ Donation: a strategic approach to solving the organ donor shortage

CAROL L. BEASLEY, JESSICA D. BLAUSTEIN

Introduction   389
Strategic overview   390
The future for organ donation   397
Appendix   398
References and notes   399

Introduction

The Partnership for Organ Donation is a private non-profit organization working to close the gap between the number of organ transplants that are possible and the number of organ donations that take place. The organization was created in 1990 by a group of business strategy and healthcare professionals who saw an important opportunity to save and improve lives, and who had extensive experience in solving complex organizational problems. They believed these skills could be applied successfully to the organ donor shortage.

The Partnership works to identify the most promising strategies for relieving the organ donor shortage, and joins forces with a variety of other transplant and donation organizations to bring about major improvements in the organ donation system. Oversight to the organization's work is provided by advisory boards comprising physicians, nurses, and other professionals in the fields of transplantation and critical care medicine.

Since 1990, The Partnership has been actively involved in research and implementation projects to increase organ donation in 18 regions of the USA and Canada and over 150 hospitals. The organization has also sponsored the most extensive survey on public attitudes about donation, and continually investigates new and promising avenues for solving the donor shortage. The Partnership collaborates with the Harvard School of Public Health and the Harvard Medical School in designing, analysing, evaluating and publishing the results of these efforts.

Strategic overview

The success of transplantation has given rise to its most intractable problem – the severe shortage of donated organs. While various approaches have been proposed, none has delivered the hoped-for increases in donation. Too often the field has pursued one strategy after another without a coherent framework or the sustained focus needed to bring about donation improvements. The Partnership's approach is to focus on those opportunities for change that offer the greatest benefit, address them with systematic programmes, and carefully evaluate the results.

Looking broadly at the transplant system in the USA, The Partnership believes that the donor shortage is a soluble problem. First, there is significantly more cadaveric donor potential than is currently being realized. Based on research and hospital interventions, The Partnership estimates that 12,000 to 15,000 people die each year under circumstances that would allow organ donation.[1] In 1995, however, only 5,360 people became organ donors.[2] In other words, only about one-third of potential donors actually donate.[3] Based on studies conducted in 1987[3] and 1990[4] approximately one-third of families were not offered the option to donate, and the other third declined donation. In simple terms, asking all eligible families, and finding ways to encourage more families to consent to donation could significantly increase the number of organs available for transplantation.

Knowing the amount of untapped potential for donation is only the beginning. Actually designing and implementing improvements to the system is a complex undertaking, involving hundreds of donor hospitals, thousands of health care professionals, about seventy organ procurement organizations, and the general public. In order to solve a problem of this magnitude and complexity it is necessary to look at the problem from a broad systems point of view, as detailed in Table 21.1.

We know from public attitudes research that the first condition, support for donation, has largely been attained with the American public. A recent Gallup survey showed that 85% of Americans supported donation and 69% said they would be somewhat likely or very likely to want to donate their own organs.[5] Support varies somewhat across ethnic groups, with lower support expressed by African-Americans and Hispanics than non-Hispanic whites. But, overall, donation and transplantation enjoy an extremely high level of public support.

Table 21.1 Donation would be maximized if these criteria are met

The public was predisposed to donate
+
Families knew each other's wishes regarding donation
+
All eligible families were asked to donate in an appropriate manner

The second condition, that families know each other's wishes, is currently far from being met. Of those who are likely to want to be donors, only about half reported having communicated this wish to family members,[5] even though, in the USA, the family makes the final decision about whether donation will take place. Strategically, improving family communication about donation is an important area of opportunity.

Finally, for donation to take place, the hospital must identify potential donors, and assist the family in making a decision about donation at a time of intense personal tragedy. Despite US legislation mandating that hospitals request donation of eligible families, the quality of the donation process in hospitals is extremely inconsistent. Too often families are not offered the option to donate, and even when donation is offered, there is no assurance that the process will meet their needs, or that staff working with the family will be properly trained.

FOCUS ON DONOR HOSPITALS

The donation process that takes place in hospitals is the principal focus of The Partnership's work. There are two main reasons for this. First, intervening in hospitals is efficient. Due to the severity of their injuries, all potential donors are treated in hospitals and, therefore, can be accessed through a hospital process. Furthermore, donor potential tends to be concentrated in relatively few large hospitals. The Partnership estimates that of the nearly 6,000 hospitals nationwide, fewer than 1,000 hospitals account for 70% of all potential donor cases. This means that interventions can be targeted to those hospitals where there is strong potential to increase donation, without having to invest resources in reaching *every* hospital.

Second, independent of a family's prior attitudes about donation, the in-hospital process matters to the family's decision. A poor hospital process can undermine even a supportive family's interest in donation, whereas a good hospital process can support families through the donation decision, regardless of whether they had previously considered donation.

When compared to the challenge of educating the public, there are significant advantages to focusing efforts on educating hospital professionals. Instead of reaching and influencing roughly 250 million Americans, most of whom will never be faced with a donation situation in their lives, effort and resources can be targeted to the 10–15,000 health professionals who are directly involved in the care of severely ill or injured patients to ensure that they are prepared to support families appropriately when donation becomes an option.

FAMILY NEEDS AND OPTIMAL DONATION PRACTICES

At the core of each donation situation is a bereaved family, coping with a sudden and often traumatic loss. Many experts in organ procurement, critical care, ethics, and other related disciplines have helped to shape a consensus on what donor families need. Donor family members themselves have provided invaluable insight into the process by sharing their experiences. At this point there is reasonable agreement about the typical needs of families, as well as the recommended practices for healthcare and organ procurement professionals to follow in caring for bereaved families in situations where organ donation is a possibility (see Table 21.2).

Table 21.2 Experts' recommended practices to follow in caring for bereaved families

- Hospitals should establish and facilitate a systematic, ongoing programme to identify all potential donors in order to ensure that every family is provided with the opportunity to donate all appropriate organs and tissues.
- Information regarding the patient's condition and prognosis should be discussed with the family on a continual basis in a sensitive and consistent manner.
- Death should be determined and confirmed in a timely fashion and communicated clearly, consistently and unequivocally to the patient's family.
- Health professionals should ensure that the family acknowledges that death has occurred before initiating a request for organ and tissue donation – a 'decoupled' request.
- All families should be offered compassionate assistance in coping with the patient's death and the subsequent bereavement process.
- Professionals who discuss donation with the family should be comfortable with the concept of brain death, trained in the organ and tissue donation process, and skilled in interacting with grieving families.
- Sensitivity to ethnic and cultural values is required of those engaged in the organ and tissue donation request process.
- The hospital and organ and tissue procurement organizations should work together to support the hospital's quality assurance programme.[6]

Despite general agreement among a wide range of professionals about the recommendations above, the current donation system is not succeeding in carrying out these recommendations consistently. Work done by The Partnership in four Organ Procurement Organization (OPO) regions over three years showed that fewer than one out of three potential donor cases exhibited the core characteristics of a well-run donation process, including decoupling, offering donation in a private setting, and involving the OPO in the donation request. A major challenge for the field is finding the best ways to implement consistent processes that ensure that every family is offered donation in the right time, in the right place, and by the right professionals. There is still much to learn about the needs of donor families, and ongoing research into donor family needs by The Partnership and others will help to refine our understanding about what donation practices are most helpful to families.

IMPLEMENTING OPTIMAL DONATION PRACTICES IN HOSPITALS

While every hospital is different, any initiative to improve donation in a hospital will focus on a couple of key performance outcomes. First, is the identification of all potential donor cases, ensuring that every opportunity for donation is recognized and acted upon. Second, is ensuring that a consistent donation process takes place in every case, so that every family receives a process that incorporates the characteristics listed above.

Implementing a new set of recommended donation practices requires changing the behaviour of a wide range of health professionals. It also requires that hospitals and OPOs have organizational systems to ensure that every potential donor situation is handled consistently and correctly and to provide ongoing monitoring and quality assurance for donation. Much of The Partnership's work has focused on the practical goal of bringing about donation improvements in hospitals (see Table 21.3).

Table 21.3 System for improving donation practice in hospitals

---

Diagnose current hospital performance
+
Build consensus around a new donation protocol
+
Establish a donation infrastructure
+
Conduct in-depth education of all relevant staff
+
Enact the protocol
+
Conduct ongoing quality assurance to monitor and improve the process

---

Hospital diagnosis

The diagnosis is a critical first step in improving donation, because – as with an ill patient – it is impossible to treat a hospital with an 'ailing' donation system without first understanding the symptoms and causes of the ailments. To conduct the diagnosis, The Partnership explores the hospital's past record of donation as well as the practices and attitudes of the staff, and relevant policies and procedures. The data collected are objective and subjective, quantitative and qualitative, written and verbal. By gathering data from a wide variety of sources, it is possible to develop a comprehensive evaluation of the hospital's donation process – where it is healthy, and where it needs improvement – and identify a strategy for change.

The hospital diagnosis typically starts with a medical records review (MRR) for the previous year (see the Appendix at the end of this chapter), and provides an answer to the following critical questions:

- How many potential donor cases were there in the institution?
- What was the outcome for each case?

Knowing the answers to these questions reveals the broad outlines of hospital performance, showing how consistently the hospital recognizes the potential for donation, and how often families consent to donation when asked.

The Partnership's work in hospitals has shown that this summary measure of performance – donors as a percent of medically suitable potential donors – varies widely across hospitals and that the reasons for a given level of performance can also vary significantly. In a diagnostic study of 39 large potential donor hospitals conducted in conjunction with four OPOs, hospitals realized anywhere from 12% to 68% of their underlying donor potential (Figure 21.1). By measuring donor potential, it is possible to look beyond the number of actual donors to identify which hospitals have the greatest unrealized potential and to identify the most promising opportunities to improve the situation.

A closer examination of selected individual hospitals (Figure 21.2) reveals very different performance issues among hospitals. In Hospital B, for example, most families of potential donors were not offered the option of donation, while in Hospital C, the

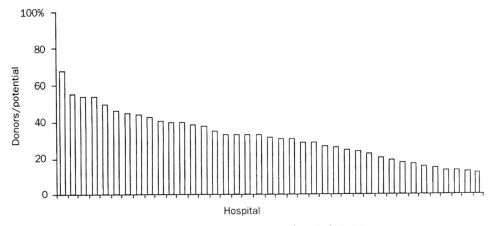

Figure 21.1 Donation rate – 39 hospitals 1990.

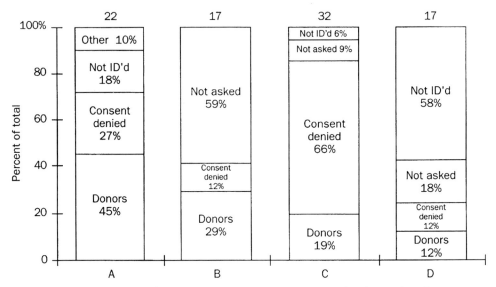

Figure 21.2 Four hospital donation situations. (Source; MRRs of 39 hospitals with average potential ≥ 10, 1990.)

vast majority of families were asked, but denied consent. In Hospital D, identification is the key issue. Understanding where donors are lost is a prerequisite to tailoring an appropriate strategy for change – one that fits the reality of each individual hospital's situation.

Qualitative data are collected from interviews as well as a hospital staff survey (see the Appendix at the end of this chapter). The purpose of these tools is to understand the knowledge, attitudes and behaviour of the hospital staff who are likely to be involved in donation situations. The data have been very revealing, typically confirming that hospital staff have extremely high levels of support for organ donation, but showing where factual and process knowledge are weak.

For example, results of a survey conducted in 1991–1992 with 1,450 hospital professionals[7] revealed strong support for organ donation (93%) but limited knowledge both of general facts about donation and transplantation, and about the donation process. Only 13% of professionals were aware of the high level of general public support for organ donation. Approximately half underestimated both the success of a kidney graft, and the size of the waiting list for an organ transplant.

Most importantly, it is clear that there is currently no consensus on standards of practice for requesting donation. When asked about the most appropriate time to introduce the donation option to the potential donor family, only 31% said 'after declaration of brain death'. The majority of respondents felt donation should be raised prior to the declaration of death or at the same time, in contrast to research results indicating that families are more likely to consent to donation when it is raised after brain death has been determined and explained.

Finally, while virtually all staff surveyed believed that training is important for the person making the donation request, only 37% had actually received such training.

Hospital interviews provide insight into the perspectives of key individuals involved in donation, and provide a mechanism for assessing commitment to donation and willingness to change. As a part of every hospital assessment, the Partnership meets individually with top administrators, physicians, intensive care nurses, social workers, chaplains and representatives from all intensive care units and relevant services to learn their perspectives on the donation process, gauge their support for organ donation and assess their commitment to improving the process within their own institution.

The diagnostic process as a whole results in a complete portrait of donation performance in a hospital, including:

- Data about all the potential donor cases occurring in the institution.
- Outcome of every case, and the patterns of non-donation specific to each hospital.
- Staff attitudes, knowledge and practices, and their perceptions about how the process can be improved.

Building consensus around new protocol

In order to ensure that a hospital, or any organization, can embrace a change initiative, it is necessary to build consensus that change is necessary and that the proposed change is appropriate and will work in that setting. Creating a sense of ownership for the donation process and the issues surrounding the donation process requires the involvement of hospital staff throughout the design and implementation of a solution.

Sharing the findings revealed in the hospital diagnosis is a first step in raising the staff's awareness of issues and opportunities to improve donation in their institution. Often, hospital staff do not realize that potential donors are being missed, or why they are missed. And few hospitals have developed structured methods for evaluating their own performance.

The consensus building process continues through the formation of a multidisciplinary donation committee typically comprising Partnership and OPO staff, physicians and nurses from units or services that care for potential donors and their families, administrators, social workers, and chaplains. This core team is responsible for overseeing the entire change process in the hospital – interpreting the findings of the hospital diagnosis, identifying the root causes for problems identified in the diagnosis, developing solutions that are suitable to the individual hospital, developing and executing an implementation plan, and monitoring the process and results. Initially they are an important engine of consensus-building, as they synthesize the diagnostic information, and gather input from staff about feasible solutions.

Establish donation infrastructure

Building a strong organizational support structure for organ donation is a complex challenge. Consider the organizational context for donation. Even in a relatively busy donor hospital, donation is a rare event – perhaps two or three potential cases might arise in a typical month. These cases may be dispersed among several intensive care units, and could involve any of dozens of hospital personnel. In most hospitals there is no locus of control for the process, so theoretically many hospital professionals could interact with the family on the issue of donation. With no current consensus on professional standards of practice for donation there is a strong likelihood that staff may end up working at cross-purposes. In the welter of demands facing critical care staff, the benefits – to the deceased's family, and to prospective transplant recipients – may not be apparent. Similarly, there are typically no consequences for staff if organ donation is handled poorly, as most donor hospitals do not track their donation performance, and staff understandably put the needs of living patients first.

An effective infrastructure for donation must include the following elements:

1. A standard organ donation protocol with clearly defined roles and responsibilities for each individual involved in donation.
2. A hospital-based team with specific training and responsibility to carry out the donation protocol, ensuring that the protocol is followed in every case.
3. Explicit hospital support for those involved in the donation process.
4. A measurement system which will track organ donation potential, track the process for each potential donor situation, and provide data to the donation committee and other relevant staff.

The hospital-based donation team is critically important as its members manage the communication with the family. By centralizing the responsibilities for donation within a core group – typically composed of critical care nurses and physicians – the hospital guarantees that the individuals working with the families are trained in organ donation, will be sensitive to the family's needs, and will understand and implement the hospital's donation process effectively.

Education

Once the protocol has been determined, it is important to educate all hospital staff involved in potential donor situations. This education – conducted for all intensive care units, the emergency room, residents, critical care staff, social workers, and chaplains – will ensure that all relevant hospital staff understand the protocol and have the information necessary to perform their role in the process.

Education is best conducted in concentrated training sessions which take place just before the protocol launch, and ideally is accompanied by internal publicity and a visible show of top administration support. The training sessions provide participants with a general overview of organ donation, detailed information about their own hospital's performance, each step of the new protocol and their specific role in the donation process. For members of the core donation team, the education is more intensive, in-depth, and participative. In a two-day programme, this group learns the details of the donation process and practises a variety of family communication scenarios.

A number of educational programmes exist that can be used for the purpose of training a core donation team. While The Partnership has developed a programme that incorporates all of the key elements of the change process, programmes such as Making the Critical Difference,[8] or the European Donor Hospital Education Programme (EDHEP)[9] cover similar material. The crucial factor is to provide in-depth training for the small designated team of donation experts and to provide broader education to the rest of the staff directing them to call upon the donation team when donation is a possibility. This role distinction is important to ensure that the core team develops a strong base of experience with donation enabling them to master the variety of skills involved, while ensuring that the rest of the hospital staff can carry out clear, simple instructions regarding their role in the process.

Enact new protocol

While enactment of the a new donation protocol might seem like a mere formality, focusing organizational attention and energy on the start of a new set of practices helps the transition to a new system go more smoothly. It is also an important way for hospital administration to re-emphasize institutional commitment to donation.

Ongoing monitoring, quality assurance

After the new protocol has been implemented in a hospital, it is essential to monitor the quality of the process and the implementation of the protocol. Information from on-going medical records reviews as well as prospective tracking of the donation process in each potential donor situation enables the donation committee to monitor donation performance and identify opportunities for improvement on an ongoing basis. Of particular importance, the group tracks the key elements of the donation process: Were all patients identified? Were all families offered the option of donation? Was it offered in at the right time? In an appropriate setting? By the right people?

The role of the organ procurement organization

While the locus of activity and responsibility for donation resides primarily within the individual hospital's core donation group, the local organ procurement organization plays a critical role. The OPO co-ordinators apply their experience by educating hospital staff about donation, making the actual donation request, clinically managing organ donors, and co-ordinating the recovery and placement process. The OPO, together with The Partnership, is often a key motivating force behind a hospital's initiative to improve its management of the donation process.

The future for organ donation

If all the recommendations outlined above were adopted in all the relevant hospitals in the United States, would the donor shortage be resolved? Perhaps not, because increases in donation may themselves fuel the demand for transplantation, so we may be living with a shortage for many years to come. However, improving hospital practice has the potential to increase donation significantly, perhaps as much as doubling it. This would mean shorter waits, fewer deaths, and better clinical outcomes for recipients. Realistically, we must continue to pursue and evaluate all the good opportunities available to maximize the availability of organs for those who need them. The Partnership believes that the same systematic approach to improving the hospital process for cadaveric brain-dead donation can also be applied to other donation processes, whether living donation, non-heart beating donation, or modes of donation currently unknown.

APPENDIX

TOOLS AND METHODS FOR IMPROVING DONATION IN HOSPITALS

The Partnership has developed diagnostic, analytic, and educational tools to support and enhance the process for improving organ donation within hospitals. Descriptions of the most essential tools follow.

Medical records review

The medical records review (MRR) is the richest source of quantitative information in the diagnostic process. This method entails systematic review of the medical records of all deaths that occurred within a hospital over a defined period of time, usually at least a year. Ruling out only cases that would be clearly disqualified from donation – such as patients with metastatic cancer or HIV infection – the remaining charts are reviewed and key clinical and process information relevant to donation is abstracted.

Donor tracking tool

The 'donor tracking tool' is a data collection form that is used prospectively to track the process steps in each donation situation. Focusing primarily on the family communication process, the donor tracking tool records who was involved in the process, when the subject of donation was raised, and how the family responded. Donor tracking tools from a single hospital can be combined and analysed to identify where there is opportunity for improvement in the donation process, and to target individuals or units in need of education.

Hospital staff surveys

Hospital staff surveys provide insight into the attitudes and knowledge of key staff about donation and transplantation, and assess hospital staff awareness of the optimal donation process components. This information highlights potential barriers to organ donation in hospitals and assists in identifying opportunities to remove those barriers.

The survey is administered to all hospital staff involved in organ donation, typically including critical care physicians and nurses, neurosurgeons, neurologists, social workers, and clergy. The survey reveals attitudinal issues – whether or not the staff support organ donation and feel comfortable with their role in the process – as well as their knowledge of factual information, such as the success rates of organ transplantation and the steps in an optimal donation process.

Educational tools

A variety of educational tools is used to teach hospital staff about organ donation and their roles in the donation process. These tools are used in a variety of forums, such as grand rounds, one-on-one meetings, brief hospital-wide inservices, or intensive two-day training programmes for in-house donation teams. The tools include the following.

Hospital education package

The 'hospital education package' is a comprehensive collection of presentation materials on an array of topics including an overview of the donor shortage, the American public's attitudes towards donation, key steps in the donation process, minorities and organ donation, and other topics. The slides can be combined in many ways in order to meet the educational needs of various groups.

Meeting family needs

'Meeting family needs' is a full-day presentation and workshop exploring the needs of families in critical care and looking specifically at the needs of families facing organ donation. It provides information on the grief process and on how healthcare professionals can help families through the initial shock and grief of losing a loved one.

In-house co-ordinator training programme

The in-house co-ordinator training programme is an intensive two-day workshop preparing selected hospital staff to fulfil the role of a donation co-ordinator. Using highly interactive exercises, the team learns how to co-ordinate the many steps in an optimal donation process – from the referral through the request – and gains skill in family communication through structured role-playing.

References and notes

1. For the purposes of this discussion we will assume that potential donors are brain dead individuals with otherwise acceptable health histories, and satisfactory organ function. Some centres are experimenting with organ recovery from non-heart beating donors, and this may prove to be another opportunity to relieve the organ shortage.
2. *1996 Annual Report of the US Scientific Registry of Transplant Recipients and the Organ Procurement and Transplantation Network – Transplant Data: 1988–1994*. UNOS, Richmond, VA and the Division of Transplantation, Bureau of Health Resources Development, Health Resources and Services Administration, US Department of Health and Human Services, Bethesda, MD.
3. Nathan H et al. Estimation and characterization of the potential renal organ donor pool in Pennsylvania, *Transplantation* 1991, **51**: 142–149.
4. Gortmaker SL, Beasley CL, Brigham LE et al. Organ donor potential and performance: size and nature of the organ donor shortfall. *Crit Care Med* 1996, **24**: 423–439.
5. The Gallup Organization, Inc., *The American Public's Attitudes Toward Organ Donation and Transplantation*, conducted for The Partnership for Organ Donation, Boston, MA, February, 1993.
6. The Partnership for Organ Donation, Inc. and The Annenberg Washington Program in Communications Policy Studies of Northwestern University, *Solving the Donor Shortage by Meeting Family Needs: A Communications Model*, 1990.
7. The Partnership for Organ Donation, *Partnership Hospital Staff Survey*, (unpublished data, 1990–1992).
8. The National Kidney Foundation, American Association of Critical Care Nurses, *Making the Critical Difference: Education, Motivation, Donation*, 1990.
9. The Eurotransplant Foundation, *The European Donor Hospital Education Programme*.

CHAPTER 22

Education in schools

NAPIER M. THOMSON, ROY KNUDSON, GEOFF SCULLY

Introduction 400
The role of education in the formation of attitudes to social issues 401
Education in schools 402
School education programme: *Transplantation – The Issues* 403
References 411

Introduction

Numerous factors determine the availability of organs for transplantation but probably the most important is the attitude to and support of organ donation by society at large. Retrospective and prospective surveys of organ donation in Australia have found that only 30–40% of potential organ donors become actual donors.[1] The other 60–70% fail to become donors either because of refusal by the next of kin or failure of hospitals to notify transplant teams of potential donors. This latter reason is often blamed on inadequate resources in hospitals to provide cardiopulmonary support for several hours for brain-dead potential donors, but more likely reflects concerns and fears of medical and nursing staff about organ donation.[2] Whatever the cause, cadaveric organ donation could be increased two- to three-fold (or by much more in countries where cadaveric organ donation is severely limited by cultural factors) if all potential donors became actual donors and such a rate of donation would probably more than satisfy the current need for organs. The critical question is how do we maximize the number of potential donors becoming actual donors?

Strategies to improve organ donation are either legal/procedural or educational. Legal/procedural strategies include presumed consent, required request or compulsory state ownership of the dead body (see Chapters 7 and 17). However, such strategies have resulted in minimal improvement in organ donation rates and in some countries a reduced rate as societies react against perceived limitations of civil liberties.

Thus it would seem that only strategies which are designed to educate society about organ donation are likely to result in a high and sustained rate of organ donation. These educational programmes need to be directed to both the population at large and to the healthcare professionals who can influence and facilitate organ donation. Examples of education programmes for healthcare professionals include the EDHEP programme as described in Chapter 19.

Education of the community at large, whilst essential, is a vast undertaking which needs constant reinforcement and which can easily be devastated by any adverse publicity about organ donation. Such adverse publicity rarely comes about through mismanagement of an organ donation but usually as a consequence of a perceived or real lack of support for the family of the organ donor or from misconceptions about whether the donor was 'really dead'. This indicates the need to maximize the bereavement support for donor families but very importantly again indicates the importance of having a society which is fully informed about organ donation and transplantation well before the devastating event that has led to the death of a potential donor (see Chapters 3 and 15). A family which has discussed organ donation is much more likely to agree to organ donation at the time of the death of a family member and the deceased is more likely to have previously indicated personal consent to become an organ donor by such mechanisms as a donor card, vehicle licence or donor register.

Given the enormous undertaking to educate the whole community about organ donation, can a subpopulation be targeted, a population which in turn could carry the message to and influence the attitudes of the community at large? This chapter describes an education programme which targets the teenage population whilst still at school, the premise being that this age group learns rapidly, is unencumbered by many out-of-date prejudices and beliefs and will be influencing the development of beliefs and attitudes of the future (and to some degree current) society.

The role of education in the formation of attitudes to social issues

Organ donation has only been in existence in a significant way for about 30 years and many of society's attitudes and cultural beliefs have been formed and consolidated before the era of organ transplantation. Over the last few years awareness of transplantation in the community has risen considerably, largely brought about by media reports of successful cases of organ transplantation and the benefits to the recipient and his or her family. Although such positive media reports have probably facilitated organ donation in general, they rarely focus on organ donation and have probably had only little impact on community attitudes to organ donation. Thus, education of the community has largely depended on programmes formulated and delivered by healthcare professionals involved in organ transplantation with help from experts in publicity, education and information technology (see Chapter 16).

There is little doubt as to the positive value of education programmes in the formation of attitudes to social issues. In the health area examples include education in road safety, water safety, adverse effect of smoking, and prevention of communicable diseases such as AIDS. Many, but not all, of these programmes have been accompanied by legislative changes (e.g. random breath tests, compulsory seat belts, swimming pool fences) and it may be difficult to dissect out the effect of legislative from attitudinal change. However, in all these areas changes in attitudes have been demonstrated.

It is fundamental to evaluate the effectiveness of any educational programme, especially if the aim is to influence society's attitude to a social issue. Many education programmes in the area of organ donation can be criticized for lack of evaluation of effectiveness or for failure to recognize co-existent factors which could themselves have a positive benefit. Expert help in evaluation of the effectiveness of educational programmes in organ donation is thus very important.

In influencing the community's attitude to organ donation multiple factors have to be considered, factors whose importance will vary considerably depending upon the society involved. For example, many countries in the Asian-Pacific and Middle East areas of the world have cultural beliefs which make cadaveric organ donation a very unusual event. Factors which need to be considered in developing education strategies include current awareness and knowledge about organ donation, dominant cultural influence, cultural diversity, ethnic diversity, religious influences, language barriers, level of educational attainment, negative influences in the community and the laws of the country or state relevant to organ donation.

Thus, educational strategies to improve organ donation will depend on initial identification of the factors preventing a high rate of organ donation in the community. It must be recognized, however, that the major barrier is lack of information and understanding or even misinformation about many aspects of organ donation and transplantation. For example, there are only very few religions or religious sects which have been identified to have objection to cadaveric organ donation but the followers of many other beliefs are under the impression that their church does not agree with organ donation.

A firm principle of education strategies to improve organ donation must be not to brain wash society but to allow the members of the society to formulate their own view based upon correct information which is effectively and meaningfully conveyed. There will always be individuals who remain implacably opposed to organ donation and transplantation despite excellent education programmes and their opinions must be respected.

As mentioned above, it is important to target education programmes to individuals or groups in the community who have the greatest potential in their own right to influence others in the community. Examples include religious or community leaders, medical science and legal correspondents of the media, high profile people such as sporting personalities and, as discussed below, the young.

Education in schools

In the modern school education system greater emphasis has been placed upon education programmes which deal with social, health, legal and ethical issues facing society. Educators recognize the tremendous capacity of school-aged children and adolescents to understand the important principles behind these issues and that life-long positive attitudes are developed. Examples include education on racial and gender discrimination, cigarette and alcohol abuse, road safety and nutrition. However, the authors are aware of only two published reports concerning school education about transplantation: one concerning a programme of education of children aged 13–14 years from schools in a principal regional city in Victoria, Australia[3] and the Australian Kidney Foundation programme *Transplantation – The Issues*[4] discussed below. It is believed, however, that there are probably thousands of 'one-off' education sessions about organ transplantation and related issues given to school children by transplant health professionals on an ad hoc basis. One can only make a guess as to the format, topics and educational value of such events.

The programme of education about renal transplantation reported by Agar et al.[3] is a three-part tutorial programme augmented by an illustrated booklet and is directed towards children aged 13–14. The first part is classroom-based during which the children are introduced to simple renal anatomy and physiology, causes of kidney failure and principles of dialysis. The students then spend 90 minutes in a dialysis unit learning about peritoneal and haemodialysis and interacting with patients to understand better the life style restrictions associated with dialysis. In the final classroom session the concepts of organ donation, brain death and organ procurement are introduced and the life style of patients with successful transplants is contrasted with that of patients on dialysis. The effectiveness of the programme has been evaluated by assessing the students' subsequent understanding of what they have learnt and of their attitudes to dialysis and transplantation, the control group being students not exposed to the programme.

Although the programme has only been taught to a limited number of students in a provincial city in Australia, the authors believe such short, simple education courses could be valuable across the nation and internationally. They have targeted 13–14 year old school children because they believe children of this age have sufficient maturity to understand the programme and to develop long-lasting attitudes.

School education programme: Transplantation – The Issues

HISTORY

In 1989, the almost chance meeting in Melbourne, Australia, between a transplant co-ordinator and a school teacher whose child had become an organ donor, led to the idea of the development of an education programme for secondary school students to cover all aspects of organ donation and transplantation. The idea was based on the concept that informing students about this important issue at a time when they are developing a greater awareness of social issues would allow them to think through the issues, make up their own minds and then promulgate discussion within their families and pass on their awareness and opinions to subsequent generations. It was appreciated that to develop such an educational programme, professional help was essential to prepare material and direct it in an appropriate educational manner. A development committee was thus formed, composed of transplant professionals, educationalists from the Science Teachers' Association of Victoria and officers of the Australian Kidney Foundation. Important contributions also came from the various Australian special societies representing intensive care physicians, transplant physicians and surgeons and transplant co-ordinators (Figure 22.1). The programme was completed in 1991 and launched nationwide in early 1992.

FORMAT OF THE PROGRAMME

The education package consists of a video and a book containing student activities, all photocopiable masters and teachers' notes (Figure 22.2). The whole programme provides for a semester's work. It is not simply directed to the area of organ donation but covers all aspects of transplantation. While the core of the material is science-based, the activities can be used in many subject areas – legal studies, economics, health education and the English language. An important component of the programme is

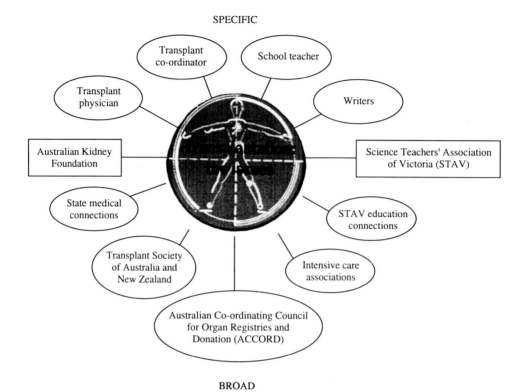

Figure 22.1 Individuals and expert groups who contributed to the developing of the school education package *Transplantation – The Issues*.

learning about ethics and a number of the activities require students to consider several of the ethical issues of organ and tissue transplantation.

The units of the programme are listed in Table 22.1. Each unit contains basic information about the topic as well as illustrative articles from the media, stories from patients or extracts from resource texts (e.g. Codes of Practice). The video is of seven sections and contains illustrative stories largely taken from television programmes.

It is ideal that the whole programme be undertaken either within a single subject or across a number of subjects. However, individual units or activities can be used successfully as part of other programmes, e.g. the unit on ethics can be used as part of a legal studies course.

The programme facilitates: (a) open discussion; (b) the study of ethical and social issues; (c) economic considerations of organ transplantation; (d) biological understanding of the body and its immune system; (e) co-operative learning; (f) individual decision making; and (g) greater awareness of the roles and responsibilities of the many people associated with transplantation, including the patient.

PHILOSOPHY OF THE EDUCATIONAL PROCESS AND DESIRED OUTCOME

The programme facilitates understanding and discussion by immersing the students in the issues, not by preaching facts and figures to them. Students' current perceptions are

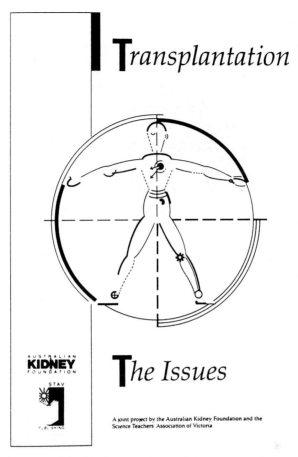

Figure 22.2 The cover of the book containing student activities, photocopiable masters and teachers' notes, for the school education programme *Transplantation – The Issues*.

sought and then challenged. Students are encouraged to look at the positive and negative aspects of all the issues raised and to decide for themselves the stances they wish to take on any particular issue (Figure 22.3). The recognition that they are unable to decide which stance to take is a legitimate position which is to be respected. Students' rights to hold, but not to express in public, their point of view is also respected; personal diary entries are encouraged in these circumstances.

Teaching strategies that reflect current knowledge of how students learn best are used and encourage students to take more control over their own learning. Contexts are used to relate science, technology, economics, legal studies and social studies to the everyday life of the students. The ethical and social issues are dealt with through a range of co-operative learning tasks as well as individual decision-making tasks. Modern teaching techniques, such as the development of 'concept' and 'consequences' mapping (Figures 22.4a and b), facilitate the teaching and learning processes.

The students who are exposed to the materials in *Transplantation – The Issues* will be more informed and active members of a future society in which the issues related to

Table 22.1 Abbreviated table of contents of the book of the education package, *Transplantation – The Issues*

Teaching and learning about ethics

Unit 1: Organ donation – what do you know?

Unit 2: The transplant equation
    From donor to recipient

Unit 3: The transplant experience
    The recipient
    The staff
    The donor and the donor's family
    The living related donor

Unit 4: Defining and determining health

Unit 5: Human bodies

Unit 6: Living with a terminal illness
    Heart disease
    Kidney function and failure
    Comparison and review

Unit 7: Transplant technology

Unit 8: Is it legal? Is it ethical?

Unit 9: The economics

Unit 10: What does the future hold?

About the activities: a guide for teachers

---

the developing technologies associated with transplantation will continue to challenge ethical and social values currently held in our society.

IMPLEMENTATION OF THE PROGRAMME

The successful deployment of any education programme requires: (a) awareness of the existence of the programme and of its merits; (b) distribution to as many outlets as possible; (c) in-service education of those who will teach the programme, and (d) mechanisms to perpetuate the use of the programme once the initial phase of enthusiasm has passed. To achieve this for *Transplantation – The Issues*, the 'co-ordinating' committee formed a partnership with the community service club 'Rotary' and developed a strong infrastructure of outlets for the programme. This consisted of state branches of the Australian Kidney Foundation and a national distributor of health education material combining to create a system of in-service education of a key group of secondary school teachers from all states of Australia. Crucial to the programme was the appointment of a national education officer whose role has been to co-ordinate all these implementation strategies.

Figure 22.3 'Brainstorming'. One of the techniques used in the school education programme *Transplantation – The Issues* to allow students to discuss the issues in transplantation.

Rotary, through its many clubs throughout the country, has progressively taken on the task of advertising the programme within schools covered by each club as well as facilitating the purchase of the programme by the schools and attendance of teachers at in-service education sessions. Rotary has produced a *Rotary Transplant Education Portfolio* which consists of information about the programme, a promotional video and a strategy to implement the programme in schools. Each club nominates an interested club member to be responsible for 'Transplant Education' and he/she educates the whole club about the programme using the promotional video and the education kit itself. The club then facilitates school implementation by identifying the schools within the club's boundary, contacting key personnel at each school (including the health co-ordinator, legal studies teacher, curriculum co-ordinator, personal development teacher and religious studies teacher), and, if necessary, finding sponsors to purchase kits for the school and to have teachers trained in the effective use of the kit. Rotary members have also been encouraged to discuss the issues of organ donation within their own home. As one Rotarian stated, 'it was one of the easiest and most effective projects that we have implemented'.

In-service education programmes for teachers from schools taking up *Transplantation – The Issues* are held regularly throughout the country. Up to twenty teachers are taken through a *Pathway Document for Transplantation* which details the content of the kit and how it may be used either in a focused manner, to provide a complete study of transplantation, or very broadly to develop in each student the knowledge, skills and attitudes needed to understand, value and lead healthy and fulfilling life styles.

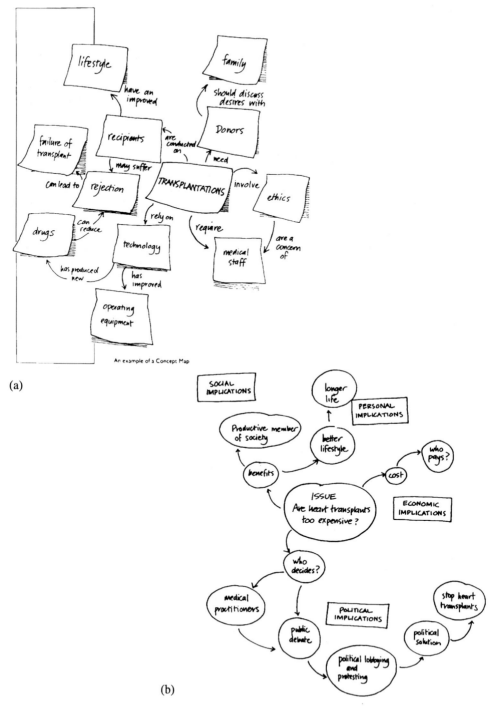

Figure 22.4(a,b) 'Concept' and 'consequences' mapping as used in the school education programme *Transplantation – The Issues*.

FUNDING OF THE PROGRAMME

The development of the programme was funded by government, industry and charitable trust grants. A small charge was introduced for the purchase of each kit to aid the funding of the awareness campaigns, in-service education and an evaluation of the programme of education. A charge for the kit was also thought to be important in focusing each school's attention on the programme and continued use of it as a 'valued' educational resource.

EVALUATION OF THE PROGRAMME

In the three years since the programme was launched the kit has been introduced into about 600 (26%) of the 2,300 Australian secondary schools and this introduction is continuing at a steady rate (Figure 22.5a and b).

It was recognized that the simple introduction of the kit into a school and in-service education of the teachers did not necessarily mean that the programme of education was either being used or used effectively. Implementation of curriculum materials is a complex social process.[5] A study was therefore undertaken in 1994 in which 55 schools in Victoria which had participated in the teacher training programme since 1992, were asked to participate in a questionnaire/evaluation of how the education programme had been established into their school curriculum.[6,7] The questionnaire was devised with the intention of determining if teacher training was a viable option taking into account the cost of teacher release and administration, as well as ascertaining whether the kit was being used in the classroom and in what capacity. Forty-six schools (84%) responded to the questionnaire and only four of these had not introduced transplant education into their curriculum on an ongoing basis. The programme was being taught mainly in year 11 (16–17 year olds) but also in years 7, 8, 9, 10, and 12. Most teaching was within the Science and Biology courses but also in English, Religious Education, Legal Studies, Human Development and Society, Mathematics, Psychology and Social Education.

Training only 1–2 teachers from each school to use the kit effectively seemed to facilitate the use of the kit by many other teachers within each school. Within the schools surveyed, a total of approximately 5,500 students were being taught, to some degree, the topic of transplantation. The potential is thus that the messages are reaching 5,000 families every year and that over five years these few teachers have the capacity to reach 30,000 families.

A survey was also undertaken in late 1992 to evaluate the effectiveness of the transplant education package. The evaluation was focused on a small number of schools and used a case study approach. The schools were selected across the spread of educational regions and represented both the independent and government school systems. Students' knowledge of transplantation and attitudes towards organ donation were assessed before and after the introduction of the transplant education programme. Teachers were interviewed and asked to maintain a diary of the activities used. Class discussions were conducted after exposure to the programme to further probe attitudes and limited family interviews were also held. Overall, the kit was very well received by teachers and students and had the potential to achieve the objectives as set out in the material. Data from both students and teachers pointed to the role of the teacher as being crucial. Teachers themselves needed assistance in how best to use the various activities. Most students could not recall what they had done unless the teacher had actively introduced the activities and directed how they should be used. Ongoing

Figure 22.5 Year 10 secondary school students (aged 15–16 years) participating in the school education programme *Transplantation – The Issues*. (a) An 'activity' in which students are attempting to determine the relative frequency of organ and tissue transplantation. (b) Viewing one of the seven television programmes contained on the video component of the education package.

in-service education was identified as essential. The greatest single activity recalled by the students and which had the most impact was the single visit to each school by the programme's education officer. The development of a positive attitude of students to organ donation and transplantation was evident. Parents commented that the topic was an emotive one and that this often precluded the discussion of the issues at home. However, they were delighted that such an important issue was handled at school and that their children were being given the opportunity to reach an informed decision about transplantation in a 'neutral' environment. Evidence collected during the study suggested that the kit did not initiate a great deal of home discussion but that most students discussed the issues amongst themselves.

FUTURE OF THE TRANSPLANT EDUCATION PROGRAMME AND RELEVANCE TO OTHER COUNTRIES

It is appreciated that the continued effectiveness of the programme will depend on the participation of as many schools as possible, regular updates of in-service education of teachers, periodic (five-yearly) updating of the resource material and teaching methods (e.g. transfer of material to CD ROM) and regular evaluation of the effectiveness of the programme. Although some of the material used in *Transplantation – The Issues* has an Australian flavour, the kit could be easily modified to suit the needs of another country both in content, education methods and language.

Although results of the education programme in the form of organ donations will not be seen immediately, the broad base of the community who will be educated in the many issues surrounding transplantation will increase on a yearly basis, allowing potential donors the chance to have made an informed decision about organ donation and transplantation, thus ensuring that decisions at the time of organ donation will be made both easier and more likely in favour of donation.

References

1. Hibberd AD, Pearson IY, McCosker CJ et al. The potential for cadaveric organ retrieval in New South Wales. Brit Med J 1992, 304: 1339–1343.
2. Mohacsi PJ, Herbertt KL, Thompson JF. Donations and retrieval of cadaveric organs in Australia. Accepting the challenge. *Med J Aust* 1993, **158** (2): 121–124.
3. Agar, JWM, Simmons R, Heintz R. An education program promoting organ donation for 13–14 year old school children. Nephrology 1995, 1: 157–161.
4. Transplantation – The Issues. Australian Kidney Foundation and Science Teachers' Association of Victoria, STAV Publishing Pty, Ltd, 1992.
5. Fullan M. In The Meaning of Educational Change. New York, Teachers' College Press, 1982, page 30.
6. Knudson R, Thomson N, Scully G, Wragg F, Keaney L, Angus J, Marshall A, Warrick A. *Transplantation – The Issues. School education extended to the community.* Proc 11th meeting of the Transplantation Society of Australia and New Zealand, 1993, 67.
7. Thomson NM, Scully G, Knudson R, Wragg F, Keaney L, Angus J, White G, Marshall A. (Eds) Transplantation – the issues: A cross curriculum programme for secondary schools. *Transplant Proc* 1993, **25**: 1687–1689.

CHAPTER 23

Publicity and marketing strategies

MIRIAM RYAN

The role of public attitudes in influencing donation rates   412
The role of public education in determining public attitudes   414
Organ donation and audience segmentation   418
Identifying motivational messages   419
Presenting motivational messages   421
Importance of evaluation in shaping programmes and measuring
    their effectiveness   422
The role of the mass media in donation promotion   423
Is public education cost-beneficial?   425
Conclusion   427
References   427

The role of public attitudes in influencing donation rates

It is generally agreed that public attitudes to organ donation and transplantation are of prime importance since prior consent of the donor or of a close relative, at the time of death, forms the basis for cadaveric organ donation in most countries.

Despite often high levels of awareness and positive attitudes towards organ donation, as expressed through general public opinion surveys, people's attitudes about donating their own or a close relative's organs vary enormously. The percentage of families who refuse to donate their relatives' organs can offer an insight into public attitudes in general although refusals may also be due to a range of other factors including the manner in which relatives are approached about donation.

In the USA the average refusal rate is around 50% but amongst certain ethnic minority populations this rises to 70%. A similar variance in family refusal rate is seen in a number of European countries. Spain, for example, despite having one of the highest number of cadaveric renal transplantation rates in Europe (with 38.4 per million population (pmp) in 1993) has a family refusal rate of 70% in some more remote parts of the country.

Public confidence in donation is easily undermined and any drop in public confidence is often rapidly reflected in subsequent decreased donor referral rates or increased family refusal rates.

Public attitudes are notoriously susceptible to adverse publicity concerning organ donation and transplantation issues and in this respect the media have an important role to play. Annual UK donor rates, for example, fell in the wake of a respected prime-time television programme in October 1980 which raised doubts about the validity of brain stem death criteria. The 8% drop in the number of kidneys transplanted in 1981 was largely attributed to the negative consequences of this one programme.

From time to time stories about the kidnapping of children for organ retrieval and sale arise in the media in different parts of the world. This became a serious national issue in Argentina in the mid-1980s. Public support for donation declined significantly with every new 'case' claimed by the press, resulting in a significant rise in the rate of potential donor family refusals.[1] The situation became so critical that the health minister was obliged to call a special conference of senior members of the police and justice departments together with senior doctors and members of parliament to dispel the rumours and plan a communications campaign to rebuild public trust and confidence. It was five years before donor rates began to recover.

More recently doubts raised over brain death, stories of the growth of organ trafficking in Eastern Europe and worries about organ allocation, as reflected in the media, are believed to be the main contributors to the rise in donor refusal rates in Germany from 21% in 1992 to 32% in 1994.[2]

Although the media is often guilty of exaggerating the negative aspects of such stories and encouraging a false sense of alarm among the population, events themselves can also precipitate a fall in public confidence and a series of adverse events can have a devastating cumulative effect, as has been the case in France in recent years.

In 1991, with 33.4 cadaveric renal transplantations per million, France stood at the top of the European league. By 1994 this rate had fallen to 26 pmp with an increasing number of families refusing to donate their relatives' organs, despite the existence of presumed consent legislation. In the intervening years, France had been rocked by a major blood products scandal, with four senior physicians, including the Director General of the National Blood Transfusion Centre, being charged with deliberately supplying HIV-infected blood to haemophiliacs. As the full facts of this scandal were emerging, a case hit the headlines of a family from the south of France who, unaware of the legal position regarding organ donation, were horrified to discover that their young daughter's corneas had been removed without their permission. With the whole area of donation and transplantation under close scrutiny, the high number of non-residents on French transplant waiting lists became a cause of national concern, further undermining public confidence.

Although public opinion often appears to express itself rapidly and sometimes dramatically in the wake of negative events and publicity, the process of rebuilding public confidence and addressing misconceptions is much slower and any achievements in this area are considerably harder to document. The positive effects of good public education, often achieved over a number of years, can also be badly damaged by a wave of negative publicity.

The fact is that effective public education requires long-term commitment, a high level of campaign planning and development and sufficient resources. Although public education may be slow to show results and is often expensive to conduct and evaluate, reading public opinion and attempting to influence attitudes and behaviour is essential to the creation of an effective donation programme, no matter what legislative system is in place.

The role of public education in determining public attitudes

In order to determine how effective public education has been in influencing attitudes and behaviour towards donation, it is important to look back at the role of public education in donation and transplantation.

In the early 1980s the main thrust of education campaigns was that 'transplantation was a great medical success able to transform the quality of people's lives' and the general message was – donate your organs after death because it is a good and right thing to do. In tandem with this many countries ran campaigns to get people to make public their stated wishes by carrying donor cards, agreeing to donation on driving licences, or recording their wishes either by opting-out or opting-in to whatever registration system was in place.

According to numerous public opinion surveys and polls conducted over more than ten years, public education campaigns undertaken in many parts of the developed world have been quite effective in increasing public awareness of transplantation. A 1992 Gallop survey in the US revealed an 85% awareness of and support for organ donation.[3] This represented a significant increase in awareness since 1985.[4] In the late 1980s public opinion research in Australia found that an extraordinarily high 98% of those questioned were aware of organ donation.[5] Similarly a major survey of public opinion in France in 1992 revealed 89% of people to be aware of and in favour of organ donation.[6]

Although these studies show a high level of awareness and often favourable attitudes towards organ transplantation, there remained a striking disparity, in many countries, between the number of people who purport to be aware and in favour of organ donation and those who actually donate. Even as early as the mid-1980s it was acknowledged that 'attitudes towards organ donation are often cited as an impediment to the procurement of organs for transplantation'.[7]

It is clear in fact that public education campaigns can claim far less success when it comes to influencing people's attitudes and, more importantly, their behaviour in favour of donating their own or a loved one's organs. Interestingly enough, when it comes to the issue of stated willingness to donate one's own or a loved one's organs, there is much less of a disparity between stated and actual behaviour.

In Britain, for example, 70% of people say they would be willing to donate their organs after death. The average UK family donation refusal rate of around 30% suggests a consistency between stated and actual behaviour. Similarly surveys in the USA during the 1980s, which found that only 45–50% of Americans were willing to donate their own organs is consistent with the average US family refusal rate of 50%.[8] In Australia between 60–65% of those surveyed expressed a willingness to donate their own organs[9] with a national family refusal rate of around 25%.[10]

An analysis of the results of public education campaigns which aim to get individuals to make public their wishes to be a donor by carrying donor cards, agreeing to donation on driving licences, or registering their wishes either by opting-in or opting-out suggest that the effectiveness of these education campaigns has been limited.

In the UK, for example, the donor card was first introduced in 1971. Twenty-two years later, however, only 26% of adults had obtained a card and only 18% acknowledged actually carrying one. This is despite the distribution of an estimated 10 million cards each year since the scheme's inception and despite a willingness, regularly reaffirmed in opinion polls since 1983, of 70% of the population to be donors after death.[11] In the USA the data on percentage of donor card carriers varies between 8%

and 39%[12] and in many European countries, the percentage of donor card carriers is similarly low. In Holland 18% of the population have donor cards although only 9% always carry them, while in Germany only 2% of the population have cards.[13]

A variety of explanations for the relatively low percentage of donor card carriers, particularly among teenagers and young adults, have been offered.[14] Among these are:

- That people do not fill out donors cards because to do so would make them aware of their own mortality.
- People's positive views about donation are outweighed by their fears concerning donation.
- Donor cards are not where the public wants them.

The effectiveness of campaigns to encourage people to sign up to opting-in registers, which operate on the same principle as the donor card, have been similarly disappointing. A significant publicity campaign conducted in Wales in the UK between 1990 and 1993, promoting a computerized opting-in system, resulted in only about 300,000 out of a population of 2.8 million registering their willingness to donate.[15]

The question remains, why is there such disparity between people's general attitudes towards donation and their individual decision to act positively and can public education help to close the gap and lessen the ambivalence which people so clearly feel? For an insight into this we need to look at some of the communication theory behind behaviour change and to determine the elements that characterize successful intervention.

A THEORETICAL BASIS FOR SUCCESSFUL PUBLIC EDUCATION

Public awareness and favourable attitudes towards donation are only the beginning of a chain of internal events which need to take place if the individual is to act. As healthcare communications specialist Elaine Arkin has pointed out:

> For an individual to take positive action, he or she must become interested in organ donation, be convinced not only of its value but also of its personal relevance and know what to do about it. There must also be a supportive environment and increasing organ donation will require strategies that extend beyond educating and motivating individuals – thus social support and health system support need to be in place to reinforce individual decision-making. [she argues] ... some of the factors that are influential in facilitating or blocking behaviour changes include an individual's knowledge, attitudes, behaviour, beliefs and values; the structure of the environment that facilitates or presents obstacles to change and the positive or negative effects of adopting the behaviour (including social acceptance and support).[16]

A misunderstanding of brain death may be an example of a belief which prevents an individual from acting positively towards the issue of donation. In a Spanish study to determine the reasons for the high rate of family donation refusals in south-west Spain, 31 families who had refused to donate the organs of their brain dead relative were asked to complete a questionnaire designed to elicit their reasons. One in four families stated that they had not understood the concept of brain death and this inability to accept that their relative had died was the reason for their withholding of permission.[17]

An environmental barrier to behaviour change may be the difficulty of obtaining donor cards. This issue, for example, is likely to be a major barrier amongst people aged 16 or over in Japan.

Donor cards were first introduced in Japan in 1977 but, by 1991, only 0.36% of people aged 16 years and over carried cards. This low take-up rate is less a reflection of unfavourable attitudes amongst the Japanese to organ donation and more to do with the complex donor card issuing system which involves registering with a prefectural kidney bank. Indeed a number of opinion surveys have revealed that 50% of Japanese are willing to donate their organs after death, a percentage similar to that of the USA. This survey revealed that, when asked, if donor cards were distributed freely and registration to carry a card was not required, over half of respondents, which included 400 high school students and 600 workers, replied that they would be willing to carry cards.[18]

Finally the positive or negative effects of adopting the behaviour, including social acceptance and support may be a inhibition to action. An elegant illustration of this can be seen in the results of many years of promoting the organ donation message in Australia.

With a national organ donation rate, which has hovered around the 13 pmp mark since 1989, Australia has one of the lowest donation rates in the developed world. This is not to say that Australians have an unfavourable attitude towards donation, in fact 60–65% of the population regularly express a willingness to donate their own organs. Although the percentage of Australians expressing their willingness to donate a close relative's organs drops to 38%, this is no different to the situation in Holland, which has a donor per million rate three times that of Australia. As Phillip Dye, the National Publicity Officer for the Australian Co-ordinating Committee on Organ Registries and Donation points out:

> Since the promotion of organ donation began in Australia, the public have been constantly told that we have one of the lowest rates of organ donation in the developed world. We are constantly reminded of the suffering on waiting lists due to our 'poor' donor rate and have been confused and made to feel guilty by a myriad of condemning statistics.

Campaigns which aim to motivate behaviour change through inspiring guilt or fear, in the target audience, are rarely effective and although there are likely to be a variety of reasons for Australia's low donor rate, an important contributory factor may have been the adoption, over a number of years, of a flawed promotional strategy which fails to take account of accepted communication theories of behaviour change. One of these theories suggests that people adopt new behaviours at different rates, depending on a range of socioeconomic factors. A new behaviour, therefore, is likely to be first adopted by a small percentage of people who are known as *innovators* and another group called the *early adopters*. In many countries the people most likely to identify themselves as potential donors will fall into these two categories and they tend to be the better educated and the more affluent. These two groups are followed by the *early* and *late majority* who make up the bulk of the population – thus transforming the idea or behaviour from an innovation to a social norm. Finally there are two other segments of the population called the *late adopters* and the *resistors* who are most likely to be socially or economically disadvantaged and represent the groups which are most difficult to reach and influence.

To return to our Australian example for a moment, as Phillip Dye points out:

> In promoting the organ donation message in Australia, the group we must target is the middle majority, who respond less to mass media and more to interpersonal messages. This is the group who will usually only adopt a new behaviour if others in the groups are adopting it.

However, as Dye rightly concludes, if the message this group is hearing is that Australia has the lowest rate of organ donation in the developed world, why should this group change their behaviour when what is effectively being presented is that this particular behaviour change is neither popular nor mainstream. Far from motivating people to change their behaviour therefore these promotional messages are reinforcing the status quo.

Where many public education initiatives have come unstuck in developing effective strategies to motivate behaviour change, is that they have focused on the intrinsic value of organ donation rather than on identifying and relating to the needs, wants and values of the target population.

As Prottas and Batten pointed out in 1991:

> Public education about donation has up to now offered a broad message to the public. The theme has been that donation is good, it offers the gift of life, it encourages altruism. The majority of the public appears to concur and agrees that, if the tragedy that makes donation an option should happen, they would donate the organs of their kin. However, for substantial segments of the public, the central messages of public education fail to address their concerns and needs.[19]

TARGETING THE RIGHT MESSAGE TO THE RIGHT AUDIENCE

In recent years there has been much discussion of the application of commercial marketing approaches to social issues, an area of communication theory and practice which has come to be known as social marketing.

In the late 1960s, two of the orginators of the term, Kotler and Levy, argued that marketing should be seen as a universal human activity and its principles applied not only to conventional products but also to people, organizations and even ideas. Kotler originally defined social marketing as:

> the design, implementation and control of programmes seeking to increase the acceptability of a social idea, cause or practice in a target group(s). It utilizes market segmentation, consumer research, concept development, communication, facilitation incentives and exchange theory to maximize target group response.[20]

The most fundamental aspect of marketing (and social marketing) is knowing the consumer. It follows therefore that any effective public education must be able to identify the perceptions, concerns and needs of the target audience to be addressed. This in turn has implications for the nature of the message to be communicated and the manner of its communication. Effective health education, for example, already draws on the lessons of social marketing.[21]

Health educators recognize the importance of segmenting the broad population into particular target groups among whom behaviour change is sought. For example, in the area of AIDS/HIV prevention, the message of 'Safer Sex' needs to be communicated, for example, to adolescents as well as to business travellers, but to be effective, the message needs to be presented in different ways.

Target audiences also need to be prioritized according to their potential for behaviour change, particularly where resources are limited. Again in the HIV/AIDS area those who perceive themselves to be most at risk, for example, those people practising high-risk behaviour, are more likely to change their behaviour than those who do not perceive themselves to be at risk. It may, for example, be more cost-effective to target educational messages at practising homosexuals, who have a higher risk of contracting HIV than, for example, at sexually active women.

Effective health education programme design is based on a thorough knowledge of individual target audience needs and concerns, which in turn are elicited through a variety of market research techniques. Finally programme design includes mechanisms to ensure that education messages and materials are both pretested, to ensure maximum effectiveness, and their impact evaluated, to monitor the programmes progress and demonstrate results.

Organ donation and audience segmentation

The 1986 US Government's Task Force on Organ Transplantation cited local, uncoordinated programmes with a high degree of inconsistency and redundancy as one of the main problems associated with current public education programmes. They added:

> Public education programmes are often conventional or unimaginative and targeted groups are frequently those already convinced of the benefits of organ donation. With few exceptions, public education activities in organ transplantation are neither developed nor implemented by education specialists.[22]

Throughout the 1980s, the situation with regard to public education efforts in many countries was not very different. Public education initiatives in the field of organ and tissue donation tended to be piecemeal, fragmented and under-resourced. Government commitment to supporting public education was often minimal. Australian government spending on the promotion of organ donation throughout the 1980s, for example, totalled less than $A200,000.[23] Campaigns were conducted for the most part by voluntary or charitable groups with limited resources, or by transplant co-ordinators or organ procurement or exchange agencies as part of their wider duties. As a result, activities have tended to be short-term, insufficiently planned and often lacking in focus.

As the donor rate in many countries began to plateau or fall at the end of the 1980s, so more serious attempts were made to implement a planned and targeted approach to public education on organ donation. One area to benefit from this more focused approach (and identified as a priority by the US government's Task Force in 1986) was the building of greater public awareness and participation in donation among ethnic minorities.

A brief review of one of the programmes which have been undertaken in this area illustrates the advantages of conducting detailed research on the attitudes and beliefs of the target audience and crafting a communications programme which takes account of those attitudes and beliefs.

US studies have shown that while the variables of both race and education are important in determining attitudes to donation, race has an independent effect on attitudes, separate from education.[19] In the USA, Callender has shown that African American (AA) families have a 60% refusal rate for organ donation, representing twice that of white families.[24] The rationale for targeting this group for special attention is as follows:

- Although AAs make up only 9% of the US population, the incidence of end-stage renal disease among this group is significantly higher than other ethnic groups, resulting in AAs comprising around a third of the cadaveric renal transplant waiting list.[25]
- In 1991, less than 10% of solid organ donor nationally were AAs.
- Because a better HLA match should be achieved if donor and recipient race are identical, it follows that AAs may have reduced waiting time and better long-term graft survival with an increase in the number of AA donors.[26]

A considerable body of information on black attitudes towards donation had been collated by Callender and the Howard University Group, through focus group sessions and face-to-face interactions, resulting in an identification of the critical success factors defining a successful black donor programme. These included ethnically sensitive messages to be delivered by appropriate messengers, in tandem with a strong mass media effort and a grass roots approach.[27]

Utilizing this excellent groundwork, the Mid-America Transplant Association supported the development of a organ donation educational programme, targeted at the AA community, in order to better serve a large AA population in St Louis, Missouri. In order to ensure that the right messages were communicated by the appropriate messengers, the programme invited participation from key sections of the black community – the church, educators, medical professionals, organ recipients and donor families and engaged a local AA-owned public relations agency. A symposium on 'Black Issues in Organ Donation' was held and an advertising print and poster campaign was launched, together with broadcast media support.[28]

Despite increasing awareness of donation in the black community and generating a 77% increase in referrals of potential AA donors, the campaign was deemed to have had no effect on the number of families consenting to organ donation. However, in the wake of this community campaign, a decision was made to appoint two AAs to the role of approaching all AA family referrals to make the donation request. Within two years, the consent rate amongst AA donor families rose from 31% in 1988 to 48% in 1992, with the major rise occurring in 1992. The study's authors conclude, 'whether the educational programme or the use of specific AA organ requesters separately or jointly accounted for the increased consent rate cannot be known from the present study'.[29] Apart from demonstrating that tangible benefits can be derived from a planned, focused campaign which takes account of audience perceptions, this example also illustrates the complex evaluation required to determine the precise role of the mass communication component versus the personal component (i.e. the AA organ requesters) in influencing the shift in target group behaviour. This experience suggests that the effects of public education in raising awareness and creating a positive climate of opinion, need to be complemented by equally careful attention to the manner in which donor families are approached.

Identifying motivational messages

As the situation regarding the donor shortage became more critical during the late 1980s and early 1990s, so attention has focused more closely on why the messages traditionally employed to increase organ donation have not worked. These messages have tended to revolve around the need for organs such as 'don't take your organs with you, heaven knows we need them here' or the moral worth of donation, such as 'offer the gift of life'.

Attention has focused instead on trying to reduce one of the most identifiable barriers to donation – the rate at which relatives refuse donation. A steep rise in the rate of relatives' refusals in some countries has made this a priority. France, for example, saw refusal rates rise, from 47% in 1991 to 62% in 1992, and this from a country with traditionally one of Europe's highest number of transplants pmp.

Public education efforts, relegated to a secondary role during the 1980s, a period during which the supply of organs continued to grow, suddenly became important when donor rates began to slip. For many countries the beginning of the 1990s was the time to take a closer look at what the public was thinking and feeling about organ donation, regardless of the national laws regarding organ procurement. At the same time increasing efforts were made in trying to identify the reasons for family refusals. Although, as before, many surveys continued to show high levels of public support for donation, the surveys also revealed worrying areas of public concern, fear and ignorance. It also became clear that individuals may be committed to donation but if their families are not aware of their feelings or decision, should they become a potential donor, their individual wishes may not be fulfilled.

In France, for example, a major public opinion survey conducted in 1992/93 revealed that an astonishing 98% of the population were unfamiliar with the national law of presumed consent.[6] This law, ratified in 1976, allows medical staff to retrieve organs without a relative's permission if the potential donor was over 18 years of age and had not objected to organ donation during their life. This level of public misunderstanding is perhaps less surprising when it becomes apparent that many medical professionals apply the law in different ways. In one study conducted in 144 public and private hospitals around Paris, physicians asked for family consent to donation in 50% of cases and in only 18% of cases was the family notified without being questioned on either their feelings or those of the deceased.[30]

It was against this confusing background that a number of cases hit the headlines during the early 1990s where families, unaware of the legal position, were appalled to discover that their relatives' organs had been removed without their permission. It is highly likely that cases such as these (together with a long-running national blood products scandal) contributed to the significant rise in the percentage of donor family refusals which occurred during this period. With French public confidence in organ donation at an all time low, it was imperative that the aim of any public education was to restore confidence. In order to minimize the possibility of families not knowing their relatives' wishes regarding donation, France's Organ Donation Committee launched a public awareness campaign around the theme 'Express your opinion and let it be known'.

Similarly the findings of opinion surveys conducted in the USA during the early 1990s revealed widespread misunderstanding about the role of a donor card and the fact that few people had discussed their wishes with regard to donation with their close relatives. A Gallup survey conducted by the Partnership for Organ Donation in 1993 revealed that 8 out of 10 people questioned, wrongly believed that a signed donor card is required for a person to be a donor and less than half of those interviewed had told a family member of their decision. The survey also found that while 93% of people would be willing to donate a relative's organs if they had expressed that wish prior to death, only 47% said they would consent if their relative had not told them of that wish.

A similar picture of relatives not knowing the wishes of their next-of-kin emerged from national opinion research conducted in the UK during a government-sponsored education campaign in 1993/94. This research revealed that while nearly 8 out of 10 donor card holders were confident that their family understood their wishes regarding donation, and 7 out of 10 card holders were confident that they understood the wishes of individual family members, the same was true of only 4 out of 10 non-card holders.[31] With non-card holders representing three quarters of the adult population, it is clear that the majority of people do not know their relatives' wishes with regard to donation. Recent research into the reasons for family refusals in the UK suggests that this issue is indeed an important barrier to organ donation. The first year results (1993) of a two-year national survey conducted amongst the approximate 30% of families refusing donation, revealed that 1 in 5 had refused because they did not know their relative's wishes.[32]

Presenting motivational messages

During the early 1990s, the search for a more motivational message than the appeal to altruism or the shortage of organs appears to have been based, in many parts of the industrialized world, on the theme of encouraging people to express a positive decision regarding organ and tissue donation and to communicate this to close relatives. 'Organ and Tissue Donation. Share your Life. Share your Decision', for example, is the communications theme of a major national campaign launched by the US Coalition on Donation in mid-1994.

As the lessons of commercial and social marketing demonstrate, messages that are based on the target group's knowledge, attitudes, values and behaviour are more likely to be effective. Research has shown that families are more likely to donate if they know that the donor would have wanted it and direct survey of actual donor families confirms that this is the case.[19] However, most of the time families do not know their relatives' wishes and are forced to decide in the absence of information. The challenge therefore is stimulating people to discuss this issue with their relatives and to make their wishes known.

The way in which a message is presented can crucially affect its impact. Important lessons in presentation can be learned from both the worlds of commercial advertising and effective health promotion techniques. One such lesson is that target audience perceptions about the credibility of a spokesperson or message sponsor affects message acceptance. People tend to trust sources similar to themselves and this is one reason why testimonials by individuals from the target group are often compelling.[16]

One of the most successful of seven years of HIV health promotion media campaigns in the UK, in terms of spontaneous (as opposed to prompted) awareness, involved a 'Personal Testimony' campaign. This press and TV campaign, targeted at sexually active 16–34 year-olds, featured the voices of individual ordinary people who had contracted HIV, describing their own experiences. Post-campaign evaluation revealed that two-thirds of the target population agreed that the experiences of the people in the advertisements made them 'think about their own sexual relationships' – one of the campaign's key objectives.[33]

The involvement of well-known 'personalities' can be helpful in attracting attention to a message, but experience in organ donation promotion suggests that they may also detract from the effectiveness of the message. The OPO, Lifebanc of Cleveland, Ohio, for example, developed a television advertisement featuring a well-known Cleveland Brown football player next to a skinny and weak-looking man, with the catchline 'you don't have to have a super body to be a superstar... '. Subsequent opinion surveys revealed that 'the majority felt that celebrity endorsements were not necessary – unless the celebrity had personal experience with organ donation. The "real people, real situations" theme was favoured by all'.[34]

There is also a risk that in appealing very directly to one particular target group, other sections of the population are alienated. In 1992, the Australian Co-ordinating Committee on Organ Registries and Donation launched the 'Happy as Larry' campaign, targeted at 16–25 year old males and designed to promote family discussion of donation. The campaign centrepiece was a television advertisement featuring Larry, lying dead in his coffin and smiling, apparently happy at having his wishes followed and his organ donated. Campaign follow-up revealed that although well-accepted by the principal target audience, the reaction from medical professionals was less favourable, with the image of a smiling body eliciting many complaints.[23]

A message which inspires fear or which emphasizes the negative consequences of failing to do something can also be counterproductive. One of the earliest AIDS/HIV prevention advertising campaigns run in the UK featured a granite tombstone and the slogan 'Don't Die of Ignorance'. This high profile campaign was subsequently found to have had very little impact on young audiences – one of the principal target audiences.

Strong emotional appeal, however, can be very effective in gaining the viewer or reader's attention. In 1993 and 1994 the UK Department of Health ran a television advertising campaign during 'Donor Awareness Week', designed to encourage people to make their wishes known to their next of kin and to stimulate the uptake of donor cards. The 40 second advertisement featured a young girl in a hospital bed opening a beautifully wrapped present which turns out to be an empty box. A narrator's voice-over says, 'By carrying a donor card you're offering the gift of life, but if you haven't told your next-of-kin, whose consent will be asked, it could turn out to be an empty gesture'.

Pre- and post-campaign opinion research revealed that the two campaigns had been very successful in increasing the number of people carrying donor cards – rising from 33% in February 1993 to 43% in April 1994.[31] The research also showed that while around 7 in 10 donor card holders had raised the subject of organ donation with their immediate family, this was only true of 3 in 10 non-card holders. However, the percentage of respondents raising the subject of donation with their families as a direct result of the advertising campaign was only a total of 4%.

While these campaigns may not have been sufficient in achieving the ultimate behaviour change goal – getting people to discuss their wishes regarding donation with their family – it is important to recognize the value of identifying and measuring intermediate objectives such as the rise in donor card carrying. No effective public education can expect to achieve major changes as a result of brief interventions.

Importance of evaluation in shaping programmes and measuring their effectiveness

Another criticism of public education programmes made by the US Government's Task Force on Organ Transplantation in 1986 was the absence of any method of evaluation. Public education programmes, they said, 'lacked specific operationally defined goals and objectives, with little formal effort made to evaluate programmes' effectiveness'.

To explore the importance of evaluation in shaping programmes and measuring their effectiveness, let us look once again at the practice of commercial or social marketing. These disciplines underline the vital role of evaluation not only in the post-campaign period, when surveys of audience response are common, but also the importance of pre-campaign evaluation to establish a baseline from which subsequent changes can be measured.

Formative evaluation methods, such as pre-testing education materials/messages to assess their effectiveness with target groups and tracking opinion change over the course of a campaign are also important in fine-tuning all aspects of campaign design including the choice of communication medium, message, language and images.

Evaluation is important, not only in determining the impact and assessing the effectiveness of the campaign – to know what worked and why – but also in offering further insights into the nature of the challenge which needs to be addressed. If, for example, the goal of current public education is to encourage families to talk about donation, it is useful to know a little about the internal dynamic of how families discuss sensitive issues such as this.

The UK research quoted above, for example, revealed that the family members most likely to discuss the subject of donation are husband and wife. Both pre- and post-campaign research revealed that nearly half of UK familial discussions of organ donation take place between husband and wife. One in four of those raising the subject are either a son or daughter aged over 16 years and in one in five cases, it is a parent who raises the issue. In only 9% of cases was the subject of donation brought up by a son or daughter under 16 years old.

Information of this nature may, for example, be essential in understanding why in some countries, such as Australia, it is possible to appear to increase the level of family discussion without affecting the percentage of those willing to donate a close relative's organs.

The role of the mass media in donation promotion

Over the past 10–15 years, television, radio, and the print media, have become an increasingly popular channel for delivering health promotional and prevention messages. A survey of the Australian population as far back as the late 1970s, for example, identified the mass media as a primary source of health advice.[35]

Television in particular has been identified as a major resource for increasing awareness. In US studies that asked where people heard about organ donation, television was cited as the primary resource for organ donor education.[4] UK opinion research also found that one in three adults cited television programmes or news as their source of information about donation in comparison to less than one in ten mentioning the print media.[31]

The US Government's Task Force on Organ Transplantation in 1986 identified television as the most effective educational medium for organ donation awareness but concluded that its strengths were underutilized, pointing out that:

> The use of national and local electronic media is poorly co-ordinated and uneven in quality and use of national TV for public service announcements was rare ... . Both local organ procurement organizations and voluntary health agencies try to get media co-operation in developing news and feature stories on donation but this is difficult to do effectively without staff experienced in media and public relations.

Each media discipline has particular strengths and weaknesses and effective communication strategies capitalize upon these strengths and optimize the interplay between different media. Advertising, for example, is a powerful communication medium but its main purpose is to keep an issue on the public agenda and contribute to a supportive environment thereby encouraging people to take in information from other sources – television programmes, press, leaflets, helplines and so on.

Effective management of the print media can stimulate more responsible reporting of an issue and address and target specific audiences with particular messages. Just as messages need to be targeted to specific groups, so different audiences get their information from different sources. Results from AIDS/HIV health promotion campaigns in the UK, for example, reveal that although television programmes and advertising are the most popular source of information for the general population (aged 13–34 years), it is one of the least effective channels of information for gay men.

The gay press is the principal source of information for gay men, followed by health promotion leaflets providing detailed explicit information. The 40% decrease in the practice of unprotected sex in the five years to 1991 suggests that the message of 'safer sex' from these two information sources has played an effective part in stimulating behaviour change.[33]

Although the mass media may be very effective in providing information and raising awareness, its role in stimulating behaviour change is limited. A number of studies evaluating the role of the mass media in the field of health promotion have concluded that media-alone interventions designed to alter behaviour directly have often had little impact on behaviour. However, when the media have been used in an agenda-setting role, in combination with a community component, significant changes in behaviour have been reported.[36]

The Cleveland Ohio public education programme which has been running since 1991 is a good example of the use of the media in calling the public to action, while a community component ensures that there is follow-up support for the action. The media element consists of a series of television and print advertisements which are aired periodically throughout the year with a call to ring a free telephone number for more information. The information going out to those responding, for example, encourages people to request a speaker to come and talk to groups at a more personal level, amongst other specific activities. The one thousand percentage increase in calls for information from the community following the programme's launch in 1991 was subsequently reflected in a 46% increase in the donor rate in 1992, a rate which though slipping slightly throughout 1993 and 1994, has remained considerably higher than in the days before the programme's launch.[37]

Encouraging sustained behaviour change therefore requires a range of interventions and messages and long-term commitment. A review of many health-related programmes that promote behavioural change suggest that interpersonal communication is often crucial to promoting change. Research from Australia and the USA has found that smoking, in particular, can be successfully reduced using mass media, complemented by local activity and interagency alliances.[38] A number of AIDS/HIV prevention campaigns have demonstrated that it is the interpersonal communication, in the form of small groups, workshops, counselling and outreach workers that appears to be necessary to promote behavioural change.[39]

The Finnish North Karelia Study is now a classic example of how a long-term series of media and community interventions can result in sustained behaviour change. This five-year cardiovascular risk factor intervention programme, focused on one county and involved screening, registration of hypertensive patients, GP education, health educational and local press and radio services. The programme's impact was evaluated after five years, by comparison with changes in cardiovascular risk factors, in a neighbouring 'no-intervention' county. Significant reductions across a range of variables were observed in the North Karelia group and ten years on, these reductions had persisted. The final proof of the programme's success came when the ten-year evaluation revealed evidence of the decrease in cardiovascular mortality among men and women in this one county compared to the rest of Finland.[36]

An analysis of successful public education donation programmes in the USA, conducted during the early 1990s, concluded that a critical success factor involved combining mass media communications with a grass-roots community approach, using volunteers to talk about issues on a one-to-one basis.[4]

Is public education cost-beneficial?

As the healthcare systems of many industrialized countries have come under fierce cost-containment constraints during the early 1990s, so the pressure to quantify and demonstrate cost-effectiveness, has grown in all areas of healthcare – the field of health promotion and public education being no exception.

While the cost-effectiveness of transplantation, particularly over long-term renal dialysis is undisputed, much less, if anything, is known about the economic evaluation of educational interventions designed to increase organ and tissue donation. The aim of conducting such an evaluation, however, would be to determine which educational/promotional strategies give the best value for the resources available.

There are three broad areas of activity where educational intervention can make a difference. One is in the area of medical infrastructure, where the organizational support for donation both within hospitals and between hospitals and the organ exchange agencies can influence the level of donation. The second area lies in raising the level of commitment and motivation of the medical professionals involved in the process of donor identification and referral. The third area involves addressing the negative attitudes and misconceptions held by the general public about donation and transplantation.

There has been growing interest, in recent years, in investing resources in the first and second of these areas – in improving the process of identifying and referring donors and in professional, rather than public, education.

Significant short-term increases in donation rates have, indeed, been observed as a consequence of a number of programmes that concentrate on improving the identification of potential donors and on supporting hospital medical personnel in their discussions with donor families.

By focusing on professional rather than public education, Lifelink of Tampa, Florida, for example, in the first three years of their programme, increased overall referrals by 400% (from 379 to 2154). Thanks to this programme, Florida now has a high of 33 donors per million, compared to the average US figure of 19 donors per million. Lifelink believe that professional contact and education is more critical than public education. With an approximate consent rate of 50% and studies showing an 80% willingness by the public to donate a loved one's organs, Lifelink determined in 1989 that the public already had an approval rating for organ donation they could not improve upon in the short term. Their analysis of data in the state of Florida indicated that any obstacle to donation occurred 'primarily because of hospital staff, generally nurses and physicians, not because of donor families.'[40]

The US Partnership for Organ Donation has also focused resources on improving the logistics of donation and, in recent years, a range of programmes, such as The European Donor Hospital Education Programme and Making the Critical Difference, have been developed to assist medical professionals in raising the subject of donation with bereaved families.

Clearly the task of the professional is much easier if the potential donor family is already somewhat informed about the donation possibility. Also an understanding of the likely fears and concerns of the family, gleaned through public attitude surveys, help the professional identify some of the issues that may need to be tackled during the crucial donation conversation.

The dilemma, therefore, over where to put limited resources is probably best summed up by Prottas and Batten, who concluded in 1991, that while 'the supply of organs is probably more contingent on the co-operation of medical professionals than on the attitudes of the general public, in the final analysis the success of organ procurement depends on the altruism of people'.[19]

The fact is that, for many years in many countries, public education initiatives in the field of organ and tissue donation were largely piecemeal, fragmented and under-resourced. They were conducted for the most part by a patchwork of transplant institutions, government health departments, voluntary or charitable groups, or assigned to transplant co-ordinators or organ procurement or exchange agencies, as one of their many duties. Rarely were communication professionals from advertising, marketing, public relations and the media involved in programme design or implementation.

The extent to which public education becomes a spending priority is dependent to some extent on what happens to the donor referral rate and the effectiveness of interventions in the area of professional education. Increasingly in the 1990s the issue of public education has become important because of declining rates and there are signs in a number of countries that, rather than work in isolation, interested organizations are banding together to seek professional advice in order to develop well-conceived national programmes.

The US Coalition on Donation is an example of this newly-emerging consensus approach. Founded in 1992, the Coalition represents an alliance of 47 national patient, professional, scientific, governmental, voluntary and commercial organizations together with equivalent coalitions at the State and local level. Its goal is 'to ensure that every individual understands the need for organ and tissue donation and accepts donation as a fundamental human responsibility'.

In 1993 the Coalition formed a five to seven-year partnership with the US Advertising Council, which pledged to co-ordinate creative and advertising support from volunteer advertising agencies and marketing experts, to develop a national public education campaign on organ and tissue donation employing a range of media strategies. This approach represents the first time in US transplant history that a national strategic communications plan has been developed which aims to co-ordinate activities at a national as well as at a local level.

In pooling their resources behind one unified national theme – 'Organ and Tissue Donation. Share your Life. Share your Decision' and using this as the central slogan around which to build strong local advocacy, communications professionals aim to focus the public's attention and optimize limited resources.

There are important issues of cost-effectiveness to consider in the selection of target audiences. When resources are limited there is little point targeting those sections of the population who are likely to fall into the categories of 'late adopters' or 'resistors' in terms of behaviour change. The conclusions of US research among black attitudes to blood donation, for example, echoes this sentiment. This research reveals that non-donors of blood who are also less predisposed to donating organs and tissues (either themselves or for their next-of-kin) tend to be females of low socioeconomic status. It concludes, 'might it not be more cost-effective to increase promotional efforts for blood, organs and tissues that are directed towards blacks of somewhat better socioeconomic standing'.[41]

The issue of which communication technique or combination of techniques is likely to yield the greatest cost-benefit in terms of increasing the donor referral rate, however, is a much harder question to answer. Simply measuring the net benefits of public education is difficult enough – certainly research can be conducted to measure the effects of an education campaign on the rate of donor card carrying or on the incidence of family discussions of donation – but a rise in either of these activities does not necessarily lead to an increase in donors. If an increase in the donor referral rate is to be used as the outcome upon which the cost-benefit of competing intervention policies is to be measured, the evidence to date is inconclusive.

Some public education programmes, such as that of Lifebanc in Cleveland Ohio, attribute the 46% increase in donation from 1991 to 1992 and sustained throughout 1993 and early 1994, to the US$400,000 education programme launched in 1991.[42]

In the UK, however, the donor rate dropped 4% during the first year of a two-year advertising campaign costing £2.5 million, despite increasing the number of donor cards carried and improving discussion amongst families.[43]

Inspiring and maintaining public confidence in the entire system of organ retrieval and distribution must be an important objective of any promotional strategy targeted at the general public, especially if external factors in the shape of scandals or lack of transparency in the system are undermining this confidence. Although general public education campaigns may not be able to point to obvious increases in national donor rates, their cost-benefit often lies in creating a supportive environment within which individuals can be encouraged to act positively.

Conclusion

In this chapter, through an analysis of geographically diverse examples and reference to the body of literature which exists on the subject, I have attempted to identify the critical success factors which I believe contribute to effective public education in the field of organ and tissue donation. Although there are many variables which need to be taken into account when planning such campaigns, I believe the following to be five of the critical keys to success.

1. The first of these success factors is that programmes which take the principles of behaviour change into account in their design and set realistic goals are likely to be more effective than those which do not.
2. The second involves the professional application of well-established marketing and promotional techniques to target audience selection; to an identification of their needs and the crafting and pretesting of target audience messages.
3. The third success factor involves an appreciation of the different strengths and weaknesses of the various communication disciplines, including advertising, direct mail, public relations, print and broadcast media together with an effective orchestration of any multi-media approach to impact at a local as well as at a national level.
4. The fourth critical success factor lies in a thorough and ongoing evaluation of the strengths and weaknesses of campaigns and a continuous tracking of any shifts in target audience opinion and behaviour.
5. The fifth success factor is recognition of the fact that interpersonal communication (in addition to any other form of communication) is often necessary to promote behavioural change and that such a mechanism should be built into the overall programme design.

References

1. Cantarovitch F, Castro L, Davalos M et al. Sectarianism, uncertainty and fear: mechanisms that may reverse attitudes toward organ donation. *Trans Proc* 1989, **21** (1): 1409–1410.

2. Personal communication Bernard Cohen, Eurotransplant, 1994.
3. Progress Notes, *Newsletter* of The Partnership for Organ Donation, Spring/Summer 1993.
4. Oberley E, Sacksteder P, Braun R Curtin et al. *Public Education in Organ and Tissue Donation*. Medical Media Publishing Inc., Madison, Wisconsin, USA, 1991.
5. Small F & Associates. *Public Awareness and Attitudes towards Organ Donation for Transplantation Purposes*. A Research Report funded by the Australian Kidney Foundation, Melbourne, 1987.
6. SOFRES, *Les Français et Le don d'organes*. French Health Ministry publication, Paris, 1993.
7. Evans R, Manninn D. Public attitudes and behaviour regarding organ donation. *J Amer Med Assoc* 1985, **253** (21): 3111–3115.
8. Evans R, Manninen D. Public Opinion concerning the procurement and distribution of donor organs. *Trans Proc* 1988, **20** (5): 781–785.
9. Quadrant Research Services. *Organ Donation Awareness Qualitative and Quantitative Studies*, Research Report, Melbourne 1991.
10. Personal communication, Philip Dye, ACCORD, 1994.
11. Ward E. *A Case for Opting Out*, April 1993, British Transplant Society Meeting, Exeter, UK.
12. Caplan A, Siminoff L, Arnold R et al. *Increasing Organ and Tissue Donation: What are the Obstacles, What are our Options?* Background Papers for the Surgeon-General's Workshop on Increasing Organ Donation, US Department of Health and Human Services, 1991.
13. Kokkedee W. Kidney procurement policies in the Eurotransplant region: 'opting in' versus 'opting out'. *Social Science and Medicine* 1992, **35** (2): 177–182.
14. Watts M. *How people feel about Organ Donation*. Lieberman Research Inc., 1991.
15. Williams R. Why we need to change the law on organ donation. *The Silver Lining Appeal in aid of the British Kidney Patient Association*, 20th Edition, 1993.
16. Arkin EB. *Motivating the Public: Application of Lessons Learned in Increasing Organ Donation*. Background Papers for the Surgeon-General's Workshop on Increasing Organ Donation, US Department of Health and Human Services, 1991.
17. Dominguez-Roldan JM, Murillo-Cabezas F, Munos-Sanchez A et al. Psychological aspects leading to refusal of organ donation in Southwest Spain. *Trans Proc* 1992, **24** (1): 25–26.
18. Hasegawa T, Maeda Y, Yamakawa K. Proposal for free delivery of donor cards in Japan. *Trans Proc* 1994, **26** (2): 980–982.
19. Prottas J, Batten H. The willingness to give: the public and the supply of transplantable organs. *J Health Politics, Policy and Law* 1991, **16** (1): 121–134.
20. Kotler P. *Marketing for Non-Profit Organisations*, Prentice-Hall, New Jersey, 1982, page 490.
21. Hastings G, Haywood A. Social marketing and communication in health promotion, *Health Promotion International* 1991, **6** (2): 135–145.
22. Willoughby D (Ed.). *Organ Transplantation: Issues and Recommendations*. US Department of Health and Human Services, 1986, pages 39–41.
23. Dye P. *Organ Donation Promotion in Australia: A historical overview, current directions and future principles*. ACCORD, 1993, (unpublished).
24. Callender CO. Organ donation in blacks: a community approach. *Trans Proc* 1987, **19**: 1551–1557.
25. *The distribution of organs for transplantation: expectation and practices*, August 1990, Washington DC, Office of Inspector General.
26. Yang S, Abrams J, Smolinski S et al. Organ donation and referrals among African–Americans. *Trans Proc* 1993, **25** (4): 2487–2488.
27. Toledo-Pereyra LH. The problem of organ donation in minorities: some facts and incomplete answers. *Trans Proc* 1992, **24** (5): 2162–2164.
28. Kappel DF, Whitlock ME, Parks-Thomas TD et al. Increasing African–American organ donation: the St Louis Experience. *Trans Proc* 1993, **25** (4): 2489–2490.
29. Hong B, Kappel D, Whitlock M et al. Using race-specific community programs to increase organ donation among Blacks. *Amer J Public Health* 1994, **84** (2): 314–315.

30. Benoit G, Spira A, Nicoulet I et al. Presumed consent law: results of its application/outcome from an epidemiologic survey. *Trans Proc* 1990, **22** (2): 320–322.
31. UK Department of Health, Organ Donation Campaign Evaluation Research, London, 1994.
32. Personal communication, UK Department of Health, London, 1994.
33. UK Health Education Authority HIV/AIDS Mass Media Activity 1986–1993, London.
34. Lifebanc Marketing/Education Research Project 1993, Cleveland, Ohio.
35. Better Health Commission. *Looking Forward to Better Health.* 1979, Vols 1–3, AGPS, Canberra.
36. Redman S, Spencer E, Sanson-Fisher R. The role of mass media in changing health-related behaviour: a critical appraisal of two models. *Health Promotion International* 1990, **5** (1): 85–100.
37. Personal communication, Lifebanc, Cleveland, Ohio, 1994.
38. UK Health Education Authority Annual Report, London, 1992/93.
39. Franzkowiak, Wenzel. AIDS Health Promotion for Youth, *Health Promotion International* 1994, **9** (2).
40. Campbell J, Layne J. *The Donor Dilemma: The Lifelink Foundation Approach.* Lifelink, Tampa, Florida, 1994.
41. Bayton J, Jennings P, Callender C. The role of Blacks in blood donation and the organ and tissue transplantation process. *Trans Proc* 1989, **21** (1): 1408.
42. Personal communication, Lifebanc, 1994.
43. New B, Solomon M, Dingwall R et al. A question of give and take: improving the supply of donor organs for transplantation, *Research Report 18* 1994, Kings Fund Institute, UK.

CHAPTER 24

Unrelated bone marrow donor registries

PATRICIA A. COPPO

Initial establishment of unrelated bone marrow donor registries   430
Centralized marrow donor registries   431
Registry organization   432
Alternative sources of stem cells   442
References   443

Initial establishment of unrelated marrow donor registries

In the early 1980s, as unrelated donor bone marrow transplantation (UDBMT) became an accepted therapeutic option[1-13] patients, their families and physicians soon recognized the need for a source of HLA typed volunteer marrow donors. In the United States, families of patients initially turned to members of their immediate community and requested they have their HLA typing performed to determine whether they were a match for the patient. Unfortunately, given the diversity of the HLA system and the need for an extremely close HLA match between donor and recipient, most of the individuals typed in an attempt to find a donor for a specific patient were not suitably matched. The typings were often retained by the families, creating small listings of several hundred HLA typed volunteers who could be searched for other patients. When patients needed unrelated donors their transplant physicians would send search requests to the individual families and they would manually search their files in an attempt to find potential matches for patients.

During this time, blood banks also began approaching their HLA typed platelet donors and asking them to consider volunteering to be marrow donors.[14-19] These individuals had already demonstrated their altruism through their commitment to repeated apheresis platelet collections. In addition, platelet donors receive a verbal health screening prior to each donation and their blood products are screened for infectious diseases, thus they were considered to be excellent marrow donor candidates. Recruitment of platelet donors was initiated by several blood banks in the United States and other countries but, as with the family-run registries, each of these sources of potential marrow donors had to be contacted and searched separately, thus adding time and expense to the process of finding an unrelated marrow donor.

Centralized marrow donor registries

ANTHONY NOLAN BONE MARROW TRUST

Shirley Nolan, the mother of a young boy diagnosed with Wiscott–Aldrich Syndrome, began marrow donor recruitment efforts in 1974 in an attempt to find a donor for her son, Anthony. She received support from the National Association of Round Tables of Great Britain and Ireland. Although Anthony died before a donor could be found, Shirley Nolan's efforts inspired the creation of the Anthony Nolan Research Centre, the first registry of unrelated bone marrow donors in the world. The Anthony Nolan Research Centre, started as an HLA typing laboratory in 1974, provided a model for the development of centralized marrow donor files in other countries around the world.[20]

UNITED STATES NATIONAL MARROW DONOR PROGRAM

In the United States, patients and their families approached members of Congress to request federal government funding for a centralized marrow donor registry. In 1986, Congress appropriated funds for this purpose, and a request for proposals was issued. The proposal selected for funding was submitted by three co-investigators, each representing one blood banking organization: the American Red Cross, the Council of Community Blood Centers and the American Association of Blood Banks. The collaboration of the three largest US blood banking organizations helped to ensure that the newly formed marrow donor registry, the National Marrow Donor Program (NMDP), had a broad base of support in the industry most familiar with recruiting, screening and managing blood and tissue donors. Members of the NMDP Board of Directors were selected from patients' families, marrow transplant physicians, immunologists and blood bank representatives. The NMDP began search operations in September, 1987.[21-26]

REGISTRIES AROUND THE WORLD

By the early 1990s marrow donor registries were established in many countries around the world. Often, the impetus for the initial recruitment of donors was the need of a specific patient, and families were responsible for raising funds to HLA type donors. However, as the value of marrow donor registries became evident, federal governments stepped in to help fund the registries in many countries, allowing them to continue recruiting and HLA typing additional donors.

Registry size varied widely but even the largest registries could not find matches for all their patients needing marrow transplants. In an effort to find matched donors for their patients, registries began exchanging searches, unfortunately, the reliance on paper forms sent through the mail often resulted in unacceptably slow search turn-around times. This issue was addressed in 1988 when the European Community granted a subsidy for the establishment of the European Donor Secretariat (EDS) in France. EDS provided the first opportunity for electronic transmission of search requests between registries. The European Marrow Donor Information System (EMDIS), also funded by the European Community, was subsequently developed to further enhance electronic search capabilities between countries.

Bone Marrow Donors Worldwide (BMDW), established in 1988, provided another important method for identifying sources of potential matched donors in registries around the world. BMDW is a compilation of phenotypes of all donors from participating registries. The phenotype listing, which is updated regularly, is incorporated into a computerized search programme thus allowing the user immediately to identify registries which may provide a donor. It has been estimated that the use of BMDW can reduce search times by 75%, thereby providing for fast and cost-effective location of donors with rare phenotypes.[27]

Registry organization

In general, the concept of a registry is that of a 'hub' in the centre of a wheel. That is, the registry serves to connect and co-ordinate the functions of individual transplant, collection and donor centres. Transplant centres are responsible for the treatment of patients. Collection centres are responsible for donor care from the time of hospitalization for marrow harvest until discharge, and for the quantity and quality of the collected marrow. Donor centres are responsible for donor recruitment and management, including post-collection follow-up until donors have completely recovered from the procedure.

DEVELOPMENT OF REGISTRY STANDARDS, POLICIES AND PROCEDURES

The efficient functioning of a donor registry requires established standards, policies and procedures governing all aspects of donor recruitment and management, patient eligibility, and marrow collection, handling and transport. In the case of most marrow donor registries and the World Marrow Donor Association, standards, and periodic revisions to the standards, are drafted by committees composed of individuals with relevant expertise. The proposed standards are then circulated to members for review and comment. Members' comments are considered by a committee, incorporated if appropriate, and the final document is disseminated as a reference for registry users.

Detailed procedures governing daily operations are developed by registry staff and are disseminated in manuals of operation. The manuals contain complete instructions for submission of search requests; communication between transplant, donor and collection centres during the search process; collection and shipment of blood samples for compatibility testing and research purposes; and completion of data collection forms. Policies (e.g. policy on approaching marrow donors for a second donation for the same patient) serving to clarify the intent of the standards and apply them to daily operations are developed as needs arise.

TRANSPLANT CENTRES

As marrow donor registries developed the question arose as to who would be eligible to search their files and obtain marrow from unrelated donors. Many registries felt a need to protect donors from undergoing a marrow collection procedure if a successful outcome for the recipient seemed unlikely. It was felt that a successful outcome was less likely at inexperienced transplant centres; for the treatment of certain diseases; and with less than perfect matches. Each registry developed its own criteria for who would be allowed access to its donor file and for which patients unrelated donor marrow would be provided.

Several registries developed a formal transplant centre accreditation procedure designed to assess each centre's experience, staffing levels, facilities and support services. Only accredited transplant centres are able to conduct searches through these registries, however, once a centre is accredited the centre is responsible for determining patient eligibility for unrelated marrow transplantation through its own standard or experimental treatment protocols. The second edition of the WMDA standards document[28] provides suggested transplant centre accreditation criteria for evaluation of centres wishing to obtain marrow from a donor resident in another country.

Other registries follow less formal transplant centre accreditation procedures but establish strict patient eligibility and HLA match criteria, and only provide marrow for patients whose diagnosis and degree of HLA match meet the registry's established criteria. The second edition of the WMDA standards lists diagnoses for which unrelated donor marrow transplantation is considered standard treatment and others for which it should be considered experimental.

COLLECTION CENTRES

Only institutions with acceptable support facilities and experienced marrow collection staff should be considered for the harvesting of marrow from unrelated donors (see Chapter 13). A few registries work with only one or two collection centres and bring donors from all parts of the country to these institutions for their harvests. Other registries establish marrow collection centres close to most donor centres thus minimizing the need for donors to travel.

The final nucleated cell count of the collected marrow, anticoagulant used, method of processing, if any, and other factors which have an impact on the suitability of the marrow for the recipient must be discussed by the transplant and collection centre in advance of the harvest. The outcome of these discussions, as well as the date and time of marrow harvest, should be documented in writing.

DONOR CENTRES

Most registries rely on a network of donor centres (usually blood banks) to recruit, perform health screening, HLA type and contact donors for additional testing. As a means of protecting donor confidentiality in these cases, donor identifying information (i.e. names, addresses and phone numbers) are maintained only at the local donor centre. The donor centre provides the registry with the donors' HLA type, sex, date of birth and other information required for the search and donor selection process but donors are identified to the registry only by coded numbers. When donors are requested for further testing the registry notifies the donor centres which are then responsible for contacting the donors and obtaining blood samples or performing the pre-collection donor work-ups.

In a few countries the registry is responsible for all donor recruitment and management, including contacting donors for further testing. These registries maintain complete information on their donors in one central file but only identify donors by coded numbers to transplant centres and other outside groups.

Donor centre accreditation criteria vary between registries. Some require donor centres to meet specific criteria and document their experience during a formal accreditation process similar to that used for the accreditation of transplant centres. In other countries virtually all blood banks are asked to participate in providing marrow donors to the registry.

DONOR RECRUITMENT

A 1993 survey of 15 WMDA participating 'Hubs' indicated that, although all registries were engaged in donor recruitment activities to some extent, the groups being targeted for recruitment and the methods employed to reach each group varied considerably between registries. For example, Australia, The Netherlands, Ireland and Switzerland were primarily recruiting blood donors, while registries in most other countries were recruiting both blood and 'community donors'. Community donors are defined as individuals who respond to an appeal for marrow donors and are tested at community-based donor drives. Although the US NMDP data indicate that blood and community donors are equally likely to respond positively to requests for additional testing on behalf of patients, registries recruiting primarily blood donors maintain that the blood centres' frequent contacts with, and health and infectious disease screening of, blood donors result in faster contact and fewer deferrals when their donors are requested for further testing as potential marrow donors.

For recruitment purposes potential donors may also be grouped according to racial or ethnic groups. It has long been recognized by blood banks that members of certain racial and ethnic groups are reluctant to donate blood. In some cases their reluctance is based on religious beliefs, in others it is based on their distrust of the medical community. Marrow donor registries encountered the same reluctance when members of these groups were first approached and asked to consider becoming donors. The registries which are successfully recruiting within these groups have established relationships with influential community leaders and are relying on them to convey the need for donors to members of their own communities. This effort also requires the development of culturally sensitive recruitment materials and translation of brochures, consent forms and other materials into a variety of languages.

PROMOTION OF RECRUITMENT ACTIVITIES

Community-based recruitment activities are more successful when the national and/or local media support and promote recruitment efforts through television coverage, the airing of public service announcements, and the printing of newspaper articles. Registries may increase media coverage in advance of donor recruitment activities by preparing and distributing their own news releases and public service announcements. Press conferences may also be used to notify the news media of special activities of potential interest. For example, several registries have held press conferences to announce the first time one of their donors provided marrow for a patient in another country.

Although many successful recruitment efforts, including those leading to the establishment of the Anthony Nolan Research Centre in the United Kingdom, have been built on the need of a specific patient, focusing on the need of a single individual may result in the recruitment of donors willing to donate only for that person. It is important that potential donors understand that they are being recruited for registries which will be searched by patients from many countries, and that they should only register and be tested if they would be willing to donate for any patient with whom they matched. In an effort to reduce potential donor identification with the needs of a specific patient, many registries produce and distribute recruitment materials featuring several patients, either from a single (e.g. Asian) or different (e.g. Asian, Caucasian and Hispanic) racial backgrounds.

It is also important that patients' families wishing to initiate recruitment activities understand that the probability of their finding a donor through their own recruitment efforts is very low, rather they will be adding to the registry to help future patients. Patients and their families should also be counselled to raise funds for their own search and transplant costs, if these costs are not covered by national or private health insurance, before assisting registries in raising funds for donor recruitment activities.

DONOR INFORMED CONSENT

Both blood and community donors must be provided with written materials and/or videotapes explaining the time and commitment required to be a marrow donor, including sufficient detail on the marrow harvest for the donor to understand that it is a surgical procedure. Potential donors should also be provided with an opportunity to ask questions of a designated, knowledgeable, individual before they provide informed consent. Most registries require that donors read and sign written consent forms prior to blood sample collection for HLA typing.

DONOR HEALTH SCREENING

All donors should be given a verbal health history screening similar to the one administered to blood donors. The screening is designed to eliminate those individuals with a history of illness which would place them at increased risk during the marrow harvest or recovery period. The screening should also eliminate potential donors whose health history or life style may indicate that their marrow could place potential recipients at increased risk for infectious or other diseases which could be transmitted by the marrow.

Some registries perform ABO and/or infectious disease testing on donors from blood samples taken at the time of initial recruitment. Although testing at the time of recruitment may eliminate from further consideration donors who are positive for certain disease markers, infectious disease testing must be repeated at several stages in the search process, and testing at recruitment increases the cost of adding donors to the registry.

COLLECTION OF DONOR DEMOGRAPHIC AND CONTACT INFORMATION

At the time of recruitment it is essential that donors provide their current address, home and work phone numbers as well as contact information for family and friends who will know how to reach them in the future should they move without informing the donor centre. Some registries require that donors also provide their Social Security number, Health Service number or other government issued identification number to be used as a means of tracking donors who are lost to contact by other methods.

Donors must also provide their date of birth, sex and, in some cases, race. Although laws prevent the collection of race data in some countries, registries which are not legally prevented from collecting donor race data should collect and provide it, in addition to donor sex and age information, to assist in the selection of donors for further testing for specific patients, since tissue types are more likely to match within ethnic groups rather than across racial boundaries.

All registries have established donor age limits though the limits vary between registries. The 1993 WMDA Hub survey indicated the minimum age for marrow donation varied between 17 and 20 years while the maximum age at which donors could donate varied between 50 and 70 years.

HLA TYPING OF NEW DONORS

Virtually all registries HLA-A and B type new donors though the level of resolution varies. Some registries also routinely type all new donors for HLA-DR antigens. Over 90% of the donors in these registries are completely HLA-A,B and DR typed thus reducing search time and costs for patients. Unfortunately, many registries, especially those recruiting large numbers of new donors, have neither the funding nor the laboratory capacity to HLA-DR type all new donors. Instead, most now store a blood sample on newly recruited donors and use this for molecular Class II typing.

Typing of stored samples can either be performed as donors are requested for patient-directed typing or in advance of patient need (prospective typing). The donors' HLA-A,B typing are performed before they are placed in the file and, therefore, the HLA-A,B type can be used to select stored samples for prospective typing. By selectively Class II typing donors it is possible to increase the diversity of completely typed donors, and increase the probability of new patients finding HLA-A,B and DR matched donors at the time of their initial search.

DONOR DATA MANAGEMENT

Following recruitment, donor information must be merged with a registry's centralized listing of donors. The data may either be entered directly onto the central registry's computer system, or added to a local donor centre system and transmitted to the central registry electronically or by computer disk. In either case, all donor data required to match patients and donors, and select potential donors for further testing, must be maintained by the central registry and made available to physicians. The format and type of donor data presented varies between registries but most registries produce search reports listing individual potential donors and providing each donor's HLA type, sex, age, and other available information which may influence donor selection.

The central registry is also responsible for removing from the search process donors who can not be contacted; are no longer eligible to donate due to age or health; or who wish to be removed because they are no longer willing to be marrow donors. In most cases local donor centres make direct contact with donors and obtain this information, and it is their responsibility to inform the central registry of the donors' change in status.

Donors may also be temporarily unavailable for a period of time for reasons of health (e.g. pregnancy) or other personal considerations. If it is known that a donor is temporarily unavailable, this information and the date after which the donor will be available should be conveyed on the search report, or through follow-up contact with the physician, to avoid the potential for delays on a patient's search.

DONOR AND PATIENT HISTOCOMPATIBILITY MATCHING

Registries are responsible for matching patients and donors and providing this information to physicians in a manner which promotes rapid identification and selection of the most closely matched donors for each patient. Computerized matching algorithms must take into account whether the patients' and donors' antigens are broadly defined or split, and produce lists of those donors who may be potential matches once the donors' and/or patients' antigens are definitively typed. As more donors and patients are typed by molecular methods rather than serology, the matching algorithm must also take into account the typing method used in each case and incorporate this information in the donor selection process.

Given the size of many registries' donor files, most patients will have more than one potentially matched donor available on their search reports, especially if the search results also list donors who are one antigen mismatches with the patient. To promote the rapid identification of the most closely matched donors, search reports should be ordered to provide listings of donors according to the following criteria.

1. Potential six antigen matched donors should be listed before donors who are mismatched at one HLA locus with the patient.
2. Donors who are one antigen mismatched with the patient should be listed in the same section of the report as other donors who are mismatched at the same locus (i.e. donors who are a one antigen mismatch at the HLA-A locus should be listed with other donors who are also mismatched at the HLA-A locus).
3. Donors whose antigens have been split, and whose split antigens match those of the patient, should be listed before potential donors with unsplit antigens.
4. Completely HLA-A,B and DR typed donors should be listed before donors who are equally compatible at the HLA-A and B loci but who have not been HLA-DR typed.

Although degree of HLA match is the primary consideration in the selection of potential marrow donors, given two or more donors who are equally well HLA matched with the patient, physicians will use other donor information to select the best potential marrow donor. Donor age, sex, CMV status, race and other characteristics have all been used by transplant physicians to select donors, however, the relative importance assigned by physicians to each of these characteristics in the selection of the final marrow donor varies. For this reason, donors should be listed according to degree of HLA match with the patient, and additional donor information should be provided, if available, on the search report without regard to ordering donors according to these characteristics unless the registry's computer system will allow those searching to establish their own sort criteria.

EXCHANGING SEARCHES BETWEEN REGISTRIES

When matched donors are not available in the first registry searched by a patient, the search should be extended to other registries around the world. It is recommended that a search of Bone Marrow Donors Worldwide (BMDW) be first undertaken to identify other registries which already list donors with the same phenotype as the patient. If phenotypically identical donors are listed, the registries listing the donors should be contacted first. If no matched donors are listed in BMDW, patient search information should be transmitted to other registries directly, on the chance that newly recruited donors, not yet listed in BMDW, may be matches for the patient. Patient search requests may be transmitted to other registries by facsimile or electronically by EDS, EMDIS or direct file transfer.

FACILITATING THE DONOR SEARCH PROCESS

As mentioned previously, registries serve to co-ordinate the functions of individual donor, transplant and collection centres thus eliminating the need for transplant centres to contact multiple donor centres to request donor testing for their patients. Registries may also serve to centralize the payment of donor centres and the billing of transplant centres in countries where marrow donor search services and marrow procurement are not covered by national health plans.

In general, registries facilitate the following steps in the search process.

Preliminary search

Physicians submit their patients' HLA antigens, diagnosis, date of birth and other relevant information to the registry and request that a search of the donor file be initiated. The preliminary search indicates whether any potentially matched donors are available for a patient, and helps physicians evaluate the feasibility of UDBMT for the patient. The results of preliminary searches are usually available within a few days of the registry receiving the search request.

HLA-DR typing

If the potentially matched donors available for a patient are matched at the HLA-A and B loci but have not yet been HLA-DR typed, the first step in the search process will be to HLA-DR type selected donors. The transplant centre, often with guidance provided by registry staff, selects donors for HLA-DR typing, and the registry communicates the typing request to the donor centres managing the donors. Typing is performed either on a fresh blood sample obtained from a donor or on a stored sample. When typing is complete results are transmitted to the registry which updates the donors' files and informs the transplant center of the typing results. Donors who are suitably matched with the patient may be requested for further testing; those who do not match will be released from the search and become available for other patients.

Confirmatory typing

When suitably matched donors have been identified for patients based on the HLA typing information provided by the registry, blood samples from the donors may be requested for confirmatory typing. The registry transmits the requests to the appropriate donor centres. The donor centres contact the donors, arrange for collection of the blood samples and shipment of the samples to the transplant centre or designated laboratory. The confirmatory typing sample is used to verify the donors' HLA type, perform infectious disease testing and other tests. Final HLA matching of patients and donors typically includes allele level typing of HLA Class II antigens, and repeat HLA-A,B typing by serology or DNA technology. The final patient and donor HLA typing results are then compared to ensure a close match.

Given the diversity of HLA antigens, and the even greater diversity of DNA defined HLA alleles, most patients and their potential donors are mismatched at some level. The challenge facing transplant physicians is to select donors whose mismatch is least likely to cause post-transplant complications including graft rejection and graft-versus-host disease. Different transplant centres have developed different criteria for donor selection based on their perceptions of the effects of mismatches at various HLA loci. The policy of most registries is to require transplant centres to perform confirmatory typing, and to report the typing results to the registry, but to allow transplant centres to select the best matched donor, according to their own criteria.

Donor pre-collection work-up

Donors selected by the transplant centre as suitably matched with patients may be requested for pre-collection work-up. Donor work-ups generally include an information session at which the donors are given more detailed information on the marrow collection procedure, potential risks associated with marrow collection and anaesthesia, and the potential benefit to the patient. Following the information session, donors who agree to proceed are given a physical examination to ensure they are in good health.

When donors have been determined to be physically able and willing to donate, donor centres inform the registry which relays the information to the transplant centres. At this point transplant centres and donor centres may begin direct communication to select transplant dates and negotiate marrow volumes and cell counts. Once finalized, this information should be confirmed, in writing, by the transplant and donor centres and transmitted to the collection centres.

Marrow collection and transport

Marrow harvests are performed by collection centres which are responsible for the donors' care during and after the collection procedure. Donor medical problems should be communicated to the donor centre and the registry. The marrow volume, nucleated cell count, type of anticoagulant and transport conditions should have been agreed upon in advance of the collection. Any deviations must be communicated to the transplant centre and the registry.

Following collection, marrow must be transported by a qualified courier as rapidly as possible to the transplant centre. Couriers must be experienced travellers and be aware of their responsibility for the safety of the marrow.

Patient and donor tracking throughout the search process

Each patient actively searching for a marrow donor, and all donors requested for testing on behalf of each patient, should be tracked through-out the search process by the registry. Registries monitor search activity in different ways but each registry's computer system should maintain individual patient search files listing the donors who have been requested for HLA-DR typing, confirmatory typing and work-up.

In addition, most registries monitor donor recovery following marrow collection and patient progress post-transplant. These data may also be stored on a registry's computer system and be made available for research and quality assurance purposes.

Maintaining donor and patient confidentiality

Most registries limit the patient information provided to donors during the search and pre-collection work-up process to the patient's age, sex and diagnosis. Patients are usually not provided with any personal donor information; in most cases they are not even informed of the region of the country in which the donor lives. This extreme confidentiality is necessary for several reasons but is primarily based on the concern that the urgency felt by patients and their families to find a donor may lead them to place undue pressure on donors. However, both patients and donors have the same rights to privacy and confidentiality and these rights must be respected and protected by registries. Throughout the search process many individuals will have contact with donors, patients, or their medical records. It is imperative, therefore, that registries develop policies and procedures to strictly maintain patient and donor confidentiality at each step in the search process. This includes providing detailed instructions to marrow couriers not to disclose the location of the donor nor marrow collection site to the patient, nor to inform the donor of the destination of the marrow.

In the early days of UDBMT, patients and donors frequently exchanged names, addresses and other personal information at the time of transplant, perhaps even meeting each other before the marrow collection. Although this contact increased the emotional reward for many donors, it became clear over time that the policy of providing an opportunity for patients and donors to become personally involved around the time of transplant often resulted in extreme emotional distress for those donors whose patients did not do well following transplant. In some cases it also lead to direct requests from patients' families for donor platelet support or additional marrow for a second transplant.

For these reasons most registries now restrict the amount of personal patient and donor information which may be exchanged before transplant, and for a period following the transplant. Any communication between patients and donors which occurs during that time, for example thank-you notes and small gifts, are passed through the respective transplant and donor centre personnel who screen the materials to ensure that identifying information is not included. Complete protection of donor and patient confidentiality usually extends for up to one year post-transplant. After that time, provided both the donor and patient are agreeable and understand the potential consequences of becoming personally involved, most registries will facilitate the exchange of personal information thus allowing the patient and donor to correspond directly. Some patients and their donors arrange for personal meetings after this time.

SUBSEQUENT DEMANDS ON DONORS

Although most donors report they have fully recovered within three weeks of their marrow collection,[29,30] ideally, donors are deferred from donating marrow a second time for a period of one year. Several registries actually remove the donor from their files for one year following marrow donation and, at the end of this period, ask whether the donor would like to be placed back on the registry. However, there are cases in which marrow recipients require peripheral blood stem cells, leukocytes or additional marrow, and the original donor is felt by the transplant physicians to be the best source of these cells. In these situations most registries will allow consideration of the request, though usually the appropriateness of each request is anonymously reviewed by a panel of physicians who determine whether the donor may be approached for a second donation. In the NMDP when reviewing each request the panel takes into account the donor's physical and emotional condition (provided by the donor centre), the patient's clinical condition (provided by the transplant centre), and the appropriateness and availability of alternate donors (provided by the registry).

If the panel determines the original donor may be approached, the donor centre is requested to contact the donor and give the donor the opportunity to make an informed decision by providing him or her with additional information on the patient's clinical condition and the likelihood of improvement with the donation of the requested blood product or marrow. Those donors agreeing to proceed may require another physical examination, repeat infectious disease testing and/or the storage of autologous blood, depending on the blood product or volume of marrow being requested, and the length of time since the first donation.

Increasingly, transplant centres are requesting that donors be given growth factors (e.g. G-CSF) to stimulate the production and release of stem cells into the peripheral blood prior to collection via apheresis. Although many registries and individual donor centres are concerned about giving growth factors to normal volunteer donors since the long-term effects of these substances has not been established, these potential risks must be weighed against the effects of a second marrow donation and the risks associated with anaesthesia. Prior to administration of growth factor to donors, the donor centre's or registry's Ethical Review Board should approve both the protocol for the use of the substance in normal donors and a consent form which clearly explains the potential risks and side-effects associated with the growth factor. Transplant centres seeking to receive stimulated stem cells for their patients should also be required to have an Ethical Review Board approved protocol for the therapeutic use of these blood products.

FINANCIAL CONSIDERATIONS FOR ESTABLISHING AND MAINTAINING A MARROW DONOR REGISTRY

The cost associated with establishing and maintaining a registry include recruiting and HLA typing volunteer donors; developing and maintaining a computer system to match patients and donors and track patient searches; communicating electronically with donor, transplant and collection centres and registries around the world; hiring and training personnel; producing materials for donor recruitment and education; travel costs associated with recruitment, training, and other functions; and purchasing equipment, supplies and office space. In addition, when services are requested for specific patients the registry must cover the costs associated with providing the services (e.g. blood sample collection, shipping and testing) whether the registry provides the services directly or indirectly through donor centres. The costs associated with marrow collection and transport must also be covered by the registry.

Although most registries receive some level of government funding and charitable contributions, virtually all registries must also charge fees for their services to make up the difference between their costs and the income they receive from other sources. In most cases registries have a different fee schedule for searches which are submitted by foreign registries, or transplant centres in foreign countries, than they do for searches originating within their country. These differences in fees may reflect coverage of domestic patients by a national health system which also funds the registry, and/or the increased costs associated with transporting blood samples and marrow to transplant centres in foreign countries.

There is considerable variability in the actual costs of establishing and maintaining a registry and providing services in different countries. Unfortunately, this may result in large differences in the search fees charged by registries which, in turn, may limit access to foreign donors for some patients. For this reason registries are now considering alternative fee schedules or the development of a credit systems to promote the exchange of marrows between countries.

FUTURE DIRECTIONS FOR MARROW DONOR REGISTRIES

Long-term disease-free survival varies according to the degree of match between patient and donor as well as the patient's diagnosis, stage of disease, age, and other factors.[31-39] Unfortunately, for many patients transplantation with haematopoietic stem cells following high dose chemotherapy and/or radiation remains their only potentially curative therapeutic option. Since most of these patients do not have an acceptable related donor, there will be a continued need for registries to maintain centralized listings of potential stem cell sources and to facilitate transplants.

As registries around the world continue to recruit more donors, many emphasizing the need for more non-Caucasians, scientists have begun to evaluate the size of the donor pool required to find matches for patients from different racial and ethnic groups.[40-43] Although the number of patients who find matched donors continues to grow, and the number of transplants facilitated by registries continues to increase, there are practical limitations to the effective size of donor registries. The limitations on registry size are financial (i.e. costs associated with recruiting and HLA typing new donors) as well as technical (i.e. managing and updating data bases and tracking systems).

Fortunately, research to advance understanding of the correlation between degree of HLA match and transplant outcome, and improvements in the management of graft rejection and graft-versus-host disease, may reduce the need for donors and patients to be as closely matched as is now considered necessary for successful post-transplant outcome. If marrow from donors who were one or two antigen mismatches with patients could be successfully transplanted, the size of donor registries could be maintained at current levels or even reduced through donor attrition, and virtually all patients could still receive unrelated donor marrow transplants.

Alternative sources of stem cells

Stem cells collected from the umbilical cords of new born infants may prove to be an effective alternative to marrow stem cells for transplantation. Cord blood stem cells (CBSC) are collected from umbilical cords and placentas, completely HLA typed, frozen and stored in liquid nitrogen. CBSC HLA information is made available through registries in the same manner as unrelated donor information. Once a potential CBSC match is identified through a preliminary search, the transplant centre may move rapidly to transplantation because the need to contact donors and arrange for further testing and marrow collection is eliminated, thus substantially reducing search times and costs for many patients. Registries benefit through increased diversity of their donor files, and the elimination of recruitment and donor management costs, although the costs associated with CBSC processing and long-term storage offsets these savings.

Peripheral blood may also serve as a source for stem cells for transplantation, provided the risks associated with the use of growth factors (e.g. G-CSF) is considered acceptable for unrelated donors. As mentioned previously, the use of G-CSF in unrelated donors and the administration of stimulated products to patients must first be approved by Ethical Review Boards at both the donor centre and transplant centre. Although the collection of peripheral blood stem cells through apheresis would eliminate the risks associated with anaesthesia and marrow collection, the potential risks and side-effects associated with the use of growth factors, and the time and discomfort required for repeated apheresis collections, may discourage many donors from agreeing to this alternative stem cell collection procedure. It is the responsibility of registries to ensure that their donors are provided with adequate information before they are asked to give their informed consent for the procedure. Registries must also ensure that growth factors are administered and apheresis collection of stem cells are carried out only by experienced institutions using approved protocols.

Both of these alternative stem cell sources provide new challenges and opportunities for registries. Transplants using unrelated donor stem cells, whether obtained from marrow, peripheral blood or umbilical cords, will continue to provide patients with the opportunity for a cure for otherwise fatal diseases. Registries will continue to be responsible for providing patients and their transplant physicians with accurate information on the availability of donors and CBSC, and to facilitate the delivery of these cells in a manner which protects the health and safety of patients as well as donors.

References

1. Beatty PG, Ash R, Hows JM, McGlave PB. The use of unrelated bone marrow donors in the treatment of patients with chronic myelogenous leukemia: experience of four marrow transplant centers. *Transplantation* 1989, **4**: 287–290.
2. Phillips GL. The use of unrelated donors (UD) for allogeneic bone marrow transplantation (BMT): a pilot study of the Canadian BMT group. *Bone Marrow Transplantation* 1991, **7**: 52–53.
3. Howard MR, Hows JM, Gore SM, Barrett J, Brenner MK, Goldman JM, Gordon-Smith EC, Poynton C, Prentice HG, Whittaker JA, Bradley BA. Unrelated donor marrow transplantation between 1977 and 1987 at four centres in the United Kingdom. *Transplantation* 1990, **49**: 547–553.
4. Speck B, Zwan FE, van Rood JJ et al. Allogeneic bone marrow transplantation in a patient with aplastic anemia using a phenotypically HLA-A identical unrelated donor. *Transplantation* 1973, **16**: 24–28.
5. Hansen JA, Clift RA, Thomas ED et al. Transplantation of marrow from an unrelated donor to a patient with acute leukemia. *N Eng J Med* 1980, **303**: 565–567.
6. O'Reilly RJ, Dupont B, Pahwa S et al. Reconstitution in severe combined immunodeficiency by transplantation of marrow from an unrelated donor. *N Eng J Med* 1977, **297**: 1311–1318.
7. Horowitz SD, Bach FJ, Groshong T et al. Treatment of severe combined immunodeficiency with bone marrow from an unrelated, mixed-leukocyte culture nonreactive donor. *Lancet* 1975, **2**: 431–433.
8. Foroozonfar N, Hobbs JR, Hugh-Jones K et al. Bone marrow transplant from an unrelated donor for chronic granulomatous disease. *Lancet* 1977, **1**: 210–213.
9. Lohrmann HP, Dietrich M, Goldmann SF et al. Bone marrow transplantation for aplastic anemia from a HLA-A and MLC identical unrelated donor. *Blut* 1975, **31**: 347–354.

10. Gordon-Smith EC, Fairhead SM, Chipping PM et al. Bone marrow transplantation for severe aplastic anemia using histocompatible unrelated volunteer donors. *Br Med J* 1982, **285**: 835–837.
11. Duquesnoy RJ, Zeevi A, Marrari M et al. Bone marrow transplantation for severe aplastic anemia using a phenotypically HLA identical, SB compatible unrelated donor. *Transplantation* 1983, **35**: 566–571.
12. Hows JM, Yin JL, Marsh J et al. Histocompatible unrelated volunteer donors compared with HLA nonidentical family donors in marrow transplantation for aplastic anemia and leukemia. *Blood* 1986, **68**: 1322–1328.
13. Buskard NA, Smiley RK, Perrault RA, Messner H. Unrelated bone marrow transplantation in Canada, the national experience. *Bone Marrow Transplantation* 1993, **II**: 45–48.
14. Beatty PG, Atcher C, Hess E, Meyer DM, Slichter SJ. Recruiting blood donors into a local bone marrow donor registry. *Transfusion* 1988, **29**: 778–782.
15. McCullough J, Scott EP, Halagan N, Strand R, McGlave P. Effectiveness of a regional bone marrow donor program. *JAMA* 1988, **259**: 3286–3289.
16. Stroncek DF, Strand R, Hofkes C, McCulough J. The changing activities of a regional marrow donor program. *Transfusion* 1994, **34**: 58–62.
17. McElligott MC, Menitove JE, Aster RH. Recruitment of unrelated persons as bone marrow donors. *Transfusion* 1986, **26**: 309–314.
18. McCullough J, Rogers G, Dahl R et al. Development and operation of a program to obtain volunteer bone marrow donors unrelated to the patient. *Transfusion* 1986, **26**: 315–323.
19. McCullough J, Bach FH, Coccia P et al. Bone marrow transplantation from unrelated volunteer donors: summary of conference on scientific, ethical, legal, financial and other practical issues. *Transfusion* 1982, **22**: 78–81.
20. Cleaver S. The Anthony Nolan Research Centre and other matching registries. In *Bone Marrow Transplantation in Practice*, Traleaven J, Barrett J (Eds), Churchill Livingstone, Edinburgh, 1992, pages 361–366.
21. Champlin R, Coppo P, Howe C. National Marrow Donor Program: progress and challenges. *Bone Marrow Transplant* 1993, **1**: 41–44.
22. Stroncek DF, Bartsch G, Perkins HA, Randall BL, Hansen JA, McCullough J. The National Marrow Donor Program. *Transfusion* 1993, **33**: 567–577.
23. McCullough J, Hansen JA, Perkins H, Stroncek D, Bartch G. Establishment of the national bone marrow donor registry. In *Bone Marrow Transplantation: current controversies*. UCLA Symposium of Molecular and Cellular Biology, new series, Gale RP, Champlin R (Eds), Alan R. Liss, New York, 1989, pages 641–658.
24. Zumwalt ER, Howe CWS. The origins and development of the National Marrow Donor Program. *Leukemia* 1993, **7**: 1122.
25. McCullough J, Hansen JA, Perkins HA, Stroncek DF, Bartsch G. The National Marrow Donor Program: how it works, accomplishments to date. *Oncology* 1989, **3**: 63–72.
26. Perkins HA, Kollman C, Howe CWS. Unrelated-donor marrow transplants: the experience of the National Marrow Donor Program. In *Clinical Transplants 1994*, Terasaki P (Ed.), Terasaki Reference, 1995, pages 295–301.
27. Oudshoorn M, van Leeuwen A, Zanden HGM, van Rood JJ. Bone Marrow Donors Worldwide: a successful exercise in international cooperation. *Bone Marrow Transplantation* 1994, **14**: 3–8.
28. Goldman J. A special report: bone marrow transplants using volunteer donors – recommendations and requirements for a standardized practice throughout the world – 1994 update. *Blood* 1994, **84**: 2833–2839.
29. Stroncek DF, Holland PV, Bartsch G, Bixby T, Simmons RG, Antin JH, Anderson KC et al. Experiences of the first 493 unrelated marrow donors in the National Marrow Donor Program. *Blood* 1993, **81**: 1940–1946.
30. Stroncek DF, Strand R, Scott EP et al. Attitudes and physical condition of unrelated bone marrow donors immediately after donation. *Transfusion* 1989, **29**: 317–322.
31. Beatty PG. Results of allogeneic bone marrow transplantation with unrelated or mismatched donors. *Seminars in Oncology* 1992, **19**: 13–19.

32. Stroncek DF. Results of bone marrow transplants from unrelated donors. *Transfusion* 1992, **32**: 180.
33. Lanino E, Lamparelli T, Dini G, Alessandrino EP, Aversa F, Calori E, Medugno D et al. Bone marrow transplantation from unrelated donors: the Italian experience. *Bone Marrow Transplantation* 1993, **11**: 88–89.
34. Phillips GL, Barnett MJ, Brain MC, Chan KW, Huebsch LB, Klingemann HG, Meharchand J, Reece DR, Rybka WB, Shepard JD, Spinelli JJ, Walker IR, Messner HA. Allogeneic bone marrow transplantation using unrelated donors: a pilot study of the Canadian Bone Marrow Transplant Group. *Bone Marrow Transplantation* 1991, **8**: 477–487.
35. Beatty PG, Anasetti C, Hansen JA, Longton GM, Sanders JE, Martin MJ, Mickelson EM, Choo YS, Petersdorf E et al. Marrow transplantation from unrelated donors for treatment of hematologic malignancies: effect of mismatching for one HLA locus. *Blood* 1993, **81**: 249–253.
36. Hows JM, Bradley BA, Gore S, Downie T, Howard MR, Gluckman E. Prospective evaluation of unrelated donor bone marrow transplantation. *Bone Marrow Transplantation* 1993, **12**: 371–380.
37. McGlave P, Bartsch G, Anasetti C, Ash RC, Gajewski J, Kernan NA. Unrelated donor marrow transplantation therapy for chronic myelogenous leukemia: initial experience of the National Marrow Donor Program. *Blood* 1993, **81**: 543–550.
38. Kernan NA, Bartsch G, Ash RC, Beatty PG, Champlin R, Filipovich A, Gajewski J, Hansen JA et al. Analysis of 462 transplantations from unrelated donors facilitated by the National Marrow Donor Program. *N Eng J Med* 1993, **328**: 593–602.
39. Armitage JO. Bone marrow transplantation. *N Eng J Med* 1994, **330**: 827–838.
40. Sonnenberg FA, Eckman MH, Pauker SG. Bone marrow donor registries: the relation between registry size and probability of finding complete and partial matches. *Blood* 1989, **74**: 2569–2578.
41. Takahashi K, Juji T, Miyazaki H. Determination of an appropriate size of unrelated donor pool to be registered for HLA-matched bone marrow transplantation. *Transplantation* 1988, **45**: 714–718.
42. Beatty PG. The world experience with unrelated donor transplants. *Bone Marrow Transplantation* 1991, **7**: 54–58.
43. Beatty PG, Mori M, Milford E. Impact of racial genetic polymorphism upon the probability of finding an HLA-matched donor. *Transplantation* 1995, **6**: 778–783.

CHAPTER 25

Xenotransplantation – a solution to the donor organ shortage

DAVID J.G. WHITE

Introduction   446
A strategy for discordant xenotransplantation   449
The scientific realities today in xenotransplantation   450
Conclusion   454
References   455

Introduction

Organ transplantation has become an established form of therapy for patients suffering from end-stage disease that is not amenable to other forms of treatment. So successful is this treatment that since the first successful kidney transplant by Lawler in 1950[1] there has been a steady growth in the number of transplants performed. This has been particularly marked since the introduction of cyclosporin A as an effective immunosuppressive agent.[2]

As transplantation has become more successful in terms of survival,[3] quality of life[4,5] and cost benefit[6] the demand for donor organs is now greater than ever before and for some organs this demand is now significantly greater than supply. For example, there were 454 thoracic organ transplants performed in the UK in the year ending December 1992, but by the end of the same year the number of patients on the waiting lists for cardiac and pulmonary transplantation had grown to 763.[7] Thus even if no more patients were accepted on to the lists, it would take nearly two years to clear the backlog of potential recipients. The flaw in this argument is that many of these potential recipients will die on the waiting list before suitable organs become available. It is worthy of note that the patients who are accepted for transplantation represent the tip of the iceberg and, for every patient who is accepted, there are two or three who are rejected but who might have benefited from transplantation if there were a limitless donor pool.

The indications for transplantation are widening and while kidney, liver, heart and even lung transplantation is now seen as routine, the necessary skills are being developed to transplant other organs such as the small intestine and pancreas, either as islet cells or as a whole organ. Clearly this stretches the donor pool substantially beyond its limit.

Another solution to the donor shortage must be sought if transplantation is to be extended to treat all those in need. One obvious source of some organs is to use living related donors and, while renal transplant surgeons have used this resource for a long time,[8,9] the use of livers and lungs from live related donors has only recently been explored.[10,11] The potential hazards for the donor of such a procedure have stimulated fierce ethical debate.[12,13] Living related donation will never solve the problem entirely and the fact that such drastic measures can be considered, and indeed put into practice, underlines the severity of the donor organ shortage.

ALTERNATIVES TO CONVENTIONAL TRANSPLANTATION

Organ transplantation may be supplemented, or even replaced in due course, using totally artificial organs. The only implantable device which finds clinical use at present is the artificial heart.[14,15] Fundamental problems such as power supply, thrombosis, infection and bio-compatibility of mechanical surface–blood interfaces remain and these obstacles must be overcome to allow long-term function. While forms of dialysis can treat kidney failure, the internal replacement of organs with complex metabolic functions is more difficult and complete replacements for the kidneys, lungs and liver are still a long way distant.

Hybrid technology devices also offer some hope[16] and it appears likely that the next organ to be replaceable with such a device is the pancreas. Islet cells are encapsulated inside a semi-permeable chamber through which blood passes but which protects the islets from contact with the host immune system. The cells are still able to respond to varying blood glucose levels and to produce insulin in an appropriate response.[17] The obvious drawback to this device is that a source of donor cells is still required, but it seems likely that this approach will achieve clinical usefulness in the future.

OBSTACLES TO TRANSPLANTATION BETWEEN SPECIES

In the face of limited therapeutic alternatives, an increasing number of centres are considering the use of animal organs as an alternative source. The idea of transplantation between species (xenotransplantation) is not new for initial attempts at transplantation in man used the organs of other animal species.[18–20] However, when the immunological basis of rejection was established, scientific interest in xenotransplantation waned and, despite a brief period of revised interest in the 1960s, this area has been largely unexplored until recently. The use of organs from an appropriate animal species would overcome any deficiency in numbers since suitable breeding programmes could be established to meet demand, and it would perhaps allow the expansion of current indications for transplantation so that a larger population could benefit.

The major scientific obstacle to the use of animal donor organs is the rejection reaction. This reaction varies in severity depending on the species combination used and in 1969 Calne distinguished between the two major xenorelationships on the basis of this behaviour.[21] The rejection of distantly related species combinations (hyperacute rejection) is rapid, resembling the second set allograft reaction, an observation which lent support to the role of preformed natural antibody in hyperacute xenograft rejection. This species combination was termed discordant, while a species combination in which the rejection process resembled first set allograft rejection occurring in days rather than in minutes was termed concordant. An example of a concordant relationship is baboon to man, while a discordant combination is pig to man. Unfortunately the terms have become distorted in use to imply the presence or absence of naturally occurring antibody to another species and, although this is not necessarily the case, the importance of, so called, heterophile preformed natural antibody is assumed in most studies of discordant xenograft rejection.

In clinical practice the use of concordant animal donors for man is appealing in the sense that hyperacute rejection will not occur, but it necessitates the use of non-human primates as donors. Some primate species are endangered and their level of intelligence and social structure makes their use ethically unacceptable to many.[22] Goodall and other research workers have highlighted the distress that removal of an individual causes to their social groups, or the distress if they are purpose bred in captivity.[23,24] There are practical problems too, with slow breeding rates, low parity and the risk of transmission of disease.[25,26]

The hurdle of hyperacute rejection in discordant combinations prevents the study of events beyond the first few minutes and it is not clear what immune processes may be important subsequently. Once hyperacute rejection can be overcome it seems likely that a more chronic cellular or antibody mediated process may be important, but this remains speculative.

The clinical use of xenografts has been limited to only 33 cases[27-29] and is summarized in Table 25.1. Until the last few months, all clinical attempts at xenotransplantation had used concordant donors and most were transplants of the kidney,[30-32] although hearts[33,34] and livers[21] have also been attempted. Recently a pig liver was used to support a patient for a number of hours until a human liver was available for transplantation, and a number of pig hearts were used in an attempt to transplant a man for whom there was no human heart available. However, there was no scientific indication that the organs could be expected to have anything but a very short survival. Several patients have undergone xenogeneic cross circulation for hepatic coma with both concordant[35] and discordant[36] species.

Although no long-term survivors have emerged from the clinical xenograft experience a great deal of useful information has been revealed. It is clear that animal organs can function in a human environment and that they can perform their own appropriate tasks. Certainly it seems likely that a cross species heart or kidney graft could reasonably be expected to perform adequately if transplanted into a human. The liver is somewhat different since many animals have different levels of products made by the liver and it is unclear whether or not the synthetic function of a baboon liver will be sufficient, or that the proteins will be sufficiently similar, to maintain a human in health for a prolonged period.

Studies of comparative physiology are relatively scarce, but one such study examines the similarities between man and pigs.[37] Renal morphology, blood flow and glomerular filtration rates are similar between the two species, while plasma electrolytes and acid-base balance are also similar. In addition the cardiovascular parameters suggest that a pig heart should be able to support a man with similar blood pressures, and response to exercise.[23,24]

Table 25.1 The clinical application of xenotransplantation

Year	Organ	Donor	Cases	Survival
1964	Kidney	Chimpanzee	12	<9 months
1964	Kidney	Monkey	1	10 days
1964	Kidney	Baboon	1	4.5 days
1964	Kidney	Baboon	6	<2 months
1964	Heart	Chimpanzee	1	2 hours
1968	Heart	Sheep	1	0
1968	Heart	Pig	1	4 minutes
1969	Heart	Chimpanzee	1	Short time
1969–73	Liver	Chimpanzee	3	<14 days
1977	Heart	Baboon	1	5 hours
1977	Heart	Chimpanzee	1	4 days
1984	Heart	Baboon	1	20 days
1992	Liver	Baboon	1	70 days
1992	Heart	Pig	1	<24 hours
1993	Liver	Baboon	1	30 days

A strategy for discordant xenotransplantation

The greatest hope for the future of organ transplantation lies with discordant xenotransplantation. Animals already bred in captivity as a source of food may be more acceptable as organ donors. Breeding colonies could be easily established with no threat to the species. In addition an animal such as the pig has high numbers in each litter, breeds well, grows rapidly and is cared for relatively easily. Although the idea may still be distasteful to some it is hard to see how the majority of people in meat-eating society could object to this plan on moral grounds. Animals are already used to provide society with other medical materials, and the use of porcine insulin or porcine heart valves has been embraced with enthusiasm because of the benefits they offer patients. A further benefit from the purpose of breeding donor animals is that it may offer an opportunity to modify the donor organ in such a way as to render it less likely to be rejected by a human recipient.

Complement is known to be important in discordant xenograft rejection. Down regulation of complement activity is produced by cell surface proteins such as membrane cofactor (MCP) and decay accelerating factor (DAF).

Thus a potential strategy for the clinical application of xenotransplantation is the production of pigs transgenic for human complement regulators (RCA) (see below).[38] *In vitro* studies[39–41] have demonstrated that such regulators protect animal cells from lysis by human complement, although whether this approach would be sufficient to protect an organ from hyperacute rejection is at present unclear. There are a number of RCAs which would be potential candidates for such a strategy. Cell membrane bound RCA offer the advantage over those active in the circulating blood[42] in that they exert only a local effect and thus do not induce systemic inhibition of complement. This local activity would reduce the immunosuppressive load placed upon the recipient and hence reduce the risk of infection. Most research groups investigating the use of membrane bound RCA in order to prevent hyperacute xenograft rejection have chosen to work with a combination of the RCA's Decay Accelerating Factor (DAF or CD55), Membrane Cofactor Protein (MCP or CD46) and CD59 (sometimes called homologous restriction factor or HRF). My own research is directed towards the production of pigs transgenic for human DAF (hDAF) and the testing of hearts from these animals in an *ex vivo* perfusion circuit with fresh whole human blood.

The scientific realities today in xenotransplantation

PRODUCTION OF TRANSGENIC PIGS

Pro-nuclear stage embryos were collected through a midline laparotomy from the oviducts of 7–10 month old mature gilts 49–52 hours after super ovulation, which had been inseminated 20 and 26 hours earlier. Visualization of the pro-nucleus is an essential step in making transgenic pigs not found necessary in the production of transgenic mice. This was achieved by centrifugation of the ova at 15,000 g for 3 minutes. These ova were then micro-injected with a mini gene[43] for hDAF using an inverted microscope with Nomarski optics and micro-manipulators. DNA was injected until the diameter of the ovum had increased by approximately 50%. Injected ova were transferred to both oviducts of recipient gilts whose oestrous cycles were either synchronous with or 24 hours behind those of the donors. The hDAF mini gene construct consisted of approximately 4 kb of genomic DNA between two Hind III sites and 2 kb of cDNA between the Hind III site and the EcoR1 site. The genomic fragment which includes 5' untranslated and signal peptide sequence, the first exon and 0.4 kb of the first intron, extends to the Hind III site of the second exon. The Hind III to EcoR1 fragment includes 41 adenosine's constituting the mRNA poly A tail.

A total of 2,432 ova micro-injected with the hDAF mini gene were transferred to 85 recipient gilts. Forty-nine (65%) of these became pregnant and produced 311 offspring of which 28 were either still born or died shortly after birth. Forty-five (15.9%) of the 283 surviving offspring and 4 (14.8%) of the fatalities were identified by dot blot analysis using DNA isolated from ear biopsy samples, as having incorporated into their genome from 1 to 30 copies of the hDAF mini gene per cell. Expression of the transgene in PBMC was determined by RT-PCR using primers designed to span an intron/exon boundary and by Northern analysis.[44] Messenger RNA for hDAF was detected in PBMC of 31 (69%) of the 45 live transgenic pigs. The level of RNA varied from pig to pig and was found to be independent of sex, copy number or fertility.

EXPRESSION OF hDAF ON TISSUES OF TRANSGENIC PIGS

The levels of hDAF protein expression on the tissue of the transgenic pigs was determined using a double capture RIA technique. Briefly, anti-DAF 1A10 mAb[45] was coated to 96-well microtitre plates at 10 μg/ml. After saturating with 1% PBS-BSA, plates were incubated for 4 hours at 4 °C with transgenic pigs tissues lysed in 1% NP40 buffer. Subsequently plates were incubated with 100 ng of iodinated anti-DAF BRIC 216 mAb,[46] for 1 hour and 30 minutes at room temperature. After washing, the remaining radioactivity in each well was determined by counting in a Packard gamma counter.

Production of transgenic lines requires that founder animals be retained for breeding and that the offspring be appropriately back crossed to produce homozygotes. Therefore, tissue analysis of founder animals has been restricted to blood samples and ear biopsies, however, tissues from F1 heterozygotes can be analysed. Northern analysis demonstrated that transcription could occur in every tissue analysed although levels of hDAF RNA can vary from organ to organ and founder to founder. One founder transgenic pig was sacrificed because repeated RT-PCR had demonstrated conclusively an absence of RNA in its PBMC. However, RNA were detected in every other organ tested (heart, lung, kidney, spleen and liver). This demonstrates the danger of relying on any single tissue sample as being indicative of expression in any other tissue or organ. Thus, as a matter of policy, all founders should be bred for at least one generation to allow total analysis of expression levels in all tissues without the loss of the line. The level of hDAF protein expressed also varied from line to line (Figure 25.1) and from tissue to tissue within one line. Comparison of the level of hDAF protein in the tissues of these pigs with that seen in the corresponding human tissues demonstrated that even heterozygous transgenics were able to express levels of hDAF comparable to or greater than that detected in equivalent human tissues (Figure 25.2).

Immunohistological staining of the tissues from these pigs demonstrated a wide variety of tissue distribution. However, pigs with the highest expression levels as quantified by RIA were more likely to have hDAF expression on endothelium.[47]

PROTECTION OF PIG HEARTS FROM HUMAN COMPLEMENT

Since hDAF is designed for the down regulation of human complement, testing its efficiency is by the nature of the transgene restricted. To investigate the protective effect which might be conferred by possession of hDAF, a mock circulation was designed such that pig hearts were both perfused with human blood and were required to 'work' by

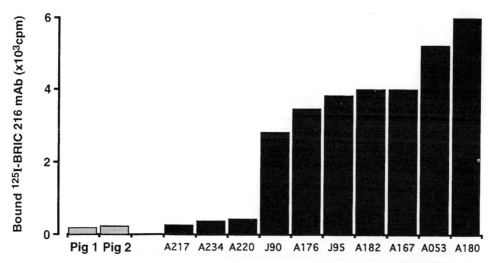

Figure 25.1 Expression of human DAF on a range of transgenic pigs, all offspring of different founders selected to illustrate the variation in levels of expression that can be produced by the same construct. Pig 1 and pig 2 are normal pig controls.

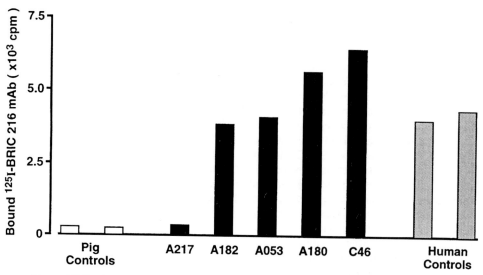

Figure 25.2 A comparative analysis of human DAF expression levels in the hearts of normal control pigs, a selection of transgenic pigs (see Figure 25.1) and man. Note that some transgenic pigs have higher levels of protein than detected in man.

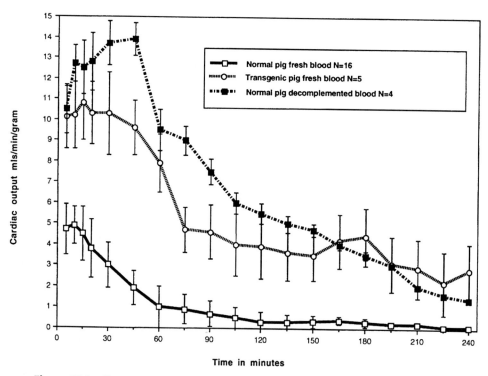

Figure 25.3 Comparison of cardiac output of pig hearts perfused with human blood. Comparison is between normal pig hearts perfused with fresh human blood ($n = 16$) (open squares), normal pig hearts perfused with decomplemented human blood ($n = 4$) (closed squares) and transgenic pig hearts perfused with fresh human blood ($n = 5$) (open circle).

pumping the blood round the circuit. Hearts were obtained from either normal or transgenic 3–4 week old piglets under halothane anaesthesia via a median sternotomy and cooled topically by immersion in haemacell. The aorta and left atrial appendage was cannulated with customized metal cannulae. An opening was made in the right ventricle which, in combination with ligation of both pulmonary arteries, caused all coronary sinus return to drain on the surface of the heart. The heart was connected to a perfusion circuit constructed as previously described.[48] In brief, clear plastic tubing of 1/16th inch internal diameter was used to connect a roller pump oxygenator, heat exchanger, filter and cardiac, atrial and venous reservoirs. The atrial reservoir was positioned 700 mm above the heart which was suspended directly over the cardiac reservoir. The circuit was primed with 1.2 litres of either fresh or decomplemented blood.

Decomplementation of fresh blood was performed with cobra venom factor. Initially the heart was perfused retrogradely to allow warming and establish sinus rhythm. After 10 minutes the circuit was converted to working mode in which the heart ejects its stroke volume against a pressure equivalent of the height of the atrial reservoir. The overflow from this reservoir represents the cardiac output. Several parameters of the function of the heart, such as survival and coronary artery flow can be measured in this model, however, the most informative is the cardiac output measured in millilitres of blood pumped per minute per gram of heart tissue against a fixed afterload.

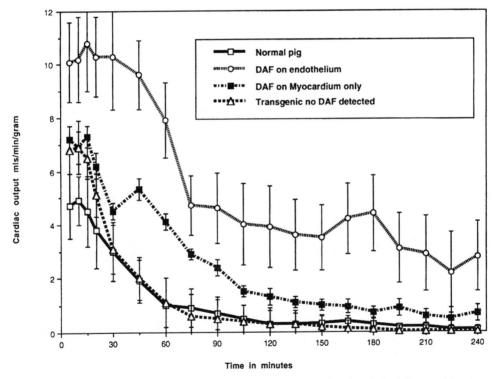

Figure 25.4 Comparison of cardiac output of pig hearts perfused with fresh human blood. Normal pigs (open square), transgenic pigs with no DAF detected (triangle), transgenic with DAF on the myocardium only (closed square), transgenic pig with DAF on endothelium (circle).

Results (Figure 25.3) show cardiac function for normal pig heart perfused with fresh human blood ($n = 16$), normal pig heart perfused with decomplemented fresh human blood ($n = 4$) and transgenic pig hearts perfused with fresh human blood ($n = 5$). These data clearly indicate that hearts from transgenic pigs perfused with fresh human blood can perform work in this model as though the blood had been decomplemented. Both of these groups showed significantly longer survival and greater cardiac output than that observed with hearts from normal pigs perfused with fresh human blood. Comparison of offspring from different founder transgenics (Figure 25.4) show there was almost no protection compared to controls in transgenic pig hearts where little or no hDAF could be detected, moderate protection in transgenes with expression of hDAF only on myocardium and excellent protection in those animals with expression on both myocardium and endothelium.

The protective effect of the transgene in these perfusion experiments can also be demonstrated by staining the hearts for deposition of C3b. Histology shows that in a normal pig heart following perfusion with fresh human blood there is profuse deposition of C3b at the time of cardiac failure. By contrast the heart from a transgenic pig showed only minimal C3b deposition.

Conclusion

WHERE NEXT?

The data presented here demonstrate that it is possible to produce viable pigs transgenic for hDAF. The success rate in producing two transgenic pigs per 100 eggs injected is substantially greater than that previously reported by other groups using different constructs (0.5%).[49,50] The factors contributing to this increase were a high pregnancy rate and large litter size. Expression of hDAF RNA and protein in the transgenic animals appeared to be dependent upon the site of integration and was independent of copy number. Our previous studies with transgenic mice[51] have demonstrated that the inclusion of the genomic promoter region enhanced the expression of the protein. Currently attempts are being made to produce transgenic pigs using yeast artificial chromosomes containing the full genomic sequence for RCA molecules. The large number of transgenic pigs produced enabled us to select by analysis of the offspring lines of pigs which have high levels of hDAF in appropriate tissues. The high level of expression seen in the tissues of some of the animals was not predicted either by our mouse studies or by analysis of levels in lymphocytes of the relevant pig. In principle, when such animals are bred to homozygosity their tissues expression levels will be doubled.

Our intention at the outset of these studies was to produce pigs transgenic for the human complement regulators DAF, MCP and CD59 as a first step towards producing animal organs suitable for transplantation into human patients. *In vitro* studies have demonstrated a linear relationship between cell surface concentration of hDAF[52] and MCP[39] and protection from human complement. If no synergistic activity exists between these different RCA in the processes of complement regulation it is possible that pigs expressing sufficient amounts of hDAF alone could become impervious to the complement mediated component of hyperacute xenograft rejection. The perfusion studies performed with fresh human blood suggest that the expression of hDAF alone on endothelium is sufficient to confer protection from complement mediated hyperacute rejection within the limits of the assay system used. However, it may be that more than one type of RCA is necessary for complete protection from hyperacute rejection. Once hyperacute rejection has been avoided what other forms of immunological damage such xenografts would attract is at present largely unknown, however, these data give hope that – by exploiting modern molecular biology – the organ shortage may be resolved by xenotransplantation.

ACKNOWLEDGMENTS

The work summarized in this chapter was only possible with the assistance of a large number of people. In particular, the author would like to acknowledge the considerable contributions of Mr J. Dunning, Mr V. Young, Mr K. Elsome, Mr R. Lancaster, Mr A. Tucker, Professor J. P. Atkinson, Dr M. Davitz, Dr C. Carrington, Dr E. Cozzi, Dr G. Langford, Dr G. Chavez and Professor C. Polge.

References

1. Lawler R, West J, McNulty Clancy E, Murphy R. Homotransplantation of the kidney in the human. *JAMA* 1950, **144**: 844.
2. Kaye M. The registry of the International Society for Heart and Lung Transplantation. Ninth Official Report – 1992. *J Heart Lung Trans* 1992, **11**: 599–606.
3. Caine N, Sharples L, English T, Wallwork J. Prospective study comparing quality of life before and after heart transplantation. *Trans Proc* 1990, **22**: 1437–1439.
4. Hoffenberg R. Working party on the supply of donor organs for transplantation. *HMSO*, London, 1987.
5. Murray J, Merrill J, Harrison J. Renal homotransplantation in identical twins. *Surg Forum* 1955, **6**: 432.
6. Najarian J, Frey D, Matas A. Renal transplantation in infants. *Ann Surg* 1990, **212**: 353–365.
7. Broelsch C, Edmond J, Whitington P, Thistlethwaite J, Baker A, Lichtor J. Application of reduced-size liver transplants as split grafts, auxiliary orthotopic grafts, and living related segmental transplants. *Ann Surg* 1990, **212**: 363–375.
8. Shaw L, Miller J, Slutsky A. Ethics of lung transplantation with live donors. *Lancet* 1991, **338**: 678–681.
9. Kormos R, Borovetz H, Armitage J, Hardesty R, Marrone G. Evolving experience with mechanical circulatory support. *Ann Surg* 1991, **214**: 471–477.
10. McCarthy P, Portner P, Tobler H, Starnes V, Ramasamy Y, Oyer P. Clinical experience with the Novacor ventricular assist systems. *J Thorac Cardiovasc Surg* 1991, **102**: 578–587.
11. Weber C, Zabinski S, Norton J, Koschitzky T, D'Agati V, Reemstma K. The future role of microencapsulation in xenotransplantation. In *Xenograft 25*, Hardy M (Ed.) Elsevier, New York, 1989, pages 297–308.
12. Maki T, Ozata H, Caretta M. Use of xenogeneic islets in hybrid artificial pancreas for treatment of diabetes without immunosuppression. *Trans Proc* 1992, **24**(2): 661–662.
13. Neuhof H. In *The Transplantation of Tissues*. Appleton & Co, New York, 1923, page 260.
14. Jaboulay M. Greffe de reins au pli du coude par soudures arterielles et veineuses. *Lyon Med* 1906, **107**: 575.
15. Unger E. Nierentransplantationen. *Klin Wsch* 1910, **47**: 573.
16. Sanfilippo E. The use of non-human primates in xenotransplantation. In *Xenograft 25*, Hardy M (Ed.), Elsevier, New York, 1989, pages 335–349.

17. Goodall J. Ethical concerns in the use of animals as donors. In *Xenograft 25*, Hardy M (Ed.), Elsevier, New York, 1989, pages 335–349.
18. Lawick-Goodall J. *In the shadow of man*. William Collins, Glasgow, 1971.
19. Chiche L, Adam R, Caillat-Zucman S, Castaing D, Bach JF, Bismuth H. Xenotransplantation: baboons as potential liver donors? *Transplantation* 1993, **55**(6): 1418–1421.
20. Ogden D, Sitprija V, Holmes J. Function of the baboon renal heterograft in man and comparison with renal homograft function. *J Lab Clin Med* 1965, **65**: 370–386.
21. Giles G, Boehmig H, Amemiya H, Halgrimson C, Starzl T. Clinical heterotransplantation of the liver. *Trans Proc* 1970, **2**: 506–512.
22. Kirkman R. Of swine and men: organ physiology in different species. In *Xenograft 25*, Hardy M (Ed.), Elsevier, New York, 1989, pages 125–131.
23. Hannon J. Haemodynamic characteristics of the conscious resting pig. In *Swine in biomedical research*, Tumbleson M (Ed.) Plenum, New York, 1985, pages 1341–1352.
24. Erickson H, Faraci F, Olsen S. Effect of exercise on cardiopulmonary function in domestic pigs. In *Swine in biomedical research*, Tumbleson M (Ed.) Plenum, New York, 1985, pages 1341–1352.
25. UKTS Users Bulletin 1993, Winter 1992/93.
26. Buxton M, Acheson R, Caine N, Gibson S, O'Brien V. *Costs and benefits of heart transplant programmes in Harefield and Papworth hospitals*. HMSO, 1985.
27. United Kingdom Transplant Service Transplant Update 1993, January 1993.
28. The National Organ Procurement and Transplantation Network. Pig liver recipient's race against time ends unsuccessfully. *UNOS Update 1992*, November 1–3.
29. O'Donnell M. Pork Futures. *International Management* 1993, 63.
30. Langer R, Vacanti JP. Tissue Engineering. *Science* 1993, **260**: 920–926.
31. Kemp E, Dieperink H, Jensen J, Kemp G, Kuhlman I-L, Larsen S, Lillevang ST. Newer immunosuppressive drugs in concordant xenografting – Transplantation of hamster heart to rat. *Xenotransplantation* 1994, **1**: 102–108.
32. White DJG. Xenotransplantation. *Transplant Proc* 1990, **22**: 2095–2096.
33. Carrington CA, Cozzi EC, Langford GA, Richards AC, Rosengard A, Yannoutsos N, White DJG. Transgenic Pigs and Xenotransplantation. In *Organ Shortage: The Solutions*, Touraine JL (Ed.), Kluwer Academic, Netherlands, 1995, pages 309–316.
34. Bailey LL, Nelson-Canerella SL, Concepcion W, Jolley WB. Baboon-to-human cardiac xenotransplantation in a neonate. *JAMA* 1985, **254**: 3321–3326.
35. Saunders SJ, Bosman SCW, Walls R. Acute hepatic coma treated by cross-circulation with a baboon and by repeated exchange transfusions. *Lancet* 1968, **2**: 585.
36. Eiseman B, Liem DS, Raffucci F. Heterologous liver perfusion in treatment of hepatic failure. *Ann Surg* 1965, **162**: 329.
37. Rotenberg MO, Chow LT, Broker TR. Characterization of rare human papillomavirus type 11 mRNAs coding for regulatory and structural proteins, using the polymerase chain reaction. *Virology* 1989, **172**: 489–497.
38. Atkinson JP, Farries T. Separation of self from non-self in the complement system. *Immunology Today* 1987, **8**: 212–215.
39. Oglesby TJ, Allen CJ, Liszewski MK, White DJ, Atkinson JP. Membrane cofactor protein (CD46) protects cells from complement-mediated attack by an intrinsic mechanism. *J Exp Med* 1992, **175**: 1547–1551.
40. White DJG, Oglesby TJ, Tedja I, Lisezwski K, Wallwork J, Wang M-W, Atkinson JP. Human decay accelerating factor or membrane co factor protein inhibit cross species mouse cell lysis by human complement. *Transplant International* 1992, **5**: 648–650.
41. Dalmasso AP, Vercelloti GM, Platt JL, Bach FH. Inhibition of complement mediated cytotoxicity by decay accelerating factor. Potential for prevention of xenograft hyperacute rejection. *Transplantation* 1991, **52**: 530–533.
42. Pruitt SK, Kirk DA, Bollinger RR, Marsh HC, Collins BH, Levin JL, Mault JR, Heinle JS, Ibrahim S, Rudolph AR, Baldwin WM, Sanfilippo F. The effect of soluble complement receptor type 1 on hyperacute rejection of porcine xenografts. *Transplantation* 1994, **57**: 363–370.

43. Cary N, Moody J, Yannoutsos N, Wallwork J, White D. Tissue expression of human decay accelerating factor, a regulator of complement activation expressed in mice: a potential approach to inhibition of hyperacute xenograft rejection. *Transplant Proc* 1993, **25**: 400–401.
44. Chirgwin J, Przybyla A, McDonald R, Rutter W. Isolation of biologically active ribonucleic acid from sources rich in ribonuclease. *Biochemistry* 1979, **18**: 5294–5296.
45. Kinoshita T, Medof ME, Silber R, Nussenzweig V. Distribution of decay-accelerating factor in the peripheral blood of normal individuals and patients with paroxysmal nocturnal hemoglobinuria. *J Exp Med* 1985, **162**: 75–92.
46. Spring FA, Judson PA, Daniels GL, Parsons SF, Mallinson G, Anstee DJ. A human cell surface glycoprotein that carries Cromer related blood group antigens on erythrocytes and is also expressed on leucocytes and platelets. *Immunology* 1987, **62**: 307.
47. Rosengard AM, Cary NRB, Langford GA, Tucker AW, Wallwork J, White DJG. Tissue expression of Human Complement Inhibitor, Decay-accelerating Factor, in Transgenic Pigs – A potential approach for preventing Xenograft rejection. *Transplantation* 1995, **59**: 1325–1333.
48. Forty J, White DJG, Wallwork J. A technique for perfusion of an isolated working heart to investigate hyperacute discordant xenograft rejection. *J Thorac Cardiovasc Surg* 1993, **106**: 308–316.
49. Pursel VG, Pinkert CA, Miller KF, Bolt DJ, Campbell RG, Palmiter RD, Hammer RE. Genetic engineering of livestock. *Science* 1989, **244**: 1281–1288.
50. Brem G, Brenig B, Muller M, Krausslich H, Winnacker EL. Production of transgenic pigs and possible application to pig breeding. *Occ Publ Br Soc Anim Prod* 1988, **12**: 15–31.
51. Cary N, Moody J, Yannoutsos N, Wallwork J, White D. Tissue expression of human decay accelerating factor, a regulator of complement activation expressed in mice: a potential approach to inhibition of hyperacute xenograft rejection. *Transplant Proc* 1993, **25**: 400–401.
52. Lublin DM, Coyne KE. Phospholipid-anchored and transmembrane versions of either decay-accelerating factor or membrane cofactor protein show equal efficiency in protection from complement-mediated cell damage. *J Exp Med* 1991, **174**: 35–44.

Index

Note: page numbers in italics refer to figures and tables

ABO blood group 232, 233
ABO blood group matching 173
   liver allocation 237
   thoracic organ allocation 237, 238
abortion, induced 99
acyclovir
   CMV infection prevention 126
   herpes simplex virus 128
adenosine triphosphate (ATP) 210
adult polycystic kidney disease 172
advertising 423
age
   bone marrow donation 250
   donors in Spain 364
   live kidney donation 171
AIDS
   transmission prevention 113
   see also HIV infection
allele-specific oligonucleotide typing (ASO) 244
allograft
   European regulations 280
   infection transmission 121
   malignancy transmission 138–40
Alport's syndrome 172
altruism 426
   incentives 53
   indirect 58
   legislation 101
   medical profession 53
   minority populations 37, 39
   Muslim 30, 31
   public 35

Alzheimer's disease 99
American Association of Critical Care Nurses 383
American Association of Tissue Banks 276
amylase 156
Anatomy Act (UK, 1832) 273, 348
anencephaly 99
angiography
   cerebral 82, 83
   renal 175–6, *177*
anoxia, ischaemic 202
anterior pituitary hormone replacement therapy 83
Anthony Nolan Bone Marrow Trust 431
Anthony Nolan Research Centre 240
anti-glomerular basement membrane antibody 170
anti-HCV 132, 133
anti-HSV antibody testing 127
aortic valve
   tissue banking 272
   transplantation 271
artificial organs 447
assumptive world 305
asthma 155
   live kidney donation exclusion 172
atelectasis 181, *184*
atropine, brainstem function test 88
attitude formation to social issues 401
audience segmentation 418–19
Australasian Transplant Co-ordinators Organization 329

Australia
  donation rate 416
  legislation on obtaining tissues for
      transplantation 273
  public awareness 414
  technical regulation of tissue banking 280–2
  tissue banking organization 286–7
  Transplantation – The Issues educational
      programme 402, 403–7, *408*, 409, *410*,
      411
Australian Corneal Graft Registry 286, 287
Australian Donor Awareness Program for
    Transplantation 318, *319*
Australian Kidney Foundation 402, 403, 406
Australian Law Reform Commission report 77
Austria
  autopsy laws 349, 352, 354
  family of donor 356
autonomy
  individual 354–5
  legislation 101
autopsy
  laws 348
  teaching 349
azathioprine 163

bacterial infection
  donor 121
  positive blood cultures 121, 122
bacterial infection transmission 120–2
behaviour change 35
  sustained 424
Belgium
  family of donor 356
  legislation 104, 351, 352, 354
  registry 355
bereavement 310
  aftercare programmes 317–20
  literature 315–16
  Partnership for Organ Donation 391–2
  simulation 376, *377*
  specialist 315–16
  support 316, 401
    groups 317
    programmes 315–17
Bio-Implant Services 284–6
bioethics 49, 52
  donation rules 63
  shift to 55
Births and Deaths Registration Act (UK, 1836)
    273
blood bank 271
blood vessels
  autograft 220–1
  banking 269

bone
  allograft distribution, USA 282, *283*
  banking 284
  cancellous chips 300
  grafting 221–2
    development 271
    tissue banking 272
  tissue banking 229
    processing 299, 300–1
    retrieval 298–301
  wedges 300
bone marrow donation 239–40
  age 250
  anti-coagulation 260
  autologous blood transfusion 260
  cell count 260
  collection 439
  colony stimulating factors 264
  complications 264
  confidentiality 262, 440
  counselling 250–1
  couriers 261, *263*
  degree of matching 249–50
  donor
    recruitment 250–3, 430–1
    search 245, *248*, 249–50
    unrelated 249–50
  erythropoietin 264
  extended family search 245, *248*
  follow-up 264
  haplotype investigation *248*
  HLA-compatibility 240
  HLA-identical family member 240, 241–2,
      249
  HLA-matched unrelated 240, *241*, 242–3
  infectious disease 252, 253
  informed consent 250, 251, 253, *254*–8
  insurance 261
  labelling 261
  marrow collection 260
  process 253, 255, *259*, 260–2
  recipient protection 251–3
  recovery 263
  registries 240
  repeat 263
  tissue typing techniques 243–5
  transport 261, *262*, *263*
  unrelated harvest procedure 253, 260
  volume 260
  young minor to other family member
      250
bone marrow donor registries 229, 431–2
  collection centres 433
  confirmatory typing 438–9
  contact information 435–6

data management 436–7
demographic information 435–6
donor
  centres 433–4
  confidentiality 440
  pre-collection work-up 439
  recruitment 434
  tracking 439
finance 441–2
growth factors 441
health screening 435
HLA typing 436
informed consent 435
organization 432–42
patient
  confidentiality 262, 440
  histocompatibility matching 437
  tracking 439
policies/procedures 432
recruitment promotion 434–5
search
  exchange 437–8
  process 438–40
standards 432
subsequent demands on donors 440–1
transplant centres 432–3
Bone Marrow Donors Worldwide 249, 432
  searches 438
bone marrow transplantation 221, 239–40
  allografts 221
  autologous 221
  development 9
  parental consent 98
bone marrow transport 439
bone tissue banking 269
  Australia 286
  cadaveric donors 291
brain *71*
  cell necrosis 74
  function loss 82
  injury 74
    diagnostic investigations 82
    outcome 4, *6*, 7
    traumatic 71
  neoplasms 139
  vascular occlusion 72–3, 74
    mechanisms 73–4
  vascular supply *73*
brain death 69
  apnoea testing 81
  caution 89
  cell survival 82–3
  confirmation 70
  confusion 313
    with vegetative state 80
continuing hypothalamic/pituitary function 83–5
criteria 28
  agreement 63
definition 74–80
diagnosis 69–70
EEG function persistence 85
evoked potential persistence 86
family of donor 306
  understanding 309–10
function persistence 85–6
guidelines 75–8
injuries 71
intracranial blood flow objective assessment 86–7
intracranial vessel filling 86–7
legislation 70
Making the Critical Difference programme 385
minority population views 38
misunderstanding 415
Muslim concept 32
objective tests 89
organ donation 312–14
pathophysiological processes 71
prediction 70
process 90
public attitudes 40, 41, 413
safeguards 87–9
somatic reflex responses 81
Spanish legislation 361
spinal reflex responses 81
time from initial trauma 306
transplant programmes 227
whole 79–80, 88
brainstem
  auditory evoked potential 86
  death 78–9
  function 88
  response 75, 76
  traumatic injury 82
Bretschneider's HTK solution *206*, 207
Britain *see* UK
Buddhism 24
burden to benefit ratio 58

cadaveric donation 3–4, *5*
  futures market 58
  potential 390
  potential increase 400
  rate 388
  Roman Catholic Church 27–8
  see also donation
cadaveric donors 152
  abdominal organs 157, 158

cadaveric donors (*continued*)
  cannula insertion 159
  cardiac 153–4, 157, 158
  core cooling 159
  cross-clamping 159
  evaluation 152–6
  flushing 159
  hemodynamic status 157
  hepatitis C virus infection 133–4
  HLA matching 165
  liver *154*, 155
  lung 154–5, 157, 158
  management 156–7
  organ removal 159
  organ retrieval 157–60
  organ specific tests *154*
  pancreas *154*, 155–6
  renal *154*, 155, *164*
  resuscitation 156–7
  small bowel 156
  Spain 362
  technique difference from non-heartbeating donors 159–60
  tissue banking 287–92
    contraindications *289*
    next of kin 288, 289–90
    relatives' consent 288–90
    serological testing 291–2
    tissue suitability criteria 290–1
  see also donors
calcium paradox effect 218
Canada, Making the Critical Difference programme 387–8
Canadian Association of Transplantation 329
cancer *see* malignancy
capital punishment 63, 348
cardiac donation 153–4
  persistent hypotension 157
  retrieval 157, 158
    for tissue banking 292–6
  see also heart transplantation; heart valve
cardiac output monitoring 153
cardiac standstill, inevitability 80–1
cardioplegic solutions 218
cardiovascular tissue banking 269
Carolina rinse solution 213
cartilage grafts 222
catecholamines, brain death 81
Centers for Disease Control 275
central venous pressure monitoring 156
cerebellar haemorrhage 82
cerebral death, Scandinavia 353
cerebrospinal fluid 72
  diaphragm sella 84

cerebrovascular accident
  cadaveric kidney donors 3–4
  donors in Spain 365, 369
chemoradiation/chemotherapy 239
child
  kidnapping stories 413
  organ donation 98
  organ shortage 39
  parental loss 311
cholangiography 188
Christianity 25–9
Church of England 25–7
circle of Willis 73, 84
  circulation 87
citrate solution *206*, 207
civil law 96
  application outside boundaries 106
civil liberties 104
CNS malignancy spread 139–4
cold perfusion, *in situ* 214
Collins' solution 205, *206*, 207, 208
  see also EuroCollins' solution
colloids 208–9
  machine perfusion 211
colony stimulating factors 264
Columbia University solution 219
commercial market legislation 97, 98
commercial transaction 25–6
commercialism
  Ethics Committee of Transplantation Society 66
  fear of 55–7
  international response *114*
communication
  donor families 41–2
  family of donor 379
  follow-up 337
  oral 346
community donors 434
compensated donation 58
complement
  activity 449
  pig heart protection 451, *452*, 453–4
  regulators 449, 454, 455
composite tissues, preservation 220
concept mapping 405–6, *408*
confidentiality 105–6
coning 73
consent
  express 100, 101–3
  family member 102
  geographical differences 346–9
  legislation 100–1
  recording refusal 101, 102
  relatives 8

consent, informed 28, 176–7, 344–5
  bone marrow donation 250, 251, 253, 254–8
  bone marrow donor registries 435
  efficiency 349–54
  embalming 347
  family of donor 356
  medical profession 357
  organ donation rate 350, 351–2
consent, presumed 40–1, 100, 101, 103–5, 344, 345–6
  efficiency 349–54
  European Donor Hospital Education Programme (EDHEP) 379
  family of donor 356
  France 420
  legislation 275, 347, 355
  organ donation rate 350, 351–2, 354
  public opinion 356
consequences mapping 405–6, *408*
continuous perfusion 208–11
cord blood
  harvest 266
  stem cells 442
core cooling 159
cornea
  evaluation 297–8
  presumed consent law in USA 354
  removal 296–7
  retrieval for tissue banking 296–8
corneal graft banking 12, 229, 269, 271–2
  Australia 286
  cadaveric donors 290–1
  UK 284
  USA 284
corneal grafting, development 270–1
Corneal Grafting Act (UK, 1952) 273
Corneal and Tissue Grafting Act 1955 (NSW) 273
counselling
  bone marrow donation 250–1
  grief 318
  private 317
CPK isoenzyme 153
cranial trauma, donors in Spain 364, 365
creatinine clearance 155
Creutzfeldt–Jakob disease (CJD) *126*, 135–6, 251
  pituitary gonadotrophin 270
  transmission 135
  transplantation policy 135–6
cross-clamping 159
cryoprotectants 201
cultural awareness 36
cultural beliefs 23, 24, 402

current awareness 402
cyclosporine 15, 163
cytomegalovirus 122–7
  bone marrow donation 243, 250
  clinical syndromes 124–5
  distribution 122–3
  hyperimmune globulin (CMVIG) 126, 127
  renal allograft recipients 125
  strategies to reduce infection 126–7
  tests 123
  transmission 123–4
  transplant policies 125–6
cytotoxic T-lymphocyte precursor frequency 243

dead body, religious concepts 24, 347–8
death
  nature of 312
  sudden 306
  violent 312
decay accelerating factor 449, 450, *453*, 454
deep vein thrombosis 172–3
  donor nephrectomy 184
denial 306
developing countries 29
dextrans 209
diabetes insipidus 156, 157
diabetes mellitus
  corneal graft tissue banking 290, 291
  kidney/pancreas transplantation 10
  live kidney donation exclusion 172
  organ donor contraindication 153
  pancreas transplantation 193
dialysis 13, *14*
disease
  infectious and bone marrow donation 252, 253
  Muslim concept 30
disease transmission 120
  bacterial infection 120–2
  parasites 136–8
  risk evaluation 152
  viral infection 122–36
dissection 348
distress, family of donor 309
doctors
  gatekeeper role 382–3
  see also medical profession
domino procedure 193
donation
  approaching relatives 379
  brain death 312–14
  compensated 58
  demand 388
  with incentive 58

donation (*continued*)
  increasing *47*
  infrastructure establishment 396
  living donor 28
  living related 64
  loss of potential 373
  optimal practice 391–2
  process in Making the Critical Difference programme 385
  public perception 382
  rate 349–50
  requests 395
  shortage 35, 374, 390
  voluntary 26
  see also cadaveric donation; paid donation
donor card 39, 42, 101, 346
  campaigns 422
  Japan 416
  next of kin consent 354
  UK 414
  USA 414–15, 420
Donor Tissue Bank of Victoria 287–92
donors
  actual 400
  age 8, 306
  audit of potential 7
  brain-dead 8
  co-ordination organization 11–12
  conventional 9
  detection in Spain 369
  guidelines 140
  identification
    by transplant co-ordinator 325–6
    of potential 8
    protocols 275
  management 334–5, 369
  marginal 140
  maximum potential rate 8
  neoplasm screening 139
  payments to remove disincentives 58
  potential 8, 400
  procurement agencies 12
  prospective
    conditions 102–3
    recipient discrimination 103
  records 336–7
  shortage 35, 374, 390
  suitability 333–4
    potential living 339–40
  tracking tool 398
  transplant unit availability 11
  see also cadaveric donors; living donor
driving licence 346
  donation card 101
  donor intent indication 42
  registry 43
  see also motor vehicle agency
dura mater tissue banking 269

early adopters 416
echocardiogram 153
education
  audience
    segmentation 418–19
    target 426
  commitment 413
  community 401
  cost-benefits 425–7
  evaluation 409, 411, 422–3
  factors for consideration 402
  funding 409
  hospital staff 396–7
  motivational messages 419–22
  nurses 383–8
  parental response 411
  pre-testing of materials 422
  professional 331–2, 338, 425
  programme
    effectiveness 422–3
    evaluation 401–2, 422–3
    implementation 406–7
    organ exchange organization (OEO) 229
  public 36, 39, 331, 337–8
    attitudes 414–18
    awareness 415–17
  resource pooling 426
  in schools 400–11, 402–7, *408*, 409, *410*, 411
  spending priority 426
  targeting 417–18
  teaching strategies 405–6
  tools 399
  Transplantation Society 64
  see also European Donor Hospital Education Programme (EDHEP)
educational attainment 402
educational materials, organ recipients 328
EEG, function persistence in brain death 85
electrocardiogram 153
electroencephalography *see* EEG
electrolytes 209–10
embalming 347
emotional appeal 421
empathy 39
empty sella syndrome 84
End Stage Renal Program (US) 48
erythropoietin 264
ethics
  change 57
  Church of England 26

debate 52
decision-making 53
living donor renal transplantation *51*
societal 49
ethnic diversity 402
EuroCollins' solution 207, 208, 219, 220
see also Collins' solution
Europe
  anatomy with human dissection courses 348–9
  legislation 350–3
    obtaining tissues for transplantation 273, 274
  Making the Critical Difference programme 387–8
  tissue banking
    organization 284–6
    technical regulation 278–80
European Donor Hospital Education Programme (EDHEP) 229, 373
  adaptation 376–7
  content 375–6
  current practice effects 379, *380*
  development 274–5
  evaluation 377, *378*, 379, *380*
  Grief Response and Donation Request 375–6
  implementation 376–7, *378*
  Meeting the Donor Shortage 375
  national working groups 376
  objectives 375–6
  quality assurance 377
  self-confidence of participants 379, *380*
  trainers 376
  workshop moderators 376, *377*
European Donor Secretariat (EDS) 431–2
European Marrow Donor Information System 432
European Transplant Co-ordinators Organization 329
Eurotransplant International Foundation (ET) 11, 229, 230, 373
  Acceptable Mismatch Programme 234, 235
  European Donor Hospital Education Programme 376
  finance 232
  Highly Immunized Trial (HIT) protocol 234, 235
  kidney allocation 234, 235–6
  liver allocation 237
  organ allocation 234, 235–8
  patient priority 233
  recipient pool 235
  Special Urgency 237
  thoracic organ allocation 237–8

evoked potentials, brain death 86
execution 63
Eye Banking Association of America 284

facilitator role 384
fairness issues 38
family
  decision 39
  discussion 41–2, 43, 423
  issues 37–8
  knowing each other's wishes 391
  Partnership for Organ Donation 391–2
family of donor 304–5, 306–7
  acceptability of law 356–7
  aftercare 314–20
  approaching 379
  brain death 306
    understanding 309–10
  communication 379
  consent 310–11, 333, 354, 355
  declining option 390
  distress 309
  forewarning 306
  grief 311–14, 314–20
    counselling 317, 318
  immortality of loved one 313
  information 308–9
    recipients 313
  knowledge of wishes 420
  nature of death 312
  needs 315–17, 391–2
    immediate 307–10
    meeting 310–11, 399
  objections 97
  offer of option 390
  perspective in Making the Critical Difference programme 385
  premature restitution 314
  refusal 373
    rate 412, 419–20
  response
    from recipients 313
    to loss 305
  Spain 370–1
  stress 41
  sudden loss 306, 311–12
  support 307–8, 310–11
    groups 317
  support programmes 317–19
    efficacy 319–20
  visiting 308
  see also bereavement
family member
  coercion 103
  consent 102, 103

family member (*continued*)
  inducement 103
  objection 102
  well-being 102
fear 422
felons, hanged 348
fetal tissue transplantation 26–7, 99
Finland, behaviour change 424
flush solutions
  Bretschneider's HTK 207
  cardioplegic 218
  citrate *206*, 207
  Collins' 205, *206*, 207, 208
  Columbia University 219
  continuous perfusion 208–11
  electrolytes 209–10
  EuroCollins' solution 219, 220
  hypothermic storage 205, *206*
  impermeable solutes 209
  single-pass 208–11
  sodium lactobionate-sucrose *213*
  University of Wisconsin (UW) *206*, 208, 211, 216, 217, 218
Food and Drugs Administration (FDA) 275, 276, *277*
France
  donation law (1994) *104*
  legislation 107, 109, *110*, *111*, *112*, 350–1
    obtaining tissues for transplantation 273–4
  marrow donor registry 431–2
  presumed consent 420
  public attitudes 413, 420
futures market 57, 58

ganciclovir 126–7
gift relationship 25
glioblastoma multiforme 139
glomerular filtration rate, renal donor 175
glucose 156
graft, iatrogenic contamination 120–1
graft-versus-host disease
  control 242
  cord blood transplant 266
  donor age 266
  HLA one antigen mismatch 240
  unrelated bone marrow transplantation 243
granulocyte colony stimulating factor (G-CSF) 265, 442, 443
grey basket concept 57–8
grief 306
  counselling 317, 318
  family of donor 311–14, 314–20
growth factors, marrow donors 441, 443

Guidelines for the Determination of Brain Death in Children (USA) 77
Guidelines for the Determination of Death (USA President's Commission) 77

haematoma removal 72
haemoglobinopathy 172
haemopoietic stem cells 266
haemorrhage, donor nephrectomy 182
Health Act (Victoria, 1958) 280
heart preservation 218–19
heart transplantation
  development 9
  living donor 192–3
  thoracic organ allocation 238
  Toxoplasma gondii 137
  see also cardiac donation
heart valve
  graft distribution in USA 282, *283*
  tissue banking 269, 291, 293–6
heart–lung transplant
  preservation 219–20
  thoracic organ allocation 237, 238
helper T-lymphocyte precursor frequency 243
hepatectomy, reduction 186
hepatectomy, donor 189–92
  perfusion 191
  resection plane 189–91
  vascular grafts 191–2
hepatic failure, acute fulminant 187
hepatitis B e antigen (HBeAg) 129, 130
hepatitis B immune globulin 129
hepatitis B surface antigen (HBsAg) 129–30
  carriers 130–1
  tissue banking serological testing 291
hepatitis B virus 129–31
  infection *126*
  living donor renal allograft recipients 171
  living partial liver donation 187
  seropositivity 140
  transmission 129, 130, 131
  transplantation policies 130–1
hepatitis C virus 131–5
  infection *126*
  infection markers 133–4
  living partial liver donation 187
  pre-transplantation infection 134–5
  screening 134
  seropositivity 140
  tests 132, 134
  tissue banking serological testing 291
  transmission 132–3
hepatitis transmission prevention 113
herpes simplex virus *126*, 127–8

Highly Immunized Trial (HIT) protocol 234, 235
Hinduism 24
histocompatibility matching 437
HIV infection *126*
  bone marrow donation 251
  donor screening 129
  seroconversion 128
  tissue banking 270
    serological testing 291, 292
  transmission 128, 275, 276
    prevention 113
  transplant policies 129
HLA matching 165, 173, 232–3
HLA mismatch 234
HLA one antigen mismatch 240
HLA polymorphism 243
HLA typing 188, 232–3
  bone marrow donors 436
  donor centre screening 433
  methods 243–4
  organ exchange organization (OEO) 228
  unrelated marrow donor registries 430
HLA-A typing 245, *246–7*
HLA-antigen sharing 236
HLA-B27 antigen 243
HLA-B typing 245, *246–7*
HLA-compatibility, bone marrow donation 240
HLA-DR typing 245, *246–7*
  bone marrow donor matching 436, 438
HLA-sensitization 233
hospitals
  concensus for change 395–6
  diagnosis 393–5
  donations
    improving 398–9
    infrastructure establishment 396
    optimal practices 392–7
    requests 395
  education package 399
  education of staff 396–7
  in-house co-ordinator training programme 399
  Partnership for Organ Donation 391, 392–7
  performance 393–4
  records 346
  staff survey 398
HTLV-1
  seropositivity 251
  tissue banking serological testing 292
human decay accelerating factor (hDAF) 450–1, *452*, 454
human growth hormone 251

human immunodeficiency virus *see* HIV infection
Human Organ Transplant Act (Singapore, 1987) 104–5
human pituitary gonadotrophin 251
human rights treaty, international multilateral 107
Human Tissue Act (UK, 1961) 273, 346, 347
Huntington's disease 99
hybrid technology 447
hydroxyethyl starch 209
hypothalamic function 83–5
hypothermia 203
hypothermic storage, simple 205, *206*

immunosuppression 162–3
  graft-versus-host disease 242
  patient care protocols 328
India
  cadaveric donation development 50
  paid donation 48, 49, 50
information, family of donor 308–9
initiator role 384
innovators 416
intensive care units 7, 8
intracranial blood flow, objective assessment 86–7
intracranial pressure 72–3, 74
  filling absence 87
intracranial vessel filling, brain death 86–7
intracranial volume 72
invasive treatment for body material utility 97
ischaemic/hypoxic damage 202
Islam 29–33

Japan
  donor cards 416
  religious beliefs 24
Jews 24
Journal of Transplant Coordination 330
jurisdiction, extent 106–7
justice principle 232, 233

keratoplasty 271, 272
kidney
  allocation 235–6
  bacterial contamination 121–2
  cadaveric donation *154*, 155
  CMV secondary infection in recipients 124–5
  donor registry 345
  persufflation during storage 210
  preservation 205, 215–16
  selling in India 48
  simple cold storage 205

kidney donation, live 163, *164*, 165, *166*, 167, 169–86
  anatomy 175–6
  angiography 175–6, *177*
  complications 181–2, *183*, 184–5
  donor
    evaluation *170*
      medical suitability 174
      potential 171
      risk 177
      screening 174
  donor–recipient matching 173
  exclusion criteria 171–3
  graft survival *173*
  HBV infected recipients 171
  infants 176
  informed consent 176–7
  left side 175
  lifestyle adjustment 185
  long-term renal function 184–5
  recipient 169–71
  renal function assessment 175
  right side 176
  swap programme 185
  transplantation 12, 13, *51*, 55
  unrelated 186
  see also nephrectomy, donor
kidney transplantation 3, 12–14, *15*
  cadaveric donors 12, 13, *154*, 155
  chronic graft dysfunction 14, *15*
  development 9, 12
  donors 3, 4
    living 12, 13
  identical twin 162, 163
  live 163, *164*, 165, *166*, 167
  living donor 55
    ethics *51*
  rate *4*, 12–13, 163, *164*
  related 165, *166*
  success rate 14
  survival *166*
  waiting list 167
kidney/pancreas transplantation 10
Kupffer cells 211, 212

language barriers 402
late adopters 416
legislation 56–7, 95–6, 402
  acceptability 355–7
  brain death 70
  civil law 96
  commercial initiatives 97, 98
  confidentiality 105–6
  consent 100–1, 344–6
  countries where pending *109*

  current configuration 107–19
  European 273, 274, 350–3
  express consent laws 101–3
  France 107, 109, *110*, *111*, *112*
  geographical differences 346–9
  individual autonomy 354–5
  key issues 96–107
  known 107, *108*
  living donors 97–100
  non-statutory measures *109*
  obtaining tissues for transplantation 272–5
  parents 98
  Portugal 110–11, *112*
  post-mortem donation 100–7
  presumed consent laws 103–5, 275
  records 106
  religious law 96
  Russian Federation 111, 113, 116–19
  Spain 361
  Turkey 100
  United States of America 113, 274–5
leucocytes, peripheral blood 265
leukapheresis 265
life support 81
ligament tissue banking 269
limbs, preservation 220
lipase 156
liver
  allocation 237
  cadaveric donation *154*, 155
  donor hazards 168–9
  live transplant 167–8
  orthoptic transplantation 186
  partition 167
  preservation 216–17
  reperfusion injury 212
liver donation, living partial (LRLT) 168–9, 186–92
  anatomy 188–9
  blood grouping 188
  cholangiography 188
  co-morbid conditions 187
  donor hepatectomy 189–92
  HLA typing 188
  imaging 188
  induction immunosuppression 188
  liver function assessment 188–9
  morbidity 192
  mortality 192
  potential donor 187–8
  recipient 186–7, 188
  relationship to recipient 188
  screening 188
  volume estimation 188, 189
liver, heart and lung programmes 12

living donor 28, 162–3
  ethics 163
  legislation 97–100
  liver 167–8
  lung 192–3
  non-related 54–5
  renal transplantation *51*, 55
  spousal 55
  see also kidney donation, live; liver donation, living partial (LRLT)
locked-in syndrome 80
lung
  donation 154–5
    retrieval 157, 158
  living donor transplantation 192–3
  preservation 219–20
  thoracic organ allocation 238

machine perfusion
  colloid 211
  extracorporeal organ storage 204–5
  flush solutions 209
  reperfusion injury 212
Making the Critical Difference programme 383–8
  Canadian applications 387–8
  European applications 387–8
  outcome 386
  outcomes study 386–7
  workshop design 384–6
malignancy
  bone marrow donation 252
  donation contraindication 153
  transmission 138–40
    risk evaluation 152, 153
mandated choice 355
mandatory exchange category 234
marketing, social 417
mass media, promotion 423–4
medical profession
  acceptability of law 357
  public attitudes 53
medical records 398
Medical Royal Colleges and their Faculties in the UK Conference 75–6
medical utility principle 232
medulloblastoma 139
membrane cofactor 449
Memorandum on the Diagnosis of Death 76
memorial programs 316–17
mental competency 97, 98–9
MHC antigen sequencing 245
minority populations 36–40
  attitude measurement 37–8
  defining attitudes 38–40

  inhibiting factors 38
miscarriage, spontaneous 99
mixed lymphocyte cultures 242, 243
motivational messages 419–20
  presenting 421–2
motor vehicle agency 42–3
  registry of donation intention 43
  see also driving licence
multi-organ donation 12
multi-organ retrieval 157, 158, 159–60
  en bloc 158–9
  procurement *214*
  sequential 158
Munich Conference (1990) 52
Muslims
  informed consent laws 347
  Singapore legislation 104–5

Na/K ATP-ase 202
National Corneal Service (UK) 284
National Donor Family Council (USA) 318
National Kidney Foundation (USA) 318
  Making the Critical Difference programme 383
National Marrow Donor Program (USA) 240, 242, 431
National Organ Transplant Act (USA, 1984) 329, 330
National Organ Transplantation Act (USA) 97
nephrectomy, donor 178–81
  approach 178, 179
  kidney perfusion 180, *182*
  late complications 184–5
  morbidity 181–2, *183*, 184
  mortality 182, 184
  patient position 178
  post-operative nursing 181
  retroperitoneoscopic 186
  vessel clamping 179, *180*
nephropathy, reflux 172
nerve injury 182, *184*
Netherlands legislation 351, 352
neurodegenerative diseases 99
neuronal cells, intracranial 73, 82
nitric oxide 219
non-heartbeating donors 159–60
North American Transplant Center Organizations 383
North American Transplant Coordinators Organization 329
nurses
  education 383–8
  gatekeeper role 382–3
  see also medical profession
nutrient substrate, organ storage 210

O_2 challenge 154
organ
  cadaver
    availability 163
    payment 63, 64
  definition 97
  distribution 334
  exchange balance 233
  maximum placement 227
  optimal usage 228
  purchase 47–8
  retrieval rate 349
  sharing 329
  shipment 231
  surgical recovery 335–6
  transparent allocation 228
  waste reduction 227
organ allocation
  central office 231
  centre-oriented 234
  Eurotransplant International Foundation 234, 235–8
  highly sensitized patients 234
  mandatory exchange category 234
  medical factors 232–3
  model 232–3
  non-medical factors 232, 233
  organ exchange organization (OEO) 231, 232–4
  patient-oriented 234
  transparent 228
organ exchange organization (OEO) 226, 227–8
  allocation 231, 232–4
  bone marrow donor registry 229
  development 229
  donation 231
  educational programmes 229
  finance 232
  infrastructure 229–30
  internal operational structure 230–1
  organization 229–32
  policy development 230
  reference tissue typing 228
  research function 228
  tissue banking 229
  transplantation 231
organ preservation 200–2, 336
  applied strategies 213–15
  clinical transplantation 215–20
  cold storage 215
  hypothermia 203
organ procurement 12, 201, 204
  optimal 213–15
  organizations (OPOs) 231, 328, 396

tissue bank relationship with agencies 284
transplant co-ordinator 325–6
see also Partnership for Organ Donation
organ storage
  cold 215
  duration 201
  extracorporeal 204–5, *206*, 207–8
  hypothermic 205, *206*, 211, *212*
  injury 211
  machine perfusion 204–5, 211
  nutrient substrate 210
  oxygen supply 210
  temperature 201, 211, *212*
  viability loss 202
Organización Nacional de Trasplantes (ONT) 11, 362
  network development 362–4
  training programmes 364
organization of organ allocation/donation 226–7
Ottawa Conference (1989) 50
oxygen
  free radical scavengers 211
  free radicals 202
  persufflation 210
  supply in organ storage 210

paediatric organs
  donation 39–40, 98
  shortage 39
paid donation 46, 63, 64
  arguments 53–4
  development 47–8
  grey issues 48, *49*
  model 48
  to remove disincentives from potential donors 58
pancreas 193–4
  donation *154*, 155–6
  hybrid technology 447
  living donor transplantation 192, 193–4
  preservation 217
  reperfusion injury 212
pancreatic islet transplantation 221
paralytic ileus 181
parents
  child organ donation 39–40
  education programme response 411
  legislation 98
  live liver transplant 168–9
  loss of child 311
  see also family of donor
Parkinson's disease 99
Partnership for Organ Donation 389
  consensus for change 395–6

donation
    improving in hospitals 398–9
    infrastructure establishment 396
    optimal practice implementation in hospitals 392–7
donor hospitals 391
education of hospital staff 396–7
family needs 391–2
hospital diagnosis 393–5
monitoring 396
new protocol 396
organ procurement organization 396
organ shortage 390
quality assurance 396
Pathology Services Accreditation Act (Victoria, 1984) 280
patient care protocols 328
patient priorities 62
perfluorochemical polyols, synthetic 210
perfusion
    cold 205
    machine 204–5
    solutions 205
perfusion techniques 201
    machine 209
peripheral blood stem cells 443
    harvest 264–5
personal belief assessment 384
personal decisions 41
personal loss, simulation 385
personal testimony 421
personalities, well-known 421
philanthropy, mandated 58
pig heart protection from human complement 451, *452*, 453–4
pigs, transgenic 449, 450, 454
    hDAF expression 450–1
pituitary
    arterial supply 84
    function 83–5
        retention 84, 85
    gonadotrophin 270
placenta 99
    see also cord blood
plastic surgery, reconstructive 220
platelet donors 430
pneumonectomy, live donor 193
pneumothorax 181, *184*
policy for obtaining tissues for transplantation 272–5
polyethylene glycol 209
Portugal, legislation 110–11, *112*
post-mortem donation legislation 100–7
potassium bolus release with limb reimplantation 220

Practice Parameters for Determining Brain Death in Adults 78, 89
prevascularization rinse 212–13
print media 423–4
privacy legislation 104
professional roles 57–8
promotion, mass media 423–4
public attitudes 34–6, 390
    acceptability of law 355–6
    brain death 40
    donation rates 412–13
    education 35–6
    family discussion 41–2
    France 420
    medical profession 53
    motor vehicle administrations 42–3
    target audience 426
    see also education, public
public awareness campaigns 39
public confidence 413
    promotional strategy 427
    rebuilding 413
public health, national structures 227
public perception 382
public understanding in UK 420
pulmonary disease, chronic 155
pulmonary embolus 182, *184*
purchase of organs 47–8
    see also paid donation
pyrimethamine prophylaxis 137–8

quality adjusted life year (QALY) 17, *18*

reasoned action theory 35
recipient
    confidentiality 105–6
    donor family
        prolonged attachment 313–14
        response to 313
    kidney transplantation 3
    outpatient care 342
    post-operative care 341–2
    suitability 340–1
    travel by potential 106
records, legislation 106
reference tissue typing 228
registries 345
    see also bone marrow donor registries
rejection
    hyperacute 448, 455
    xenotransplantation 447–8
religious influences 402
religious law 96
religious view 23–4
renal allograft recipients, CMV infection 125

renal evaluation 155
renal failure, end-stage 48
renal graft failure 14, *15*
renal transplantation *see* kidney transplantation
reperfusion injury 211
　after reimplantation 202
　hypothermia *203*
　prevascularization rinse 212–13
　prevention 212–13
resistors 416
restitution, premature 314
road traffic accidents
　cadaveric kidney donors 3
　deaths in Spain 367, 369
Roman Catholic Church 27–9
Rotary club 406, 407
Russian Federation, legislation 111, 113, 116–19

sale of organs 47
Scandiatransplant 11
Scandinavia 353–4
　acceptability of law 357
schmerzensgeld 58
scleral tissue banking 269
sella turcica 84
shared interest/purpose 53
Shariah 31–2
Shinto concept 24, 347
sick euthyroid state 83, 85
single-pass flush solutions 208–11
skeletal tissue retrieval for tissue banking 298–301
skin graft, split thickness 221
　retrieval for tissue banking 301–2
skin tissue banking 229, 269, 271
　cadaveric donors 291
　retrieval for 301–2
skin transplantation 221
　development 270
skull vault 72
small bowel
　donation 156
　preservation 217–18
　transplantation 194
smoking 154
sodium lactobionate-sucrose solution *213*
sodium pump, cell membrane 202
solutes, impermeable 209
somatic evoked potential 86
source of organs 62
Spain 361–2
　approach to family 370–1
　brain death misunderstanding 415
　donor detection 369

donor management 369
family of donor 356–7
legislation 351, 354, 361
media relations 363–4
numbers of donors *366*, 367
organ donor evolution 364–5, *366*, 367
organ transplantation 368
Special Urgency, organ allocation 237
spinal cord 81
spinal reflexes 81
splenectomy 182, *184*
Staphylococcus aureus septicemia 121
Statement and Guidelines on Brain Death and Organ Donation – 1993 78
stem cells
　alternative sources 442–3
　haemopoietic 266
　harvest 264–5
　mobilization 265
steroid immunosuppression 163
stillbirth 99
stress, donor families 41
sucrose solution *206*, 207
support
　family of donor 307–8, 310–11
　groups 317
　programmes 317–20
supporter role 384
supratentorial function 78, 79, 80
　brainstem function absence 82
Swan–Ganz catheter 157
Swan–Ganz monitoring of cardiac output 153
swap programme for live kidney donors 185
Sydney Declaration 75
syphilis, tissue banking serological testing 292

target groups 421
television 423
temperature
　bone marrow transport 261
　organ storage 211, *212*
　tissue storage 201, 203
Tennessee Donor Services 318, 319
territorial principles of jurisdiction 106–7
tertiary healthcare costs 53
Therapeutic Goods Act (Australia, 1989) 280–2
Therapeutic Goods Adminstration (Australia) 281
thoracic organ allocation, ET 237–8
tissue
　availability 275
　obtaining 272–5
　preservation 200–2
　typing 173

viability 203
tissue banking 12, 268–70
  assessment of potential to donate 287
  cardiac tissue retrieval 292–6
  development 271–2
  donor selection 287–92
  infection risk 270
  organization 282–7
  technical regulation 275–82
Toxoplasma gondii 136–8
toxoplasmosis
  clinical features 137–8
  prophylaxis 138–9
  transmission 136–7
training for trainers 376
transgenic animals 26
transplant co-ordinators 231, 325–6
  clinical 327–8
    outcome criteria 337–42
  donor issues 328
  educational catalyst 328
  management function 327
  Organización Nacional de Trasplantes (ONT) 363, 364
  patient care 327–8
  practitioner certification 330
  procurement 326–7
    outcome criteria 331–7
  professional organizations 329
  professional relationships 327
  professionalism 329–30
  recipient issues 328
transplantation
  alternatives 447
  between species 447–8
  centres 10–11
  cost-effectiveness 425
  costs 16–18
  development of clinical science 9–10
  donor numbers 3–4, 5, 7–9
  global numbers 15, 16
  indications 15, 17
  medical profession benefits 53
  Muslim draft law 32
  outcome 1
  programmes 231
  public awareness 34
  qualities conducive to in Islam 30–1
  rate 1, 2–4, 5, 7–12
    factors 2–3
  Spain 368
  survival 3
  Transplantation–The Issues educational programme 403–7, *408*, 409, *410*, 411

Transplantation Society
  commerce response 55–6
  commercialism debate 48–50
  education strategy 64
  Ethics Committee 62, 63
    commercialism 66
    policy statement 65–6
    payment for cadaver organs 63, 64
  transplantation tourism 106–7
  tumour necrosis factor alpha 211
Turkey, legislation 100

UK
  legislation on obtaining tissues for transplantation 273
  public awareness 414
  public understanding 420
  technical regulation of tissue banking 276–8
  tissue banking organization 284, *285*
UK Tissue Banking Review 276–8
UK Transplant 345
Uniform Anatomical Gift Act (USA, 1968) 99, 274, 347
  irrevocability 354
Uniform Anatomical Gift Act (USA, 1971) 326
United Kingdom Transplant Support Service Authority (UKTSSA) 3, 11–12, 229
  finance 232
  organ allocation 234
  patient priority 233
United Network for Organ Sharing (UNOS) 11, 229, 230
  organ allocation 234
  patient priority 233
universal crime 107
University of Wisconsin (UW) solution *206*, 208, 211, 216, 217, 218
unrelated marrow donor registries 430–1
  organization 432–42
US Coalition on Donation 426
USA
  African American families 418–19
  consent legislation 354
  corneal graft banking 284
  donor cards 414–15, 420
  donor screening routines 283
  legislation 113
    obtaining tissues for transplantation 274–5
  National Marrow Donor Programme 431
  professional education 425
  public awareness 414
  public education programmes 424

USA (*continued*)
  technical regulation of tissue banking 275–6, *277*
  tissue banking organization 282–4

vascular endothelial cells, reperfusion injury 212
vegetative state 80
ventilation, artificial 70
ventriculosystemic shunt 140
vertebrobasilar artery occlusion 82
video scenarios 384
virus transmission 122–36
Vital Connections 229

waiting list time 233
warm ischaemia 213–14

Wiscott–Aldrich syndrome 431
World Health Organization
  commerce issues 56–7
  guiding principles *114–15*
  known legislation 107, *108*
World Marrow Donor Association 432, 433, 436

xenotransplantation 64–5, 446–51, *452*, 453–5
  clinical use 448, *449*
  concordant 448
  discordant 448, 449–50
  Muslim view 33

Yorkshire Regional Tissue Bank 284, *285*